SCIENCE AND MEDICINE IN THE
SCOTTISH ENLIGHTENMENT

Science and Medicine in the Scottish Enlightenment

Edited by

Charles W. J. Withers and Paul Wood

TUCKWELL PRESS

First published in Great Britain in 2002 by
Tuckwell Press Ltd
The Mill House
Phantassie
East Linton
EH40 3DG

Copyright © The editors and contributors severally

All rights reserved
ISBN 1 86232 285 6

A catalogue record for this book is available
on request from the British Library

Typeset by Hewer Text Ltd, Edinburgh
Printed and bound by Bell & Bain Ltd, Glasgow

In memory of Roy Porter
(1946–2002)
colleague, Londoner, historian

Contents

List of Illustrations ix
List of Tables xi
Notes on Contributors xiii
Acknowledgements xv

1 Introduction: Science, Medicine and the Scottish
   Enlightenment: An Historiographical Overview   1
  Paul Wood and Charles W. J. Withers

2 'Feasting my eyes with the view of fine instruments':
   Scientific Instruments in Enlightenment Scotland, 1680–1820   17
  A. D. Morrison-Low

3 Situating Practical Reason: Geography, Geometry and
   Mapping in the Scottish Enlightenment   54
  Charles W. J. Withers

4 Science and Enlightenment in Glasgow, 1690–1802   79
  Roger L. Emerson and Paul Wood

5 Maclaurin and Newton: The Newtonian Style and the
   Authority of Mathematics   143
  Judith V. Grabiner

6 The Burden of Procreation: Women and Preformation
   in the Works of George Garden and George Cheyne   172
  Anita Guerrini

7 William Smellie and Natural History: Dissent and Dissemination   191
  Stephen W. Brown

8 Charles Elliot's Medical Publications and the International
   Book Trade   215
  Warren McDougall

9 Reading Cleghorn the Clinician: The Clinical Case
    Records of Dr. Robert Cleghorn, 1785–1818   255
    Fiona A. Macdonald

10 Appealing to Nature: Geology 'in the Field'
    in Late Enlightenment Scotland   280
    Stuart Hartley

11 Late Enlightenment Science and Generalism: The Case
    of Sir George Steuart Mackenzie of Coul, 1780–1848   301
    Charles D. Waterston

12 Afterword: New Directions?   327
    Charles W. J. Withers and Paul Wood

    Index   337

List of Illustrations

(BETWEEN PAGES 176–7)

1. An air pump by James Crichton. (Photograph courtesy of the Collins Gallery, University of Strathclyde)
2. A barometer by James Watt. (Photograph courtesy of the National Museums of Scotland)
3. Class cards belonging to William Dean. (By Permission of University of Edinburgh Library Special Collections)
4. The title page of John Gray's *The Art of Land Measuring Explained* (1757). (Photograph courtesy of University of Edinburgh Photographic Services).
5. A portrait of William Cullen after William Cochrane. (By permission of the Royal College of Physicians, Edinburgh)
6. A detail of Alexander Bryce's 'Map of the North Coast of Britain' (1744). (By permission of the Trustees of the National Library of Scotland)
7. 'Sir Archibald Grant and Family' by an unknown artist. (In the collection of Sir Francis Grant's Accumulation and Maintenance Trust and with the permission of the Scottish National Portrait Gallery)
8. James Hutton, John Clerk and Hutton's dog in the field. (From facsimiles in the care of Jean Jones and with the permission of Sir John Clerk of Penicuik)

List of Tables

2.1 Markets for Instruments during the Eighteenth Century   31
4.1 The Growth of Glasgow and Its Resident Scientific Community   130
4.2 Glasgow's Men of Science (by birth cohorts)   131
4.3 Places of Origin of Glasgow Men of Science   137
4.4 Fathers' known Occupation   138
4.5 Initial Occupations of Glasgow Men of Science   138
4.6 University Attendance   139
4.7 Characteristics of Glasgow Men of Science   140
4.8 Principal Occupation of Glasgow Men of Science by Birth Cohort   141
4.9 The Changing Occupational Composition of the Glasgow Scientific Community by Ten-Year Intervals   142
8.1 Short-Title List of Charles Elliot's Medical Publications   237

Notes on Contributors

STEPHEN BROWN is a member of the English Department at Trent University where he is also Master of Champlain College. He has published over twenty papers on William Smellie, science and publishing in Enlightenment Edinburgh, and has just completed a biography of Smellie. He has also indexed Smellie's archives for the Society of Antiquaries of Scotland and is a member of the History of the Book in Scotland project.

ROGER L. EMERSON retired as a Professor of History at the University of Western Ontario in 1999. He is the author of many essays on eighteenth-century Scotland, including a number on the Philosophical Society of Edinburgh, on Scottish universities and on Scottish patrons. He is currently working on a biography of the 3rd Duke of Argyll.

JUDITH V. GRABINER is the Flora Sanborn Pitzer Professor of Mathematics at Pitzer College, one of the Claremont Colleges in California. She is the author of two books, *The Origins of Cauchy's Rigorous Calculus* (1981) and *The Calculus as Algebra: J.-L. Lagrange, 1736–1813* (1990), as well as many articles in the history of mathematics. Her current research interest is Colin Maclaurin and his relationship to the Newtonian tradition and to Scottish culture.

ANITA GUERRINI is an Associate Professor of Environmental Studies and History at the University of California, Santa Barbara. She is the author of *Obesity and Depression in the Enlightenment: The Life and Times of George Cheyne* (2000). Her current research interests centre on public anatomy in the eighteenth century.

STUART HARTLEY recently completed his PhD at the Science Studies Unit, University of Edinburgh, on the history of the earth sciences. His thesis is entitled 'Robert Jameson, Geology and Polite Culture, 1796–1826: Natural Knowledge Enquiry and Civic Sensibility in Late Enlightenment Scotland'.

FIONA MACDONALD is a Research Fellow at the Wellcome Trust Centre for the History of Medicine at University College London. She is currently working on a history of Scottish medical journals in the period 1733 to 1832. Her publications include (with Johanna Geyer-Kordesch) *Physicians and Surgeons* in Glasgow: The History of the Royal College of Physicians and Surgeons of Glasgow, 1599–1858 (1999).

ALISON MORRISON-LOW has been Curator of Scientific Instruments and Photography at the National Museums of Scotland since 1980. After co-authoring books on Scottish and Irish scientific instrument making, she recently completed her doctorate on the instrument trade in provincial England during the Industrial Revolution. Her most recent work has been concerned with the evolution of Scottish weights and measures before the imposition of the Imperial system in 1826.

WARREN MCDOUGALL is an Honorary Fellow of the English Literature Department, University of Edinburgh, and of Champlain College, Trent University. He is writing a book on Charles Elliot and is editor of the volume covering the eighteenth century in the History of the Book in Scotland project.

CHARLES WATERSTON is a Fellow and past General Secretary of the Royal Society of Edinburgh and former Keeper of Geology at the Royal Scottish Museum. His research interests in the history of science have concentrated upon nineteenth-century geologists and the development of scientific institutions in Scotland. His *Collections in Context: The Museum of the Royal Society of Edinburgh and the Inception of a National Museum for Scotland* was published in 1997.

CHARLES W. J. WITHERS is Professor of Historical Geography at the University of Edinburgh. He has research interests in the historical geographies of science. He is the co-editor (with David Livingstone) of *Geography and Enlightenment* (1999) and author of *Geography, Science and National Identity: Scotland since 1520* (2001).

PAUL WOOD is a member of the History Department and Director of the Humanities Centre at the University of Victoria (Canada). He has published *Thomas Reid on the Animate Creation* (1995) and *The Correspondence of Thomas Reid* (2002) as part of the Edinburgh Reid Edition, and is engaged in writing a biography of Thomas Reid.

Acknowledgements

A number of friends and colleagues provided us with their advice and assistance during the preparation of this volume. We would like to thank: Rosemary Barlow, Chris Fleet and Andrew Grout, who solved various technical problems; and Roger Emerson, who read and commented on a number of the chapters. We greatly appreciate the help with illustrations provided by Jean Jones and Alison Morrison-Low, as well as staff at Edinburgh University Special Collections and the National Library of Scotland. We are also grateful to the following individuals and institutions who permitted us to reproduce our illustrations: Sir John Clerk of Penicuik; Laura Hamilton, Curator of the Collins Gallery, University of Strathclyde; the National Museums of Scotland; the Scottish National Portrait Gallery; and the University of Edinburgh.

1
Introduction: Science, Medicine and the Scottish Enlightenment: An Historiographical Overview

PAUL WOOD AND CHARLES W. J. WITHERS

The study of the history of science and medicine in eighteenth-century Scotland came of age in the 1970s. This is evident in the appearance in 1974 of a special issue of the journal *History of Science* devoted to 'Science in the Scottish Enlightenment' and in numerous other writings. The empirical study of eighteenth-century Scottish science and medicine blossomed as never before and, by 1975, a commonality of interpretation had emerged which challenged the tidy division then current among philosophers and some historians between the so-called 'internal' and 'external' factors affecting the growth of scientific knowledge.[1] It is our intention here both to explain this significant 'moment' and to review the main features of work subsequently published on science and medicine in the Scottish Enlightenment. Through such a review, however partial, we hope to place this collection in historiographical context.

Science and Medicine in Eighteenth-Century Scotland: Initial Approaches

One of the distinctive characteristics of the literature on eighteenth-century Scottish science and medicine published during the 1970s was its recognition that, however we may theorise the connections between the cognitive and the social, natural and medical knowledge is socially situated. Some historians drew such connections in terms of the theoretical framework known as the 'strong programme' in the sociology of scientific knowledge then being developed by members of the Science Studies Unit at the University of Edinburgh.[2] Others used variants of Marxist theory, while many either showed no allegiance to any theoretical school or were theoretically eclectic. Despite such divisions,[3] the consensus was that the flourishing of the natural sciences and medicine in eighteenth-century Scotland could only be explained in terms of the contemporary state of Scottish society.

The combination of the cognitive and the social can be seen, for example, in work on the history of chemistry by Arthur Donovan. Having completed his PhD on William Cullen and Joseph Black in 1970, he published a major monograph in 1975 on the chemical tradition initiated by Cullen and Black. Donovan related their teaching and research not only to the theoretical legacies of earlier chemists and the Newtonians, but also to their place within the Scottish universities and their engagement with efforts at economic improvement.[4] A similar sensitivity to social context is to be found in the writings of J. B. Morrell, including his now classic essays on the teaching of science in Edinburgh during the latter part of the eighteenth century.[5] Attention to the cultural settings and the politics of scientific knowledge was likewise a feature of Steven Shapin's work on the early history of the Royal Society of Edinburgh, while parallel concerns can be seen in the discussion of the birth of the Edinburgh medical school by Christopher Lawrence and others in 1975.[6]

Furthermore, the debates over scientific methodology spawned by the rise of the wave theory of light in the early nineteenth century were contextualised in terms of the intellectual traditions, curricula and teaching practices of the Scottish universities by G. N. Cantor, who also reconstructed the activities of one of the many ephemeral eighteenth-century student societies in Edinburgh, the Academy of Physics.[7] Questions of methodology, as well the philosophical aspects of scientific inquiry, figured in the writings of Richard Olson, who was perhaps less interested than other scholars to locate ideas socially.[8] Like Cantor, Olson traced the impact of eighteenth-century Scottish thinkers on the practice of nineteenth-century British physics and, although details of Olson's argument were questioned, his book *Scottish Philosophy and British Physics* was a timely reminder of the role played by the Scots in shaping the pursuit of science and medicine in the Victorian era.[9] The 1970s also saw the first major steps taken by D. J. Bryden and others towards the recovery of a central aspect of the material culture of the natural sciences and medicine, namely the historical study of scientific instruments and instrument makers.[10]

Arguably, the most influential essays to appear in the 1970s were by John Christie.[11] Christie's interpretation of science and medicine in eighteenth-century Scotland wove together the analytical themes and empirical findings of much of the literature surveyed above. Christie himself observed in 1975 that the work of Cantor, Olson, Morrell, Shapin and Nicholas Phillipson (who will be discussed below) 'refus[es] to fall into any neat classification along "internal" and "external" lines, [which is] a direct consequence ... of the manifest social role and status which knowledge-seeking achieved in eighteenth-century Scotland'.[12] Christie's own analysis cut across the internal-external demarcation in seeking to explain the efflorescence of scientific activity in terms of the economic, political and cultural priorities of Scottish landed and urban elites during the 'long' eighteenth century. Foregrounding the significance of the Parliamentary Union of 1707 in the formation of a 'distinctive philosophical and educational programme' which informed the pursuit of the natural sciences and medicine in the period, he echoed Donovan and others in emphasising the importance of the ideology of improvement in both the economic and cultural spheres.[13] Christie underlined too the crucial role played by the universities, clubs, student societies and more formal bodies like the Royal Society of Edinburgh in the cultivation of natural knowledge. In discussing the universities, Christie followed the lead of George Davie in highlighting the centrality of moral philosophy in the curriculum, and consequently paid close attention to the philosophical dimensions of scientific practice.[14] Christie was also sensitive to the cognitive implications of pedagogical change in higher education, as can be seen in his comments on the tension between the specialisation engendered by the professorial system and the need for prospective academics to maintain a competence in a number of fields due to the limited number of positions available in academe.[15] Like Donovan, Christie saw chemistry as being central to the achievements of Scottish men of science in the eighteenth century, writing that '[m]uch of the history of Scottish science between 1760 and 1800 can be written in terms of the research traditions which Black initiated in pneumatic chemistry and heat'.[16] For Christie, the study of chemistry was an especially revealing site for historical enquiry because the researches of chemists from Andrew Plummer onwards displayed the full range of institutional, ideological and intellectual factors shaping Scottish science and medicine more generally. In his later work, Christie returned to Cullen and Black in order to explore further the social and cultural matrix within which natural knowledge was created in Scotland.[17]

Placing Science and Medicine in the Scottish Enlightenment

Christie's essays were not written with only historians of science and medicine in mind. They also engaged with the historiography of the Scottish Enlightenment, a field of study itself undergoing a renaissance at precisely the same moment as the blossoming of the literature on the history of eighteenth-century Scottish science and medicine.[18] Renewed interest in the Scottish Enlightenment was at least partly prompted by the polemical address on the subject given by Hugh Trevor-Roper (later Lord Dacre) at the Second International Congress on the Enlightenment held at St. Andrews in 1966.[19] Although Trevor-Roper mentioned a number of Scottish men of science and medicine in the course of his discussion, for him the intellectual focus of the Scottish Enlightenment was the study of the human rather than the natural sciences. He maintained that the 'real intellectual pioneers' amongst the Scots were Francis Hutcheson, David Hume, Adam Ferguson, William Robertson, Adam Smith and John Millar, and it was this group that constituted the Scottish Enlightenment.[20]

A decade later, Trevor-Roper was even more explicit in his exclusion of science and medicine from what he took to be 'the Scottish Enlightenment'. He claimed that:

> What foreigners admired in late eighteenth-century Scotland, what they found, in a peculiar form, there and nowhere else, was not art or literature or medicine or natural science or even philosophy. No doubt all these things could be found there, but they could be found elsewhere too, and perhaps better elsewhere . . . Where Scotland was unique, and

not merely improved from its former state or sharing the general advance of England and Europe, was in two areas: in the organisation of university teaching and in a particular branch of study. That branch of study is historical explanation, the historical analysis of human progress: what would come to be called "political economy".[21]

By 1977, then, Trevor-Roper had become even more emphatic that the Scottish Enlightenment had nothing to do with science and medicine, and that it was to be defined exclusively in terms of the emergence of a distinctive form of historical understanding involving an account of human progress in terms of four stages of economic development.[22]

While Trevor-Roper's interpretation effectively excluded historians of science and medicine from the field, Nicholas Phillipson's series of influential articles on the Scottish Enlightenment that began to appear in 1974 informed the historiographical orientation of the work of John Christie among others. Unlike Trevor-Roper, Phillipson has never denied that the pursuit of science or medicine was a facet of the Scottish Enlightenment, but the cultivation of natural knowledge occupies a somewhat ambiguous place in his view of the Enlightenment in Scotland. In Phillipson's classic study of Edinburgh as an exemplar of eighteenth-century provincial culture, he recognised the role played by Colin Maclaurin in shaping the cultural priorities of the enlightened in Edinburgh during the 1730s and 1740s and pointed to the formation of the Honourable the Society for Improvement in the Knowledge of Agriculture as marking a decisive moment in the history of the ideology of improvement espoused by Edinburgh's 'provincial oligarchy'.[23] In his later essay in *The Enlightenment in National Context*, however, he moved to a position closer to Trevor-Roper's, arguing that after the Union Scottish thinkers were preoccupied with forging an 'alternative language of civic morality' to replace that of classical republicanism and that, by the 1760s, 'a new language of civic morality had been created which provided the Scots with a new understanding of civic virtue and that "sociological" understanding of the Science of Man which is the unique contribution of the Scots to the philosophy of the Enlightenment'.[24] According to Phillipson, then, the Scottish Enlightenment was, first and foremost, an extended meditation on the problem of political identity in a commercial polity that was no longer an independent state. He can thus be read as implying that society not nature was the primary focus of intellectual activity in the Scottish variant of the Enlightenment.[25]

By contrast, in 1976 Anand Chitnis found a place for both the human and the natural sciences in his *The Scottish Enlightenment*, the first book-length synthesis of research in the field. The story of the Scottish Enlightenment, as recounted by Chitnis, was one in which the desire for improvement played itself out in the institutions of the law, the Kirk and the universities, and which encompassed the nascent social sciences, as well as medicine, chemistry and the various branches of natural history.[26] An interpretation of the place of science and medicine in the Scottish Enlightenment akin to that found in

Chitnis was also advanced by John Christie. His avowed historiographical debts to Phillipson notwithstanding, Christie saw the culture of enlightenment in Scotland as embracing '[m]edicine, natural philosophy, idealist empiricism, polite literature, and agricultural improvement, pursued in an institutional and social environment constituted by university and polite society'.[27]

In his attention to the relations between science, medicine and enlightenment, Roger L. Emerson has most directly challenged the historiographical perspectives of Trevor-Roper and Phillipson. Going beyond the qualified claims of Chitnis and Christie, Emerson has argued that the pursuit of natural knowledge was central to enlightened culture in eighteenth-century Scotland. Rejecting the view of Phillipson and others who see the Scottish Enlightenment as a cultural and institutional response to the Union of 1707, Emerson has looked for the origins of the enlightenment in Scotland in the late seventeenth century and he has stressed the continuity of outlook between *virtuosi* and medical men active after 1660, such as Sir Robert Sibbald, with later Scottish savants. According to Emerson, the *virtuosi* wanted to improve Scotland through the cultivation of history, polite letters, moral philosophy, the natural sciences and medicine, as well as through the creation of institutions that would further their improving aims. Moreover, the *virtuosi* were inspired by the achievements of the Scientific Revolution, and sought to advance human knowledge in both natural and moral philosophy through the use of empirical methods. The reforming programme of the *virtuosi* was gradually applied in practice in the Scottish universities (with the assistance of patrons like the 3rd Duke of Argyll), the many clubs which flourished through the eighteenth century, and in learned bodies like the Philosophical Society of Edinburgh. Science and medicine thus gave shape and structure to the Scottish Enlightenment. But with increasing specialisation, the professionalisation of medicine, shifts in patronage and the fractures in the republic of letters that emerged during the troubled decade of the 1790s, the cultural world of the *virtuosi* disintegrated and the moment of the Scottish Enlightenment passed.[28]

Although Emerson's interpretation of the place of science and medicine in the Scottish Enlightenment has recently been challenged,[29] he has undoubtedly done more than any other individual scholar of the past two decades to illuminate the scientific and medical world of the *virtuosi* and their enlightened successors. He has reconstructed the outlook of the *virtuosi* through an analysis of the contents of their libraries.[30] He has examined in detail the institutional milieux of the *virtuosi* in studies of, variously, Sir Robert Sibbald's plan to found a Royal Society of Scotland, the Philosophical Society of Edinburgh, the Edinburgh Society for the Importation of Foreign Seeds and Plants and university patronage.[31] He has also helped to initiate one of the more significant shifts in the literature on eighteenth-century Scottish science and medicine that occurred in the 1980s, namely the move away from an exclusive focus on Edinburgh. In 1973, Phillipson stated that 'there is an important sense in which the history of the Scottish enlightenment *is* the history of Edinburgh', a comment later echoed by Chitnis.[32] Emerson, however, has consistently maintained a broader geographical focus and, in his contribution to the volume in which Phillipson's 1973 essay appeared, he sketched out a comparison between the urban contexts of Edinburgh, Glasgow and Aberdeen which has served as the starting point for much of the research subsequently published on science and medicine in the Aberdeen and Glasgow enlightenments.[33]

Consolidating the Historiographical Tradition

Over the course of the past two decades, the literature on science and medicine in the Scottish Enlightenment has built on the historiographical foundations laid in the 1970s, and our empirical knowledge of the period has grown considerably. In addition to Emerson's work on the universities, a number of studies have appeared on the teaching of the natural sciences in Scottish academe, while the development of the Edinburgh medical school from its inception in 1726 has been charted.[34] Other institutions associated with the natural sciences that have received attention include the Aberdeen Philosophical Society and the Royal Society of Edinburgh.[35] Histories of the major Scottish infirmaries of the period have been produced, along with accounts of Glasgow's Town's Hospital and Faculty of Physicians and Surgeons.[36] The ideology of improvement has likewise been examined, particularly with reference to agriculture.[37] The applications of natural knowledge in the technological sphere have similarly been considered, largely in terms of the career of James Watt.[38] The rise of Newtonianism in Scotland at the turn of the eighteenth century has been traced, and the various branches of the sciences have also been explored, most notably mathematics, chemistry, astronomy, optics and geology.[39] So too have facets of medical theory.[40] The ensemble of subjects which made up natural history in the Enlightenment has finally been given its due, as has the connection between natural history and the science of man.[41] The relations between philosophy, science and medicine have been analysed in studies which have shown that the structure of the Scottish universities in the eighteenth century nurtured close connections between the human and natural sciences.[42]

To some extent, the literature on eighteenth-century Scottish medicine and science published in the 1980s and 1990s has reflected broader historiographical trends in the history of science and medicine. Ludmilla Jordanova's pioneering essay on the illustrations to William Hunter's *The Anatomy of the Human Gravid Uterus*, for example, posed complex questions about gender, and about the interpretation of images in relation to medicine and science more generally.[43] Following on from Steven Shapin's treatment of the audience for science in eighteenth-century Edinburgh, Jan Golinski, Paul Wood and Charles Withers have discussed science as a form of 'public culture' and probed the meaning of the 'public sphere' in the Scottish Enlightenment.[44] Prompted by Shapin, Withers has taken up the role of trust in the context of natural historical and geographical inquiry, while Wood has used Shapin's work as his starting point for explorations of the identities of John Anderson and Thomas Reid.[45] Withers has utilised constructivist notions of the making of natural knowledge in his writings on geography and cartography, as has Judith Grabiner in her account of the meanings of Colin Maclaurin's attempt to apply mathematics to contemporary social problems in Scotland in the 1730s.[46] Lastly, historians such as Richard Sher have turned to the history of the book in order to delineate more fully the context in which science and medicine were pursued.[47]

The theoretical perspectives deployed in the literature have thus undergone a degree of change since the 1970s, even if it is also true that the field as a whole has remained fairly traditional in its historiographical approaches. It is precisely to expose these approaches to further scrutiny, and, in so doing, to identify new questions and point to new directions, that this collection has been assembled.

Rethinking Science and Medicine in the Scottish Enlightenment

The essays which follow engage with the body of work surveyed above and attempt to further our understanding of eighteenth-century Scottish science and medicine at both the empirical and conceptual levels. Generally speaking, students of Scottish science have hitherto paid insufficient attention to the history of scientific instruments, despite the existence of a rich literature on the subject. In her chapter, Alison Morrison-Low presents a wide-ranging synthesis of current knowledge about instruments and instrument-making in the Scottish Enlightenment that provides the details necessary for historians to fashion a more informed view of the material basis of scientific practice in the period. Moreover, Morrison-Low's classification of the different markets for instruments in the eighteenth century gives the reader an invaluable analytical resource for thinking about the place of instruments in enlightened culture, and her discussion points to important connections with recent work on consumerism and consumption in the eighteenth century.[48] In his chapter on map-making, Charles Withers investigates the social locations and cognitive dimensions of geography in the Scottish Enlightenment. Withers here turns to an aspect of geography linked to the subject of Morrison-Low's essay, namely the use of mathematical instruments to create more accurate maps of Scotland. He shows how the utilisation of such instruments in the field was socially mediated through networks of patronage, the ascription of trust and the attainment of credibility. Withers' chapter addresses, therefore, some of the same theoretical questions regarding representation that were earlier posed by Jordanova.

In their study of science and medicine in eighteenth-century Glasgow, Roger Emerson and Paul Wood take up the contentious issue of the relationship between enlightenment and the pursuit of scientific and medical knowledge. They profile Glasgow's scientific community and identify the major areas of theoretical and practical interest to Glasgow men of science in the period. They also shed light on the institutions and values which sustained public science in eighteenth-century Glasgow, and underline the differences between Glasgow's enlightenment and those not just of Edinburgh and Aberdeen but also those provincial enlightenments in England discussed by historians such as Roy Porter and Arnold Thackray.[49]

From the depiction of a community through time, we move in Judith Grabiner's chapter to the portrait of one of the leading figures of the Scottish Enlightenment, Colin Maclaurin. Grabiner surveys diverse facets of Maclaurin's career to reveal how his inquiries in fields ranging from moral theory to physics were given unity through his application of a style of mathematical reasoning derived from the works of Newton. Moreover, in looking at the social processes through which Maclaurin's form of mathematical reasoning gained authority in Scotland, she touches on themes similar to those found in Withers's essay. Both chapters can be seen to be making the same general point: viewed historically, mathematics is as socially and culturally situated as any other science.

With Anita Guerrini's study of George Garden and George Cheyne, we turn away from mathematics to the realm of natural history and medicine, and, at least initially, move north to Aberdeen. Aberdeen differed from Edinburgh and Glasgow as a centre for enlightenment partly because of the strength of Episcopalianism, as well as Catholicism and Quakerism, in the north-east of Scotland. It is in this complex religious context that Guerrini finds the key to understanding the theories of generation proposed by both Garden and Cheyne. Guerrini's chapter is a valuable reminder of not only the importance of generation as a topic for investigation in the eighteenth century, but also of the role of gender in scientific and medical thought, and the prominent part played by expatriate Scots like Cheyne in the making of the English Enlightenment.[50] Further, her comments on the relationship between religion and enlightenment highlight an historiographical problem which has yet to be addressed in relation to the Scottish Enlightenment.

The chapters by Stephen Brown and Warren McDougall draw on the insights of book historians to illuminate the careers of two men who made signal contributions to the print culture of eighteenth-century Scotland, William Smellie and Charles Elliot. Smellie is a fascinating figure whose historical significance is only now appreciated, and Brown offers a picture of him as a social and intellectual outsider who managed to combine a chequered career as a printer with a lifelong passion for medicine and natural history which he managed to channel into print. But if Smellie was somewhat fickle in his business dealings, McDougall shows Elliot to have been an astute entrepreneur who succeeded in marketing medical books by such authors as William Cullen across the Atlantic world. Elliot's acumen as a publisher ensured that the Scottish Enlightenment was exported to Europe and to America, and we see in the network of contacts in the book trade that Elliot built up on both sides of the Atlantic the routes through which medical knowledge and enlightened thought circulated in the latter part of the eighteenth century.

Through his publishing Elliot was instrumental in consolidating Edinburgh's reputation as a 'hotbed of genius'. As Emerson and Wood suggest in their chapter, Glasgow was also notable as a centre for medical knowledge, and Fiona Macdonald takes up their story in her examination of the notebooks of the Glasgow physician Robert Cleghorn. Like other recent historians of medicine, Macdonald is interested in using such notebooks to recover the details of physician-patient relationships and to shed light on the development of clinical medicine in the eighteenth century. Significantly, the notebooks show that unlike many of his medical contemporaries, Cleghorn actually examined his patients physically, and that he worked closely with a number of local surgeons, a fact which underscores a distinctive feature of medical culture in eighteenth-century Glasgow. As Macdonald suggests, Cleghorn's notebooks raise a number of epistemological questions of broader import having to do with trust and the reliability of historical records.

The last two chapters, by Stuart Hartley and Charles Waterston, take us into the period when the Scottish Enlightenment was at an end. Hartley deals with the subject which dominated Scottish science in the late eighteenth century, namely geology, but looks at it by focusing on the practice of field work, rather than on purely theoretical concerns or on leading figures like James Hutton. Like Macdonald, he uses archival materials along with printed sources to document the different approaches to working in the field adopted by Robert Jameson, Sir James Hall and Sir George Stuart Mackenzie. Hartley provides fascinating insights into the ways in which these men related theory and evidence in their practice, and poses further questions about the role of trust in scientific inquiry. Mackenzie is also the subject of Charles Waterston's chapter, which provides the first detailed delineation of the full range of Mackenzie's interests as a man of science. Waterston's discussion of Mackenzie's career marks a fitting close to the collection, because it tells the story of someone who was, to use John Christie's phrase, a multi-competent intellect, and who was thus more at home in the Enlightenment than in the new cognitive and social regime of specialisation and professionalisation of the nineteenth century.

Notes

The references in the notes which follow are in no way exhaustive; they are intended to serve as a reasonably detailed guide to the literature on eighteenth-century Scottish science and medicine produced in the period *c.* 1970 to 2001.

1. 'Internal' factors were considered to be the purely intellectual aspects of scientific inquiry, whereas the 'external' factors were taken to encompass the political, social and economic conditions in which the pursuit of science takes place. Religious and

philosophical ideas were also sometimes treated as being external to science. For an influential discussion of this distinction, see A. Rupert Hall, 'Merton Revisited, or Science and Society in the Seventeenth Century', *History of Science* 2 (1963), 1–16.
2. Christopher Lawrence, 'The Nervous System and Society in the Scottish Enlightenment', in Barry Barnes and Steven Shapin (eds.), *Natural Order: Historical Studies of Scientific Culture* (Beverley Hills and London, 1979), 19–40. For classic statements of the strong programme, see Barry Barnes, *Scientific Knowledge and Sociological Theory* (London and Boston, 1974); *idem, Interests and the Growth of Knowledge* (London, Henley and Boston, 1977); David Bloor, *Knowledge and Social Imagery* (London, Boston and Henley, 1976).
3. That there were sharp disagreements over theoretical issues in the period can be seen in the exchange between G. N. Cantor and Steven Shapin published as 'Phrenology in Early Nineteenth-Century Edinburgh: An Historiographical Discussion', *Annals of Science* 32 (1975), 195–256.
4. Arthur Lovekin Donovan, 'The Origins of Pneumatic Chemistry: William Cullen, Joseph Black and the Unification of Natural Philosophy' (unpublished Ph.D. thesis, Princeton University, 1970); *idem, Philosophical Chemistry in the Scottish Enlightenment: The Doctrines and Discoveries of William Cullen and Joseph Black* (Edinburgh, 1975).
5. J. B. Morrell, 'The University of Edinburgh in the Late Eighteenth Century: Its Scientific Eminence and Academic Structure', *Isis* 62 (1970), 158–71; *idem*, 'Professors Robison and Playfair, and the *Theophobia Gallica*: Natural Philosophy, Religion and Politics in Edinburgh, 1789–1815', *Notes and Records of the Royal Society of London* 26 (1971), 43–63; *idem*, 'The Leslie Affair: Careers, Kirk and Politics in Edinburgh in 1805', *Scottish Historical Review* 54 (1975), 63–82.
6. Steven Arthur Shapin, 'The Royal Society of Edinburgh: A Study of the Social Context of Hanoverian Science' (unpublished Ph.D. thesis, University of Pennsylvania, 1971); *idem*, 'Property, Patronage and the Politics of Science: The Founding of the Royal Society of Edinburgh', *The British Journal for the History of Science* 7 (1974), 1–41; R. G. W. Anderson and A. D. C. Simpson (eds.), *The Early Years of the Edinburgh Medical School* (Edinburgh, 1976).
7. G. N. Cantor, 'Henry Brougham and the Scottish Methodological Tradition', *Studies in History and Philosophy of Science* 2 (1971), 69–89; *idem*, 'The Academy of Physics at Edinburgh, 1797–1800', *Social Studies of Science* 5 (1975), 109–34.
8. Richard Olson, 'The Reception of Boscovich's Ideas in Scotland', *Isis* 60 (1969), 91–103; *idem*, 'Scottish Philosophy and Mathematics, 1750–1830', *Journal of the History of Ideas* 32 (1971), 29–44; *idem, Scottish Philosophy and British Physics, 1750–1880: A Study in the Foundations of the Victorian Scientific Style* (Princeton, NJ, 1975).
9. For criticism of Olson's book, see G. N. Cantor's review in *The British Journal for the History of Science* 10 (1977), 81–84, and John R. R. Christie, 'Influencing People', *Annals of Science* 33 (1976), 311–18.
10. See, *inter alia*, D. J. Bryden, 'James Short, M.A., F.R.S., Optician Solely for Reflecting Telescopes', *University of Edinburgh Journal* 24 (1969–70), 251–61; *idem, Scottish Scientific Instrument-Makers, 1600–1900* (Edinburgh, 1972); R. G. W. Anderson, *The Playfair Collection and the Teaching of Chemistry at the University of Edinburgh, 1713–1858* (Edinburgh, 1978).
11. John R. R. Christie, 'The Origins and Development of the Scottish Scientific Community, 1680–1760', *History of Science* 12 (1974), 122–41; *idem*, 'The Rise and Fall of Scottish Science', in Maurice Crosland (ed.), *The Emergence of Science*

in Western Europe (London and Basingstoke, 1975), 111–26. Unfortunately, some of Christie's more provocative papers (including that given to the King's College, Cambridge seminar on the Scottish Enlightenment) remain unpublished.
12. Christie, 'Rise and Fall', 111.
13. Christie, 'Scottish Scientific Community', 127–28, 133–34, and 'Rise and Fall', esp. 114–18.
14. Christie, 'Rise and Fall', 118–19, 121–22. Much of the work done on Scottish science in the 1970s was influenced by George Davie's *The Democratic Intellect: Scotland and her Universities in the Nineteenth Century* (Edinburgh, 1961).
15. Christie, 'Scottish Scientific Community', 126–27.
16. Christie, 'Rise and Fall', 120.
17. John R. R. Christie, 'Ether and the Science of Chemistry', in G. N. Cantor and M. J. S. Hodge (eds.), *Conceptions of Ether: Studies in the History of Ether Theories, 1740–1900* (Cambridge, 1981), 85–110; *idem*, 'Joseph Black and John Robison', in A. D. C. Simpson (ed.), *Joseph Black, 1728–1799: A Commemorative Symposium* (Edinburgh, 1982), 47–52; *idem*, 'The Culture of Science in Eighteenth-Century Scotland', in Andrew Hook (ed.), *The History of Scottish Literature, Volume 2: 1660–1800* (Aberdeen, 1987), 291–305; *idem*, 'William Cullen and the Practice of Chemistry', in A. Doig, J. P. S. Ferguson, I. A. Milne and R. Passmore (eds.), *William Cullen and the Eighteenth Century Medical World* (Edinburgh, 1993), 98–109.
18. For a recent survey of the literature on the Scottish Enlightenment which is relevant to the concerns of this introduction, see Paul Wood, 'Dugald Stewart and the Invention of "the Scottish Enlightenment"', in Paul Wood (ed.), *The Scottish Enlightenment: Essays in Reinterpretation* (Rochester, NY, 2000), 1–35.
19. Hugh Trevor-Roper, 'The Scottish Enlightenment', *Studies on Voltaire and the Eighteenth Century* 58 (1967), 1635–58.
20. Trevor-Roper, 'The Scottish Enlightenment', 1639.
21. Hugh Trevor-Roper, 'The Scottish Enlightenment', *Blackwood's Magazine* 322 (1977), 371–88, at pp. 373–74.
22. While Trevor-Roper's approach has found little favour among historians of the Scottish Enlightenment, his student John Robertson has advanced a similar interpretation; see, for example, John Robertson, 'The Scottish Contribution to the Enlightenment', in Wood, *The Scottish Enlightenment*, 37–62.
23. N. T. Phillipson, 'Culture and Society in the Eighteenth Century Province: The Case of Edinburgh and the Scottish Enlightenment', in Lawrence Stone (ed.), *The University in Society: Studies in the History of Higher Education*, 2 vols. (Princeton, NJ, 1974), 2:407–48, esp. pp. 436, 440–41; see also Phillipson's 'Towards a Definition of the Scottish Enlightenment', in Paul Fritz and David Williams (eds.), *City and Society in the Eighteenth Century* (Toronto, 1973), 125–47.
24. Nicholas Phillipson, 'The Scottish Enlightenment', in Roy Porter and Mikuláš Teich (eds.), *The Enlightenment in National Context* (Cambridge, 1981), 19–40, at p. 22.
25. George Davie advanced a similar interpretation in his pamphlet *The Scottish Enlightenment* (London, 1981), subsequently reprinted in his *The Scottish Enlightenment and Other Essays* (Edinburgh, 1991), 1–50.
26. Anand Chitnis, *The Scottish Enlightenment: A Social History* (London, 1976), 1–10. For Chitnis's later critique of Phillipson and Davie, see Anand C. Chitnis, 'Agricultural Improvement, Political Management and Civic Virtue in Enlightened Scotland: An Historiographical Critique', *Studies on Voltaire and the*

Eighteenth Century 245 (1986), 475–88; 'The Eighteenth-Century Scottish Intellectual Inquiry: Context and Continuities versus Civic Virtue', in Jennifer J. Carter and Joan H. Pittock (eds.), *Aberdeen and the Enlightenment* (Aberdeen, 1987), 77–92.
27. Christie, 'Culture of Science', 294.
28. Roger L. Emerson, 'Natural Philosophy and the Problem of the Scottish Enlightenment', *Studies on Voltaire and the Eighteenth Century* 242 (1986), 243–91; *idem*, 'Science and the Origins and Concerns of the Scottish Enlightenment', *History of Science* 26 (1988), 333–66.
29. Robertson, 'Scottish Contribution to the Enlightenment', 40–41; Richard B. Sher, 'Science and Medicine in the Scottish Enlightenment: The Lessons of Book History', in Wood, *Scottish Enlightenment*, 99–156, esp. pp. 99–112.
30. Roger L. Emerson, 'Scottish Cultural Change, 1660–1710, and the Union of 1707', in John Robertson (ed.), *A Union for Empire: Political Thought and the Union of 1707* (Cambridge, 1995), 121–44; *idem*, 'Catalogus Librorum A.C.D.A.: The Library of Archibald Campbell, Third Duke of Argyll (1682–1761)', in Paul Wood (ed.), *The Culture of the Book in the Scottish Enlightenment* (Toronto, 2000), 13–39.
31. Roger Emerson, 'Sir Robert Sibbald, Kt, the Royal Society of Scotland and the Origins of the Scottish Enlightenment', *Annals of Science* 45 (1988), 41–72; *idem*, 'The Philosophical Society of Edinburgh, 1737–1747', *The British Journal for the History of Science* 12 (1979), 154–91; *idem*, 'The Philosophical Society of Edinburgh, 1748–1768', *The British Journal for the History of Science* 14 (1981), 133–76; *idem*, 'The Philosophical Society of Edinburgh, 1768–1783', *The British Journal for the History of Science* 18 (1985), 255–303; *idem*, 'The Scottish Enlightenment and the End of the Philosophical Society of Edinburgh', *The British Journal for the History of Science* 21 (1988), 33–66; *idem*, 'The Edinburgh Society for the Importation of Foreign Seeds and Plants, 1764–1773', *Eighteenth-Century Life* 7 (1982), 73–95; *idem*, 'Lord Bute and the Scottish Universities, 1760–1792', in Karl W. Schweizer (ed.), *Lord Bute: Essays in Reinterpretation* (Leicester, 1988), 147–79; *idem*, *Professors, Patronage and Politics: The Aberdeen Universities in the Eighteenth Century* (Aberdeen, 1992); *idem*, 'Medical Men, Politicians and the Medical Schools at Glasgow and Edinburgh, 1685–1803', in Doig, Ferguson, Milne and Passmore, *William Cullen*, 186–215; *idem*, 'Politics and the Glasgow Professors, 1690–1800', in Andrew Hook and Richard B. Sher (eds.), *The Glasgow Enlightenment* (East Linton, UK, 1995), 21–39.
32. Phillipson, 'Towards a Definition', 125; Chitnis, *Scottish Enlightenment*, 5.
33. Roger L. Emerson, 'The Enlightenment and Social Structures', in Fritz and Williams, *City and Society*, 99–124. See, for example, Paul Wood, 'Science and the Aberdeen Enlightenment', in Peter Jones (ed.), *Philosophy and Science in the Scottish Enlightenment* (Edinburgh, 1988), 39–66; Charles W. J. Withers, 'Toward a Historical Geography of Enlightenment in Scotland', in Wood, *Scottish Enlightenment*, 63–97.
34. Christine M. Shepherd, 'Newtonianism in Scottish Universities in the Seventeenth Century', in R. H. Campbell and Andrew S. Skinner (eds.), *The Origins and Nature of the Scottish Enlightenment* (Edinburgh, 1982), 65–85; John S. Reid, 'Patrick Copland, 1748–1822: Aspects of His Life and Times at Marischal College', *Aberdeen University Review* 50 (1984–85), 359–79; *idem*, 'Late Eighteenth-Century Adult Education in the Sciences at Aberdeen: The Natural Philosophy Classes of Professor Patrick Copland', in Carter and Pittock, *Aberdeen*

and the Enlightenment, 168–79; A. D. Boney, *The Lost Gardens of Glasgow University* (London, 1988); Paul Wood, *The Aberdeen Enlightenment: The Arts Curriculum in the Eighteenth Century* (Aberdeen, 1993); Christopher Lawrence, 'Medicine as Culture: Edinburgh and the Scottish Enlightenment' (unpublished Ph.D. thesis, University of London, 1984); *idem*, 'Ornate Physicians and Learned Artisans: Edinburgh Medical Men, 1726–1776', in W. F. Bynum and Roy Porter (eds.), *William Hunter and the Eighteenth-Century Medical World* (Cambridge, 1985), 153–76; *idem*, 'The Edinburgh Medical School and the End of the "Old Thing", 1790–1830', *History of Universities* 7 (1988), 259–86; Rosalie M. Stott, 'The Incorporation of Surgeons and Medical Education and Practice in Edinburgh, 1696–1755' (unpublished Ph.D. thesis, University of Edinburgh, 1984); Lisa Rosner, *Medical Education in the Age of Improvement: Edinburgh Students and Apprentices, 1760–1826* (Edinburgh, 1991); L. S. Jacyna, *Philosophic Whigs: Medicine, Science and Citizenship in Edinburgh, 1789–1848* (London and New York, 1994).

35. H. Lewis Ulman (ed.), *The Minutes of the Aberdeen Philosophical Society, 1758–1773* (Aberdeen, 1990); Charles D. Waterston, *Collections in Context: The Museum of the Royal Society of Edinburgh and the Inception of a National Museum for Scotland* (Edinburgh, 1997). Under the rubric 'Scotland's Cultural Heritage', a project sponsored by the late Eric G. Forbes produced three volumes of short biographies of early Fellows of the Royal Society of Edinburgh edited by Sheila Devlin-Thorpe in 1981–82.

36. Guenter B. Risse, *Hospital Life in Enlightenment Scotland: Care and Teaching at the Royal Infirmary of Edinburgh* (Cambridge, 1986); Derek A. Dow, *Paisley Hospital: The Royal Alexandra Infirmary and Allied Institutions, 1786–1986* (Glasgow, 1988); Iain D. Levack and Hugh A. F. Dudley, *Aberdeen Royal Infirmary: The People's Hospital of the North-East* (London, 1992); Jacqueline L. M. Jenkinson, Michael S. Moss and Iain F. Russell, *The Royal: The History of the Glasgow Royal Infirmary, 1794–1994* (Glasgow, 1994); Fiona A. Macdonald, 'The Infirmary of the Glasgow Town's Hospital, 1733–1800: A Case for Voluntarism?', *Bulletin of the History of Medicine* 73 (1999), 64–105; *idem*, 'The Infirmary of the Glasgow Town's Hospital: Patient Care, 1733–1800', in Wood, *Scottish Enlightenment*, 199–238; Johanna Geyer-Kordesch and Fiona Macdonald, *Physicians and Surgeons in Glasgow: The History of the Royal College of Physicians and Surgeons of Glasgow, 1599–1858* (London and Rio Grande, 1999).

37. For an important discussion of the concept of utility, see J. V. Golinski, 'Utility and Audience in Eighteenth-Century Chemistry: Case Studies of William Cullen and Joseph Priestley', *The British Journal for the History of Science* 21 (1988), 1–31. On agricultural improvement, see Charles W. J. Withers, 'A Neglected Scottish Agriculturalist: The "Georgical Lectures" and Agricultural Writings of the Rev. Dr. John Walker (1731–1803)', *Agricultural History Review* 33 (1985), 132–46; *idem*, 'William Cullen's Agricultural Lectures and Writings and the Development of Agricultural Science in Eighteenth-Century Scotland', *Agricultural History Review* 37 (1989), 144–56; *idem*, 'On Georgics and Geology: James Hutton's "Elements of Agriculture" and Agricultural Science in Eighteenth-Century Scotland', *Agricultural History Review* 42 (1994), 138–49.

38. For a suggestive treatment of Watt, see David Philip Miller, ' "Puffing Jamie": The Commercial and Ideological Importance of Being a "Philosopher" in the Case of the Reputation of James Watt (1736–1819)', *History of Science* 38 (2000), 1–24.

39. Anita Guerrini, 'The Tory Newtonians: Gregory, Pitcairne and their Circle', *Journal of British Studies* 25 (1986), 288–311; Ian Tweddle, *James Stirling: 'This about series and such things'* (Edinburgh, 1988); Niccolò Guicciardini, *The Development of Newtonian Calculus in Britain, 1700–1800* (Cambridge, 1989); Judith V. Grabiner, 'A Mathematician among the Molasses Barrels: Maclaurin's Unpublished Memoir on Volumes', *Proceedings of the Edinburgh Mathematical Society* 39 (1996), 193–240; idem, 'Was Newton's Calculus a Dead End?: The Continental Influence of Maclaurin's *Treatise of Fluxions*', *American Mathematical Monthly* 104 (1997), 393–410; Simpson, *Joseph Black, 1728–1799*; Carleton E. Perrin, 'A Reluctant Catalyst: Joseph Black and the Edinburgh Reception of Lavoisier's Chemistry', *Ambix* 29 (1982), 141–76; idem, 'Joseph Black and the Absolute Levity of Phlogiston', *Annals of Science* 40 (1983), 109–37; David Gavine, 'Astronomy in Scotland, 1745–1900', 2 vols. (unpublished Ph.D. thesis, Open University, 1982); D. J. Bryden, 'The Edinburgh Observatory, 1736–1811: A Story of Failure', *Annals of Science* 47 (1990), 445–74; Kurt Møller Pedersen, 'Roger Joseph Boscovich and John Robison on Terrestrial Aberration', *Centaurus* 24 (1980), 335–45; Geoffrey Cantor, *Optics after Newton: Theories of Light in Britain and Ireland, 1704–1840* (Manchester, 1983). The literature on Scottish geology remains focused largely on James Hutton. For important studies of Hutton, see: Arthur Donovan and Joseph Prentiss, *James Hutton's Medical Dissertation* (Philadelphia, 1980); Jean Jones, 'James Hutton and the Forth and Clyde Canal', *Annals of Science* 39 (1982), 255–63; idem, 'James Hutton: Exploration and Oceanography', *Annals of Science* 40 (1983), 81–94; idem, 'The Geological Collection of James Hutton', *Annals of Science* 41 (1984), 223–44; idem, 'James Hutton's Agricultural Research and His Life as a Farmer', *Annals of Science* 42 (1985), 573–601; idem, 'James Hutton', in David Daiches, Peter Jones and Jean Jones (eds.), *A Hotbed of Genius: The Scottish Enlightenment, 1730–1790* (Edinburgh, 1986), 116–36; Dennis R. Dean, *James Hutton and the History of Geology* (Ithaca, NY and London, 1992). For a study of John Walker, see M. D. Eddy, 'Geology, Mineralogy and Time in John Walker's University of Edinburgh Natural History Lectures (1779–1803)', *History of Science* 39 (2001), 95–119.

40. Andrew Cunningham, 'Sydenham versus Newton: The Edinburgh Fever Dispute of the 1690s between Andrew Brown and Archibald Pitcairne', *Medical History*, Supplement 1 (1981), 71–98; Anita Guerrini, 'James Keill, George Cheyne and Newtonian Physiology, 1690–1740', *Journal of the History of Biology* 18 (1985), 247–66; idem, 'Archibald Pitcairne and Newtonian Medicine', *Medical History* 31 (1987), 70–83; Akihito Suzuki, 'Psychiatry without Mind in the Eighteenth Century: The Case of British Iatro-Mathematicians', *Archives Internationales d'Histoire des Sciences* 48 (1998), 119–46; Guenter B. Risse, 'Hysteria at the Edinburgh Infirmary: The Construction and Treatment of a Disease, 1770–1800', *Medical History* 32 (1988), 1–22; Michael Barfoot, 'Brunonians under the Bed: An Alternative to University Medicine in Edinburgh in the 1780s', in William F. Bynum and Roy Porter (eds.), *Brunonianism in Britain and Europe* (London, 1988), 22–45. Essays on medical theory are also to be found in Bynum and Porter, *William Hunter*, and Doig, Ferguson, Milne and Passmore, *William Cullen*.

41. A. G. Morton, *John Hope, 1725–1786: Scottish Botanist* (Edinburgh, 1986); Charles W. J. Withers, 'The Rev. Dr. John Walker and the Practice of Natural History in Late Eighteenth Century Scotland', *Archives of Natural History* 18 (1991), 201–20; idem, 'Natural Knowledge as Cultural Property: Disputes over

the "Ownership" of Natural History in Late Eighteenth-Century Edinburgh', *Archives of Natural History* 19 (1992), 289–303; *idem*, 'Geography, Natural History and the Eighteenth-Century Enlightenment: Putting the World in Place', *History Workshop Journal* 39 (1995), 136–63; Paul Wood, 'Buffon's Reception in Scotland: The Aberdeen Connection', *Annals of Science* 44 (1987), 169–90; *idem*, 'The Natural History of Man in the Scottish Enlightenment', *History of Science* 28 (1990), 89–123.

42. See especially Michael Barfoot, 'James Gregory (1753–1821) and Scottish Scientific Metaphysics, 1750–1800' (unpublished Ph.D. thesis, University of Edinburgh, 1983); *idem*, 'Hume and the Culture of Science in the Early Eighteenth Century', in M. A. Stewart (ed.), *Studies in the Philosophy of the Scottish Enlightenment* (Oxford, 1990), 151–90; *idem*, 'Philosophy and Method in Cullen's Medical Teaching', in Doig, Ferguson, Milne and Passmore, *William Cullen*, 110–32; John P. Wright, 'Metaphysics and Physiology: Mind, Body and the Animal Economy in Eighteenth-Century Scotland', in Stewart, *Scottish Enlightenment*, 251–301; Paul Wood (ed.), *Thomas Reid on the Animate Creation: Papers Relating to the Life Sciences* (Edinburgh, 1995).

43. L. J. Jordanova, 'Gender, Generation and Science: William Hunter's Obstetrical Atlas', in Bynum and Porter, *William Hunter*, 385–412.

44. Steven Shapin, 'The Audience for Science in Eighteenth Century Edinburgh', *History of Science* 12 (1974), 95–121; Jan Golinski, *Science as Public Culture: Chemistry and Enlightenment in Britain, 1760–1820* (Cambridge, 1992); Paul Wood, 'Science, the Universities and the Public Sphere in Eighteenth-Century Scotland', *History of Universities* 13 (1994), 99–135; Charles W. J. Withers, 'Towards a History of Geography in the Public Sphere', *History of Science* 37 (1999), 45–78.

45. Charles W. J. Withers, 'Reporting, Mapping, Trusting: Making Geographical Knowledge in the Late Seventeenth Century', *Isis* 90 (1999), 497–521; *idem*, 'Travel and Trust in the Eighteenth Century', in John Renwick (ed.), *L'Invitation au voyage: Studies in Honour of Peter France* (Oxford, 2000), 47–54; Paul Wood, ' "Jolly Jack Phosphorus" in the Venice of the North; or, Who was John Anderson?', in Hook and Sher, *The Glasgow Enlightenment*, 111–32; *idem*, 'Chi era Thomas Reid?', in Antonio Santucci (ed.), *Filosofia e cultura nel Settecento Britannico*, 2 vols. (Bologna, 2000), 2:275–300.

46. Charles W. J. Withers, 'How Scotland came to know Itself: Geography, National Identity and the Making of a Nation, 1680–1790', *Journal of Historical Geography* 21 (1995), 371–97; *idem*, 'Geography, Science and National Identity in Early Modern Britain: The Case of Scotland and the Work of Sir Robert Sibbald (1641–1722)', *Annals of Science* 53 (1996), 29–73; Judith V. Grabiner, ' "Some Disputes of Consequence": Maclaurin among the Molasses Barrels', *Social Studies of Science* 28 (1998), 139–68.

47. Richard B. Sher, 'William Buchan's *Domestic Medicine*: Laying Book History Open', in Peter Isaac and Barry McKay (eds.), *The Human Face of the Book Trade: Print Culture and Its Creators* (Winchester and New Castle, DE, 1999), 45–64; *idem*, 'Science and Medicine in the Scottish Enlightenment'; Wood, *The Culture of the Book in the Scottish Enlightenment*, 99–156.

48. This work informs John Brewer's *The Pleasures of the Imagination: English Culture in the Eighteenth Century* (London, 1997), which provides a useful context for considering the broader implications of Morrison-Low's chapter.

49. Arnold Thackray, 'Natural Knowledge in Cultural Context: The Manchester Model', *American Historical Review* 79 (1974), 672–709; Roy Porter, 'Science,

Provincial Culture and Public Opinion in Enlightenment England', *British Journal for Eighteenth-Century Studies* 3 (1980), 20–46.
50. Guerrini has also discussed the role of the Scots in the metropolis in ' "A Scotsman on the Make": The Career of Alexander Stuart', in Wood, *Scottish Enlightenment*, 157–76.

2
'Feasting my eyes with the view of fine instruments': Scientific Instruments in Enlightenment Scotland, 1680–1820

A. D. MORRISON-LOW

My acquaintance with Mr Watt began in 1758. I was then a Student in the University of Glasgow, and was then studying the Science [natural philosophy] which I now profess to teach . . . Mr Watt came to settle in Glasgow as a Math[ematica]l and Phil[osophica]l Instrument maker, and was employed to repair and fit up a very noble collection of Instruments bequeathed to the University by Mr. McFarlane of Jamaica, a Gentleman well known to the scientific world. Mr Watt had apartments, and a workshop within the College . . . [I was taken by various eminent professors] into Mr Watt's shop, and when he saw me thus patronized and introduced, his natural complaisance made him readily indulge my curiosity. After first feasting my eyes with the view of fine instruments, and prying into every thing, I conversed with Mr Watt. I saw a workman, and expected no more, but was surprized to find a Philosopher as young as myself, and always ready to instruct me.[1]

Thus wrote John Robison, the Edinburgh Professor of Natural Philosophy, in 1796 of his introduction while still a student to the engineer James Watt. After the death of Joseph Black, Robison explained to Watt his reasons for undertaking an edition of Black's lectures for publication, and mentioned how he had first encountered the famous chemist: 'My first acquaintance with Dr Black began in your Rooms, where you was rubbing up McFarlanes Instruments. Dr Black used to come in, and, standing with his back to us, amuse himself with Birds Quadrant, whistling softly to himself'.[2] These 'fine instruments' were a part of Watt's general merchandise, which included musical instruments as well as those used in natural philosophy, astronomy and chemistry.[3] 'Scientific instruments' is, in fact, a singularly inappropriate term for the surviving material culture which was once used in a number of loosely connected, quasi-technical spheres. The term is, perhaps, both broader and narrower than it might suggest, but despite its inherent anachronisms, 'scientific instrumentation' remains an important component of Scotland's contribution to the making of the modern world. This essay will sketch out what is meant by 'scientific instruments' today, what they were and what they were not during the long eighteenth century, and what instruments were to be found in Scotland in the period. It will also look at the markets for such devices, and how these changed and grew during the Scottish Enlightenment. Then, as now, scientists needed instruments, particularly ones which enhanced the five senses, or could be used to measure observed phenomena. The conduct of science needed what became known in the late nineteenth century as 'scientific instruments'. These cover the wide range of instruments used by 'mathematical practitioners' – the people who used some form of measuring or optical devices in their everyday jobs. Mariners, land surveyors, architects and teachers of these subjects all required different tools for various aspects of their work. In particular, the role of the extramural teacher, whether itinerant or resident, proved to be a catalyst for promoting the use of instrumentation, especially in parts of the market not hitherto discussed in the scholarly literature. This essay will explore the development of the use of such devices in Scotland during the long eighteenth century, their historiography, their users and the genesis of an indigenous trade. It will become apparent that there is still considerable room for further investigation in this subject.

Since 1974, when pioneering papers by Jack Morrell, John Christie and Steven Shapin defining the scientific component of the Scottish Enlightenment were published, a historiographical problem has emerged.[4] Just as instrument historians have paid little attention to other historians in related fields – notably to historians of science, since 'issues important to instrument historians were not embraced until recently by historians of science, whose post-war work focused on issues more closely akin to those of philosophers and historians of idea'[5] – historians of the Scottish Enlightenment appear to diverge, depending on their particular interest in the huge range of issues covered by the subject. In particular, the stress has fallen on the philosophical nature of the period, rather than its scientific content, and this has meant that the case for the practical scientific component – the investigation of the natural world as opposed to the exploration of the human sphere – is underplayed.[6] Paul Wood has put this bluntly: 'many eminent writers on eighteenth-century Scotland still tend to exclude science from their purview when defining the characteristics of the Scottish Enlightenment'.[7] If such a large and important topic as science has been excluded, it is small wonder that its tools have been overlooked.

The Scottish Enlightenment was predominantly an urban phenomenon, and was centred around the three largest cities – Edinburgh, Glasgow and Aberdeen – and in all these cases the universities of each city had a major role to play. Recent work has looked at the history of these institutions, their patronage systems, their teaching and their networks of knowledge and friendship. In some instances, this has been done with regard to the practice of science. The role of Edinburgh's provost, George Drummond, in placing 'proved specialist scholars, preferably of international reputation, who were appointed to bring renown to the University through their specialised published research' was highlighted in the mid-1970s, while more recent work (including Chapter 4 below) has shown similar patterns emerging in Glasgow and Aberdeen.[8]

Who were the men who helped these 'specialist scholars' realise their researches? The shadowy figure of what amounted to a forerunner of today's university technician could not have supported himself in this role alone: he had to diversify in order to survive economically. By developing his expertise and applying his skills, he was able to make items which appealed to a broad and growing range of customers. Just how the instrument trade operated on a day-to-day basis remains largely a mystery: no substantial papers or accounts from any Scottish eighteenth-century business survive.[9] What does emerge from surviving information demonstrates that there was a small, but strongly rooted, indigenous trade in Scotland from about the mid-eighteenth century, and that it differed in character from that found in London, which by then was dominating the world market for instruments.[10]

By the beginning of the eighteenth century, Scotland had four universities, or six colleges, all of which taught some form of science as a part of the undergraduate arts curriculum. Ronald Cant dates the acquisition of instruments at St. Andrews to 1673, at Edinburgh to 1674, at Glasgow to 1658; and we shall see that Marischal College, Aberdeen, had a bequest of instruments as early as 1613. It was at Glasgow in 1712–13 that a special 'course of experiments' was begun, with similar courses initiated at St. Andrews in 1714 by Charles Gregory, at Marischal College in 1726 by Gregory's former pupil Daniel Gordon, while Glasgow's course was put on a more regular basis by Robert Dick I in 1727. Meanwhile, Colin Maclaurin, having graduated from Glasgow in 1713, was appointed Professor of Mathematics at Marischal in 1717 but moved to the chair at Edinburgh in 1725.[11]

Instruments were, of course, used in other scientific subjects, particularly chemistry and medicine, but the greatest variety were called for in the natural philosophy classroom. Several university teachers of these subjects were, by the mid-eighteenth century, undertaking consultancy work in industry and commerce, and needed assistance in this area, as well as having apparatus made for their more orthodox classroom work.[12] Edinburgh, in particular, required active craftsmen, for although there had been no Parliament since 1707, the city was the headquarters of post-Union Boards run by placemen: Excise, Manufactures, Fisheries. It was also the main venue for the law, the Church, various learned societies, schools, the Northern Lighthouse Board and a vibrant publishing trade, to say nothing of the building of the New Town from 1760. Potential patrons and customers were there for enterprising craftsmen to attract and, indeed, the period saw rapid and interesting developments in advertising techniques.[13] The broad output of any Scottish workshop was proportionately tiny, however, when it came to advancing 'science'; but the broader category of what we now call 'scientific instruments' made a larger impact qualitatively and quantitatively on the Scottish way of life, in particular for the upwardly mobile consumer.

The Nature of Scientific Instruments

In making the case for seeing the Enlightenment in Scotland as a logical extension of the Scientific Revolution in that country, Paul Wood has described the subject of the Scientific Revolution in Scotland as 'an historiographical minefield'.[14] Understanding the tools of that revolution is no less fraught with difficulties. The first of these is defining the term 'scientific instrument', which, as a catch-all description, has come to include various classes of instrument and to exclude others. Most of the instruments are not by any stretch of the imagination 'scientific'. Coined towards the end of the nineteenth century as antiquarian interest was aroused in these historic artefacts, and curiously not as an extension of interest in the history of technology, the term 'scientific instrument' has come to mean the material culture survivals of a former, if not also a more attractive, time.[15] 'Scientific instruments' are items currently prized for their external beauty, brassy glitter and, above all, for the name of the workshop which produced them. By the eighteenth century this was usually a London maker, but in an earlier period, it would have been a continental *atelier*. Their use – certainly as perceived by most contemporary collectors, antique dealers and even by museum curators – has now become secondary to their appearance and to the cultural significance invested in them by these interested parties as symbols of a more elegant but bygone age. In contrast, today's scientific instruments are, especially to the layman, the externally unromantic, functional and dreary-looking black boxes of Nathan Rosenberg (which prove to be exciting economic engines within),[16] but that is another essay. When 'scientific instruments' become historic, they take on a respectability which they may not have had when they were working or up-to-the-minute instruments. Then they were the practical tools which extended the human senses; now, they are valuable and beautiful artefacts.[17]

For the period under discussion, there were no such classifications as 'scientific instrument' or 'precision instrument'. These terms have taken on specific meanings for contemporary historians of instruments. 'Scientific instruments' were made in London, as well as elsewhere in England, and, as we shall see, in Scotland. The manufacture of 'scientific instruments' provided the everyday trade, which produced the basic expertise and financial reserves for the production of 'precision' items. By contrast, 'precision instrument' now means a finely divided, usually London-made, probably large astronomical or first-order surveying item made by a leading craftsman that would have proved expensive in terms of labour, skill and materials.

Contemporaries classified instruments by their use as 'mathematical', 'optical' and 'philosophical' instruments, and they described themselves as makers of such in street directories, in advertisements and on trade-cards. Historically, 'mathematical' instruments were the first to be developed, and included such items as sundials, logarithmic scales and sighting instruments. Any device which had a graduated scale and which was used to measure angles or distance or perform calculations could be regarded as a mathematical instrument. For such an instrument to be useful, both maker and user had to be numerate and literate, and the form of knowledge sought transferable between them through the instrument's markings. 'Optical' instruments were developed after the discovery, probably made in the Netherlands at the end of the sixteenth century, that two or more lenses could enlarge distant or very small objects; this discovery led to the construction of the telescope and of the microscope. There is still controversy about the identity and nationality of the individuals who actually invented the telescope and the microscope.[18] It would seem that improvements in glass production allowed spectacles to be made from about 1300, but the combining of lenses for magnification did not occur to anyone until much later. In fact, optical glass quality remained generally poor, inhibiting the performance of both telescopes and microscopes until the mid-nineteenth century, so much so that for the eighteenth century, the larger aperture telescopes had polished metal mirrors rather than glass optics. Optical components replaced open sights on various mathematical instruments, such as the octants and sextants used in navigation and the theodolites and levels used in surveying, to improve pointing accuracy.

The third grouping, 'philosophical' instruments, dates from the mid-seventeenth century. New instruments were developed to investigate or demonstrate naturally occurring phenomena. For example, magnetism, although long used at sea in the mariner's compass, was newly investigated when it was realised that the earth itself behaved like a giant lodestone, and that the magnetic field had changing vertical and horizontal components.[19] The barometer, invented in Italy, was used to measure air pressure and predict changes in the weather, whereas the air pump demonstrated to those willing to pay attention – and there were many – how a guinea and a feather fell at the same rate in a vacuum; or, as dramatically portrayed by the artist, Joseph Wright of Derby, how life itself could be extinguished without God's great gift of air.[20] Other apparatus illustrated the action of static and dynamic forces and electrical phenomena, the latter being one of the most popular of eighteenth-century demonstrations, believed to have beneficial medical properties.[21]

What were *not* 'scientific instruments'? There are items which the layman might consider to be 'scientific instruments', but which are, by custom and usage, excluded from contemporary definition. For instance, watches, clocks and chronometers were manufactured by specialist makers, had a distinct trade and have a historic and technical literature of their own. More is known, for example, about the structure and economics of horology than about instrument-making.[22] This trade was centred in Prescot in Lancashire, Coventry and the Clerkenwell area of London, although there were examples of specialist clockmakers in all the major population centres (and some outside these) capable of making complex mechanisms from the late seventeenth century.[23] Specialist precision clockmaking was an essential adjunct to positional astronomy, just as the invention of the marine chronometer was the key to finding a ship's longitude at sea. This trade was again concentrated in the hands of a few – mostly London-based – craftsmen.[24] Many domestic clock and watch retailers and repairers were to be found in Scotland, however, and as with many instruments, the engraved signature often indicates the point-of-sale, and not the workshop of manufacture.

Barometers, which have already been mentioned as philosophical instruments, rapidly became items of domestic furniture. Some were undoubtedly made and sold by instrument makers, such as the air barometer developed and patented by the Edinburgh maker, Alexander Adie, in 1818 and devised specifically for marine use, although it could also be useful in experiments on mountains.[25] The Glasgow Professor of Mathematics, George Sinclair (discussed below in Chapter 4), used the barometer shortly after its invention for scientific purposes, namely to measure the heights of mountains and the depths of mines. Nicholas Goodison comments that Sinclair's discussion in his *Hydrostaticks*, first published in 1672, 'concentrates on the mercurial barometer as an instrument for the measurement of heights, and the only weather-glasses he mentions contain water'.[26] In Sinclair's second edition of 1683, as Thoday observes, 'the first details on how to interpret the changes in the mercury height for foretelling the weather' were published.[27] David Bryden claims that Sinclair's teaching and writing 'did much to make Scots aware of the barometer both as a philosophical instrument and as an aid to forecasting the weather, [and he] had to turn his hand to making for sale the instruments for which his activities had stimulated a demand'.[28] Indeed, in January 1684 the Faculty of Advocates authorised its treasurer 'to pay Mr George Sinclair for his Barometer and Higroscope ten dollars [sic]'.[29] By 1687, the almanac-compiler James Paterson was including weather glasses amongst the mathematical instruments he had for sale in his Edinburgh shop. It is not clear whether Paterson made these himself, or merely retailed them. A flyer on 'The USE of the BAROSCOPE', which has notes of cash payments on the reverse dated May 1697, to be found in the papers of one of the early Edinburgh glasshouses in the National Archives of Scotland, says towards the bottom of the first column:

> Inhabitants of every Countrey ought to consider the Quarters from whence [come fairest] and foulest Weather, and make . . . Observations on the Weather-Glass accordingly. For example, Here at Edinburgh (where it is found that the North-East is a moist Quarter) though the Mercury be risen a little, while the Wind is about the North-East, it may notwithstanding prove foul, because these Winds from the Sea are ordinarily attended with moisture . . .[30]

By the mid-nineteenth century, the large majority of barometers appear to have been wholesaled in very large numbers and distributed throughout the kingdom with many different signatures, in particular those of the itinerant Italians who sold them.[31]

Medical instruments – apart, perhaps, from the thermometer – are not usually classified as scientific instruments: the cutlers and metal workers who produced them led separate, parallel existences, and were grouped in locations where the iron and steel trades were important, such as Birmingham and Sheffield. In Scotland, the specialist makers of medical instruments similarly do not appear to be linked with the brass-workers who manufactured 'scientific instruments'.[32] Weights and measures have been largely ignored by instrument historians, but recent work by Allan Simpson and Robin Connor has demonstrated that precision weighing and measuring, particularly for national standards and those enshrined in legislation, did involve instrument makers. Capacity measures, for instance, raised from the ancient Scots pint under the direction of John Robison, Professor of Natural Philosophy at Edinburgh, were probably made by James Milne, an Edinburgh copper-smith who had links with the Adie family of instrument makers; examples were retained as burgh standards at Peebles, Forfar, Cupar and Edinburgh.[33] Scale-makers, on the other hand, appear largely to have been producing blacksmith's work for the market place, and these – with a few honourable exceptions, including makers of precision balances – have been disregarded as 'scientific instruments'.[34]

Seventeenth-Century Beginnings

How early did instruments arrive in Scotland? To answer this question, it is essential to refer briefly to developments in the better-documented case of England, and, specifically, to consider London, given that activities there had an impact on Scotland from an early date. The current received wisdom on the subject is that 'the skills of printing, and the engraving of book illustrations, maps, charts, and brass scientific instruments came to England from Continental Europe, and more particularly from the Low Counties', and that the major figure responsible was a skilled engraver named Thomas Gemini, who settled in London from Flanders in about 1540.[35] The second recorded maker of instruments in England was Humphrey Cole, who claimed to be 'a English man born in ye north', who worked at the Mint in the Tower of London for twenty years.[36] At this time, the demand for instruments in London was so slight that neither man had workshops, apprentices or a direct trade succession. Such developments had to wait for further concomitant economic expansion and commercial growth, of which a later generation of London-based Elizabethan instrument makers was able to take advantage. In these early years, people migrated to London in order to develop their skills and to find ready markets in the capital and close to the court. Subsequently, people with such skills moved into the provinces and into established and growing population centres, as new and local markets were to be found during the Stuart period.[37]

The first scientific instruments to be used in Scotland came from abroad. This is unsurprising, when the travels of the educated Scot before the 1680s are taken into account, as much of that education was undertaken on the Continent. Duncan Liddell, who founded the chair of mathematics at Marischal College, Aberdeen, with a bequest in 1613, 'was the prototype of the kind of scholar who became characteristic of Scottish intellectual activity in the later seventeenth century'. Born and educated in Aberdeen, he had 'wide-ranging interests in mathematics, philosophy and medicine – all of which he had either studied or professed in a whole sequence of German universities for a full quarter-century'.[38] He also bequeathed his books and instruments to the College, and these were supplemented by further mathematical instruments listed as being in the library by 1670, which probably came from the first occupant of the mathematics chair, William Johnstone of Caskieben.[39] None of these early pieces is known to have survived.

Perhaps the oldest surviving instrument in Scotland (outside those pieces found on archaeological digs or through marine archaeology) is the unsigned Gothic astrolabe dating from the early fifteenth century, attributed to the French workshop of Jean Fusoris. Astrolabes were used as calculating devices by astronomers, mathematical geographers and surveyors to solve problems of time, position and trigonometry; they were also outstandingly beautiful pieces of luxury craftsmanship. The Fusoris astrolabe has been engraved at a later date on the rim with an owner's name, that of the chorographer Robert Gordon of Straloch, Aberdeenshire. Gordon's major contribution to Scottish cartography, mapmaking, and to early modern chorography (the geography of regional description) was his extensive revision of Timothy Pont's earlier manuscript maps for publication in the *Atlas Novus* of Johannes Blaeu in Amsterdam in 1654.[40] The astrolabe has undergone some alteration: two projection plates have been added for the latitudes of Edinburgh and Straloch, and a date of 1597 has been engraved on the reverse. The instrument was probably reconditioned that year for Gordon at the time of his early studies in Paris.[41]

Instruments with provenances as long as this are rarities, and Scotland has only a very few. A Portuguese-made mariner's astrolabe, dating from about 1550, was at one point owned by a Dundee skipper, Andrew Smyton, who stamped it with both his name and date of ownership, 1688.[42] The group of three early London-made instruments – two by Humphrey Cole (a large astrolabe of 1575 and a nautical hemisphere dated 1582), and one by Elias Allen (a large mariner's astrolabe dated 1616) – may have formed part of the collection of secondhand equipment intended for an observatory at the University of St. Andrews assembled in 1673 by the Aberdeen-born astronomer, James Gregory, who had been appointed to the Regius chair of mathematics at St. Andrews in 1668. This observatory was abandoned after Gregory's early death in 1675; just who the original clients were for such large and prestigious instruments has never been resolved.[43] As inventor of the Gregorian telescope and one of the early teachers of Newtonian philosophy, Gregory was an important figure, but 'his comparative isolation from his mathematical contemporaries, his reluctance to publish his work, and finally his early death, all conspired to restrict his contribution to mathematics'.[44] There are other pieces dating from the late seventeenth century at the

University of St. Andrews, which may have been negotiated from London makers and designers on Gregory's behalf by his London correspondent, the mathematical intelligencer John Collins.[45] One further item has been in the collection at St. Andrews since the seventeenth century: an unsigned twelve-inch diameter dip circle – used to measure the angle of variation in the earth's magnetic field – inscribed 'Archibaldi Arskini Armigeri Londini'. It was presented by a former student; the maker of the instrument and the circumstances of its use then remain unknown.[46]

These brief remarks give some idea of the paucity of documented instrumentation in Scotland before 1700. This evidence is supported by the comments in 1662 of the Glasgow-born mathematician, almanac compiler and teacher of geography, James Corss:

> I have oftentimes lamented with myself to see so many Learned Mathematicians to arise in sundry parts of the world, and so few to appear in our Native Country. In other things we are parallel with (I shall not say in a superlative degree far above) other Nations; but in Arts and Sciences Mathematical, all exceed us. And had not that thrice Noble and Illustrious Lord, viz. John Lord Nepper, Baron of Merchiston &c. preserved the honour of our Nation by his admirable and more than mortal invention of Logarithms, we should have been buried in oblivion, in the memories of Forraign Nations.[47]

For David Bryden, writing in 1972, the poor standing of Scots in mathematics and astronomy meant that 'the instrument-maker's trade could not become economically viable, except within an established framework of scientific endeavour, which could provide a market for his goods and skills'. This suggests 'only a limited infrastructure of scientific activity in Scotland at that time'.[48] In the intervening thirty years, Bryden's claim has been modified only partly by further research and by the discovery of more extant instruments.

Bryden discusses the attempts of Edinburgh Town Council to attract craftsmen to set up and train apprentices in new skills, and cites the 1647 statement from the Burgh Records, when the Council

> grants libertie and licence to Robert Devinport Englishman, maker of mathematical instruments to exercise his trade and calling of making all the saids mathematical instruments and making or mending of watches within this burgh . . . and to keep his abode and residence heir for serveing of his Majesties Leidges and teaching his trade to such uthers as shall enter his service.[49]

For Bryden, 'the fact that a stranger, and an Englishman at that, was allowed to trade and invited to take apprentices . . . is a strong indication that there were no native craftsmen actually making instruments in the Burgh'.[50] A single double-sided brass instrument with Davenport's signature is known.

One side has the maker's signature 'Rob ∴ Dauenport / fecit' above a dialling instrument constructed for use at 56° N, the conventional latitude for Edinburgh at that time. This horizontal instrument – a sophisticated form of sundial – is made to the published design of William Oughtred, and the reverse has a logarithmic calculator, known as 'The Circles of Proportion', also designed by Oughtred.[51] Oughtred had worked with the London instrument maker Elias Allen, who constructed and modified a number of Oughtred's designs, and Davenport was one of Allen's apprentices.[52] Given that a number of Oughtred's double horizontal dials designed for Scottish latitudes have been recorded in the past thirty years, and that some of them were constructed by engravers resident in Scotland, the earlier suggestion that they had all been made by London craftsmen from a particular trade succession now looks doubtful.[53]

Stone sundials have long been a traditional feature of the formal gardens of Scottish landowners, and the complex polyhedral dials first attracted the attention of the late nineteenth-century architect and antiquary Thomas Ross.[54] His work was updated in the 1980s by Andrew Somerville, who catalogued 330 surviving complex stone sundials, mostly located in the central belt between Glasgow and Edinburgh (a total which far exceeds those that survive in other European countries).[55] Somerville believes that the 'main dial making period was from 1623 to about 1731, so it was a relatively short-lived fashion'.[56] He summarised the reasons for this phenomenon as

> first of all material: more stable government and increased prosperity led to the building of mansion houses with pleasure gardens. Secondly, the Calvinist philosophy of the time frowned on decoration for its own sake and required function as well. Thirdly, interest in science was increasing, along with the Renaissance interest in re-discovering the esoteric knowledge of the ancients. In addition, freemasonry may have had its beginnings in Scotland in the late sixteenth century and certainly assumed considerable importance in the seventeenth; this is unlikely to be mere co-incidence and the dials probably had a significance which went well beyond simple time-keeping.[57]

Horizontal brass sundials are a different tradition. Scottish clients for these – mainly landowners – appear to have initially ordered these from London instrument makers, and only subsequently from Scots makers, before Scots makers found it more economic to import London-made pieces and inscribe their own names on them. Two mid-seventeenth century complex horizontal dials made by William Havart are known, one example dated 1660, the second dated 1662.[58] Both of these have been made for use at an Edinburgh latitude, and their maker, 'GULIELMO HAVARTO', describes himself as 'Cellae Cuthberti', and may have been a priest travelling north from Durham.

A further two double horizontal dials to Oughtred's pattern are known. These were made by the London-trained copperplate engraver James Clark, who lived in Edinburgh from about 1680 until his death in 1718.[59] Clark was employed as Engraver to the Scottish Mint, and also had other uses for his engraving skills. In 1681 he engraved a portrait of Charles II for an edition of the *Acts of Parliament of Scotland*, and around this time began engraving a series of five maps for John Adair's *A True and Exact Hydrographical Description of the Sea Coast and Isles of Scotland*. Although Adair began his project in the 1680s, and Clark dated the Montrose survey 1693, the complete work was not published until 1703. Adair's problems were shared with his patron, Sir Robert Sibbald, whose political influence was on the wane after the Glorious Revolution in 1688, and various projects, including Sibbald's own proposed atlas *Scotia Illustrata*, were never to be completed.[60]

The fact that sundials continued to be purchased after this early period — there survives, for example, a massive brass sundial gnomon covered with tables and calendrical calculations that is inscribed 'G. Fullarton *Designant et Soli fulgenti Consecravit* Anno 1744' and 'R: Craig Automar Scul. Excu.' — shows that interest in this sort of garden sculpture continued well into the eighteenth century. Its designer can be identified as William Fullarton of Fullarton House in Ayrshire, a gentleman with several scientific interests, including astronomy. William Fullarton, who died in 1758, was educated at the universities of Edinburgh and Leyden. He was making extensive alterations to his country house in 1744, and this may have been when he ordered the construction of his sundial from Robert Craig, a local clockmaker.[61] By the early nineteenth century, horizontal brass sundials were being retailed by the major Edinburgh, Glasgow and Aberdeen firms, such as the example probably ordered between 1822 and 1835 by Sir Andrew Agnew for Lochnaw Castle in Wigtownshire from Alexander Adie, the Edinburgh instrument maker.[62]

The principal Scottish clients for sundials were from the gentry and aristocracy, who had an interest in their gardens. The changes in fashions from the formal to the landscaped garden allowed for this sort of ornamental inclusion, and sundials proved to be items which were acquired steadily throughout this period, despite the country's poor reputation for the favourable weather conditions upon which this instrument relies. Sundials, however, can be seen as more of a luxury commodity than as an instrument and, like their purchasers, they formed only a small part of the larger social picture. If instruments were, then, commodities, who were their consumers?

The Markets for Scientific Instruments During the Eighteenth Century

In 1972 David Bryden examined the character of the Scottish scientific instrument trade and considered the 'maker-customer relationship' with reference to the 1964 analyses of Silvio Bedini, who looked at the *ante-bellum* American instrument trade and defined it in terms of three broad groups of customers, with a fourth sub-group.[63] First were the men of science, the philosophers who performed experiments and who needed new apparatus to help them in their investigations. We should think in this regard of James

Gregory as an active designer of instruments who was obliged to turn to London craftsmen for assistance with his experiments in improving the reflecting telescope. Scientific customers were to be found increasingly within the universities throughout the long eighteenth century. Although a proportionately small part of the market, this important relationship between maker and user was on a one-to-one basis, and the technology thus developed helped to increase scientific understanding, while the scientific ideas brought technological improvements. Examples of this particularly close maker-customer relationship in Scotland during this period include James Watt with Joseph Black and James Crichton with Andrew Ure in Glasgow. In Edinburgh, further examples can be found with James Short and Colin Maclaurin, Miller & Adie with John Leslie, Alexander Adie with David Brewster, John Adie with William Swan, Patrick Adie with Sir Thomas Makdougall Brisbane, and John Dunn with Thomas Graham. Other instances are to be found elsewhere in the country.[64]

A second, and somewhat larger group, was the dilettanti, that is, gentlemen with sufficient disposable income to be able to buy the latest fashionable microscope, for example, either as an amusement or to entertain and impress friends. The garden sundials already mentioned, which continued to be produced throughout the period, fall into this category of ornamental display or status symbol. Interested and educated gentry often formed small cabinets of curiosities that contained contemporary instruments in order to demonstrate their virtuosity.

Bedini's third and largest group can be identified as the so-called 'mathematical practitioners', those who required instruments for practical use in everyday life. These included navigators, surveyors, and architects (see Chapter 3). A further sub-group, the teacher-demonstrators, is a category which expanded rapidly during the long eighteenth century. The number of teacher-demonstrators who were not affiliated with the Scottish universities grew steadily as the century progressed, and there were significantly more Scots involved in extramural lecturing than Bryden has noted. Moreover, as Steven Shapin observed, 'the itinerating paths of scores of scientific lecturers traced back to Edinburgh classrooms', and although Edinburgh was by no means the exclusive source of all Scottish popularisers of science during the period, the city probably produced more than its share of such characters. Furthermore, many itinerant English lecturers included Scottish towns on their circuits.[65] As Withers has shown, many of those persons teaching geography employed instruments – globes, astrolabes and orreries – in their classes, some shop-bought, others handmade by themselves.[66]

This division of the markets for scientific instruments into three or four components has been adhered to until recently by most instrument historians, particularly those doing any analysis of a period in a given locality.[67] The only monograph other than Bryden's on the history of Scottish scientific instrument making takes a different form. The acquisition of a large collection amounting to almost 200 items of Scottish scientific instruments mostly dating from the late nineteenth century afforded the opportunity to publish much of the National Museums' information gathered since Bryden's 1972 monograph. *'Brass & Glass'* allowed its authors to consider instrument history by focussing on business histories of the firms represented in the collection, rather than emphasising a particular period. Although this allowed the model of a family-run business enterprise for the instrument trade to emerge, such work – unlike Bryden's 1972 monograph – purposefully produced no synthesis of the material offered in the book beyond what appeared in the introduction, allowing different aspects of the market for Scottish instruments to be presented.[68]

More recently, the historian of science Richard Sorrenson has attempted to construct a model of the market for scientific instruments in eighteenth-century Britain which fits the consumer society thesis initially proposed by J. H. Plumb, and taken up by Neil McKendrick and John Brewer, among others.[69] Having sketched the supply side of the eighteenth-century trade, Sorrenson sums up the demand side

> domestic, colonial and European consumers purchasing marine, surveying, and household instruments; natural philosophers in Britain and abroad ordering experimental and observational instruments; popular lecturers and schoolteachers buying a whole range of demonstration instruments to explicate the new natural philosophy; and finally . . . the British state itself, buying gauging instruments for the customs and excise, marine instruments for the navy, astronomical instruments for the Royal Greenwich Observatory, and surveying instruments for the Board of Ordnance, as well as offering prizes for navigational instruments through the Board of Longitude.[70]

Sorrenson has four market categories, merging the former 'scientific' customer into a 'Special Market' which he sees appearing in the eighteenth century, namely the British state. His contention is that different markets for different types of instruments emerged over time. According to Sorrenson, a 'Natural Philosophical' market appeared in the late seventeenth century, absorbing newly invented instruments such as pendulum clocks, telescopes, air pumps and microscopes as well as older mathematical instruments, such as quadrants and magnetic compasses, used in the new context of demonstration. Two entirely new markets, the 'Natural Philosophical Lecturing' and the 'Household', were created in the late seventeenth and early eighteenth centuries as applied science moved down the social scale 'into the newly burgeoning commercial market-place of consumers'.[71] The rather all-encompassing category of 'Marine, Astronomical, Surveying, Weights and Measures', which comprises the 'British state' market, Sorrenson dates as existing from before the Scientific Revolution. Into it he places most practical instruments (which Bedini had characterised as 'practical' or 'professional'), but adds the rider that 'the most important newcomers in the eighteenth century were chronometers, dividing engines, large theodolites, and achromatic lenses, all of which became commonly available only after they had been first developed in response to the demands of the British state'.[72] There are distinct problems with Sorrenson's analysis of the market, especially when he combines the demands of science and the state, although his attention to state demand is a welcome advance in conceptions of instrument history.

Sorrenson's analysis is difficult to apply to the Scottish case, not least because it is unclear what he means by 'Britain'.[73] His 'British state' is supplied by a small group of pre-eminent eighteenth-century London instrument makers – amongst whom, after 1738, Edinburgh-born James Short would be included. Of these, he discusses in particular John Harrison, George Graham and Jesse Ramsden (all, incidentally, Yorkshiremen by birth):

> With Ramsden's theodolite, as with Graham's quadrant and Harrison's chronometer, the British navy and army demanded and were provided with instruments that mapped with the greatest available perfection the land, the heavens and the sea . . . by encouraging both the construction of unique large instruments and the mass production of smaller instruments, the state through such organs as the Royal Greenwich Observatory, the Board of Longitude, and the Ordnance Survey, ensured that instruments of near perfection became readily available both to its servants and its subjects.[74]

These makers – world leaders in their craft – and 'state bodies' had no real equivalent in post-Union Scotland. There was no Scottish national observatory until 1822, and no Astronomer Royal for Scotland until 1833: few of the instruments were then made by Scottish makers.[75] There was no Scottish Board of Longitude, although there were several Scottish clockmakers whose innovations made an enormous impact on precision time-keeping in astronomical observatories: Alexander Cumming, William Hardy (both of whom moved to London), Thomas Reid and Robert Bryson. The Ordnance Survey eventually covered the whole of the British Isles and, as with all these bodies and the 'British state' itself, Scotsmen became contributors to them all.[76] The re-ordering of the Bedini/Bryden classification gives undue emphasis to the class which Sorrenson sees as pre-eminent: the 'British state'. It is not a model which can hold for Scotland.

By building on the analyses offered by Bedini and Sorrenson, and incorporating evidence from the provincial English trade, Ireland, and, importantly, from that in Scotland, we can create a new characterisation of the instrument trade, with six market categories which might fit a less special case than that of London (see Table 2.1). This shows how demand changed over time and reveals how new markets came into existence. Clearly, the products across all these markets changed with time. Those aimed at the first category – for instance, the dilettante market – were generally demand-led luxuries, and probably formed a larger proportion of the entire instruments market in the

Table 2.1: Markets for Instruments during the Eighteenth Century

Market	Contents	Examples	Size and character
1. Dilettante	Any bespoke item used to demonstrate its owner's virtuosity, cleverness or wealth	Clockwork automata, items in precious metals, but might be extravagantly decorated sundials or outrageously rococo microscopes	Small, often metropolitan
2. Practical	Surveying instruments: exploration, mapping, engineering. Navigation instruments: foreign and coastal trade	Compasses, chronometers, tables, telescopes, gunnery and fortification devices, levels, slide rules	Growing with the Empire, peripatetic
3. Teaching	Lecture/demonstration material	Orreries, globes, electrical machines, magic lanterns, air pumps, chemical apparatus, magnets, mechanical models	Large growth in late 18th century with growth in popular scientific lecturing; metropolitan and subsequently provincial
4. Domestic	Items that became everyday	Clocks, spectacles, telescopes, microscopes, opera glasses, camera obscuras, thermometers, barometers, 'toys'	Small, but spreading with the rise of a consumer culture
5. Scientific	Precision pieces, used for specific experiments	Optical apparatus commissioned by Brewster from 4 Scottish opticians; glass equipment by Black from Watt. More usually commissioned from London makers than from provincial/Scottish makers.	Very small, often expensive; usually London-based or generated, but extremely influential nationally and internationally.
6. State	Precision pieces, sometimes made to match legislation as national standards	Standard yard for Royal Society; chronometers and dividing engine for Board of Longitude; pendulum clocks, quadrants sectors, telescopes for Royal Greenwich Observatory; hydrometers for Customs and Excise, navigation instruments for Royal Navy; standard weights and measures for burghs	Extremely large in England; less so in Scotland

early eighteenth century than in 1820. Returning gentry needed status symbols: the Grand Tour rapidly became a fixture for the education of young gentlemen. The design of such instruments might include a certain amount of cultural involution, or change for change's sake, but it also served the purpose of increasing the skills of the maker and delighting the fashion sense of the owner, a relevant example of this being the silver microscopes of John Clark of Edinburgh, produced between 1749 and 1776.[77] There was not yet an 'industrial' category, because during the eighteenth century any instruments designed for product control on the factory floor would either have been the 'practical' instruments used in measurement or engineering, or have fallen into the 'State's' market, where standard instruments were required by legislation. Not until the late nineteenth century – beyond the period of concern here – could an 'industrial' category said to have come into existence.[78] The greatest growth area during the eighteenth century was in the 'teaching' market (which equates to Sorrenson's third class of 'natural philosophical lecturing'), which, outside the Scottish universities, led to new emerging audiences for science and customers for instruments, especially women, younger people and children. These created in turn a small but growing 'domestic' market, mostly confined to the upper and middle classes. It is to these areas of market growth, particularly in teaching, that we now turn.

Instruments and Market Growth During the Scottish Enlightenment

During the last thirty years, the cultural and social effects of the diffusion of a scientific culture during the eighteenth century have received considerable attention from historians of science.[79] Although difficult to quantify, the understanding of scientific principles, it has been argued, may have been associated with industrial advance.[80] In particular, the audience for science throughout Britain grew enormously during the eighteenth century. The early itinerant lecturers, spreading the Newtonian gospel, found eager listeners first in London, and subsequently in the provinces.[81] As Larry Stewart has written, 'The community of experimenters, the instrument makers, and self-styled engineers with their varying degrees of dependence on Newton's principles, and the devotees of the public lectures, constructed a broad bottom for natural philosophy'.[82] These first London-based lecturers at the start of the eighteenth century were closely allied with the Royal Society, and were usually university-educated men who collaborated with an instrument maker: examples include James Hodgson with Francis Hauksbee II, and Benjamin Worster with William Vream.[83] A later generation, including the prominent figures of Benjamin Martin and the Scots-born James Ferguson, combined these roles in the same person. At first, only the syllabi of the lecture courses were published, but eventually the lecture text, together with illustrations of the demonstration apparatus – much of which had evolved specifically for lecturing purposes – was published in 'textbooks which often gave details that would help someone wanting to replicate the equipment'.[84] The increase in natural philosophy lecturing by mid-century – particularly in England – has been linked with the successful establishment of a popular press, and the ability of the lecturers to attract their audience through newspaper advertisement. This fell away as the audience was diverted by other attractions, including the foundation of the Society of Arts and Manufactures in London in 1752, and other institutions which offered a more stable environment with extensive facilities.[85]

Itinerant natural philosophy lecturers appeared in Edinburgh as early as 1721, when John Theophilus Desaguliers visited the Scottish capital in connection with his involvement in freemasonry, but he also assisted the Town Council in a practical fashion by getting the air out of the water-supply pipes.[86] Desaguliers was perhaps the most important populariser of his generation of Newtonian philosophy for the layman. In Edinburgh in 1724, the Swede Mårtin Triewald and his English friend, John Thorold, together gave a course of twenty-four lectures, which used 'numerous and costly Apparatus'.[87] An engineer and natural philosopher, Triewald had been in London in the spring of 1719, and had attended the physical experiments of Desaguliers in the dome of St Paul's Cathedral. Triewald took his 'comprehensive collection of scientific instruments' back to Sweden with him in 1726, and an inventory list was made in 1736 when 327 items were presented to the University of Lund.[88] The only recognisably Scottish piece was a twelve-foot telescope signed by 'William Barclai' [Barclay] of Edinburgh, a maker of telescopes and domestic barometers, contemporary with James Short, the pre-eminent maker of reflecting telescopes.[89]

The travelling lecturer often put together a collection of apparatus, but few such collections have survived. The nucleus of George III's own collection of instruments was formed by the King's tutor, Stephen Demainbray, himself previously an itinerant lecturer. The orphaned Demainbray had been brought up in London by the family of J. T. Desaguliers, and, after a spell on the Continent, he was to be found in Edinburgh in the mid-1740s running a young ladies' boarding school. In 1748–49, he began lecturing on natural philosophy, first in Edinburgh and the north of England and, subsequently, in London. As Morton and Wess show, Demainbray added instruments to his collection as he travelled, and bought several in Edinburgh. These include a terrestrial telescope signed 'W.R. fecit', and an unsigned Gregorian telescope possibly by the same maker.[90]

Demainbray had obtained his royal appointment in 1769 through the patronage of John Stuart, 3rd Earl of Bute. Bute had arranged that Demainbray should give a lecture course to the royal children, including the future king, in 1755. Bute himself had a collection of instruments, which he kept at his English seat at Luton Hoo, just as his uncle, Archibald Campbell, Earl of Ilay and 3rd Duke of Argyll, had kept his at his seat at Whitton on Hounslow Heath.[91] Although their personal collections can be seen as forming part of the dilettanti market, both uncle and nephew played an important political role in the appointment of key figures at the Scottish universities. For Roger Emerson, 'there is a very real sense in which the Scottish Enlightenment was the product of the patronage of the anglified and progressive Scots [like Bute and Argyll] who controlled and managed patronage throughout most of the eighteenth century'.[92]

The later eighteenth-century lecture-demonstrators, of whom Benjamin Martin and James Ferguson are the best known, encouraged that continuing interest in natural philosophy through the spread of literacy by publishing their own populist works.[93] Ferguson, a farm labourer's son from Banffshire, made mechanical models while keeping sheep as a young boy. He taught himself the elements of surveying, horology, astronomy and portraiture after a fairly rudimentary education at Keith Grammar School. Under the patronage of Colin Maclaurin, Ferguson went to London in 1748, where he lectured before the Royal Society and was elected a Fellow in 1763. Ferguson was a skilled designer of clocks and astronomical models, and he lectured throughout the country using his own demonstration apparatus. He also wrote and published assiduously, expounding Newtonian ideas at a popular level on all these topics. His designs were later marketed by other instrument makers.[94] Popular science was thus disseminated to polite and domestic audiences through lectures, through the ownership of apparatus or 'scientific toys', and through books which illustrated and explained their use.[95]

Two examples from the end of the later eighteenth century, in chemistry, demonstrate that instrumentation was required as part of that subject's stock-in-trade: the examples are 'Blind' Henry Moyes and his assistant, William Nicol. Details of other Scottish lecturers, particularly in demonstrating elements of natural philosophy, peregrinating around Scottish centres including Glasgow and Aberdeen, have been recorded.[96] Kirkcaldy-born Moyes was blind from about the age of three, yet he started a career in lecturing with an assistant demonstrator, apparently in Edinburgh, before setting off on the English lecture circuit in 1779. In 1784 Moyes left for America, lecturing to great acclaim in Boston, Philadelphia, Baltimore, Princeton and Charleston, South Carolina, before returning to England in 1786, where he continued to lecture. In 1789, he visited Dublin to give two courses of lectures. In Newcastle-upon-Tyne, where he had lectured in 1780 and 1793, Moyes was elected in 1795 amongst the first of the honorary members of the local Literary and Philosophical Society. Similarly, in Hull, Moyes was in 1792 a prime mover in the formation of a 'Society for the Purpose of Literary Information', and was elected its first President: by 1804, this society had become moribund. He died in the middle of a lecture course in Doncaster, in 1807, aged 57. Because Moyes never published, his reputation faded with his memory. Recent increasing interest in the rise of science in British provincial culture has led to a reassessment of Moyes's role. He is now seen as one of a group of itinerant lecturers who helped the Literary and Philosophical movement's ideas in England permeate further down the social spectrum, thereby contributing to the beginnings of the Mechanics' Institutes.[97] This also exposed larger numbers of people of the middling sort outside London to scientific ideas and lecture-demonstration apparatus. It is significant that the best-known portrait of Moyes shows him with an air pump, and surviving lecture syllabi reveal that friction electrical machines were used in his demonstrations.[98] William Nicol, Moyes's assistant at the time of his unexpected death, was born at Humbie in East Lothian in about 1771. Nothing is known of Nicol's childhood or of his education, but it appears that at about the age of fifteen, in 1786, Nicol became Moyes's assistant. After Moyes's death in 1807, Nicol purchased his apparatus (valued for probate at £11 8s 2d) and manuscripts, and continued his master's lecturing circuit of English provincial centres in his own right well into the 1820s. He is recorded as having given the same type of course as Moyes, usually of twenty or twenty-two lectures, in provincial centres such as Sheffield, Nottingham, York, Derby, Leeds, Lincoln and Edinburgh. Retiring to Edinburgh in the late 1820s, Nicol was also the inventor of the prism which bears his name, and he pioneered the grinding of thin sections of minerals for microscopy. Nicol was, thus, not solely a user of instruments, but also someone who improved them through practical use.[99]

Having outlined one area of the market – extramural teaching – where supply was being vigorously promoted at this time, in the next section we look at the demand for instruments during the Scottish Enlightenment in the three main geographical locations in which they might be found, and the broad market categories associated with them.

Instruments in the Scottish Enlightenment

During the long eighteenth century, the instrument trade was most vigorous in Edinburgh. One of the most prominent Edinburgh makers was James Short, who was an important figure in Scottish instrument making; he proved to be skilful, a good networker and an astute businessman.[100] Short's career has been described as 'the almost unblemished success story of the archetypal lad o'pairts'.[101] He was fortunate to attend the classes of one of Provost Drummond's appointees, Colin Maclaurin, whose career and scientific style are analysed below in Chapter 5. Maclaurin allowed Short to use one of the college rooms as a workshop for experimenting in practical optics, and Short apparently acted as class demonstrator for a time under Professor Matthew Stewart.[102] Short taught himself so well by trial and error that, by 1734, Maclaurin was able to write of his protégé's success to Robert Smith, Professor of Astronomy and Experimental Philosophy at Cambridge. Smith published Maclaurin's letter praising Short's work in his influential *A Compleat System of Opticks* in 1738. Fellows of the Royal Society were testing the quality of Short's telescopes as early as 1736, and he was summoned that summer to London by Queen Caroline, who wanted Short to instruct her son William, Duke of Cumberland. The following February, a group of Scottish astronomers observed an annular eclipse. All those interested, including Short, were friends of Maclaurin – they then went on to form the Society for Improving Arts and Sciences and particularly Natural Knowledge, a precursor of the Royal Society of Edinburgh.[103] Although Short was not to remain in Edinburgh beyond 1738, 'the broadly-based Society, with members drawn across the professions and the aristocracy, was to provide [him] with a framework of patronage and influence which extended across Europe'.[104]

Short left Edinburgh because the lure of London as the scientific marketplace for the rest of the world was too strong. If the scientific community in Scotland at this point proved too small, too impoverished or lacking in diversity and communications for Short, he would find a global market in London. There he joined an élite corps of instrument makers, who were also Fellows of the Royal Society, and, to some extent, he bowed out of the Scottish Enlightenment within Scotland itself.[105] Yet each of the Scottish universities purchased examples of his telescopes, either from him when he was working in Edinburgh, or later, when he was in London.[106] By the time of his death, Short had sold over 1300 telescopes – of which he made only the optical components – to customers in St. Petersburg, Uppsala, Paris and Harvard, as well as other less exotic locations.[107] Scots took Short's instruments with them when they went abroad and, in one instance, the instruments were returned from Jamaica to Glasgow.[108] However, it would be revealing to know more about other instruments that Maclaurin used, both in a private capacity as well as professor in Aberdeen and in Edinburgh. He probably purchased for the University of Edinburgh the impressive quadrant now attributed to the London maker John Rowley, which was likely taken to the Orkney Islands by Murdoch MacKenzie in 1748–49 in order for him to fix the latitudes of Kirkwall and North Ronaldsay.[109] Maclaurin was also the probable purchaser of the surviving telescope micrometer dated 1736, described in Smith's *Opticks*, from the London maker Thomas Heath.[110] Maclaurin was thus a key figure in the development of the use of instrumentation and of instrument making in Scotland. The link between instrument makers and the natural philosophy class at Edinburgh was consolidated in the second half of the eighteenth century, when, in 1754, John Miller I manufactured and adjusted two large high-precision capacity measures for the county of Stirling raised from the ancient Scots pint.[111] This work was carried out under the supervision of Dr. John Stewart, Professor of Natural Philosophy.[112] Later, John Robison undertook comparable work in the 1780s and early 1790s. A similar role was undertaken by the Aberdeen professor Patrick Copland in 1811.[113] These investigations reveal another aspect of Scottish Enlightenment thinking: the concern for antiquarian truth, particularly where revealed in surviving artefacts and legislation.[114] It also demonstrates another Enlightenment characteristic: the quest for precise quantification, which led in turn to ever-more exact forms of measurement as the basis for scientific truth (see also Chapter 3).[115]

Miller went on to become an assistant in class demonstrations to John Robison who was appointed Professor of Natural Philosophy in 1774. Miller was the first of a family firm of scientific instrument makers which dominated the Edinburgh scene for over a century and a half: his son, John Miller II, his nephew Alexander Adie, and Adie's four sons have been described as 'talented and creative . . . Although it may ultimately be concluded that Edinburgh was no more than a provincial centre of instrument production dominated from England, the history of the Adie business makes it clear that such a judgement needs to be heavily qualified'.[116] John Miller II had begun in business just at the end of a period of rapid agricultural change, during which the number of land surveyors working in Scotland increased from about ten to about seventy, and the range of surviving surveying instruments from this period indicates how important this section of the market must have been in underpinning the more specialised 'scientific' orders.[117] That there remained a growing and diverse dilettante market in Edinburgh is demonstrated by the growing demand for domestic – as well as special-purpose – barometers, globes and special drawing instruments aimed at the amateur market.[118] In Edinburgh, it seems particularly difficult to decide when the Enlightenment ended, and David Brewster has been characterised as an 'ambitious Edinburgh intellectual in the age of the late Enlightenment'.[119] His work with instrument makers was part patronage, part scientific design. Among Scottish makers, he favoured the Adies, but he also patronised James Veitch of Inchbonny, near Jedburgh, as well as Peter Hill and William Blackie, both of Edinburgh. His 1813 publication, *A New Treatise on Philosophical Instruments*, for example, in tandem with his own investigative work in optical mineralogy, promoted new devices for measuring the angles of crystals.[120]

Recent work by Roger L. Emerson has shown that the way in which many professors at all the Scottish universities obtained their academic positions was due in large measure to the machinations of enlightened and politically active patrons such as Archibald Campbell, Lord Ilay, the 3rd Duke of Argyll.[121] Emerson has produced a list of twenty Glasgow University appointments dating between 1728 and 1761, made with the 'approbation of Lord Ilay'. He also writes that 'it was largely accidental that this great patron was also a lover of learning, an amateur scientist, and an enthusiastic improver who liked the company of intellectuals'.[122] Ilay had a substantial library, a section of which was dedicated to the current literature concerning developments in instrumentation, and was himself enthusiastic collector of instruments; his collection included a large and impressive standing dial made by Richard Glynne of London that incorporated his coat-of-arms in its design.[123] He apparently kept this collection at his English seat, while he was rebuilding the family seat at Inveraray in Argyll, to the design of the Adam brothers.[124] Ilay encouraged a bright young craftsman named Alexander Cumming, whose origins are obscure. By 1752 Cumming was set up at Inveraray as a watchmaker and he was employed to build an impressive organ (destroyed in a fire in 1877). He apparently designed Inveraray's pioneering water closet system in 1756, and in that year restored instruments belonging to Ilay which had been damaged by sea water in a stranded ship. Like James Short, Cumming had moved to London by 1763; in 1765 he made the first of his clock-barographs for George III, for which he was paid £1178 and an annuity of £150 for its maintenance.[125]

As discussed below in Chapter 4, one of Ilay's appointees at Glasgow University was Alexander Wilson, originally a typefounder with the renowned Foulis Press. Ilay had not supported the candidature of John Anderson for the chair of natural philosophy, and when Anderson was appointed, Ilay 'got a chair created for his protégé Alexander Wilson, who shared his interests in chemistry and who had long supplied him with scientific instruments'.[126] Wilson was appointed Glasgow's first Professor of Practical Astronomy in 1760, shortly after the university had received the bequest of astronomical instruments from a former graduate, Alexander Macfarlane, who had emigrated to Jamaica. Sadly, most of these instruments – which included a large transit, a zenith sector, and the astronomical quadrant by John Bird so admired in James Watt's shop by Joseph Black – have not survived.[127] Wilson had been making instruments (thermometers and barometers) since the 1730s when he had assisted George Martine in his experiments on heat. A single stick barometer signed 'Alex. Wilson' and dated 1774 has been recorded in a continental collection; and the National Museums of Scotland have a thermometer signed by Wilson and dated 1782, which came from the natural philosophy class at the University of Edinburgh.[128] Wilson had also been in a short-lived partnership in 1758 with James Watt and Joseph Black (another of Ilay's Glasgow University placemen). Both Wilson and Watt supplied Black with thermometers during the 1760s while he was still at Glasgow doing his pioneering work on the nature of heat. Wilson offered to produce thermometers for Black's students, priced at a guinea or a guinea and a half in 1768.[129]

Robert Anderson's work on Joseph Black's use of instruments and apparatus and the teaching of chemistry at the University of Edinburgh between 1713 and 1858 has allowed him to construct a history of the use of chemical equipment through six successive professorships, and to link this sequence of use to surviving apparatus.[130] Similar work has been undertaken for natural philosophy teaching at Aberdeen, although no apparatus there survives from before 1769.[131]

In contrast, only two papers appear to have been written about the teaching of natural philosophy at Glasgow at this period (although see below, Chapter 4).[132] Robert Dick I, first Professor of Natural Philosophy, was appointed in 1726, and a list of departmental apparatus dated 1727 survives. By 1730 Henry Drew had been appointed to look after the instruments and help with the class experiments, but unlike John Miller in Edinburgh, Drew was not an instrument maker himself. Drew, and his successor George Jarden, both metalworkers, have been identified by Bryden as possible candidates for the unknown master with whom James Watt trained for the year before he went to London.[133] A number of surviving pieces can be identified from the 1727 list, including the famous model Newcomen engine, subsequently repaired by James Watt.[134] James Watt was first employed by the University to unpack and refurbish the astronomical instruments from the Macfarlane bequest in 1756. After a year's training as an instrument maker with a master in Glasgow, he then spent a year in London gaining experience in the workshop of John Morgan before returning north. Glasgow University appointed Watt as its mathematical instrument maker in 1757, and he continued in this trade until at least 1771. Like Short and Cumming before him, Watt took the road south, to the proverbial fame and fortune with Matthew Boulton at the Soho Works.[135] Few items appear to have survived from Watt's instrument-making career, apart from a single domestic barometer.[136] Watt appears to have trained John Gardner I not only as an instrument maker but as a land surveyor. Gardner began as Watt's apprentice, and became his journeyman in 1769. Subsequently, Gardner set up independently, becoming assistant to the official City of Glasgow land surveyor in 1789, and succeeding him in 1792 as well as sustaining his thriving instrument business. However, over the next 75 years this family firm, which produced and sold a wide variety of merchandise, went bankrupt four times, demonstrating that the instrument trade was a risky business venture, involving large investments of capital held in stock, which, in a period of economic slump, might not sell. The technical competence of any instrument maker might be put at risk through poor business practice. Only with changes in company law, particularly with the introduction of limited liability, were businesses like the Gardners' more secure.[137]

John Anderson was Professor of Natural Philosophy at Glasgow University from 1757 to 1796. After Anderson's death, his will provided money and apparatus for 'Anderson's Institution', which became the Technical College and, later, the University of Strathclyde. Some of the instruments at the University of Strathclyde may have belonged to Anderson himself.[138] A number of these, and some items remaining in the University of Glasgow's collection, were made or designed by a local instrument maker named James Crichton. These include a bimetallic thermometer described by Crichton in 1803, and a hydrometer. At Strathclyde University surviving demonstration apparatus, presumably used in the classroom, includes a magnetic compass, a magnetic dip-circle and an air pump (see Figure 1). Besides producing items for the teaching market, Crichton also worked for 'scientific' customers: four thermometers with his signature are to be found in the surviving apparatus of Antoine Lavoisier, and a precision balance reputedly once the property of James Watt is in the museum at Greenock, Watt's birthplace. Crichton also worked with Andrew Ure and others in producing new Imperial standard capacity measures for Glasgow in 1826: his work and patrons evidently warrant further investigation.[139]

In St. Andrews, despite a decline in prestige and student population, a course in 'Experimental Philosophy' with specially bought apparatus was begun in 1714, and was conducted by the mathematics professor, Charles Gregory. Ronald Cant has commented, however, that 'it is doubtful whether the St. Andrews project was immediately successful', but he notes that two of the colleges there established 'fixed professors' of natural philosophy in the mid-1720s, and obtained further instruments.[140] These items were amalgamated with the university collection, but no survivors today can be identified.

Subsequently, some pieces of apparatus were acquired during David Young's professorship between 1747 and 1759. These included a Grand Orrery by Benjamin Cole of Fleet Street, London, and a reflecting telescope made by James Short. Short's less-skilled – and possibly less honest – brother, Thomas, maintained instruments at the University in 1748 and 1757.[141] Some chemical equipment may have been acquired through the Edinburgh instrument maker Alexander Allan from the extramural lecturer Thomas Thomson, who gave up his private practical chemistry course in Edinburgh in 1811.[142] Other material (which survived until the mid-nineteenth century) was listed in an inventory by a later Professor of Natural Philosophy, William Swan, in 1859.[143] However, although there appear to have been competent assistants to repair apparatus damaged in the ordinary wear-and-tear of everyday use, St. Andrews never sustained a local instrument-making business.

In Aberdeen, where Newtonianism was early established as instrumentation and the teaching of natural philosophy went in tandem, much surviving material comes from a relatively late date, notably from the time that Patrick Copland taught natural philosophy at Marischal College, between 1775 and 1822.[144] Of the earlier teaching and research apparatus used until the late eighteenth century, almost nothing was constructed locally. Rather, most of the instrumental hardware was bought from London: for instance, instruments made by pre-eminent London makers such as Francis Hauksbee and George Hearne figure in a list of instruments dating from the 1790s.[145] Copland was particularly good with his contacts, patrons and efforts at fundraising. In 1780, he raised almost £400 through a public appeal for instruments, and with this he bought prestigious London instruments for an astronomical observatory established on Castlehill the following year. These included a large quadrant with a two-foot radius, made by Kenneth McCulloch and divided by Edward Troughton, and two impressive pieces paid for by John Stuart, 3rd Earl of Bute, then the Chancellor of Marischal College. Bute's gifts were a transit telescope by Jesse Ramsden, and an equatorial instrument originally constructed by Jonathan Sisson, but altered in 1781 at Bute's expense by Ramsden. These two instruments apparently cost the Earl some 500 guineas.[146]

John Reid has written that, 'what particularly attracted the attention of Copland's contemporaries . . . was his extensive use of demonstration equipment, employing the largest collection of apparatus in the country, much of it made by himself'.[147] In 1783 Copland managed to obtain a grant of £50 for three years from the Board of Trustees for the Encouragement of Manufactures in Scotland (a government body) and used this to employ an assistant, John King, who had been trained as a watchmaker. King remained attached to the collection, which finally amounted to about fifty models, until about 1790. Only three items from this collection can now be positively identified, one of which is an astronomical clock to the design of James Ferguson.[148] As Reid comments, however, 'there was no Scottish supplier capable of providing the range of philosophical instruments demanded by the Scottish universities.

This was not due to any want of indigenous skills, as can be seen by examining the surviving instruments in the Scottish universities and elsewhere, constructed by Copland, King or Charles Lunan in Aberdeen, Miller or Adie in Edinburgh, or Crichton in Glasgow'.[149] From 1785, every other year or so, as well as teaching in college, Copland gave an extramural evening course in experimental philosophy. This predated the adult education classes of George Birkbeck in Glasgow by some fifteen years. The models constructed by Copland and King were used in these classes as well as the college classes to demonstrate the principles outlined in the lectures to the audience. As Reid summarizes, 'The motivation for the use of so much apparatus was almost entirely its effectiveness in teaching: it aroused the student's curiosity; it riveted the facts in his mind; it clarified the principles by demonstrating their application; it emphasised the practical aspects of science and awakened the student's spirit of discovery; it was a bonus, not particularly emphasised on the course, that apparatus was also central to many real advances being made in physical knowledge'.[150]

Conclusion

This chapter has looked in some detail at two particularly important areas of the market for instruments in Enlightenment Scotland: the teaching market, both within and without the universities, and the allied but different and smaller 'scientific' market. These areas were the two which encouraged local skilled artisans, whereas the other markets depended more on instruments bought in from the south, where indeed items with a London-made cachet could be considered the more desirable in consumer terms. There were, of course, exceptions to these distinctions, and it is apparent that the difficult-to-estimate 'practical' market underpinned the economics of the 'teaching' and 'scientific' markets.

The making of instruments as an indigenous trade came late to Scotland. The first sustained attempt was by James Short in Edinburgh between 1734 and 1738. A specialist maker of the optical components of reflecting telescopes only, Short had no workshop, workmen or trade successors. James Short's specialisation meant that he had to move to London to survive economically. Later, Scottish instrument makers did diversify their range of wares, buying in where necessary to remain financially secure, but actively making where they could:

> I have lately examined the Therm[omete]rs which you blew for me and have found them most charmingly sized – I shall clean them and fill them some of these days and then I shall be upon you again to divide them . . .[151]

James Watt was at the beck and call of his scientific patron, Joseph Black, but even his range of skills was not enough to keep him in Glasgow. Enlightenment instrument making rarely paid its way: Bryden has concluded that well into the nineteenth century other lines of business – silver, clock-retailing and jewellery, to name but three – kept the local trade for instruments afloat.[152]

In 1984, the following comment was made concerning scientific instrument making in Scotland:

> With apologies to Scottish and Irish nationalists, instrument making in the British Isles outside London was a provincial activity at least until the latter years of the nineteenth century. It was provincial in that London was *the* centre of activity, with manufacture in other towns and centres largely a peripheral activity.[153]

This chapter has suggested that this interpretation must be treated with considerable caution. The history and geography of instrument making and instrument use in the eighteenth century may, it is true, be a story of English metropolitan dominance. Yet it is also a story of local variation, different markets and divergent uses. More work needs to be done on specific instruments and apparatus belonging to key individuals, on the instrumental patrons, on the instrument-using practitioners, and on how, in given contexts, instruments were used in association with instructional texts, laboratory experiments and the philosopher's unmediated eye. Such research will give a much clearer idea about the genesis of the indigenous Scottish trade and, consequently, a better picture of how it worked. By looking at the history of the trade elsewhere, in England and in Europe, accusations of provincialism might be better expressed as questions of geographical and of social difference. Discussions between social historians and historians of science would help further our understanding of the issues that have hitherto separated instrument history from social history and unjustifiably distanced the artefact from its context.

Notes

1. Robison to Watt, 1796, in Eric Robinson and Douglas McKie (eds.), *Partners in Science: Letters of James Watt and Joseph Black* (London, 1970), 256–57. My thanks to Professor Donald MacIntyre for presenting me with this book.
2. Robison to Watt, 18 December 1799, in Robinson and McKie, *Partners in Science*, 321. For Robison and Black, see J.R.R. Christie, 'Joseph Black and John Robison', in A.D.C. Simpson (ed.), *Joseph Black, 1728–1799: A Commemorative Symposium* (Edinburgh, 1982), 47–57.
3. Peter Swinbank, 'James Watt and His Shop', *Glasgow University Gazette* 59 (1969), 5; D. J. Bryden, *Scottish Scientific Instrument-Makers, 1600–1900* (Edinburgh, 1972), 21–22.
4. J. B. Morrell, 'Reflections on the History of Scottish Science', *History of Science* 12 (1974), 81–94; Steven Shapin, 'The Audience for Science in Eighteenth Century Edinburgh', *History of Science* 12 (1974), 95–121; John R. R. Christie, 'The Origins and Development of the Scottish Scientific Community, 1680–1760', *History of Science* 12 (1974), 122–41.

5. A. D. Morrison-Low, 'The Scientific Instrument Trade in Provincial England during the Industrial Revolution, 1760–1851' (unpublished D.Phil. thesis, University of York, 1999), 29.
6. Peter Jones (ed.), *Philosophy and Science in the Scottish Enlightenment* (Edinburgh, 1988); Peter Jones (ed.), *The 'Science of Man' in the Scottish Enlightenment: Hume, Reid and Their Contemporaries* (Edinburgh, 1989); Paul Wood (ed.), *The Scottish Enlightenment: Essays in Reinterpretation* (Rochester, NY 2000).
7. Paul Wood, 'Science and the Aberdeen Enlightenment', in Jones, *Philosophy and Science*, 58.
8. Roger L. Emerson, *Professors, Patronage and Politics: The Aberdeen Universities in the Eighteenth Century* (Aberdeen, 1992); idem, 'Politics and the Glasgow Professors, 1690–1800', in Andrew Hook and Richard B. Sher (eds.), *The Glasgow Enlightenment* (East Linton, UK, 1995), 21–39; Jack Morrell, 'The Edinburgh Town Council and Its University, 1717–1766', in R. G. W. Anderson and A. D. C. Simpson (eds.), *The Early Years of the Edinburgh Medical School* (Edinburgh, 1976), 12. See also R.G.W. Anderson and A.D.C. Simpson, *Edinburgh and Medicine: A Commemorative Catalogue of the Exhibition* (Edinburgh, 1976).
9. The only business records relating to an eighteenth-century Scottish instrument firm are those at Birmingham Reference Library which relate to James Watt's Glasgow business, and are currently being used by Dr. R. L. Hills towards a biography of James Watt. My thanks to Dr. Hills for allowing me to see drafts of his early chapters.
10. G. L'E. Turner, 'The London Trade in Scientific Instrument-Making in the Eighteenth Century', *Vistas in Astronomy* 20 (1979), 173–82; W. D. Hackmann, 'The Nineteenth-Century Trade in Natural Philosophy Instruments in Britain', in P. R. de Clercq (ed.), *Nineteenth-Century Scientific Instruments and Their Makers* (Leiden and Amsterdam, 1985), 53–91.
11. Ronald G. Cant, 'Origins of the Enlightenment in Scotland: The Universities', in R. H. Campbell and Andrew S. Skinner (eds.), *The Origins and Nature of the Scottish Enlightenment* (Edinburgh, 1982), 42–64.
12. R. G. W. Anderson, 'Industrial Enterprise and the Scottish Universities in the Eighteenth Century', in Jenni Calder (ed.), *The Enterprising Scot: Scottish Adventure and Achievement* (Edinburgh, 1986), 59–67.
13. See, for instance, Dorothy Davis, *A History of Shopping* (London, 1966); M. A. Crawforth, 'Evidence from Trade Cards for the Scientific Instrument Industry', *Annals of Science* 42 (1985), 453–554.
14. Paul Wood, 'The Scientific Revolution in Scotland', in Roy Porter and Mikuláš Teich (eds.), *The Scientific Revolution in National Context* (Cambridge, 1992), 263.
15. Discussions of this problem are to be found in: J. V. Field, 'What is Scientific about a Scientific Instrument?', *Nuncius: Annali dell'Istituto e Museo di Storia della Scienza* 3 (1988), 3–26; D. J. Warner, 'What is a Scientific Instrument, When did It Become One, and Why?', *British Journal for the History of Science* 23 (1990), 83–93; A. J. Turner, 'Interpreting the History of Scientific Instruments', in R. G. W. Anderson, J. A. Bennett and W. F. Ryan (eds.), *Making Instruments Count: Essays on Historical Scientific Instruments Presented to Gerard L'Estrange Turner* (Aldershot, UK, 1993), 17–26.
16. See, for instance, Nathan Rosenberg, *Inside the Black Box: Technology and Economics* (Cambridge, 1992), and his *Exploring the Black Box: Technology, Economics and History* (Cambridge, 1994).

17. See the articles in *Journal of the History of Collections* 7 (1995), devoted to the history of collecting scientific instruments.
18. Discussed by A. van Helden, 'The Invention of the Telescope', *Transactions of the American Philosophical Society* 67 (1977), Part 4. More recently, see Colin Ronan *et al*, 'Was there an Elizabethan Telescope?', *Bulletin of the Scientific Instrument Society* no. 37 (1993), 2–10.
19. Stephen Pumfrey, '"O Tempora, O Magnes!": A Sociological Analysis of the Discovery of Secular Magnetic Variation in 1634', *British Journal for the History of Science* 22 (1989), 181–214; Patricia Fara, *Sympathetic Attractions: Magnetic Practices, Beliefs and Symbolism in Eighteenth Century England* (Princeton, N.J., 1996); for instruments, see Anita McConnell, *Geomagnetic Instruments before 1900: An Illustrated Account of Their Construction and Use* (London, 1980); Anita McConnell, *Geophysics & Geomagnetism: Catalogue of the Science Museum Collection* (London, 1986).
20. For the barometer, see Jan Golinski, 'Barometers of Change: Meteorological Instruments as Machines of Enlightenment', in William Clark, Jan Golinski and Simon Schaffer (eds.), *The Sciences in Enlightened Europe* (Chicago and London, 1999), 69–93; for the air pump, see Steven Shapin and Simon Schaffer, *Leviathan and the Air-Pump: Hobbes, Boyle and the Experimental Life* (Princeton, N.J., 1985); for the development of the instrument, see A. C. van Helden, 'The Age of the Air-Pump', *Tractrix* 3 (1991), 149–72; A. C. van Helden, 'Theory and Practice in Air-Pump Construction: The Co-Operation between Willem J. 'sGravesande and Jan van Musschenbroek', *Annals of Science* 51 (1994), 477–95; for Joseph Wright, see Judy Egerton (ed.), *Wright of Derby* (London, 1990).
21. W. D. Hackmann, *Electricity from Glass: The History of the Frictional Electrical Machine, 1600–1850* (Alphen aan den Rijn, ND, 1978); Simon Schaffer, 'The Consuming Flame: Electrical Showmen and Tory Mystics in the World of Goods', in John Brewer and Roy Porter (eds.), *Consumption and the World of Goods* (London, 1993), 489–526. For electricity in medicine, see D. Fishlock, 'A History of Electricity in Medical Treatment', *GEC Review* 9 (1994), 113–24, and W. D. Hackmann, 'The Induction-Coil in Medicine and Physics, 1835–1865', in C. Blondel, F. Parot, A. J. Turner and M. Williams (eds.), *Studies in the History of Scientific Instruments* (London, 1989), 235–50.
22. See David S. Landes, *Revolution in Time: Clocks and the Making of the Modern World* (Cambridge, MA, and London, 1983); Leonard Weiss, *Watch-Making in England, 1760–1820* (London, 1982).
23. See John Smith, *Old Scottish Clockmakers, 1453–1850*, 2nd edn (Edinburgh and London, 1921).
24. William J.H. Andrewes (ed.), *The Quest for Longitude* (Cambridge, MA, 1996).
25. T. N. Clarke, A. D. Morrison-Low and A. D. C. Simpson, *'Brass and Glass': Scientific Instrument-Making Workshops in Scotland* (Edinburgh, 1989), 35–37.
26. Nicholas Goodison, *English Barometers, 1680–1860*, 2nd edn (Woodbridge, UK, 1977), 32; for Sinclair's mining work, see Richard J. Howarth, 'Measurement, Portrayal and Analysis of Orientation Data and the Origins of Early Modern Structural Geology (1670–1967)', *Proceedings of the Geologists' Association* 110 (1999), 273–309. My thanks to Dr Mike Taylor for this reference.
27. A. G. Thoday, *Barometers* (London, 1966), 2.
28. Bryden, *Scottish Scientific Instrument-Makers*, 7.
29. John M. Pinkerton (ed.), *The Minute Book of the Faculty of Advocates, 1661–1712* (Edinburgh, 1976), 65. My thanks to Dr. Iain Gordon Brown for this reference.

30. National Archives of Scotland, MS GD 6/2173. My thanks to George Dalgleish for this reference.
31. Clarke *et al.*, 'Brass and Glass', 102–3, 203–9.
32. Elisabeth Bennion, *Antique Medical Instruments* (London, 1979).
33. R. D. Connor and A. D. C. Simpson, *Metts, Merchants and Markets: The Weights and Measures of Scotland*, forthcoming.
34. For example, the work of John White of Auchtermuchty, for which firm see Edwin J. White, 'Whites of Auchtermuchty', *Auchtermuchty Festival Programme 1984*, 12–15. However, both the instrument makers Alexander Adie of Edinburgh and James Crichton of Glasgow are known to have constructed precision balances; J. T. Stock and D. J. Bryden, 'A Robinson Balance by Adie and Son, Edinburgh', *Technology and Culture* 13 (1972), 44–54. A balance made by Crichton is at the McLean Museum Greenock; Mary Holbrook, R. G. W. Anderson, and D. J. Bryden, *Science Preserved: Directory of Scientific Instruments in Collections in the United Kingdom and Eire* (London, 1992), 135. John Gardner of Glasgow designed and manufactured a coin-scale in the 1770s; D. J. Bryden, 'John Gardner's Coin-Scale: A Scottish Gold Coin-Scale of 1773', *Equilibrium* no. 1 (1990), 1303–11; Roger Davies, 'A What's-it Identified', *Tool and Trades History Society Newsletter* no. 30 (1990), 46–48.
35. Gerard L'E. Turner, *Elizabethan Instrument Makers: The Origins of the London Trade in Precision Instrument Making* (Oxford, 2000), 5.
36. Silke Ackermann (ed.), *Humphrey Cole: Mint, Measurement and Maps in Elizabethan England* (London, 1998), 15–18.
37. Philip Staynred (fl. 1621–69) of Bristol apparently had devised and made his own measuring rulers and gauging rods; E. G. R. Taylor, *Mathematical Practitioners of Tudor and Stuart England, 1485–1714* (Cambridge, 1954), 208. The brothers John and Robert Roscoe were active in Liverpool from about 1696; Gloria Clifton, *Directory of British Scientific Instrument Makers, 1550–1851* (London, 1995), 237. An hour-glass maker, Nicholas Cosens, obtained his freedom in York in 1638; Brian Loomes, *The Early Clockmakers of Great Britain* (London, 1981), 166.
38. Cant, 'Origins of the Enlightenment', 44; A. George Molland, 'Duncan Liddell (1561–1613): An Early Benefactor of Marischal College Library', *Aberdeen University Review* 51 (1985–86), 485–99.
39. Betty Ponting, 'Mathematics at Aberdeen: Developments, Characters and Events, 1495–1717', *Aberdeen University Review* 48 (1979–80), 26–35; idem, 'Mathematics at Aberdeen: Developments, Characters and Events, 1717–1860', *Aberdeen University Review* 48 (1979–80), 162–176; Paul B. Wood, *The Aberdeen Enlightenment: The Arts Curriculum in the Eighteenth Century* (Aberdeen, 1993), 14.
40. Willem and Joannis Blaeu, *Theatrum Orbis Terrarum sive Atlas Novus*, 6 vols. (Amsterdam, 1635–55), esp. vol. 5 for maps of Scotland and Ireland; for an overview, see D. G. Moir (ed.), *The Early Maps of Scotland to 1850*, 3rd edn, 2 vols. (Edinburgh, 1973), 1:37–53.
41. National Museums of Scotland inventory no. T.1947.27; Angus Macdonald and A. D. Morrison-Low, *A Heavenly Library: Treasures from the Royal Observatory's Crawford Collection* (Edinburgh, 1994), 25. For Fusoris, see A. J. Turner, *Early Scientific Instruments: Europe, 1400–1800* (London, 1987), 31–32.
42. A.D. Morrison-Low, 'Early Navigational Instruments in Scotland: Icons and Survivals', in Anderson *et al*, *Making Instruments Count*, 218–31.
43. For the Cole instruments, see: Ackermann, *Humphrey Cole*, 33–38; Turner,

Elizabethan Instrument Makers, 149–54, 166–68. For the Allen instrument, see: Alan Stimson, *The Mariner's Astrolabe: A Survey of Known, Surviving Sea Astrolabes* (Utrecht, 1988), 78–79; Morrison-Low, 'Early Navigational Instruments', 221.

44. A. D. C. Simpson, 'James Gregory and the Reflecting Telescope', *Journal of the History of Astronomy* 23 (1992), 77–92.

45. These include: an unsigned example in brass of a horary quadrant, 'the Panorganon, or a Universall Instrument', described in 1672 by William Leybourne; a series of three clocks by Joseph Knibb; a new plate for the Cole astrolabe engraved by John Marke for the latitude 56° 25′; and an example of William Oughtred's double horizontal dial by Hilkiah Bedford. On these instruments and their makers, see: Taylor, *Mathematical Practitioners*, 230–1, 384; R. A. Lee, *The Knibb Family Clockmakers* (Byfleet, UK, 1964), 20, 154–57; A. J. Turner, 'William Oughtred, Richard Delamain and the Horizontal Instrument in Seventeenth Century England', *Annali dell'Istituto e Museo di Storia di Scienza Firenze* 6 (1981), 99–125; idem, 'A Note on the Life of Hilkiah Bedford', *Bulletin of the Scientific Instrument Society* no. 9 (1986), 3–5. Further London makers are mentioned in Gregory's correspondence; Herbert Westren Turnbull (ed.), *James Gregory Tercentenary Memorial Volume* (London, 1939). For the surviving instruments, see Holbrook et al, *Science Preserved*, 17, 201–3.

46. D. W. Waters, *The Art of Navigation in England in Elizabethan and Early Stuart Times* (London, 1958), pl. 63; Holbrook et al, *Science Preserved*, 202.

47. James Corss, *Uranoscopia: The Contemplation of the Heavens in a Perpetual Speculum, or General Prognostication for Ever* (Edinburgh, 1662), sig A3r, quoted in Bryden, *Scottish Scientific Instrument-Makers*, 1.

48. Bryden, *Scottish Scientific Instrument-Makers*, 2.

49. M. Wood (ed.), *Extracts from the Records of the Burgh of Edinburgh, 1642–1655* (Edinburgh, 1938), 25, quoted in Bryden, *Scottish Scientific Instrument-Makers*, 3.

50. Bryden, *Scottish Scientific Instrument-Makers*, 3.

51. National Museums of Scotland inventory no. T.1972.252; D. J. Bryden, 'Scotland's Earliest Surviving Calculating Device: Robert Davenport's Circles of Proportion of c.1650', *Scottish Historical Review* 55 (1976), 54–60; for Oughtred, see Turner, 'William Oughtred'.

52. Joyce Brown, *Mathematical Instrument-Makers in the Grocers' Company, 1688–1800* (London, 1979), 25. For Elias Allen, see Hester Higton, 'Elias Allen and the Role of Instruments in Shaping the Mathematical Culture of Seventeenth-Century England' (unpublished Ph.D. thesis, University of Cambridge, 1996).

53. These include a dial by Henry Wynne of London (dated 1692) at Drumlanrig Castle (seat of the Dukes of Buccleuch) which is discussed in Andrew Somerville, 'The Ancient Sundials of Scotland', *Proceedings of the Society of Antiquaries of Scotland* 117 (1987), 233–64, fiches 3: A2–G14 (also issued separately with the fiche printed in Andrew Somerville, *The Ancient Sundials of Scotland* (London, 1990), fiche 70 and 96); another by the same maker was inscribed as having been made for Richard, Earl of Lauderdale in 1695, on which see Bryden, *Scottish Scientific Instrument-Makers*, 16 n76; three have been recorded by John Marke of London – one undated at Stobhall in Perthshire, and two dated 1679 at Drummond Castle, Perthshire – and are listed in Somerville, *Ancient Sundials*, fiche 93; a dial by Henry Sutton in 1660 for the latitude 56° 30′, which was clearly made for a Scottish latitude somewhere between Dundee and Arbroath, and which was recently acquired by the National Museums of Scotland (inventory no.

T.1993.110) at Sotheby's, 7 October 1993, Lot 787; the latter dial may have been made for the same client who once owned a horary quadrant, also made by Sutton, at the same date and for the same latitude, which is now in the Whipple Museum of the History of Science, Cambridge (inventory no. 738) and described in D. J. Bryden, *Catalogue 6: Sundials and Related Instruments* (Cambridge, 1988), item 282.

54. Thomas Ross, 'The Ancient Sundials of Scotland', *Proceedings of the Society of Antiquaries of Scotland* 24 (1890), 161–273; David MacGibbon and Thomas Ross, *The Castellated and Domestic Architecture of Scotland from the Twelfth to the Eighteenth Centuries*, 5 vols. (Edinburgh, 1887–92), 5:357–514.
55. Somerville, 'Ancient Sundials' and *The Ancient Sundials of Scotland*.
56. Somerville, 'Ancient Sundials', 234.
57. Somerville, 'Ancient Sundials', 233.
58. Bryden, *Scottish Scientific Instrument-Makers*, 16 n76. National Museums of Scotland inventory no. T.1979.24.
59. One is signed 'Ja Clark Dudonens fecit'; this was acquired from a Scottish source by the National Museums and Galleries on Merseyside, inventory no. 1988.379. My thanks to the late Martin Suggett for this information. The second is signed 'Jacobus Clark fecit / Edenburgi'; National Museums of Scotland inventory no. T.1988.19. On Clark, see D. J. Bryden, ' "Mr Clerk the Graver": A Biographical Study in the Cultural Infrastructure of Early Modern Scotland', *Review of Scottish Culture* 11 (1998–99), 13–31. Thomas Ross's publications mention a dial at Niddrie Marischal outside Edinburgh, where the dial was inscribed with the arms of the Wauchopes of Niddrie and the signature 'Jacobus Clark, Dundee, fecit'; Ross, 'Ancient Sundials', 254–55; MacGibbon and Ross, *Castellated and Domestic Architecture*, 495. The house was demolished in the 1920s, and the dial, on its richly carved sandstone pedestal with the Wauchope arms in relief, was removed into the grounds of a house in north Edinburgh. Shortly after being recorded by a National Museums of Scotland photographer, the sundial, torn from its pedestal, was stolen in November 1990.
60. A. D. C. Simpson, 'Sir Robert Sibbald – the Founder of the College' in R. Passmore (ed.), *Proceedings of the Royal College of Physicians of Edinburgh Tercentenary Congress 1981* (Edinburgh, 1982), 59–91; idem, 'John Adair, Cartographer, and Sir Robert Sibbald's Scottish Atlas', *Map Collector* no. 62 (1993), 32–36; C. W. J. Withers, 'Geography, Science and National Identity in Early Modern Britain: The Case of Scotland and the Work of Sir Robert Sibbald, 1641–1722', *Annals of Science* 53 (1996), 29–73; John N. Moore, 'John Adair's Contribution to the Charting of the Scottish Coasts: A Reassessment', *Imago Mundi* 52 (2000), 43–65.
61. National Museums of Scotland inventory no. T.1988.18. Other Scottish clockmakers of this period known to have made sundials include John Russell of Falkirk, and George Jamieson of Hamilton.
62. Clarke *et al*, 'Brass and Glass', 68–69.
63. Bryden, *Scottish Scientific Instrument-Makers*, 7; Silvio A. Bedini, *Early American Scientific Instruments and Their Makers* (Washington, DC, 1964), 3–13, subsequently expanded and revised as S. A. Bedini, *Thinkers and Tinkers: Early American Men of Science* (New York, 1975).
64. Bryden, *Scottish Scientific Instrument-Makers*, 9–11; Clarke *et al*, 'Brass & Glass', 25–84, 89–95.
65. Steven Shapin, ' "Nibbling at the Teats of Science": Edinburgh and the Diffusion of Science in the 1830s', in Ian Inkster and Jack Morrell (eds.), *Metropolis*

and *Province: Science in British Culture, 1780–1850* (London, 1983), 152; John A. Cable, 'The Early History of Scottish Popular Science', *Studies in Adult Education* 4 (1972), 34–45; David Gavine, 'Astronomy in Scotland, 1745–1900', 2 vols. (unpublished Ph.D. thesis, Open University, 1981).

66. C. W. J. Withers, 'Towards a History of Geography in the Public Sphere', *History of Science* 37 (1999), 45–78; this article draws on some of the data gathered by Gavine, 'Astronomy in Scotland'.
67. For instance, Hackmann, 'Nineteenth-Century Trade'; J. E. Burnett and A.D. Morrison-Low, *Vulgar and Mechanick: The Scientific Instrument Trade in Ireland, 1650–1921* (Dublin, 1989), 19; Turner, *Early Scientific Instruments*, ch.5.
68. Clarke *et al*, *'Brass and Glass'*; the only follow-up to this was A. D. Morrison-Low, 'The Road to Ruin? Bankruptcy and some Legal Consequences for the Instrument Maker in Nineteenth Century Britain', in Giorgio Dragoni, Anita McConnell and Gerard L'E. Turner (eds.), *Proceedings of the XIth Scientific Instrument Symposium, Bologna 1991* (Bologna, 1994), 53–59.
69. Neil McKendrick, John Brewer and J.H. Plumb, *The Birth of a Consumer Society: the Commercialization of Eighteenth-century England* (London, 1982); John Brewer, *The Pleasures of the Imagination: English Culture in the Eighteenth Century* (London, 1997); Richard Sorrenson, 'The State's Demand for Accurate Astronomical and Navigational Instruments in Eighteenth Century Britain', in Ann Bermingham and John Brewer (eds.), *The Consumption of Culture: Image, Object, Text* (London, 1995), 263–71.
70. Sorrenson, 'State's Demand', 264.
71. Sorrenson, 'State's Demand', 265.
72. Sorrenson, 'State's Demand', 265.
73. The 'Scotland' or 'Britain' terminology problem still has not been resolved: see, for instance, for some explanations, Linda Colley, *Britons: Forging the Nation, 1707–1837* (New Haven, CT, 1992); Angus Calder, 'The Enlightenment', in Ian Donnachie and Christopher Whatley (eds.), *The Manufacture of Scottish History* (Edinburgh, 1992), 31–50; Marinell Ash, *The Strange Death of Scottish History* (Edinburgh, 1980).
74. Sorrenson, 'State's Demand', 270.
75. For background, see: Gavine, 'Astronomy in Scotland'; Hermann A. Brück, *The Story of Astronomy in Edinburgh from Its Beginnings until 1975* (Edinburgh, 1983), 11–21; D. J. Bryden, 'The Edinburgh Observatory, 1736–1811: A Story of Failure', *Annals of Science* 47 (1990), 445–74.
76. C. Close, *The Early Years of the Ordnance Survey* (Newton Abbot, 1969); Yolande O'Donoghue, *William Roy, 1726–1790: Pioneer of the Ordnance Survey* (London, 1977); J. R. Millburn, 'The Office of Ordnance and the Instrument-Making Trade in the Mid-Eighteenth Century', *Annals of Science* 45 (1988), 221–93.
77. D.J. Bryden, 'Three Edinburgh Microscope Makers: John Finlayson, William Robertson and John Clark', *Book of the Old Edinburgh Club* 33 (1972), 165–76; Bryden, *Scottish Scientific Instrument-Makers*, 16–20.
78. See the chapter on 'James White', in Clarke *et al*, *'Brass and Glass'*, 252–66.
79. See, for instance, Roy Porter, 'Science, Provincial Culture and Public Opinion in Enlightenment England', *British Journal for Eighteenth-Century Studies* 3 (1980), 20–46, and, more recently, Roy Porter, *Enlightenment: Britain and the Creation of the Modern World* (London, 2000).
80. Ian Inkster, 'The Development of a Scientific Community in Sheffield, 1790–1850: A Network of People and Interests', *Transactions of the Hunter Archae-*

ological Society 10 (1973), 99; Margaret C. Jacob, *Scientific Culture and the Making of the Industrial West* (Oxford, 1997); Larry Stewart, *The Rise of Public Science: Rhetoric, Technology and Natural Philosophy in Newtonian Britain, 1660–1750* (Cambridge, 1992).

81. The first number of the *British Journal for the History of Science* 28 (1995) contains articles devoted to this topic.
82. Stewart, *Rise of Public Science*, 384.
83. Alan Q. Morton and Jane A. Wess, *Public and Private Science: The King George III Collection at the Science Museum* (Oxford, 1993), 52; Stephen Pumfrey, 'Who did the Work? Experimental Philosophers and Public Demonstrators in Augustan England', *British Journal for the History of Science* 28 (1995), 131–56.
84. Morton and Wess, *Public and Private Science*, 56.
85. Morton and Wess, *Public and Private Science*, 72–87.
86. Stewart, *Rise of Public Science*, 235, 362–63.
87. Svante Lindqvist, *Technology on Trial: The Introduction of Steam Power Technology into Sweden, 1715–1736* (Uppsala, 1984), 207–10.
88. Lindqvist, *Technology on Trial*, 205, 212.
89. J. G. Tandberg, *Die Triewaldsche Sammlung am Physikal. Institut der Universität zu Lund und die Original-Luftpumpe Guerickes* (Lund, 1920), 3–6, 25. My thanks to Dr. Olov Amelin for this reference.
90. Morton and Wess, *Public and Private Science*, 89–127, 195, 236. For 'W.R.' [William Robertson], see Bryden, 'Three Edinburgh Microscope Makers'.
91. G.L'E. Turner, 'The Auction Sales of the Earl of Bute's Instruments, 1793', *Annals of Science* 23 (1967), 213–42; Karl W. Schweizer (ed.), *Lord Bute: Essays in Re-Interpretation* (Leicester, 1988); Roger Emerson, 'Catalogus Librorum A.C.D.A.: The Library of Archibald Campbell, third Duke of Argyll (1682–1761)', in Paul Wood (ed.), *The Culture of the Book in the Scottish Enlightenment* (Toronto, 2000), 17.
92. Emerson, *Professors, Patronage and Politics*, 103.
93. For Benjamin Martin, see John R. Millburn, *Benjamin Martin, Author, Instrument-Maker and 'Country Showman'* (Leyden, 1976); idem, *Benjamin Martin, Author, Instrument-Maker and 'Country Showman': Supplement* (London, 1986); idem, *Retailer of the Sciences: Benjamin Martin's Scientific Instrument Catalogues, 1756–1782* (London, 1986).
94. For Ferguson, see Ebenezer Henderson, *Life of James Ferguson, F.R.S., in a Brief Autobiographical Account, and Further Extended Memoir* (Edinburgh, London and Glasgow, 1867); John R. Millburn, *Wheelwright of the Heavens. The Life and Work of James Ferguson, F.R.S.* (London, 1988).
95. James A. Secord, 'Newton in the Nursery: Tom Telescope and the Philosophy of Tops and Balls, 1761–1838', *History of Science* 23 (1985), 127–51; G. L'E. Turner, 'Scientific Toys', *British Journal for the History of Science* 20 (1987), 377–98.
96. Gavine, 'Astronomy in Scotland'.
97. Inkster, 'Development of a Scientific Community', 100–1, 109.
98. J. A. Harrison, 'Blind Henry Moyes, "An Excellent Lecturer in Philosophy"', *Annals of Science* 13 (1957), 109–25.
99. A. D. Morrison-Low, 'William Nicol FRSE c.1771–1851: Lecturer, Scientist and Collector', *Book of the Old Edinburgh Club*, n.s., 2 (1992), 123–31.
100. D. J. Bryden, *James Short and his Telescopes* (Edinburgh, 1968); idem, 'James Short, MA, FRS, Optician Solely for Reflecting Telescopes', *University of Edinburgh Journal* 24 (1970), 251–61; idem, 'Note on a Further Portrait of James Short, FRS', *Notes and Records of the Royal Society of London* 24 (1969), 109–12;

G. L'E. Turner, 'A Portrait of James Short, FRS, Attributable to Benjamin Wilson, FRS', *Notes and Records of the Royal Society of London* 22 (1967), 105–12; idem, 'James Short, FRS, and His Contribution to the Construction of Reflecting Telescopes', *Notes and Records of the Royal Society of London* 24 (1969), 91–108; idem, 'Mr James Short, Optician, Founding Member of the Society', *Journal of the Royal Society of Arts* 118 (1970), 788–792; A. D. C. Simpson, 'The Early Development of the Reflecting Telescope in Britain' (unpublished Ph.D. thesis, University of Edinburgh, 1981). The chapter on Short in Clarke et al, *'Brass & Glass'*, 1–10, synthesises this literature and draws on new material.
101. Clarke et al, *'Brass and Glass'*, 1.
102. Thomas Somerville, *My Own Life and Times, 1741–1814* (Edinburgh, 1861), 14–15.
103. Roger L. Emerson, 'The Philosophical Society of Edinburgh, 1737–1747', *British Journal for the History of Science* 12 (1979), 154–91; for a discussion of the Society after its revival in the 1740s, see Emerson, 'The Philosophical Society of Edinburgh 1748–1768', *British Journal for the History of Science* 14 (1981), 133–76.
104. Clarke et al, *'Brass and Glass'*, 1.
105. See, for instance, Richard John Sorrenson, 'Scientific Instrument Makers at the Royal Society of London, 1720–1780' (unpublished Ph.D. thesis, Princeton University, 1993); Roy Porter, Simon Schaffer, Jim Bennett and Olivia Brown, *Science and Profit in Eighteenth-Century London* (Cambridge, 1985). For a critique of the latter, see J. R. Millburn, 'Trade in Scientific Instruments', *Annals of Science* 43 (1986), 81–86.
106. Bryden, *James Short*, 13–16.
107. Bryden, *James Short*, 24–25.
108. D.J. Bryden, 'The Jamaican Observatories of Colin Campbell, FRS and Alexander Macfarlane, FRS', *Notes and Records of the Royal Society of London* 24 (1970), 261–72.
109. Diana C. F. Smith, 'The Progress of the *Orcades* Survey, with Biographical Notes on Murdoch Mackenzie Senior (1712–1797)', *Annals of Science* 44 (1987), 277–88; Diana C. F. Webster, 'Mony and Diverse Ways: Surveying in Scotland Before 1820', in Finlay MacLeod (ed.), *Torgail Tir Marking Time: The Map of the Western Isles* (Stornoway, 1989), 79–88.
110. Macdonald and Morrison-Low, *A Heavenly Library*, 47.
111. See Ch. 4, 'The Adie Business', in Clarke et al, *'Brass and Glass'*, 26–87.
112. A. D. C. Simpson, '"Handle with Care": Handling Warnings on Early Scientific Instruments and Two Early Scottish Grain Measures', *Bulletin of the Scientific Instrument Society* no. 30 (1991), 3–4.
113. John S. Reid, 'Patrick Copland, 1748–1822: Connections outside the College Courtyard', *Aberdeen University Review* 51 (1985–86), 226–50.
114. This is discussed more fully in the chapter, 'Science and Measurement', in Connor and Simpson, *Metts, Merchants and Markets*, forthcoming.
115. Tore Frängsmyr, J. L. Heilbron and Robin Rider (eds.), *The Quantifying Spirit in the Eighteenth Century* (Berkeley, Los Angeles and London, 1990).
116. 'The Adie Business', in Clarke et al, *'Brass and Glass'*, 25.
117. Ian Adams, 'The Land Surveyor and his Influence on the Scottish Rural Landscape', *Scottish Geographical Magazine* 84 (1968), 248–55; idem, 'Economic Process and the Scottish Surveyor', *Imago Mundi* 27 (1975), 13–18. See also the introduction to Sarah Bendall's *Dictionary of Land Surveyors and Local Map-Makers of Great Britain and Ireland, 1530–1850* (London, 1997).

118. A. D. Morrison-Low, *Scottish Barometers*, forthcoming; A. D. C. Simpson, 'Globe Production in Scotland in the period 1770–1830', *Der Globusfreund: Journal for the Study of Globes and Related Instruments* nos. 35–37 (Wien, 1987), 21–32; idem, 'An Edinburgh Intrigue: Brewster's Society of Arts and the Pantograph Dispute', *Book of the Old Edinburgh Club* n.s., 1 (1991), 47–73.
119. Nicholas Phillipson, 'Sir David Brewster: Some Concluding Remarks', in A. D. Morrison-Low and J. R. R. Christie (eds.), *'Martyr of Science': Sir David Brewster, 1781–1868* (Edinburgh, 1984), 79.
120. See John G. Burke, *Origins of the Science of Crystals* (Berkeley, 1965), 147–175; A. D. Morrison-Low, 'Brewster and Scientific Instruments', in Morrison-Low and Christie, *'Martyr of Science'*, 59–63; A. D. Morrison-Low and A. D. C. Simpson, 'A New Dimension: A Context for Photography before 1860', in Sara Stevenson (ed.), *Light from the Dark Room* (Edinburgh, 1995), 15–28.
121. Emerson, *Professors, Patronage and Politics*, 104.
122. Emerson, 'Politics and the Glasgow Professors', 23.
123. For Ilay's library, see Emerson, '*Catalogus Librorum A.C.D.A*', 13–39; the mechanical equinoctial dial by Richard Glynne was until recently at the Time Museum, Rockford, Illinois, but was sold at Sotheby's New York, 2 December 1999, Lot 6; see A.D. Morrison-Low, 'Sold at Sotheby's: Sir John Findlay's Cabinet and the Scottish Antiquarian Tradition', *Journal of the History of Collections* 7 (1995), 197–209, and Roger L. Emerson, 'The Scientific Interests of Archibald Campbell, 1st Earl of Ilay and 3rd Duke of Argyll, 1682–1761', *Annals of Science* 59 (2002), 21–56.
124. Ian G. Lindsay and Mary Cosh, *Inveraray and the Dukes of Argyll* (Edinburgh, 1973); Mary Cosh, 'Lord Ilay's Eccentric Building Schemes', *Country Life*, 20 July 1972, 142–45.
125. Lindsay and Cosh, *Inveraray*, 411–12; Mary Cosh, 'Clockmaker Extraordinary: The Career of Alexander Cumming', *Country Life*, 12 June 1969, 1528–35.
126. Emerson, 'Politics and the Glasgow Professors', 31.
127. J. D. Mackie, *The University of Glasgow 1451 to 1951* (Glasgow, 1954), 218, 222–23; Bryden, 'Jamaican Observatories'; Holbrook *et al*, *Science Preserved*, 130. Interestingly, there is an octant signed by John Bird in the collection at the University of Strathclyde; Holbrook *et al*, *Science Preserved*, 133.
128. Bert Bolle, *Barometers in Beeld* (Lochem-Poperinge, Belgium, 1983), 30; A. D. Morrison-Low, 'The Provenance of the Natural Philosophy Collection at the University of Edinburgh' (unpublished M. Sc. dissertation, University of Leicester, 1987), 29.
129. R. G. W. Anderson, *The Playfair Collection and the Teaching of Chemistry at the University of Edinburgh, 1714–1858* (Edinburgh, 1978), 21.
130. Anderson, *The Playfair Collection*; idem, 'Joseph Black: An Outline Biography', in Simpson, *Joseph Black*, 7–12; idem, 'Instruments and Apparatus', in C. A. Russell (ed.), *Recent Developments in the History of Chemistry* (London, 1985), 217–37; J. G. Fyffe and R. G. W. Anderson, *Joseph Black* (London, 1992); R. G. W. Anderson, 'A Source for Eighteenth-Century Chemical Glass', in Dragoni *et al*, *Proceedings of the XIth Scientific Instrument Symposium*, 47–52; idem, 'Joseph Black and His Chemical Furnace', in Anderson *et al*, *Making Instruments Count*, 118–26; idem, 'Joseph Black', in D. Daiches, P. Jones and J. Jones (eds.), *A Hotbed of Genius: The Scottish Enlightenment, 1730–90* (Edinburgh, 1986), 92–114.
131. Wood, *Aberdeen Enlightenment*.
132. J. T. Lloyd, 'Item ane Shipe Skin: An Account of Early Experimentation in the

Natural Philosophy Department', *The College Courant: Journal of the Glasgow University Graduates Association* 21 (1969), 5–9; Peter Swinbank, 'Experimental Science in the University of Glasgow at the Time of Joseph Black', in Simpson, *Joseph Black*, 23–35.

133. Mackie, *University of Glasgow*, 218; D. J. Bryden, 'James Watt, Merchant: The Glasgow Years, 1754–1774', in Denis Smith (ed.), *Perceptions of Great Engineers: Fact and Fancy* (London, 1994), 9–21.
134. See Holbrook *et al*, *Science Preserved*, 131.
135. The exact nature of Watt's appointment as the University's mathematical instrument maker remains unclear. For a recent evaluation of Watt and his life, see Christine MacLeod, 'James Watt, Heroic Invention and the Idea of the Industrial Revolution', in Maxine Berg and Kristine Bruland (eds.), *Technological Revolutions in Europe: Historical Perspectives* (Cheltenham, UK, 1998), 96–116.
136. Richard L. Hills, 'James Watt's Barometers', *Bulletin of the Scientific Instrument Society* no. 60 (1999), 5–10.
137. Clarke *et al*, *'Brass and Glass'*, 164–79; Morrison-Low, 'Road to Ruin?'.
138. See Holbrook *et al*, *Science Preserved*, 131–34; A. M. Campbell, *Catalogue of the Collection of Historic Scientific Instruments in the University of Strathclyde* (Glasgow, 1980). On John Anderson, see: James Muir, *John Anderson: Pioneer of Technical Education and the College He Founded*, (ed.), James M. Macaulay (Glasgow, 1950), 62–70; John Cable, 'Early Scottish Science: The Vocational Provision', *Annals of Science* 30 (1976), 179–99; John Butt, *John Anderson's Legacy: The University of Strathclyde and Its Antecedents, 1796–1996* (East Linton, UK, 1996); Paul Wood, 'Jolly Jack Phosphorus in the Venice of the North; or, Who was John Anderson?', in Hook and Sher, *The Glasgow Enlightenment*, 111–32.
139. Holbrook *et al*, *Science Preserved*, 132, 134, 135; James Crichton, 'On the Freezing Point of Tin and the Boiling Point of Mercury; with a Description of a Self-Registering Thermometer', *Philosophical Magazine* 15 (1803), 147–48; Turner, 'The Earl of Bute's Instruments', 224; Connor and Simpson, *Metts, Merchants and Markets*, forthcoming.
140. R. G. Cant, *The University of St. Andrews: A Short History*, new edn (Edinburgh and London, 1970), 83.
141. Holbrook *et al*, *Science Preserved*, 201; for the orrery, see Macdonald and Morrison-Low, *Heavenly Library*, 40; for the telescope, see Bryden, *James Short*, 8; for Short, see Bryden, *Scottish Scientific Instrument-Makers*, 10 and Clarke *et al*, *'Brass & Glass'*, 6.
142. See Holbrook *et al*, *Science Preserved*, 201–3; Anderson, 'Source for Eighteenth-Century Chemical Glass'. For Thomson, see Jack Morrell, 'Thomas Thomson: Professor of Chemistry and University Reformer', *British Society of the History of Science* 4 (1969), 245–65.
143. See E. M. Wray (ed.), *Catalogue of Apparatus in the Museum of the Natural Philosophy Class in the United Colleges of St Salvator and St Leonard, St Andrews, prepared by W. Swan, LLD, May 1880* (St. Andrews, 1983); idem, *Historical Scientific Instruments from the Collection of the Department of Physics, University of St. Andrews: A Guide to Selected Exhibits* (St. Andrews, 1983).
144. John S. Reid, 'Eighteenth-Century Scottish University Instruments: The Remarkable Professor Copland', *Bulletin of the Scientific Instrument Society* no. 24 (1990), 2–8.
145. G. J. Thorkelin, 'The Northern Traveller (1790)', letters 3 and 4, mss. in Det Kongelige Bibliotek, Copenhagen, published in W. T. Johnstone (ed.), *Thorkelin*

and Scotland (Edinburgh, 1982), 43–50; the purchase of an air pump, a barometer, a hydrostatic balance and other items from Hauksbee is discussed by Wood, *Aberdeen Enlightenment*, 15; George Hearne's five-foot focus Newtonian telescope was repaired in 1819 but was no longer extant by 1837; John S. Reid, 'The Castlehill Observatory, Aberdeen', *Journal for the History of Astronomy* 13 (1982), 84–96.
146. Reid, 'Scottish University Instruments'; Reid, 'Castlehill Observatory'; Wood, *Aberdeen Enlightenment*, 82–84.
147. John S. Reid, 'Patrick Copland, 1748–1822: Aspects of his Life and Times at Marischal College', *Aberdeen University Review* 50 (1984–85), 359.
148. John S. Reid, 'A Select Clock', *Antiquarian Horology* 13 (1981), 45–50.
149. Reid, 'Scottish University Instruments', 4.
150. Reid, 'Scottish University Instruments', 4.
151. Joseph Black to James Watt, 20 December 1769, in Robinson and McKie, *Partners in Science*, 23.
152. Bryden, *Scottish Scientific Instrument-Makers*, 24–38.
153. D. J. Bryden, 'Provincial Scientific Instrument Making', *Bulletin of the Scientific Instrument Society* no. 2 (1984), 4.

3

Situating Practical Reason: Geography, Geometry and Mapping in the Scottish Enlightenment

CHARLES W. J. WITHERS

This chapter attempts to locate the practical concerns of geography, geometry and mapping on the map of scientific knowledge in Enlightenment Scotland. The chapter also has two more specific aims. First, I want to illustrate the public interest in geography in Enlightenment Scotland and in those other subjects with which it was allied (notably geometry), and to do so with reference to public lecture classes in geography and geometry, and to university teaching in those and related subjects. Second, I want to suggest that mapping and map-making and the public interest in geography should be accorded a more prominent place in our understanding of the intellectual character of the Enlightenment in Scotland than has been the case. Both aims reflect a longer-run intention to 'recover' the place of geographical knowledge as an intellectual and practical pursuit within the Scottish Enlightenment.[1]

Mapping and geographical knowledge were understood by Enlightenment commentators as expressions of national self-knowledge and as means by which one could secure accurate understanding of the shape and the state of the nation. For Sir Robert Sibbald, for example, a leading early Enlightenment figure and, from 1682, Scotland's Geographer Royal, accurate maps were crucial to the proper 'Geographical Description of our Countrey'.[2] In his *Essay on the Rise and Progress of Geography in Great Britain and Ireland* (1780), Richard Gough equated the progress of geographical knowledge in Enlightenment Scotland with the improved nature of its mapping. The same philosophy underpinned John Thomson's introduction to his 1832 *Atlas of Scotland*, where the author considered the improved understanding of the nature of the nation as the result of mapping endeavours between the late Renaissance work of Timothy Pont, through Blaeu and Sibbald in the seventeenth century and a host of map-makers in the eighteenth century, to his own 1832 work.[3]

What follows explores both the nature of mapping as a form of geographical knowledge in the Scottish Enlightenment, and the connections between the practices of geography and geometry as they underpinned such mapping. I understand mapping and these subjects as socially mediated and complex representational practices of the real world. I do not share the view expressed by Gough and Thomson and represented by subsequent generations of commentators that mapping (and, thus, geographical understanding) gets progressively 'better' over time. Nor do I see the making of these subjects as somehow 'above' the social world. In my own work on the history of geography, I am driven not just by an interest in knowing the facts of contemporary intellectual interest in geographical knowledge and in mapping in the Enlightenment, but also by the possibilities of locating what historians of science have called the analytic spaces and sites, the constitutive settings, in which such facts were made, debated, and consumed by their audiences.[4] Drawing upon writings in the histories of cartography and science, and on the place of geography in the public sphere(s) of Enlightenment Scotland, I will in this chapter address such issues for the period between Sibbald's claims of 1682 and Thomson's of 1832. I want to consider mapping as, at once, three things: first, a form of scientific endeavour reliant in various ways upon geography and geometry; second, a matter of situated social networks – of patronage, of institutional affiliation, and of certain social spaces; and third, the result of epistemological questions to do with accuracy and credibility, both in the map *and* in the map-maker. In order to explore Enlightenment Scotland in these ways, let me turn initially to consider the place of mapping and geographical knowledge in the Enlightenment in wider context.

Mapping and Geographical Knowledge in the Enlightenment

The term and the idea of 'mapping' were understood and imagined in a variety of ways by Enlightenment philosophers. The *Encyclopédie* (1751–65) of Diderot and d'Alembert, for example, arguably one of the greatest products of the Enlightenment, was itself reckoned a map:

> a kind of world map which is to show the principal countries, their position and their mutual dependence, the road that lies directly from one to the other. This road is often cut by a thousand obstacles, which are known in each country only to the inhabitants or to travellers, and which cannot be represented except in individual, highly detailed maps. These individual maps will be the different articles of the *Encyclopédie* and the Tree or Systematic Chart will be its world map.[5]

The idea of the 'map' was also a metaphor for the central place of geographical discovery in the Enlightenment, in bringing all the peoples of the world into view. This is what Edmund Burke understood by the map when he wrote: 'But now the Great Map of Mankind is unroll'd at once; and there is no state or Gradation of barbarism, and no mode of refinement which we have not at the same instant under our View'.[6] Burke was imagining a sort of human chronological and geographical 'chart', which revealed at a glance the stages of social development of the world's peoples.

At the same time, Enlightenment mapping was an innately practical affair.

For many European countries, new maps reflected the expansion of geographical knowledge and of colonial power. The 'Age of Enlightenment' was an age in which the establishment of national boundaries, the transformation of agricultural landscapes, and the expression of property rights in town and country were constituted in the map and embodied in the authority of surveyors, map-makers, and national mapping projects.[7] Modern scholars have recognised mapping and map-making as crucial elements within Enlightenment thinking in these terms. Mapping, with its dependence upon mathematics and geometry and upon their practical application 'in the field' in the form of survey, was part of what has been seen as the 'quantifying spirit' in the eighteenth century.[8] Mapping was a means to put the world into rational order. As J. L. Heilbron puts it, the 'quantifying spirit' of what he calls the European mathematical landscapers 'came to ground' not just in the map but in forestry management, in the exploitation of lands overseas, and in meteorology.[9] Matthew Edney regards or conceives of Enlightenment maps as encyclopaedic repositories of global reconnaissance, reflecting and constituting particular ideological and political interests.[10] Whether in the mapping of Bengal (1765–71), cadastral mapping in France, the work of military engineers and surveyors in Prussia (1767–87), in the territories of the Habsburgs (1763–87), in Saxony (1780–1825), or in North America, mapping was a symbol of political and scientific authority, a reflection of national self-identity, and a practical expression of what many contemporaries thought of as the triumph of Enlightenment reason.[11] Edney has also examined the social nature of Enlightenment map production, notably in late Hanoverian Britain. Following Eric Forbes, he has shown that Enlightenment scholars positioned the map 'at the heart of mathematical cosmography', which was the 'conceptual fusion of astronomy, geography, and their point of contact, cartography'. Consequently, for Edney eighteenth-century map-making merged without clear boundaries into other knowledges, such as geometry and astronomy.[12] Sven Widmalm, too, sees practical geometry in the form of geodetic survey and triangulation as promoting the idea of national cartography as a form of state measurement.[13]

Other scholars have considered the social nature of Enlightenment mapping as involving questions of native-explorer contact and of epistemological incommensurability.[14] Mapping provided a means to inscribe 'blank' space in the ways Daniel Clayton has documented of Vancouver's charting between 1792 and 1795 of the island that now bears his name.[15] Anne Godlewska demonstrates for France the central place of mapping and of mathematics and geometry in French geographical traditions in the eighteenth century, and the importance of mapping to the French state's sense of itself as a territory.[16] Catherine Delano-Smith and Roger Kain's work on English maps shows the social basis of map production in the context of eighteenth-century property and estate mapping.[17] Such concerns to understand what has been termed 'the capacity of the state' were in England paralleled in the work of the Excise, where the importance of mathematical measurement of casks and barrels reflected the centralising authority of the state as well as the particular promotion of discourses of political arithmetic.[18] Such statistical accountancy extended also to social questions, as Judith Grabiner shows in Chapter 5 of Colin Maclaurin's work with Robert Wallace over the Widows' Fund, and, in Scotland, such issues were most evident in Sir John Sinclair's management of the parish-based *Statistical Account of Scotland*, published in twenty-one volumes between 1791 and 1799.[19]

As for geography, historians of the subject have identified a number of the principal features in the development of geography in the Enlightenment. The second half of the eighteenth century in particular was characterised by the production of universal geographies, geographical dictionaries, and gazetteers whose form and purpose in offering descriptive lists of geographical facts mirrored other contemporary textual projects designed to order the world. Enlightenment geography texts were strongly mathematical as well as descriptive, reflecting a tradition that had classical origins in the work of Ptolemy as well as contemporary relevance in navigation and in astronomy. Geography was taught in many schools in the Enlightenment, often alongside civil history and mathematics. Studies have also been made of the teaching of geography within university curricula and in the public sphere. Whilst there was universal recognition of the subject's descriptive rationale, contemporary recognition of the subject's utility varied with context: from geography's place in military survey and campaigning, for example, to mining and survey, to 'rational geography's' place in historical explanation and in the geographical bases to Enlightenment theorisations of the stadial nature of human development. Geography understood as a means of commercial exchange also figured in the activities of the Royal Society and in the Académie des Sciences during the eighteenth century.[20]

In short, geography and mapping were commonplace elements in the Enlightenment, as textual subjects *and* as graphical practices. The first – geography, with its texts and practices – was used as a form of polite sociability and civic discourse. The second – mathematics – was understood as a basis for the practical application of geometry and geography, notably in the form of mapping. Both helped make sense of the world for contemporaries, whose conception of the nature of the terraqueous globe changed virtually with each new voyage of discovery. How, then, were these features apparent in the case of Scotland?

Mapping and Geographical Knowledge in the Scottish Enlightenment

Scotland offers considerable potential for examining the place of mapping and geographical knowledge in the Enlightenment, given the widespread interest that Scots had in the problems of progress and virtue in a commercialising society and in the improvement of the human condition.[21] As has been noted of chemistry's place as an enlightened public discourse, for example, Scotland provided a cultural and intellectual environment 'within which science could establish its civic credentials as public culture in close conjunction with the beginnings of its academic and disciplinary structure'.[22] Evidence relating to public lectures on agriculture and natural history supports this claim.[23]

We may think of mapping in Enlightenment Scotland in several ways. At one level were individuals, usually with patrons among the nobility and gentry who, as map-makers and surveyors, transformed the face of the Scottish countryside and whose work in the towns and planned villages reflected the Enlightenment's 'geometric spirit'.[24] Some of these men, but by no means all, had formal institutional support for their work, such as the mathematician and Church of Scotland minister, Alexander Bryce, who produced a map of the north of Scotland based on his own surveying work which was published under the auspices of the Philosophical Society of Edinburgh in 1744.[25]

At a level above the individual and the institutional was mapping work at regional and national scales. Notable in this respect were the activities of the Commission for the Annexed Estates between 1754 and 1769, which produced a large number of estate plans and maps.[26] More important still was the work of the Military Survey of Scotland between 1747 and 1754 whose origins lay in the political surveillance of Scotland through mapping in the wake of the 1745 Jacobite Rebellion.[27] Mapping in this sense was a political act and one that transformed the landscape itself. Widmalm has argued in Foucauldian terms that governmental surveillance played a leading role in the Military Survey in Scotland under William Roy and in eighteenth-century triangulation more generally:

> With the technology of triangulation, the militarization of cartography's geometrical grid ... was imposed on the landscape itself. Geodetic surveying quantified geographical information, which became atomized and readily presentable in tabular form. This kind of extensive, yet compressed, knowledge was an attractive resource for centralized government and military planning. In the terminology of Michel Foucault, the Ordnance Survey and similar organizations brought an expansion of military surveillance from the limited disciplinary units of the garrison and the fortress to the full physical extent of a nation.[28]

Encounters with the landscape were not just institutional and political and evident in the replicable geometry of maps. The Astronomer Royal, Nevil Maskelyne, spent four months on the slopes of Schiehallion in Perthshire in 1774 at the instigation of the Royal Society, conducting experiments on the gravitational attraction of mountains on plumb lines and, in so doing, exhibiting the topographical influences upon accurate geometrical and astronomical measurement. He was visited there by numerous Scottish professors, was assisted in the field by William Menzies, a local land surveyor, and was accompanied for some of the time by William Roy who used Maskelyne's instruments in the Schiehallion 'field station' to ascertain the height of the mountain and, thus, to report on geometrical and topographical measurement to the Royal Society in ways which corroborated Maskeleyne's work and brought further credibility to both men and to that institution through the warranted claims of instrumental accuracy, personal observation and residence in the field.[29] Such topographic mapping and geodetic enquiry were paralleled by the hydrographic and coastal charting carried out by individuals or by larger state-run projects such as the Admiralty-sponsored work of Murdoch Mackenzie and his hydrographic mapping of the northern and western isles between 1742 and 1749, and 1751 and 1757.[30]

Through the work of individuals and institutions, in state-sponsored activity and through the patronage of landlords and others, mapping in and of Scotland was a reflection of Enlightenment concerns more generally, expressed in French Physiocratic and German Cameralist traditions as well as in the estate mapping of Enlightenment England, of the utility of putting one's national territory in order through the map.[31] It is not possible here to document all such mapping activity in Enlightenment Scotland: the activities of Scotland's map makers and the chronology of map production in Scotland have been charted elsewhere.[32] I will discuss two examples in order to explore mapping in the several senses I have outlined above. The first considers the work of John Ainslie and the role of the Society of Antiquaries of Scotland in providing institutional support for mapping. The second example, with reference to the work of Aaron Arrowsmith and of John Thomson, explores the question of what I call 'mapping networks', the dependency of mapmakers upon the work of others and, on the basis of this, discusses questions to do with the 'scientific' production of maps as reliable documents. My use of these two examples and of the questions they raise is to allow us not just to consider the situated nature of geography, geometry, and mapping in the Scottish Enlightenment. I also want to suggest how we might more generally conceptualise the geographies of scientific production in the Scottish Enlightenment: as social hierarchies, as institutionalised networks, as locatable sites of practical reason. Before turning to these examples and questions, however, let me demonstrate something of the interest in geography and in geometry in Enlightenment Scotland by looking at the public and university teaching of those subjects.

Mapping Enlightenment Audiences: Geography and Geometry Teaching in Scotland, c.1680–c.1820

From a survey of newspaper advertisements and sources, it is possible to enumerate the number and the location of public lecture classes on geography and geometry in Enlightenment Scotland, and to know in which of Scotland's universities these subjects were taught. It is less easy, however, to know the cognitive content of such teaching, either in the public sphere or in the private spaces of university curricula, and it is harder still to know exactly how they were taught, to whom, and what the audiences' motivation was in attending such classes.

We know that geography lectures were given in Edinburgh University in 1679–80 as part of philosophy courses by John Wishart, successively Professor of Humanity and of Philosophy from 1654 to 1667 and from 1672 to 1680, and in King's College, Aberdeen by the regent (and later sub-principal) William Black in 1692–93 and the regent George Skene between 1701 and 1704. In St Andrews, the mathematics professor in 1684–85, James Gregory, gave 'some succinct instructions in Astronomy and Geography'.[33] The presence of geography should not surprise us since, with geometry and arithmetic, the subject was included in the *quadrivium* of the medieval curriculum, and was taught in Edinburgh, the two Aberdeen colleges, and Glasgow as early as the 1580s.[34] Wishart's geography was concerned both with cosmography (literally celestial or astronomical geography), and with regional description. Black taught a series of geographical 'propositions': how to calculate longitude, how to measure the dimensions of the globe, etc. Cosmological teaching in Scotland's universities largely followed traditional Aristotelian-Ptolemaic lines until the 1640s, and not until the 1670s and, more effectively, from the 1680s and 1690s, was Cartesian cosmology dominant.[35] Seen thus, it is difficult to know exactly what place geography in the universities played in advancing that mechanistic natural philosophy others have seen as characteristic of the Scientific Revolution more generally at this time.[36] It did not, we may surmise, have a central part. Yet it is noteworthy in terms of the connections I am charting here between geography and geometry and in regard to the place of scientific instruments more generally as discussed in Chapter 2 that the University of Edinburgh was in 1686 purchasing Ptolemaic and Copernican celestial globes and terrestrial globes from Holland,[37] and that, in 1705, James Gregory, Professor of Mathematics, was organising the purchase of globes and geometry and geography books from London suppliers to use in his teaching.[38]

Rather more certain evidence, both as to cognitive content and to the purpose contemporaries attached to geography, is to be found in the eighteenth century. The Newtonian natural philosopher and Professor of Mathematics at Marischal College, Aberdeen, John Stewart, taught geography to his students from at least 1748. To judge from notes of 1753, Stewart's geography was included within his first-year mathematics classes. The College's minute books also record that 'the Semi Year, or Second of the Course shall be spent in the most usefull parts of Natural History, in geography, and the Elements of Civil History'.[39] Geography was a formal part of the second-year curriculum in King's College at the same time, where it was taught by Thomas Reid. Reid's geography course had four main parts: the regional geography of the globe; the main seas and lands named; more detailed facts pertaining to given towns and countries; and description of the principal trade winds.[40] Later in the century, geography was taught at Marischal College by the Professor of Natural History, James Beattie, and by the Professor of Natural Philosophy and Mathematics, Robert Hamilton, who taught it alongside arithmetic, the use of the globes, and in association with navigation and advanced astronomy.

Hamilton also gave evening classes in which geography was taught in conjunction with navigation, astronomy, navigation, and other 'Subjects which are useful in different Departments of Life'.[41] In Glasgow, geography was taught by the Professor of Mathematics, James Millar, between 1802 and 1821.[42] Earlier it had also been taught in the mathematics class, with prizes for the 'best exercises in geography' being awarded from the 1784–85 academic year. Similar prizes were also given in following years until 1791–92, and although none appear to have been awarded after 1792, medals were given for essays of a geographical-mathematical nature such as that awarded in 1797 to David Warden, an Irish student at Glasgow, for 'the best historical and philosophical account of the application of the barometer to the mensuration of heights'.[43]

Understood as a subject of formal description and of measurement, geography was, then, encountered in one form or another in most of Scotland's universities during the Enlightenment. Cosmographic concerns paralleled attention to astronomy. Matters of earth measurement and practical utility were taught through geometry and navigation. Geography was also taught in relation to civil and natural history.

It is, of course, one thing to *locate* a subject on the map of Enlightenment knowledge and quite another to know its meaning and importance for contemporaries. Even so, it is clear that geography had a place in the universities, that one reason why it did was because of close connections with geometry and astronomy, and that geography's mathematical bases were apparent in the emphasis on navigation and surveying. It is also clear that students would encounter what we might think of as geographical 'principles' and questions of terrestrial and celestial measurement in other subject areas. In his 1765–66 course, for example, Nicholas Vilant, Professor of Mathematics at St Andrews, taught navigation, fortification, architecture, spherical trigonometry, projection of the sphere, use of the globes, fluxions, and astronomy, much of which material was used as the basis to his 1786 book, *The Elements of Mathematical Analysis*. Matthew Stewart, Professor of Mathematics at Edinburgh from 1747 to 1772, taught astronomy in his third mathematics class, and John Robison, Professor of Natural Philosophy at Edinburgh in 1773, had been an instructor in navigation to Admiral Knowles' son in 1759, and a surveyor of British territories along the St Lawrence River in 1760, before being appointed in that year to be a member of the Board of Longitude's committee judging Harrison's chronometer. This private and institutional engagement with the sciences of measurement – the institutional expression of 'mathematical cosmography' – was paralleled in the Scottish Enlightenment's public sphere.

Examination of Scottish newspapers reveals sixty-four different persons giving public lectures or private classes in geography in the period 1708 to *c*.1830, several of whom continued after the latter date.[44] If we include others for whom there is firm evidence of teaching in geometry, navigation, and astronomy (and related concerns such as the use of the globes and trigonometry), but for whom there is no certain evidence of their teaching geography, a further eighteen persons may be added.[45] Although we should recognise that this total of eighty-two persons must be regarded only as an approximate total for what we might think of as geography and geometry's public teachers (there were many others who professed interests in, for example, astronomical observation and instrumentation or who wrote about the practical applications of geometry without ever teaching the subject), the available evidence allows us to identify several distinguishing features of the place of geography in the public sphere.

Most public teaching in geography and in geometry was concentrated in Edinburgh – where the earliest class, the Rev. William Smart's, was begun in 1708 – and in Glasgow, with a few such courses in Aberdeen, and a scattering in the smaller towns. Most public teachers of geography and of geometry were self-styled as 'teachers' or 'lecturers'. Some men, like Smart, who taught mathematics, navigation, surveying, and the 'use of the globes celestial and terrestrial' as well as geography, had other secure employment. It is unlikely, of course, at the admissions levels of about 1/- which most persons seem to have charged, that such teaching was financially rewarding. Teaching took place in teachers' homes, in public sites such as coffee houses, and even in theatres, and, for a few, in the homes of the audiences themselves.

In terms of cognitive content, no absolute distinction was made between the teaching of geography, geometry, astronomy, navigation, and the use of the globes. Of course, such attention to what we might think of as 'practical mathematics' within geographical knowledge reflected individual competences. It also differed by gender. There is some evidence to suggest that the mathematical and geometrical content of geography teaching directed specifically at men presumed a practical utility for such knowledge, but that the use of globes for women and the inculcation of a geographical awareness amongst women was seen as a genteel accomplishment, part of the sense in which knowledge of the earth was a matter of polite sociability.[46]

For some individuals, we can see how these features came together: the use of domestic and institutional sites for teaching; the production of different classes for men and for women; the mix of subjects; a formal involvement with institutions as well as with teaching the public in private classes and spaces. Take, for example, the mathematician, John Cross, who succeeded Andrew Ure as Superintendent of Garnethill Observatory in Glasgow in 1811. Cross ran a 'mathematical academy' in Glasgow's John Street in 1812 where he taught astronomy for ladies, navigation, spherical trigonometry, and geodesic problems: the last-named with instruments. Four years later, at his home in Ingram Street in Glasgow, he was teaching mathematics, geography, geodesy, surveying, navigation, and astronomy and conducting practical classes at the Glasgow Observatory.[47] James Denholm, President of the Glasgow Philosophical Society from 1811 to 1814 and author in 1797 of *An Historical and Topographical Description of the City of Glasgow,* taught geography and astronomy at Glasgow Academy, styled himself 'professor' of drawing and taught geography, art and geometry at his 'Academy' in Dunlop Street for over fifteen years.[48] One parallel figure in Edinburgh might be the mathematician George Douglass, who taught 'navigations with calculations on the places of the planets' at 108 Stair's Close in 1791, and, in addition to such works as *The Elements of Euclid* (1776) and *The Art of Drawing in Perspective* (1805), published the polemical *An Appeal to the Republic of Letters in behalf of Injured Science* (1810), in which he railed against what he saw as the lack of scientific understanding within society. Other men used their mathematical and geographical expertise as a means to social advancement. Edmund Stone, for example, was a self-taught mathematician, the son of the Duke of Argyll's gardener and author, in 1768, of *Some Reflections on the Uncertainty of Many Astronomical and Geographical Positions with Regard to the Figure and Magnitude of the Earth, the Finding the Longitude at Sea by Watches, and Other Operations of the Most Eminent Astronomers, with Some Hints Towards Their Reformation*. He became a Fellow of the Royal Society in 1770 and contributed several articles to the *Philosophical Transactions*.

Given such variation in this period among the public teachers of geography, geometry, and related subjects, we should beware of uncritically terming them a 'scientific community'. Some had formal institutional connections that brought them together – in the Glasgow Philosophical Society, or as Fellows of the Royal Society of Edinburgh, for example – but most did not. There was no geographical body in which shared interests in earth knowledge could be discussed and shared methods with, say, astronomy and geometry debated. Information about such matters was more probably exchanged in correspondence, more through chance encounters at the book and map sellers and the globe and instrument makers than through formal associational meetings. Audiences for geographical knowledge were constituted only fleetingly, as people moved to lecturers' homes or to private academies and there engaged with geographical issues before dispersing, either to reconvene with others in subsequent weeks, to put their knowledge to appropriate use in given contexts such as marine navigation or terrestrial survey or, more privately, to pursue their interests at home in polite conversation and in silent reading. Even so, I do not wish to dismiss the shared interest in geography, geometry, and mapping evident in the university classrooms, private salons, public lecture classes, in textual production, and in practices of enquiry in the field. Rather than a 'scientific' or 'geographical community', we might think of the connections between these persons and practices as networks sustained by different bonds – of teacher-student, corresponding friends, client-patron and so on – of varying strength.[49] I return to this idea of networks in my examples below concerning the production of maps.

Occasionally, we are afforded insight into the social connections underlying these networks of Enlightenment interest in geographical knowledge. For example, in a letter of December 1765 to his friend David Skene, the Aberdeen philosopher and natural historian, Thomas Reid, notes that he is sending Skene what he calls 'the Perspective machine' (which was made by James Watt) with instructions for its use. Reid notes: 'I have seen the machine used for taking off a small map from a large one. You will easily see that this is done by making the large map the object and the small one the perspective draught of that object'.⁵⁰ Here, in correspondence, we can see something of the material and intellectual exchange between naturalists and something of the cultural exchange of scientific understanding as an instructional discourse as Reid informs Skene how to use the instrument to draw scaled-down maps. Others published books in order to inform their students. As John Wilson, author of *Trigonometry: With an Introduction to the Use of Both Globes, and Projection of the Sphere in Plano* (1714), noted, 'I was obliged to print it for the ease and satisfaction of such gentlemen as were pleas'd to attend my lessons'.⁵¹ Yet others – and we will probably never know how many – taught their audiences how to map as well as the essential facts of geography. Robert Darling, self-styled 'private teacher of geography' and 'teacher of mathematics and geography' in Edinburgh's Ramsay's Land, advertised in 1776 (and again in 1793 and 1794) that he would '[teach] Youth Writing, Book-keeping, Mathematics, and Geography, and Gentlemen to Measure and Plan their own estates'.⁵² Such intentions were shared by institutions as well as by individuals.

'A Centre of Geographical Calculation'?: John Ainslie, Geographical Enquiry and Map-Making in the Society of Antiquaries of Scotland

The Society of Antiquaries of Scotland, founded in 1780 in Edinburgh by the leading antiquarian and improvement thinker David Erskine, Earl of Buchan, has been the subject of some attention by historians of the Scottish Enlightenment.⁵³ Relatively little attention has been paid to the fact that under Buchan's guidance the Society was concerned with enquiries into Scotland's geography.⁵⁴

A central objective of the Society at its foundation was the comparative examination of 'the ancient state of Scotland' (by which was understood its antiquities and history studied as a matter of enlightened rational enquiry) with the modern state of the nation. Study of the 'Geography, Hydrography and Topography of the Country' played a key role in achieving this objective. In 1781, Buchan proposed what he termed 'a general parochial survey' of the country under seven headings.⁵⁵ The results of this survey would be displayed in the Society's museum. As he wrote to the topographer, John Nichols:

> Parochial histories in Scotland, Gaelic topography, with maps, charts, and views of places, will form another separate department, and will have a more extensive demand in the literary market.
>
> To this last part of my plan I propose to dedicate a room in our Museum, fitted with a separate repository for each of the parishes of Scotland, accompanied with specimens of mines, minerals, and every thing that is politically useful to the community; and from whence future proprietors

of the soil may be able to draw every information that can tend to their profit or local curiosity.[56]

Buchan noted of this proposed survey and museum that 'I flatter myself that the example of enriching the Museum in a Scotch Atlas *General & Particular* will be followed: by which means the repository maybe come usefull & interesting to the publick in that as in other departments connected with a general Survey of the Country'.[57]

Although the intended usefulness of such geographical enquiry and its representation in certain spaces is not in doubt, the extent to which Buchan's geographical 'centre of calculation' actually incorporated maps is unclear.[58] His plans for the Society and for its museum were certainly restricted by the Royal Society of Edinburgh, founded in 1783.[59] Only a handful of parishes seem to have been surveyed and publication was delayed until 1792, by which time Buchan's plans had been overtaken by Sinclair's *Statistical Account of Scotland*.

Contemporary reaction to the first volume of the *Transactions of the Society of Antiquaries of Scotland* was far from complimentary. A reviewer in the *Critical Review* wrote that 'The idle titles of seal-engraver, geographer, topographer, etc, etc, to the Society are ostentatious and unbecoming the modesty of a literary body'. Whilst noting that the paper by the mineral surveyor John Williams on plans for a royal forest of oak in the Highlands was 'a good one', the reviewer remarked: 'but what connection it, and many others in the volume have with antiquities, we are utterly at a loss to discover'.[60]

Closer examination of the Society's records and of surviving correspondence, however, reveals that mapping *was* central to the Society's activities. Furthermore, the archival materials illuminate something of the personal relationships sustained through mapping within the Society's activities. Williams (who was a founder member of the Society) was one of the surveyors employed by William Roy in the Military Survey of Scotland from 1747. He was also engaged by the Commissioners for the Annexed Estates in mapping Jacobite estates from 1754 and was active in mineralogical survey and mapping in Russia and in Italy, where he died (in Verona) in 1795.[61] John Clark, land surveyor, was a founder member, together with John Bain, 'topographer'. So, too, was John Ainslie.

John Ainslie has been referred to rather grandiosely as 'virtually the Master General of Scotland's national survey'.[62] Ainslie's career as a map-maker began in 1765 with his employment as apprentice to the Geographer to the King, Thomas Jefferys, and began in Scotland in 1770 with a plan of his home town, Jedburgh. Ainslie surveyed many Scottish counties, produced numerous town and estate plans and, on 1 January 1788, published the map 'Scotland, drawn and engrav'd from a Series of Angles and Astronomical Observations'.[63] This map effectively replaced Dorret's 1750 map of Scotland. Like his maps, Ainslie's two books on surveying established a new standard. In 1802, he published *The Gentleman and Farmer's Pocket Book, Companion, and*

Assistant, and, in 1812, the *Comprehensive Treatise on Land Surveying Comprising the Theory and Practice in all its Branches in which the Use of the Various Instruments Employed in Surveying, Levelling, etc. is Clearly Elucidated by Practical Examples.*

Ainslie's patrons included many prominent Scots gentry, the Edinburgh Town Council and James Stewart Mackenzie, Lord Privy Seal of Scotland, to whom Ainslie dedicated his 1779 map of the Edinburgh district. His 1780 map of Edinburgh was dedicated to David Steuart, the city's Lord Provost. Ainslie's patronage network is illustrative of the patron-client social relationships between landowners and surveyors that underlay the accelerating pace of estate mapping and agricultural change in later eighteenth-century Scotland.[64] It was just such interests, of course, that Robert Darling hoped would fill his private classes and that John Stewart sought to instil in his students at Marischal College. In addition to being a founder member at Buchan's invitation, Ainslie's connections with the Society of Antiquaries – which could not always be directly personal given his surveying work in the field – were through the antiquary, George Paton. In a letter to Paton of November 1775, Ainslie outlined something of his own career and commented upon Jefferys' work (who, it should be recalled, had royal patronage and was Ainslie's mentor):

> Mr Jeffreys [sic] had several other people concern'd in several of the above counties whoes work I may say was committed to the flames[.] to do that Gentleman justice he always would have the work done over again if he found it did not correspond with angles that was taken with proper instruments which he had a Great Collection of for that purpose. I dont remember the names of all the People that was imployed by him whose work turned out bad but shall give you the names of them I can remember if required and what counties they were imployed in.[65]

Ainslie then proceeded to comment upon those he could remember. Of a Mr Elliot (probably the James Elliot who produced a map of Hawick in 1802),[66] Ainslie rather dismissively remarked: 'He is a millers son born in Liddesdale or Yeusdale. Teached School at Kelsoe, Jedburgh and some other places in merse of Tiviotdale and is now in Carlisle[.] Surveys some small Estates and writes for the attornies at 2 pence per padge (is not worth the Title of a Geographer)'.[67] Ainslie had heard several accounts of Dorret. He tells Paton 'how he [Dorret] was first a Barber and got into the Argyll family as one of the Duke's valet de chambre, and did take an actual survey of the County or Shire of Argyll'. After that, Ainslie continued, '[Dorret] took angles through all Scotland and compiled a map of the Kingdom which may be called a Survey by some people'.[68] Ainslie's estimation of Dorret's competence is clear. In answer to a query in 1788 from General Hutton concerning placenames in Argyll, Ainslie drew his correspondent's attention to the fact that 'A survey is just in hand at present by a Mr Langlands who lives at Cambelton in

Cantyre . . . I hear he has got a very large subscription for the County, if you would take the trouble to write that Gentleman'. Langlands was in correspondence with other Argyll gentlemen concerning mapping in 1799.[69]

Paton was himself the object of criticism. In an angry letter of 19 May 1782 to Paton, Mostyn Armstrong complained about Richard Gough's review of his work, in which Gough had called it a 'stupid, ill-digested, ignorant jumble of scraps and opinions, to [sic] contemptible for serious perusal'.[70] Mostyn Armstrong with his father, Andrew, had produced county maps of Scotland from 1771 and, in 1777, published *A Scotch Atlas or Description of the Kingdom of Scotland*. It was this work that was the subject of Gough's scorn. Armstrong's hurt feelings were not helped by Paton's own contributions to Gough's 1780 *British Topography* which were, according to him, 'totally false'.[71] Gough had indeed referred critically to the Armstrongs' *Scotch Atlas*, writing that it

> is little valued: his pretension to actual surve is entirely chimerical: he copied others, ingrafting mistakes of his own, and run over the counties in a strange cursory manner. The Atlas is indeed more neatly engraved than Kitchen's "Geographia Scotiae:" which yet is better esteemed, though erroneous. Armstrong has attended to his own and the engraver's profit more than that of the public or their information.[72]

The attention of Paton, Buchan, Ainslie, and others to mapping should be seen as part of a concern that the Society of Antiquaries take a lead in understanding the nation. The claim that there was a shared social sentiment in Enlightenment Scotland which recognised mapping and, more broadly, geographical knowledge as a means to know one's nation ought to be thought of as taking particular shape and emphasis within certain institutions.

Yet even within the Society of Antiquaries, this sense was not uniformly shared. As Buchan wrote to Richard Gough on 9 March 1787:

> Selfish sordid views on the one hand, and a rage for idle shows and sports on the other, have rendered this country so inattentive to grave pursuits, that I despair of any thing being done to answer your design but in a long course of time; which, by the institution of the Society of Antiquaries here, and the collections I am making, and causing to be made, will, in spite of every adverse circumstance, produce at last a complete and accurate survey of the country, political as well as geographical and antiquarian.[73]

The Society of Antiquaries did not, however, act as the bright centre for Enlightenment mapping that Buchan and others intended. There is no sense that the Society acted for geography and for mapping in the ways that, for example, Joseph Banks did for natural history and for economic botany in his 'centres of calculation' in Soho Square and at Kew.[74] Nevertheless, for Ainslie and for Williams, it provided an institutionalised setting in and through which to negotiate and secure patronage. It was, in a limited way, a focal point to which individual mapmakers could direct their queries. Judging from Ainslie's remarks to Paton about Jeffreys' work and assistants, it was also an institutional site and a social space in which gossip was traded about personal capacities and rivalries.

Mapping Networks: Aaron Arrowsmith, John Thomson, and the Nature of Scotland's Mapping, 1807–1832

One of the most notable features of early modern geographical enquiry is its collaborative production. The making of geography 'in the field' was a collective undertaking in several ways: voyages of discovery included subject specialists and depended upon state support and pre-planning; surveying parties depended upon the cooperation of 'natives' and upon overcoming natural difficulties as well as upon the authority of their instruments; travellers – in the absence of other information – trusted what they saw and were told.[75]

In the case of Scotland, Sibbald's geographical enquiries in the early Enlightenment were dependent upon a large number of correspondents (usually ministers of the church) scattered throughout Scotland (and further afield) who replied to Sibbald's circulated queries, some of whom were undertaking their own local geographical enquiries.[76] Survey – in natural history, in the visualisation of the landscape in painting, in the econometric and political arithmetic sense of Sinclair's *Statistical Account* as well as in the Military Survey's work – was a major form of geographical knowledge in Enlightenment Scotland.[77] As the case of John Ainslie suggests, mapping the nation in the eighteenth century was as much a matter of social networks and of patronage relationships as it was of the presumed scientific accuracy of maps themselves. The sense in which map-making as a product of social networks depended upon epistemological questions – of trust in other people and in instruments, and, thus, of there being different routes to scientific authority – is apparent in late Enlightenment Scotland in Aaron Arrowsmith's 1807 *Memoir Relative to the Construction of the Map of Scotland*, which he published as an accompaniment to and explanation of his map of the country.[78]

The textual memoir was a common partner to the map in the Enlightenment. The memoir usually provided the map reader or user with a list of the sources used in the map's compilation; as Godlewska has shown for French geographical and mathematical practices, for example, the written text worked to secure the credibility of the map as a visual approximation to the geographical truth.[79] Arrowsmith was additionally motivated to write his *Memoir* by a sense of the advantages arising to others:

> Having had an opportunity of constructing a Map of Scotland from materials unusually copious and original, I am persuaded that in justice to the public, I ought to accompany it with a Memoir describing the

authorities on which it has been founded, and the progress of the work from its commencement in August 1805 to the publication of the Map in June 1807; and I have been the more easily induced to undertake this from a constant sense of the permanent advantage which would accrue to Geography in general, if those who have it in their power to make a good Map, would at the same time publish such Notices of their authorities as might inform future generations how far to rely on it, and in what proportion further materials remain to be sought elsewhere; whether from actual Survey, or celestial Observations of Latitude and Longitude.[80]

Arrowsmith then proceeded to acknowledge the sources he had drawn upon: earlier maps, other contemporaries, actual survey or 'ground truth' as we might term it, and others' astronomical-cosmological calculations. Closer examination of extant maps, notably of the maps resulting from the work of The Commissioners for making Roads and building Bridges in the Highlands of Scotland, begun in 1803, revealed how little he could depend upon previous surveys: 'On examination however it was found that these Surveys, sufficient as they might be for their intended purpose, were not such as to form materials for a good Map, the shaded sketches of the vicinity of the Lines surveyed, proving not to be the result of actual Survey, as from their finished appearance was at first supposed'.

Roy's Military Survey mapping was also considered deficient: 'it is on so small a Scale, that the difficulty of joining the several sections without possessing the authentic Corrective of Celestial Observations, vanished in its minuteness'.[81] General Roy himself in 1785 had admitted the flawed nature of the Military Survey:

> although the Military Survey possesses considerable merit, and perfectly answered the purpose for which it was originally intended; yet, having been carried on with Instruments of the common or even inferior kind, and the Sum annually allowed for it being inadequate to the execution of so great a design in the best manner, it is rather to be considered as a magnificent Military Sketch, than a very accurate Map of a Country.[82]

What we might think of, then, as either the value or, perhaps, the utility of the map in any one context depended not simply upon its accuracy, understood in terms of any geometric or mathematically derived correspondence between the real world and its paper representation. The 'accuracy' of the map depended also upon its audiences' sense of purpose and, thus, upon different social meanings, upon who was involved in its production, and quite how they operated – by being in the field themselves, or by being elsewhere, collecting and correcting the instrumental observations of others. Not everyone could be relied upon as an appropriate source of mappable information, however, or to undertake field-based survey work to proper levels. General Roy suggests this to be the case for official government surveys, Ainslie says as much of Dorret's work. For these reasons, practical questions to do with the scientific production of maps and with their dependence upon geometry, mathematical cosmography, and other forms of geographical knowledge must also be seen as theoretical matters to do with the social status and intellectual credibility of those people who acted as part of the networks of map production. Let me illustrate this claim with reference to John Thomson's 1832 *Atlas of Scotland*.

This work begins with a 'Memoir of the Geography of Scotland' in which, as noted above, Thomson equates Scotland's geographical enlightenment with its mapping: from Pont, Sibbald and Ainslie to Arrowsmith and on to himself. Ainslie's 1779 map, 'Scotland Drawn and Engraved from a Series of Angles and Astronomical Observations' (which is dedicated to Henry Dundas), is described by Thomson as a monument of scientific achievement. Thomson cites Arrowsmith's *Memoir* on the importance of the social and authoritative network upon which map-makers draw. In relation to each county, Thomson offers a history of that county's mapping in the past and briefly remarks upon the sources for the accuracy of his own work. For Aberdeenshire, one A. J. Ross did the map and we are told that 'The names that attest the accuracy of this county are highly respectable'. Argyll was accredited by its leading landowners, Banff by 'respectable persons'. For Perthshire, Thomson is heavily reliant upon James Stobie's 'actual survey of this county', namely Stobie's 1783 map of Perth and Clackmannan. Stobie's work is judged by Thomson 'a most meritorious map, as he measured every foot of the ground, and laid it down with great accuracy and distinctness'. In contrast, for Ayrshire, 'The Map given in this Atlas is almost from actual survey. William Johnson, who superintended the construction of the Maps in the Atlas of Scotland, resided three summers in the town of Ayr, where the plans of the different estates were sent him by the proprietors, from which he laid down the map of the county. When he found these to require correction or adjustment, he went to the fields, and made such observations as enabled him to give with accuracy the various plans furnished him'.[83] For Thomson, the whole enterprise was clearly more work than he would have wished. He closed his preface thus:

> The Publisher will now take leave of the ATLAS OF SCOTLAND, – a Work which he never would have undertaken, had he known the difficulties to be encountered, the great number of people to be employed, the advance of capital, and the time necessary to carry through such an arduous undertaking, which required at least one surveyor to each county to correct the drawings, and find respectable names to guarantee their accuracy. How far this undertaking has been properly executed, he leaves it to the discernment of the Public, to whose opinion he now submits the results of his labours.[84]

As Thomson well knew, the utility of mapping rested with the user. The social networks through which the map was constructed were of concern to the map-maker.

Conclusion

This chapter has argued that understanding the place of geographical enquiry and of mapping as forms of scientific knowledge in the Scottish Enlightenment demands attention to questions of situated institutional and individual endeavour, to the nature of social networks, and a recognition of the different criteria by which maps, and the people who represented Scotland's geography in and through the map, were judged reliable.

Geography, in the several senses in which Enlightenment contemporaries understood it, was taught in the urban centres of the Scottish Enlightenment – Edinburgh, Glasgow and Aberdeen – and, later in the eighteenth century, in smaller towns. It was taught in the universities in association with geometry, mathematics, trigonometry, the use of the globes, and navigation. The exposure of relatively small numbers of university students to the idea of geographical knowledge as mathematical cosmography was paralleled both by the links that their professors made to geography's connections with civil and natural history, and by the attention given to these and other discursive connections by the teachers of geography and geometry and related subjects in Enlightenment Scotland's public sphere. We cannot, it is true, always determine the cognitive content of such public teaching, or know the motives for individual participation and, thus, discern the different meanings people may have attached to such geographical knowledge in given social contexts. Nor should we suppose that Scots and others were geographically informed citizens simply by virtue of attending public classes on the subject, or by reading geography books. In the early 1770s, in connection with an increase in emigration from Scotland to America, the Rev. William Thom of Govan remarked how informed the ordinary people of the West of Scotland were about the American colonies:

> You would wonder to hear how exactly they know the geography of North America, how distinctly they can speak of its lakes, its rivers and the extent and richness of the soils in the respective territories where British colonies are settled: for my part, did I not know the contrary, I would be tempted to think that they had lived for some time in that country.[85]

Knowing that geography *should* be placed on the map of Enlightenment knowledge in Scotland and in some detail *where* and *when* is not directly to know *how* Scots and others became geographically aware or to know the value placed by them upon geography or its practical applications.

It is clear, too, that this chapter has only lightly traced the social and scientific nature of the connections between the presence of such situated geographical knowledge, geography's audiences, and the story of Scotland's mapping. The picture afforded here is, I suggest, less one of discrete scientific communities and rather more one of webs of affiliation between map-makers, their patrons, and the institutions they served. Map production was a social enterprise dependent upon networks of people who could, in various ways, be relied upon: because they had proven themselves capable in the field by having undertaken actual survey, because they were capable (by virtue of a university education or of practical training) of working within the discourses of mathematical cosmography, because they were 'respectable persons' whose word concerning the authoritative accuracy of Scotland's place names and geometrical representation could be trusted because of *who* they were rather than simply *what* they knew or *how* they knew it.

If these claims have a wider purchase regarding the history of science and of geographical knowledge in the age of the Enlightenment, two implications at least follow. The first concerns our understanding of the nature of the Scottish Enlightenment itself. Rather than study any one 'subject' (in the disciplinary sense), for example, it may be more fruitful to consider exactly how the scientific worlds of the period were socially connected and how the conduct of given 'subjects' (in the social sense) was discursively constituted in given institutions and in particular spaces. In understanding such connections, we might want to consider not just the fact that the conduct of scientific knowledge was socially mediated, but also to know how people like Ainslie moved within their networks through a variety of social mechanisms: by appeal to personal capacities or by association with others with Royal patronage; by being found useful 'in the field'; by belittling the work of others; or, as with Sibbald, Arrowsmith, and Thomson, by being dependent upon the claims of others.

The second implication has to do not simply with recovering the place of geographical knowledge (including mapping) on the map of Enlightenment knowledge, but with the possibility of mapping the Enlightenment itself. Rather than consider mapping *in* the Enlightenment, we might be able to move towards the mapping *of* the Enlightenment, and to consider the Enlightenment as itself geographically constituted not simply on the national scale as in 'the Scottish Enlightenment', but also in relation to the movement across space of people and of ideas, to local differences in meaning, and to the historical geography of Enlightenment audiences. Mapping in that sense is not just the situated geographical and mathematical practice that has been the subject of this chapter but an essential means by which we can locate the social agencies and scientific concerns that made up the Enlightenment in Scotland and elsewhere.

Notes

1. In this context, see also Charles W. J. Withers, 'Toward a Historical Geography of Enlightenment in Scotland', in Paul Wood (ed.), *The Scottish Enlightenment: Essays in Reinterpretation* (Rochester, 2000), 63–97.
2. Sir Robert Sibbald, *An Account of the Scotish Atlas, or the Description of Scotland Ancient and Modern* (Edinburgh, 1683), 1–2.

3. John Thomson, *Atlas of Scotland* (Edinburgh, 1832), vi.
4. For an introduction to these terms and to the attention paid by historians of science to the settings in which scientific knowledge is made, see Jan Golinski, *Making Natural Knowledge: Social Constructivism and the History of Science* (Cambridge, 1998); Crosbie Smith and Jon Agar (eds.), *Making Space for Science: Territorial Themes in the History of Science* (Macmillan, 1998).
5. The quote is from d'Alembert's *Discours Preliminaire* to the *Encyclopédie*: see Charles W. J. Withers, 'Geography in its Time: Geography and Historical Geography in Diderot and d'Alembert's *Encyclopédie*', *Journal of Historical Geography* 19 (1993), 255–64.
6. Burke is here writing to William Robertson, the Scottish Enlightenment historian and cleric: for discussion of this point, see G. H. Gutteridge (ed.), *The Correspondence of Edmund Burke* 3 vols., (Cambridge, 1961), 3: 350–1; on the idea of the 'chart' of human social development generally, see Peter J. Marshall and Glyndwyr Williams, *The Great Map of Mankind: British Perceptions of the World in the Age of Enlightenment* (London, 1982); and Ronald L. Meek, *Social Science and the Noble Savage* (Cambridge, 1976).
7. See, for example, Josef W. Konvitz, *Cartography in France, 1660–1848: Science, Engineering and Statecraft* (Chicago, 1987); Anne M. C. Godlewska, 'The Napoleonic Survey of Egypt: A Masterpiece of Cartographic Communication and Early Nineteenth-Century Fieldwork', *Cartographica* 25, 1 & 2, *Monograph* 38–39 (1988); idem, 'Map, Text and Image: The Mentality of Enlightened Conquerors – A New Look at the *Description de l'Egypte*', *Transactions of the Institute of British Geographers* 20 (1995), 5–28.
8. Tore Frängsmyr, J. L. Heilbron and Robin L. Rider (eds.), *The Quantifying Spirit in the Eighteenth Century* (Berkeley and Oxford, 1990).
9. J. L. Heilbron, 'Introductory essay', in Frängsmyr, Heilbron and Rider, *The Quantifying Spirit*, 1–23, quote from p. 14.
10. Matthew H. Edney, 'Reconsidering Enlightenment Geography and Map Making: Reconnaissance, Mapping, Archive', in David N. Livingstone and Charles W. J. Withers (eds.), *Geography and Enlightenment* (Chicago, 1999), 165–198. Edney provides a summary of the nature of map-making in the Enlightenment in Part One – 'The Enlightenment Construction of Geographical Knowledge' – of his *Mapping an Empire: The Geographical Construction of British India, 1765–1843* (Chicago, 1997), 39–118.
11. Matthew H. Edney, 'Cartography: Disciplinary History', in Gregory A. Good (ed.), *Sciences of the Earth: An Encyclopedia of Events, People, and Phenomena*, 2 vols. (New York and London, 1998), 1: 81–85.
12. Matthew H. Edney, 'Mathematical Cosmography and the Social Ideology of British Cartography, 1780–1820', *Imago Mundi* 46 (1994), 101–16; see also Eric Forbes, 'Mathematical Cosmography', in George S. Rousseau and Roy Porter (eds.), *The Ferment of Knowledge: Studies in the Historiography of Eighteenth-Century Science* (Cambridge, 1990), 417–48.
13. Sven Widmalm, 'Accuracy, Rhetoric and Technology: the Paris-Greenwich Triangulation, 1784–88', in Frängsmyr, Heilbron and Rider, *The Quantifying Spirit*, 179–206.
14. Michael Bravo, 'Ethnographic Navigation and the Geographical Gift', in Livingstone and Withers, *Geography and Enlightenment*, 199–235; see also, G. Malcolm Lewis (ed.), *Cartographic Encounters: Perspectives on Native American Mapping and Map Use* (Chicago, 1998).

15. Daniel W. Clayton, *Islands of Truth: The Imperial Fashioning of Vancouver Island* (Vancouver, 2000), esp. 190–204.
16. Anne M. C. Godlewska, *Geography Unbound: French Geographic Science from Cassini to Humboldt* (Chicago, 1999). On mapping constituting the French nation state as a territory, see also Jacques Revel, 'Knowledge of the territory', *Science in Context* 4 (1991), 133–61.
17. Catherine Delano-Smith and Roger Kain, *English Maps: A History* (Toronto and Buffalo, 1999), esp. pp. 112–41.
18. John Brewer, *The Sinews of Power: War, Money and the English State, 1688–1783* (London, 1989) has seen this as the emergnce of the 'fiscal-military state'. On the role of the Excise in measuring the capacity of the state, see Miles Ogborn, *Spaces of Modernity: London's Geographies, 1680–1780* (London and New York, 1998), 158–200.
19. Charles W. J. Withers, 'How Scotland Came to Know Itself: Geography, National Identity and the Making of a Nation, 1680–1790', *Journal of Historical Geography* 21 (1995), 371–97.
20. This work is reviewed in Charles W. J. Withers and David N. Livingstone, 'Introduction: On Geography and Enlightenment', in Livingstone and Withers, *Geography and Enlightenment*, 1–28; and in Robert Mayhew, *Enlightenment Geography?: The Political Languages of Geography in Britain, 1650–1850* (London, 2000). On the role of the Royal Society in promoting geography, see Richard Sorrenson, 'Towards a History of the Royal Society in the Eighteenth Century', *Notes and Records of the Royal Society of London* 50 (1996), 29–46. Interestingly, throughout the period 1720–79, in which Sorrenson shows that the Royal Society paid attention to geography and to navigation as part of its natural history and mathematics meetings, only two brief papers were devoted to mapping problems, and then as a subject of mathematical issues underlying mapping, rather than to mapping as an intellectual question in its own terms: P. Murdoch, 'On the Best Form of Geographical Maps', *Philosophical Transactions of the Royal Society* 50 (1758), 553–62; William Mountaine, 'A Short Dissertation on Maps and Charts: In a Letter to the Rev. Thomas Birch, D. D. and Secret', *Philosophical Transactions of the Royal Society* 50 (1758), 563–68.
21. In a large literature on the nature of the Scottish Enlightenment, see, for example, Christopher J. Berry, *Social Theory of the Scottish Enlightenment* (Edinburgh, 1997); Alexander S. Broadie (ed.), *The Scottish Enlightenment: An Anthology* (Edinburgh, 1997); Wood, *The Scottish Enlightenment*.
22. John Money, 'From Leviathan's Air-Pump to Britannia's Voltaic Pile: Science, Public Life, and the Forging of Britain, 1660–1820', *Canadian Journal of History/ Annales Canadiennes d'Histoire* 28 (1993), 521–44, esp. p. 541. More generally on chemistry as an enlightened discourse see Jan Golinski, *Science as Public Culture: Chemistry and Enlightenment in Britain, 1760–1820* (Cambridge, 1992); Larry Stewart, *The Rise of Public Science: Rhetoric, Technology and Natural Philosophy in Newtonian Britain, 1660–1750* (Cambridge, 1992).
23. Charles W. J. Withers, 'William Cullen's Agricultural Lectures and the Development of Agricultural Science in Eighteenth-Century Scotland', *Agricultural History Review* 37 (1989), 144–56.
24. The phrase comes from Isabel Knight, *The Geometric Spirit: The Abbé de Condillac and the French Enlightenment* (New Haven, 1968), 19.
25. For a note on this map, see Donald G. Moir (ed.), *The Early Maps of Scotland to 1850*, 2 vols. (Edinburgh, 1973), 1: 184. The map, which was engraved by the artist Richard Cooper who, like Bryce, was trained in mathematics by Colin

Maclaurin in Edinburgh, covered only the north coast of Scotland and the adjoining parts of the east and west coasts. The intellectual and institutional context of Bryce's mapping and the involvement in it of the Philosophical Society of Edinburgh is discussed in Roger L. Emerson, 'The Philosophical Society of Edinburgh, 1737–1747', *The British Journal for the History of Science* 12 (1979), 154–91. A detail from the map is reproduced here as figure 6.

26. For the Forfeited Estates, Virginia Wills (ed.), *Reports on the Annexed Estates 1755–1769* (Edinburgh, 1973), and Annette M. Smith, *Jacobite Estates of the Forty-Five* (Edinburgh, 1982).
27. On the 1747–55 Military Survey, see Graeme Whittington and Alexander Gibson, *The Military Survey of Scotland 1747–1755: A Critique* (Historical Geography Research Group Research Monograph 18, Lancaster, 1986).
28. Sven Widmalm, 'Accuracy, rhetoric and technology', 201.
29. Ian Mitchell, *Scotland's Mountains before the Mountaineers* (Luath Press, Edinburgh, 1998), 36. On Nevil Maskelyne's fieldwork on Schiehallion, see Nevil Maskelyne, *A Proposal for Measuring the Attraction of some Hill in this Kingdom by Astronomical Observations* (London, 1776), and his *An Account of Observations made on the Mountain Schehallien for findings its Attraction* (London, 1776). Both of these short pamphlets were papers read before the Royal Society, in 1772 and on 6 July 1776, respectively. In his biography of Maskelyne, *Nevil Maskelyne: The Seaman's Astronomer* (Cambridge, 1989) Derek Howse also discusses the Schiehallion fieldwork: see esp. pp. 129–141, and, on pp. 232–4, a list of the Schiehallion equipment and instruments. Among the professors who visited Maskelyne were Thomas Reid and John Anderson from Glasgow, and John Ramsay, first Professor of Natural History at the University of Edinburgh. Patrick Copland from Aberdeen and John Playfair also visited him, as did James Stewart Mackenzie, Lord Privy Seal for Scotland. Maskelyne's field site was, thus, both a centre of calculation and a social magnet for many Enlightened Scots.
30. Diana C. F. Smith, 'The Progress of the *Orcades* Survey, with Biographical Notes on Murdoch Mackenzie Senior (1712–1797)', *Annals of Science* 44 (1987), 277–88.
31. This point about the place of mapping in the Physiocratic and in the Cameralist traditions in eighteenth-century Europe is made in Richard Drayton, *Nature's Government: Science, Imperial Britain, and the 'Improvement' of the World* (New Haven and London, 2000), 112–15.
32. The work of individual map-makers and the overall chronology of mapping in Enlightenment Scotland are most clearly discussed in the first volume of Moir (ed.), *Early Maps of Scotland*.
33. Edinburgh University Library, Special Collections, 'Dictates on Philosophy and Dictates on the Geography of Scotland' MS DK. 5. 27, ff.57–70; Aberdeen University Library, Special Collections, MS K 153, ff.164–184, 188–206; 'Tractatus Geographicus' MS 2092, ff.44–9; Roger L. Emerson, *Professors, Patronage, and Politics: The Aberdeen Universities in the Eighteenth Century* (Aberdeen, 1992), 24–6, 72–3, 90, 93, 104–5, 126, 128–9, 136, 143, 146; Bodleian Library, Ashmolean MSS (1813), f.243, 3 February 1685.
34. Charles W. J. Withers, 'Notes Towards a Historical Geography of Geography in Early Modern Scotland', *Scotlands* 3 (1996), 111–124.
35. John L. Russell, 'Cosmological Teaching in the Seventeenth-Century Scottish Universities', *Journal of the History of Astronomy* 5 (1974), 122–32, 145–54.
36. Steven Shapin, *The Scientific Revolution* (Chicago and London, 1996); Christine M. Shepherd, 'Newtonianism in Scottish Universities in the Seventeenth Cen-

tury', in R. H. Campbell and A. S. Skinner (eds.), *The Origins and Nature of the Scottish Enlightenment* (Edinburgh, 1982), 65–85.
37. Edinburgh University Library, Special Collections, 'Matriculation Receipts and Disbursements and Library Donations and Purchases, 1653–1693', MS. Da. 1. 33, f. 101.
38. Edinburgh University Library, Special Collections, 'General Book of Disbursements, Library Accession Book 1693–1719', MS. Da. 1. 34, ff.21–2, 25.
39. Paul B. Wood, *The Aberdeen Enlightenment: The Arts Curriculum in the Eighteenth Century* (Aberdeen, 1993), 20–22.
40. Thomas Reid's 'Course Outline of 1752': Aberdeen University Library, Special Collections, MS 2131/8/V/I,
41. *Aberdeen Journal*, 15 October 1781; 4 and 7 October 1782; 4, 6 October 1783.
42. National Library of Scotland, MS 2801.
43. W. Innes Addison, *Prize Lists of the University of Glasgow from Session 1777–78 to Session 1832–33* (Glasgow, 1902), 22, 31, 46, 50, 71, 80, 89. Geography does not again appear in any class prize lists until 1832–33, in which year it is included with 'Popular Astronomy' in the Natural History Class: Addison, *Prize Lists*, 357. I am grateful to Paul Wood for drawing my attention to this source.
44. Charles W. J. Withers, 'Towards a History of Geography in the Public Sphere', *History of Science* 37 (1999), 45–78.
45. It is possible to calculate these numbers and to identify something of the background to these figures using D. Gavine, 'Astronomy in Scotland, 1745–1900', 2 vols. (unpublished Ph.D. thesis, Open University, 1987), principally from Volume 2, Appendix II, 'List of establishments and teachers of astronomy and navigation up to 1900', 407–569.
46. This point, of geography's connections with astronomy, for example, is made by Alice N. Walters, 'Conversation Pieces: Science and Politeness in Eighteenth-Century England', *History of Science* 35 (1997), 121–54.
47. *Glasgow Herald*, 30 October 1812 and 17 February 1815.
48. *Glasgow Herald*, 8 February 1805; 18 October 1808; 4 February and 3 May 1811.
49. On the point of 'the strength of weak links' in the early-modern scientific world, see David S. Lux and Harold J. Cook, 'Closed Circles or Open Networks? Communicating at a Distance during the Scientific Revolution', *History of Science* 36 (1998), 179–211.
50. Thomas Reid to Dr David Skene, 20–30 [sic] December 1765, New College Library, Edinburgh, MSS. THO. 2, f.5v.
51. John Wilson, *Trigonometry: With an Introduction to the Use of both Globes, and Projection of the Sphere in Plano* (Edinburgh, 1714).
52. *Caledonian Mercury* 1 June 1776, 1a; *Edinburgh Evening Courant*, 1793–94.
53. See, in particular, Ronald G. Cant, 'David Steuart Erskine, 11th Earl of Buchan: Founder of the Society of Antiquaries of Scotland', in A. S. Bell (ed.), *The Scottish Antiquarian Tradition: Essays to Mark the Bicentenary of the Society of Antiquaries of Scotland, 1780–1980* (Edinburgh, 1981), 1–30.
54. Withers, 'How Scotland Came to Know Itself', 386–389.
55. This is discussed in William Smellie, *Account of the Institution and Progress of the Society of the Antiquaries of Scotland* (Edinburgh, 1782), 23–25.
56. John Nichols, *Illustrations of the Literary History of the Eighteenth Century* 8 vols. (London, 1831), 8: 506 [Letter of 13 February 1782 from Buchan to Nichols].
57. Nichols, *Illustrations*, 8: 508.
58. I take the term 'centre of calculation' from Bruno Latour, *Science in Action: How to Follow Scientists and Engineers Through Society* (Milton Keynes, 1987).

59. Withers, 'How Scotland Came to Know Itself', 386–89.
60. *Critical Review,* 2nd ser. 5 (1792), 403, 405. The whole review of the Society's *Transactions* is covered in pages 402–410 and 561–70.
61. Williams' life and work is discussed in Hugh Torrens, '"Mineral Engineer" John Williams of Kerry (1732–95): His Work in Britain and his Mineral Surveys in the Veneto and North Italy', *Montgomeryshire Collections* 84 (1996), 67–102.
62. Ian H. Adams, 'John Ainslie, Map-Maker' (unpublished typescript, Edinburgh, [copies held in the National Library of Scotland, Map Room]), 4.
63. Moir, *Early Maps of Scotland,* 1: 125–6.
64. On agricultural change and the work of surveyors, see Ian H. Adams, 'The Agents of Agricultural Change', in Martin L. Parry and Terence R. Slater (eds.), *The Making of the Scottish Countryside* (Beckenham, 1980), 155–76. On the maps and plans resulting from the agricultural transformation of late eighteenth-century Scotland, see Ian H. Adams (ed.), *Descriptive List of Plans in the Scottish Record Office,* 4 vols. (Edinburgh, 1966–88).
65. John Ainslie to G[eorge] Paton, 13 November 1775, National Library of Scotland, Adv. MS. 29.3.8, f.107v.
66. Adams, *Descriptive List of Plans* (Edinburgh, 1966), 1: 150.
67. NLS, Adv. MS. 29.3.8, f.108.
68. NLS, Adv. MS. 29.3.8, f.108.
69. John Ainslie to Hutton, Edinburgh, 27 November 1788, NLS, Adv. MS. 29.4.2, f.97.
70. This is the phrasing cited in the latter from Mostyn Armstrong to Paton, 19 May 1782: NLS, Adv. MS. 29.3.8, ff.83–83v.
71. NLS, Adv. MS. 29.3.8, f.83v.
72. Richard Gough, *British Topography,* 5 vols. (London, 1780), 2: 588. I am grateful to Chris Fleet for pointing out this reference.
73. Nichols, *Illustrations,* 8: 515.
74. David Philip Miller, 'Joseph Banks, Empire, and "Centers of Calculation" in Late Hanoverian London', in David Philip Miller and Peter Hanns Reill (eds.), *Visions of Empire: Voyages, Botany and Representations of Nature* (Cambridge, 1996), 21–37; John Gascoigne, *Science in the Service of Empire: Joseph Banks, the British State and the Uses of Science in the Age of Revolution* (Cambridge, 1998); Drayton, *Nature's Government,* 50–128.
75. Steven Shapin, *A Social History of Truth: Civility and Science in Seventeenth-Century England* (Chicago and London, 1994), 243–309; Charles W. J. Withers, 'Travel and Trust in the Eighteenth Century', in John Renwick (ed.), *L'Invitation au Voyage: Studies in Honour of Peter France* (Oxford, 2000), 47–54.
76. Charles W. J. Withers, 'Geography, Science and National Identity in Early Modern Britain: The Case of Scotland and the Work of Sir Robert Sibbald (1641–1722)', *Annals of Science* 53 (1996), 29–73; idem, 'Reporting, Mapping, Trusting: Making Geographical Knowledge in the Late Seventeenth Century', *Isis* 90 (1999), 497–521. For the claim that Sibbald was a leading figure of the early Scottish Enlightenment, see Roger L. Emerson, 'Sir Robert Sibbald, Kt, the Royal Society of Scotland, and the Origins of the Scottish Enlightenment', *Annals of Science* 45 (1988), 41–72.
77. Withers, 'How Scotland Came to Know Itself'.
78. Aaron Arrowsmith, *Memoir Relative to the Construction of the Map of Scotland published by Aaron Arrowsmith in the Year 1807, with Two Maps* (London, 1807).
79. Godlewska, *Geography Unbound.*
80. Arrowsmith, *Memoir,* 3.

81. Arrowsmith, *Memoir*, 8.
82. William Roy, 'An Account of the Measurement of a Base on Hounslow Heath', *Philosophical Transactions* 75 (1785), 379–97, quote from p. 385.
83. Thomson, *Atlas of Scotland* (Edinburgh, 1832), i-iii.
84. Thomson, *Atlas of Scotland*, vi.
85. W. Thom, *A Candid Enquiry into the Causes of the Late and Intended Migrations from Scotland* (Glasgow, 1771), 50–51.

4
Science and Enlightenment in Glasgow, 1690–1802

ROGER L. EMERSON AND PAUL WOOD

Despite the growth of interest in both provincial science and the different contexts of the Scottish Enlightenment, comparatively little scholarly attention has been paid to Glasgow as a site for the cultivation of natural knowledge or enlightenment during the long eighteenth century.[1] Unlike Edinburgh or Aberdeen, Glasgow has been marginalised in the literature and has typically been noticed only incidentally in relation to other topics.[2] Among the handful of scholars who have examined aspects of Glasgow in the period, Ned Landsman and Robert Kent Donovan have explored some of the religious controversies which divided the city during the second half of the eighteenth century, and have underlined the cultural dynamism of Glasgow's evangelicals. In their view, Glasgow had two competing Enlightenments: one centred on the University and inspired by the thought of Francis Hutcheson; the other evangelical, politically radical, utilitarian and devoted to trade, industry and profit. Hutcheson's Enlightenment was genteel, polite and elitist, they argue, whereas the evangelical Enlightenment found its supporters among increasingly well-off merchants and artisans.[3] Richard Sher has extended their analysis, and treated the achievements of the University men as largely owing to initiatives coming from Hutcheson and his circle.[4] Hutcheson also figures prominently in the work of James Moore and M. A. Stewart, which points to the importance of Glasgow's social and cultural links with Ireland.[5]

Finally, we have both contended that Glasgow's enlightened community was less divided than Landsman would have it and that it preserved something of the outlook of the Scottish *virtuosi* until nearly the end of the eighteenth century. We both see this as partly attributable to the way in which Scottish patronage was administered between 1690 and 1800.[6] We also believe that the natural sciences bulked far larger in the Scottish Enlightenment than is generally acknowledged.

In Edinburgh and Aberdeen, *virtuosi*, medical men, professors and clerics interested in natural knowledge first began to shift patterns of thought toward those which subsequently characterised the Scottish Enlightenment. They were the first Baconian natural historians, the first to accept the methods of the 'new science' and the first to try to apply natural knowledge in a range of improving activities. From the 1680s onwards, they founded institutions in which the ideas, methods and outlook of the 'new science' were incorporated, perpetuated and legitimated socially. Beginning in the 1680s, they also sought to reform the universities and, by 1710, they had succeeded in effecting some significant changes. Apparent initially in medicine and natural philosophy, the new ideas quickly affected logic and moral philosophy. Furthermore, Baconian natural history stimulated an interest in civil history, which was also fuelled by the political and economic crises of late seventeenth-century Scotland. Innovative ideas, patriotism coupled with shame at Scottish backwardness and a determination to improve the nation marked the generation which flourished in Aberdeen and Edinburgh from 1680 to 1710. For such men, the natural sciences had an importance unknown amongst humanists earlier.[7] Notwithstanding recent arguments to the contrary, we continue to maintain that these men produced the Scottish Enlightenment and that natural knowledge was a dynamic and an essential component of enlightened culture, in both Scotland and elsewhere.[8] Two questions thus need to be posed at the outset. First, was Glasgow like Aberdeen and Edinburgh? Secondly, if it was not, how did it differ and why? This chapter is, then, about the nature of the enlightenments in Glasgow and Scotland, and about the place of natural knowledge in both.

Glasgow's Men of Science: A Social Profile

To answer such questions we must first identify the men of science who worked in Glasgow between 1690 and 1802.[9] We begin our analysis in 1690 because it was in that year that a Parliamentary Commission of Visitation was struck to reform the Scottish universities and to remove those who remained loyal to the Jacobite and Episcopalian cause. Although the Commission had in the end only a limited impact on teaching at Glasgow and the other colleges, its formation is widely recognised as marking the beginning of a new era in Scottish academic and intellectual life. We end our discussion with the founding of the Glasgow Philosophical Society in 1802 and the Literary and Commercial Society in 1805. The establishment of these two societies signalled the disintegration of a cultural formation embodied in the proceedings of the Glasgow Literary Society that had embraced polite letters, political economy, moral philosophy and the natural sciences, and the emergence of an economy of knowledge both increasingly differentiated and specialised. In effect, 1802 can be seen as the year in which the Enlightenment in Glasgow came to a close.[10]

A 'man of science' we define as either someone who can be identified as part of the audience for science or someone who participated directly in the production of scientific knowledge. Under the rubric 'production of scientific knowledge', we include the authorship of a text, the engagement in observational or experimental practices, the development of theory or the acquisition and transmission of relevant forms of technical expertise. By 'science' we denote the fields embraced by natural history, natural philosophy, medicine, the cluster of subjects which contemporaries grouped under the heading of 'mathematics' and arts like instrument making that applied such natural knowledge. We use the term 'community' in an informal way. While some of Glasgow's chemists in the latter half of the eighteenth century arguably formed a community in the Kuhnian sense of sharing a paradigm or a set of exemplars derived from the work of William Cullen and Joseph Black, for us the community of Glasgow men of science was knit together primarily by physical proximity, personal relationships and common ideas and values.[11]

A number of the Glasgow University regents teaching before 1727 count as men of science, but we have omitted those who had little interest in natural philosophy or little knowledge of the new sciences around 1700. The University's professors of mathematics, natural philosophy, astronomy and of the medical subjects (botany, materia medica, chemistry, anatomy, medicine, midwifery) have all been listed, as have the city's known extramural lecturers in these fields. Those named as professors of science or medicine or as scientist trustees in John Anderson's will of 1796 have been listed,[12] as well as all the area's better-known instrument makers. Men who read papers on scientific subjects to the Glasgow Literary Society or to other clubs that flourished before 1802 have been added, as have some who corresponded with men of science and *virtuosi* elsewhere. Any Glasgow merchant, industrialist or artisan known to have cultivated natural knowledge or who is said to have conducted chemical experiments is included. Managers and investors have been excluded if they do not meet these criteria. On the basis of our research we have compiled a list of 168 men working (with the exception of James Corss) in Glasgow between 1690 and 1802 (see Tables 4.1 and 4.2). While this list is inevitably incomplete, it incorporates the most prominent men of science active in Glasgow during this period.[13]

The scientific community in Glasgow resembled the much smaller one in Aberdeen, but it differed in significant ways from that in Edinburgh. Men of science in Glasgow certainly knew one another. Some were related through the University, or through clubs or medical bodies. But Glasgow did not possess an 'academy' like the Edinburgh Philosophical Society or, later, the Royal Society of Edinburgh, which could give focus, status and prominence to their activities, as well as a venue for their discussions. Consequently, the community was more loosely knit than the one in Edinburgh, it was less medically oriented, and it was, in the end, more concerned with industrial matters, especially chemical ones. Nevertheless, it was a group composed of men who shared much and were in many cases related by blood, who often collaborated with one another and who all tended to see themselves as advancers of natural knowledge. There are numerous ways in which this community can be analysed. Let us begin with the question of geographical origin.

We take 'Glasgow' to encompass a fairly large area including Greenock, Dumbarton, Lanark, Kilmarnock and the other towns within a radius of roughly thirty miles, with the exception of Stirling, which looked more to Edinburgh. This makes sense because the port, the coalfields, many industrial establishments and the estates owned by Glasgow men largely fall within this district, of which Glasgow was the focus. Two-thirds of our men of science probably hailed from this region(see Table 4.3). Of those whose place of birth is known or can be surmised from their father's residence, Glasgow supplied 19%; the contiguous counties of Lanarkshire (5%), Renfrewshire (8%) and Dumbartonshire (2%) provided another 15%, while Stirlingshire (3%) and Ayrshire (5%) gave another 8%. The counties not known to be represented by any man of science include Banff, Moray, Nairn, Caithness, Orkney and Shetland in the north, Argyllshire and Sutherland in the west, and most of the border areas. No more than one or two men of science came from Dumfriesshire, Haddington, West Lothian, Angus, Bute, Inverness or Ross and Cromarty, and only two from the small midland counties of Clackmannan and Kinross. There were few Gaelic speakers. Of those with known birthplaces, 4% came from Fife.[14] St. Andrews had ties with the Glasgow University community throughout the century, as did Aberdeen.

Glasgow's recruitment area for its men of science was fractionally wider than that from which its students normally came.[15] This was a less isolated community than the one in Aberdeen, for it both attracted and lost members. Only a few foreign-born men ended up in Glasgow. The known 'foreigners' were: Joseph Black from Bordeaux; James Towers from the West Indies; the Englishmen Thomas Garnett and George Birkbeck (along with a few Scots born in England); and two Irishmen, Adair Crawford and the surgeon James Watt, both from Scottish families.[16] Most Glasgow men of science died in the region if they remained in Scotland.

Turning to the social origins of Glasgow's men of science, the members of the Glasgow scientific community were generally less genteel than their counterparts in Edinburgh (see Table 4.4). Just six of their fathers were titled, including two who were *de jure* baronets but too poor to use the title. Of those for whom information exists, twenty-three (14%) are likely to have owned land. There seem to have been few army or naval officers other than surgeons, and no important civil functionaries among them. Edinburgh men were generally better born with fathers who were better connected with brokers of political patronage; this was less true of those working in Glasgow. The Glaswegians were 'middle class' men whose fathers were not eminent professionals. Of the 101 fathers *whose occupations are known*, 27% were clerics, 8% physicians or M.D.s, 8% professors, 6% surgeons and 1% lawyers. At least twenty of the scientists (20%) had fathers who were merchants. Few parents came from the lower ranks of society. No fathers are known to have been labourers; only eight were farmers (8%), and five (3%) seem to have been artisans. The fathers from the lower orders appear mostly in the cohorts of scientists born after 1730.[17]

On their mother's side, the Glasgow men of science came from similar families. Maternal grandfathers, wives, siblings and the in-laws of the scientists confirm this picture of them as drawn from the professions. If anything, they suggest that ties to the clergy and to merchants may have been more numerous than is suggested by paternity alone.

Over the period surveyed here, there were some shifts in the recruitment pattern. Fewer men came from landed families at both ends of the period than during the middle years, although over the period 17% founded landed families of some sort. Similar mobility is probably shown by the fact that at least 12% eventually owned city properties. Those who had shops as instrument makers or printers had other capital investments. Clerical sons were most numerous in the 1740s and '50s, but they were subsequently outnumbered by the sons of merchants and industrialists. The greatest proportion of local men seem to have been present from $c.$ 1745 to $c.$ 1780, a time when the local professions could largely supply the city's needs and before industrialisation attracted large numbers of outsiders. Otherwise, there was little dramatic change in the recruitment pattern.

Glasgow's men of science typically came from the professional classes and they tended to remain within them. The initial occupations of 164 of them are thought to be known. The traditional professions accounted for 71% of them: medicine (24%), surgery (22%), teaching (14%), the Kirk (5%), military service (5%) and the law (1%). Although we cannot establish the exact number, it is clear that some of the medical men were sometimes busier with industrial concerns than with fractures and ailments. The marginal shift away from the clergy, and the clergy's lack of importance in the scientific community overall, probably points to its decline in status and income during the eighteenth century. Almost one third of our men of science (30%) are known to have followed their father in one of his principal roles. More changed their roles and added to them in order to pursue more lucrative careers, and more seem to have done so after $c.$ 1740. Most of them ended life manifestly better off than their fathers.

Income levels are notoriously difficult to calculate with any precision, but we estimate that few men other than artisans were making under £300 sterling a year in their primes and that incomes ranged from about £75 to thousands of pounds *per annum*.[18] What are better known are the sources or expected sources of income. Professional fees or stipends accounted for at least part of the incomes of 72% of these men, but after $c.$ 1750 this figure fell as more incomes came to depend upon industrial activities. For at least 48% of our group, the emoluments of offices in the church, universities, military services and civil establishments counted in incomes at the height of their careers. Incomes partially derived from business activities and investments began to increase markedly with those born between 1710 and 1730. Those sources undoubtedly figure in 33% of the incomes of those 145 men born after 1710. Because tradesmen became more numerous, their earnings were also more significant in this category. In the 1790s about 11% of the men (mostly born after 1730) had incomes derived at some time from tradesmen's work. Managerial salaries start to appear by the 1760s. Lastly, a few men began to make money in or through publishing. In 1740 only the typefounder Alexander Wilson did so. Toward the end of the century, the list includes not only Wilson's typefounder son, Patrick, but the writers of textbooks, a bookseller, a journal owner and several editors.

Glasgow's men of science came from backgrounds affluent enough to provide them with good educations (see Table 4.6).[19] One hundred and ten (65%) of them are known to have attended at least one college or equivalent school;[20] thirty-five (21%) attended at least two; ten (6%) were at three and one (1%) at more than three. Those known not to have gone to a college or trained as apprentices were generally teachers and men in industry. Most studied at the University of Glasgow with at least 53% attending at some point in their career. More probably did so since our figure is based principally on the number of matriculants and in this period many students did not matriculate. Until the 1740s their second university was likely to be Dutch, but only one man (Robert Freer) seems to have attended a Dutch university after 1750. By then, Edinburgh had replaced Leyden as the educational Mecca for medics and Utrecht for clergymen. Rheims professors were given up for Parisian and London hospitals and for schools such as that run by the Hunter brothers or by foreign chemists. Glasgow to Edinburgh to London was the typical route for medical study by the 1770s for both physicians and surgeons. Of course, many of the latter still served apprenticeships.

While it is difficult to estimate expenses, men educated for medical careers must have spent between at least £100 and £300 on their educations. Apprenticeship fees for merchants were probably lower, but bankers expected fees that were about as high.[21] Even among the artisans and wrights (a class that included the instrument makers), apprentices paid well. Many, like James Watt, could not be fully trained at Glasgow and had to go elsewhere – in his case to London. Watt's education as an instrument maker must have cost well in excess of £200.

Men who could afford such costs had inevitably travelled. At least forty-three men (26%) are known to have been partly educated outside Scotland. A few, like William Cullen and John Moore, went abroad as fledgling surgeons or as tutors to aristocratic boys. The latter was a more common pattern for university professors because their patrons could secure such positions for them. Twenty-four (14%) men had been to Europe as students or young men; twenty-nine (17%) had been in England, with the numbers growing over the period. Nine (5%) had been either to America or the Caribbean (including six already mentioned). Probably half of the men of science had been out of Scotland as young adults.

One reason for their good educations is that they were favoured not only by birth but also by birth order. Sibling order is known for seventy-six (45%) of them. Of these cases, 74% were first or second (or first or second surviving) sons. They were more likely to be well educated than the men known to be younger sons. It does look, however, as if second sons were becoming more numerous in the cohorts born after 1730. This reflects both a marginal increase in recruits from landed families, whose eldest son would stay on the land, and the increasing opportunities represented by a career promoted by natural knowledge.

Glasgow's men of science also tended to qualify quickly for their professions (see Table 4.7). In only one generation (1650–1669) was the average age of entry into a profession more than twenty-four years. The Glasgow men of science thus joined the general scientific community fairly young. They had many years in which to contribute to or to form part of the audience for natural knowledge. About half that time was spent in Glasgow, where they tended to spend their early rather than their later years. This is partly because the list contains the names of some students, but also because success in Glasgow opened doors elsewhere in Edinburgh, London, Birmingham and the colonies. The Glasgow scientific community was a relatively young one with highly mobile members. It also tended in any given year to be comparatively small.

Our list of 168 men of science is slightly larger than the known total number of Scottish members (163) of the Edinburgh Philosophical Society between 1737 and 1783. The core of the Glasgow group was made up of men with permanent places in the town, namely university professors, clerics, medical men with established practices and skilled artisans (see Tables 4.8 and 4.9). By the 1760s merchants, manufacturers and highly skilled artisans can be added to the list. By 1800 they were about as numerous as the professors. This core of professionals and businessmen was not without changes, but it was far different from the group of young medical men, tutors and divinity students who also belonged to Glasgow's scientific community. The first group was not static, but its members were relatively permanent and tended to be the leaders until the end of the period when men like the manufacturers Charles Macintosh and Charles Tennant became more important. To gauge the results of mobility it is worth considering two typical years, 1749 and 1775.

In 1749 there were twenty-nine men of science in Glasgow on our list: seven professors; three instrument makers; ten men who had been or shortly would be lecturing on anatomy, chemistry or medicine either extramurally or as professorial deputies; two students; five other medical men; and two manufacturers. Six appear in *DNB* entries: Joseph Black, William Cullen, Thomas Melvill, James Moor, Alexander Wilson and Robert Simson. Black was a student and would move to Edinburgh to do his M.D. (1752–1754), return in 1756 and permanently leave ten years later. Cullen had not long been in Glasgow and would move to Edinburgh in 1757. Melvill would die in 1753. Moor and Wilson had arrived in Glasgow in the early 1740s and the latter tried to leave with his business in the period 1759 to 1761. Only Simson spent his life in the University and city of Glasgow. By 1760, of the twenty-nine, six were dead, four disappeared from the records and two had moved on while the group had grown in absolute numbers.

The story of 1775 is much the same. The community had grown to about thirty-two men, with seven making it into the *DNB*. Alexander Wilson was one. Of the six others, five were professors who spent their careers mostly at Glasgow: John Anderson, Archibald Arthur, Thomas Hamilton, William Irvine and Thomas Reid. Hamilton would die in 1782, Irvine in 1787. Reid was busy teaching; Anderson was still a productive man of science. The other figure in the *DNB*, George Macintosh, was a manufacturer. By 1785 seven of the group had died, three had more or less retired, one had gone into the army and three had moved on. They had been replaced by seventeen men, five of whom seem to have been new to the area. Of these, only two were or were to be professors, whereas six were chemists and five were instrument makers. The turnover rates in other decades were not much different, although mortality rather than migration was probably more often the cause of changing numbers in the years before 1750. By 1790 both inward and outward migration was increasing, partly because of the growth of industry in Glasgow and in Britain generally. This also shows up in our list as an increase in total numbers in the community after 1790.

The migration of men to and from Glasgow brought with it a number of benefits. It created a community stimulated by incoming ideas. It also promoted contacts with other centres like Edinburgh, Dublin, London and, by the 1770s, Birmingham, Manchester, Newcastle and other industrial towns. These must have been useful to the industrialists and supplemented connections maintained by the Glasgow Chamber of Commerce. There was at least one other interest served by these ties. Glasgow surgeons migrating to London seem to have sought out eminent Scots in the metropolis – the brothers James and John Douglas, William Smellie, William and John Hunter, Matthew Baillie and William Cruickshank, all from the West of Scotland and all probably Glasgow alumni. The continued and pronounced interest in midwifery and anatomy in Glasgow may owe something to their practice, their teaching and their example. One negative consequence of the migration, however, was that men who had initially worked in Glasgow, like William Cullen, Joseph Black and John Robison, achieved greater renown as members of the scientific community in Edinburgh and elsewhere.

Glasgow's men of science were a respectable group whose eminence in their own day is best measured by indicators other than *DNB* or *DSB* entries. This group included most of the city's respected physicians and surgeons, that is, those who had lucrative practices, who taught, who practised in the hospitals and asylum and who ministered to the troops in volunteer regiments. Insofar as all of these posts were in some degree political, we can see that politicians had helped these men to pre-eminence, and they did so partly because they approved of what the men of science did and thought they did it well. This is particularly true of the university men for whom patronage came not from the town fathers but from political magnates such as the 2nd and 3rd Dukes of Argyll, the 1st, 2nd and 3rd Dukes of Montrose, the 5th Duke of Hamilton, the 11th Earl of Mar, the 3rd Earl of Bute, Baron William Mure of Caldwell, or Henry Dundas, the 1st Viscount Melville.[22] Most of these patrons had an eye to merit as well as to political connections. They did well by the Glasgow men of science, especially the 3rd Duke of Argyll. The appointments which he made or encouraged reflected his own interests in and knowledge of botany, materia medica, chemistry, medicine, astronomy, mechanics and mathematics. Among his university friends were Robert Simson, the Robert Dicks I and II, John Anderson, Alexander Wilson and William Cullen. These six men owed their initial appointments to his interest. Argyll's importance to the university and his solicitude for the city's businessmen, especially those involved in the manufacture of glass, pottery and linen, has been documented.[23] Argyll was a collector of porcelain who almost certainly had investments in these fields and sought the advice of expert chemists, whom he was himself competent to judge since he was an accomplished amateur chemist. He was also a notable agricultural improver. When he thought about university places and men to fill them, he wanted able individuals who shared his interests. Glasgow, of which he was an alumnus, was well served by him but he was not the only political figure who found merit in members of its scientific community. Most of the men of science were probably also burgesses of Glasgow. This status was necessary for surgeons, merchants and artisans. For others, it was an honour which the town bestowed upon those whom it had reason to respect or to whom it was grateful.

By 1752 there was another test of local eminence, namely election to the Glasgow Literary Society, which was founded in that year. Of the Society's 103 known members, thirty-four were men of science who sometimes read papers related to their research interests to this body. Because all members had to deliver formal discourses, the Literary Society also probably heard unrecorded papers on scientific subjects from men who did not hold university posts, including locals such as Patrick Carmichael, M.D., the Rev. James Couper, Major John Finlay, Thomas Jackson, Robert Marshall, M.D. and Thomas Melvill, as well as 'outsiders' such as the Rev. Dr. John Walker (who eventually taught natural history at the University of Edinburgh).[24] But the Literary Society was not the only local club which dealt with such topics. Robert Simson had a dining club from *c.* 1746 to *c.* 1765 which included at least a dozen Glasgow men of science.[25] In the early 1750s John Anderson and Thomas Melvill belonged to a group known as the Considerable Club. Dick II, the surgeons Thomas Hamilton and John Moore, and the physician Alexander Stevenson were all members of the more fashionable Hodge Podge Club, which was formally constituted in 1752 and which started out as a serious discussion circle.[26] At the turn of the nineteenth century most of the founders of the Medical Club (1798) and the Philosophical Society (1802) came from the ranks of the scientific community. Over forty of our men belonged to some Glasgow society in which natural knowledge and its applications were discussed.

Glasgow's men of science were also recognised outside their region, although (as noted above) for many recognition came after they had left the city. At least twenty-six (16%) belonged to the Philosophical Society of Edinburgh or to its successor, the Royal Society of Edinburgh.[27] Sixteen (including seven previously unnoticed) were eventually made Fellows of the Royal Society of London.[28] Alexander Wilson was honoured by the Royal Danish Academy of Sciences and Letters in Copenhagen. John Anderson was chosen as a member of the Tzarina Catherine's St. Petersburg Economic Society and, after leaving Glasgow, John Robison was given a pension by the

Tzarina and elected to the Imperial Academy of Sciences in St. Petersburg. Sixteen others held honorary degrees, including several from Columbia and Princeton. Altogether, forty-four (26%) individuals were recognised in their time and outside Glasgow. Forty-two are also included in the *DNB* and eight in the *DSB*. For most, recognition came as a result of their scientific work. While much of this was not done in Glasgow, it does tell us that even when living in that city they must have seemed more eminent than many in their fields. Finally, most of the recognition came after 1750. This was partly because the numbers grew toward the end of the period but also because the giving of honours was then more common (which was to some extent due to the significant increase in the number of societies). It was also because the Glaswegians born after 1710 included many notable figures who made lasting contributions to many branches of the natural sciences.

Institutional Sites: The University and the Ideal of the Virtuoso

Having identified who Glasgow's men of science were, we now turn to consider the physical and institutional sites where natural knowledge was produced or presented to its audiences. Central to our concern here is the University and those other bodies which eventually challenged it as the focus of scientific work and discourse.

The primary institutional venue for Glasgow's men of science in this period was unquestionably the University. Most of them had attended it and two-thirds of them had some formal connection with the College. Others sought such ties or looked to its professors for guidance and aid. Until mid-century there were no other institutions which could challenge it as a site for discussion, research and the promotion of scientific activities. The Faculty of Physicians and Surgeons functioned primarily as a guild and, while it supported the extramural study of medicine and chemistry like the Surgeons' Company in Edinburgh, whenever possible it used the facilities available at the University to advance its own corporate interests. There was thus, as Johanna Geyer-Kordesch and Fiona Macdonald have argued, a symbiotic relationship between the Faculty and the University.[29] Before 1752 there were no significant clubs or societies possessing instruments or libraries to rival the College.[30] Nor were there any sites like an observatory, botanical garden or alternative spaces such as coffee houses where men of science could find either employment or the facilities to sustain their work.[31] Only the Glasgow Town's Hospital, which opened in 1733, offered institutional support for the cultivation of natural knowledge outside of the College, and even then the Hospital's resources were limited.[32] The University was thus virtually the only place where natural knowledge was cultivated and circulated into the public sphere.[33] It is important to ask, therefore, what sort of natural knowledge the University promoted.

Despite their reorganisation after the Revolution of 1688–89, the Scottish universities were still rooted in neo-scholastic pedagogy. They were just beginning to teach the classical languages as a form of *belles lettres* for rhetorical purposes, but the level of instruction in Greek was probably rudimentary.[34] In 1690 none of the universities taught history save as an adjunct to the classics, although proposals had been aired for the founding of history chairs. Even though most of the Scottish colleges had professors of mathematics, the subject was a marginal one in the curriculum.[35] The heart of a university education was the *cursus philosophicus*, which then encompassed logic, metaphysics, pneumatics, ethics and general and special physics. The ends of education were commonly understood to be the promotion of piety and virtuous action in both the private and public spheres. A divinely created, unified and hierarchically ordered universe was taken to be the object of human knowledge, which meant that what was known had to be presented as a system of integrated truths which described God's creation as humankind was privileged to know it. To convey the outline of such a system was what the Scottish regents of 1690 were paid to do. But they were also under some pressure to add to the curriculum the findings of the new science along with the 'polite' subjects of *belles lettres*, history, modern languages and rhetoric which were regarded as being useful to a gentleman.[36] What the Scots who reorganised their universities in 1690 wanted were seminaries, professional schools to train lawyers and physicians and colleges which would produce Whig Presbyterians prepared to become *virtuosi*.

By the early 1700s progress toward these specialised ends had been made, particularly at Glasgow. The College now had lecturers on history,[37] Greek and Latin. The latter was teaching (or would soon teach) rhetoric.[38] A mathematician had been employed and discussions about a medical school had begun. What is more, several of the regents, professors and lecturers were themselves *virtuosi* for whom civil and natural history, natural philosophy and improvements based upon a knowledge of these fields were of more importance than they had been for masters a generation earlier. By 1710 the College had embraced the values associated with the 'new science' and the ideal of gentlemanly politeness, and had institutionalised the conception of natural knowledge and its place in the economy of knowledge espoused by the *virtuosi* in England and on the Continent. The College was thus to some extent training *virtuosi* and it would continue to do so until after the mid-eighteenth century.

As an institutional site for the production and dissemination of natural knowledge, Glasgow University in 1690 accepted and transmitted the outlook of men like Sir Hans Sloane, John Woodward, Edward Lluyd and Sir Robert Sibbald. Among those in the Glasgow scientific community in the period 1690 to 1709, Sibbald found at least five correspondents who all had ties with the University. Dr. Matthew Brisbane, who served as Glasgow's Rector in 1677, had been involved with the Royal College of Physicians in Edinburgh, which Sibbald had helped to found in 1681. Brisbane's involvement earned him an honorary Fellowship in the Royal College in 1695. He was in touch with Sibbald about botanical matters, and informed him about the medicinal value of foxglove.[39] Brisbane's descendants figure on the list of Glasgow men of science until the early nineteenth century.

A more important contact was William Dunlop, who was the Principal of the College from 1690 until his death in 1700. Dunlop was a minister who had also been a merchant and a militia officer in Carolina, where he spent part of the 1680s in self-imposed exile. He had responded to Sibbald's call in 1682 for county and regional surveys by writing an account of Renfrewshire. This shows him to have been an antiquary – he became Historiographer Royal in 1693 – and something of a naturalist.[40] He supplied Sibbald with miscellaneous information and discussed tides and wells with him.[41] Both were promoters of the Scottish fisheries, and Dunlop's patriotism led him to invest some of his own and the College's money in the Darien Company. The improvements in which he was successfully involved were made closer to home with the aid of his brother-in-law, William Carstares, Chaplain to the King and later Principal of Edinburgh University. Dunlop and Carstares secured more cash for the College, which was used to fund lectureships or chairs in oriental languages, humanity, history and mathematics.

The mathematician employed by Dunlop also shows up in Sibbald's notes.[42] George Sinclair had been a Glasgow regent from 1654 until 1666, when he was forced out of the College because of his religious views. Between 1666 and 1689, when he returned to the College, he made his living as a civil engineer, a public lecturer, a mathematics teacher and possibly as a tutor. His practical involvement with commercial ventures went back to 1655 when he is said to have designed a diving bell used to salvage an Armada ship, the *Florida*. His interest in hydrostatical problems continued and he published works on the subject in 1669, 1672 and 1683. Other scientific publications dealt with navigation, astronomy and coal mining. Sinclair is probably best known for *Satan's Invisible World Discovered* (1685), a book which combines an oblique defence of his witch-killing relatives with the apologetic agenda of Joseph Glanvill's *Sadducismus Triumphatus* (1666). Sinclair was concerned with the metaphysical implications of the new science, committed to the experimental philosophy and adamant in his belief that science should improve the lot of humankind.[43]

George Sinclair was succeeded in the mathematics chair by Robert Sinclair, M.D. Robert Sinclair shows up in Sibbald's papers as the inventor of a miner's lamp and possibly as a correspondent about 'the eruption of Fyres' in Italy, between Bologna and Florence, that his brother had seen. Between 1704 and 1709, Robert Sinclair also taught Hebrew at Glasgow.[44] Like George Sinclair, Robert believed that God's creation could only be known through the cultivation of natural history, mathematics and experimental philosophy, and that natural knowledge had to serve the interests of both humankind and true religion.

Another Glasgow *virtuoso* active in the 1690s and friendly with Sibbald was Robert Wodrow. Wodrow was the son of a Glasgow divinity professor and in 1697 became the University librarian.[45] For several years he collected historical manuscripts and documents for the library, which also housed some Roman artifacts. He left in 1701 to become a tutor and chaplain in the home of Sir John Maxwell of Pollock and, in 1704, became the minister of Eastwood, a few miles from Glasgow cross. By that time he had firmly established himself as a *virtuoso*. In 1703 his cabinet 'shottles' included minerals, fossils (in the two senses of the word), artifacts and curios from North and Central America, the Far East, Ireland, England and many Scottish regions.[46] He was respected as a field naturalist by *virtuosi* south of the Tweed like Sloane, Woodward, Lluyd and the Bishop of Carlyle, William Nicolson, and he was in touch with all of these men.[47] It is thus no surprise that in 1701 he was considered to be a suitable candidate for the position of keeper of the planned physic garden at the University.[48] Like Sibbald, Lluyd and Nicolson benefited from his aid and historical knowledge. Wodrow became a conscientious antiquary whose biased history is better remembered than his diligence and care as a collector of manuscripts.

To most today it seems as if Wodrow's early scientific interests were buried under parochial concerns and his historical writing. This is not likely to be the impression left by reading through his surviving correspondence. Until the end of his life he was intensely interested in all the activities of the Republic of Letters. He knew something of Newton's work, had read Clarke's edition of Rohault's physics, owned a work by Leibnitz and followed the controversy between the great Hanoverian polymath and his Newtonian antagonists. His best friend seems to have been Robert Steuart, the Edinburgh Professor of Natural Philosophy from 1708 to 1747. Wodrow wrote approvingly of Colin Maclaurin in 1718, and he was reading books on natural history and natural philosophy well into the 1720s. He kept up with the deist controversy, the Boyle lectures and modern history and read as many journals as he could. He particularly liked the *Philosophical Transactions* and the learned Dutch periodicals. He was, in short, typical of the learned Scot of his day in his choices of secular reading.[49] Like most of his Scottish contemporaries, he also expected that learning should lead men to God, to the improvement of this world and to the rational pleasure afforded by the systematic tracing out of the truths which revealed the underlying design of the creation. Wodrow's *Weltanschauung* was not so remote from that of the enlightened *literati*, whose more Erastian clerical politics arguably descended from his own. It was also shared by some of the regents teaching at Glasgow in the 1690s and early 1700s.

John Law of Ballarnock was another investor in the Darien Company and, by 1700, the College's senior regent. While he mentioned Locke in his 1691 logic lectures, he seems to have taught a relatively old-fashioned natural philosophy course.[50] But he also invented a new kind of sundial and he seems to have published a perpetual calendar, the *Calendarium lunae perpetuum*, in 1697.[51] These interests suggest a practical concern with timekeeping as well as a certain degree of mathematical competence. It is probable that along with the Sinclairs he was the chief user of the telescope acquired by the College in 1693. The following year, Law was joined at the College by Gershom Carmichael.

When Carmichael was appointed in 1694, he was sympathetic to Cartesianism, but by 1713 he had shed his Cartesian views under the combined influence of Locke, Boyle and Newton, as well as the natural law theorist Richard Cumberland. Carmichael is best known for his work on natural jurisprudence, but he also wrote textbooks on logic and natural theology, and, in 1712–13, he was the first to give lectures and demonstrations in natural philosophy using the College's newly purchased scientific apparatus.[52] Both Law and Carmichael were probably among those discussing schemes of medical education in the University. These surfaced in the 1690s, again *c.* 1701–2 when a physic garden was discussed, and in 1703 when the University awarded its first M.D. Both men would have been involved in such talks in 1711 when Robert Houston petitioned successfully to be examined for the M.D. degree and in other discussions which led to the revival of the professorship of medicine in 1712.

When the chair was revived, it went to a man also notable as a *virtuoso*, Dr. John Johnstoun, who was formally appointed in June 1714.[53] Like the irascible Dr. Archibald Pitcairne, Johnstoun was a wit with a profane sense of humour. He was also an antiquary with a collection of Roman artifacts and inscriptions, and was reputed to be a good classicist. He was a polite and learned physician who became a founder member of the Edinburgh Philosophical Society and an honorary Fellow of the Royal College of Physicians of Edinburgh.[54] While he did not teach much, his interests were certainly wider and different from those of most Scottish medical professors of the latter half of the century. Johnstoun's more closely resemble those of William Hunter, whose fortune was laid out on splendid books, coins and medals, pictures and rarities of other kinds, as well as on a museum necessary to a teacher of anatomy.

Another *virtuoso* hired in the 1710s was Robert Simson, who was both an accomplished mathematician and a serious Greek scholar. Simson was the College's first notable Newtonian, and he is also said to have been 'one of the best botanists of his time'.[55] He had trained, however, to become a minister. His surviving correspondence and his library[56] show him to have been a man of varied interests, a fact also borne out by what is known of his clubs. As much as Hutcheson, the Foulis brothers or the Greek and Humanity professors, Simson made Glasgow University a centre for the study of classical antiquity. He much improved instruction in mathematics, astronomy and mechanics, and he continued to place these subjects in a broadly humanistic context. His attitudes were adopted by many of his students, and some of them (including William Trail, James Buchanan, Matthew Stewart, John Robison, James Moor and James Williamson) perpetuated them in Glasgow and the other Scottish universities until the last decades of the century.

Glasgow's men of science and the natural knowledge they pursued were kept within the *virtuoso* framework not only by the structure of the curriculum and the perceived mission of educators but also by the patronage of outsiders who shared their ideas. The most important of these was the 1st Earl of Ilay, or as he was from 1743, the 3rd Duke of Argyll. Ilay had a huge library in London, as well as a sizeable cabinet containing many splendid scientific toys, models and instruments.[57] He had laboratories at three of his estates and travelled with a portable chemistry set. He was a serious botanist and had gardens full of exotics catalogued by men like Dr. John Mitchel. He was also a trained lawyer who dosed his friends with medicines he had made.[58] Interested in agriculture, he improved three estates, and his investments in the industries of Scotland were designed to improve them as well as strengthen his political interest. By the 1740s he and his friends ran two of the three Scottish banks with a view not only to political advantage, but also to the economic advancement of Scotland. Ilay claimed to be a reader of the classics until the end of his days, and he read five or six modern languages. With Ilay involved in making most of the appointments at the University from roughly 1722 until 1761, it is not surprising that men of science with *virtuoso* interests kept getting appointed.[59] Perhaps the individuals whose range most resembled Ilay's were James Buchanan, John Anderson and Alexander Wilson, for whom Ilay engineered the creation of the chair of astronomy in 1760. Buchanan, who died young, was a mathematician, botanist and linguist. Wilson had trained to be a surgeon-apothecary but knew some chemistry and was notable as an astronomer, typefounder and instrument maker. Ilay had patronised him since the 1730s, and employed him to make thermometers and barometers. Anderson's concerns embraced natural history, natural philosophy, chemistry, Hebrew and the making of artillery. Wilson and Anderson were active until 1786 and 1796 respectively and, by the time of their deaths, the demands of a good education had altered. Ilay and his associates also played a conspicuous part in making these alterations.

In the 1720s, factional politics in Scotland were an ever-present backdrop to all collegiate activities. In the years 1723 to 1725, John Campbell, 2nd Duke of Argyll, and his brother Lord Ilay wrested control of Scottish patronage from the Squadrone faction, to which most Glasgow professors had hitherto owed their allegiance. Because the Campbells now held the reigns of political power in Scotland, the Glasgow Visitation Commission of 1726–27 was an Argathelian body which sought to bring the University to heel. Ilay's later influence in the University was proof of the Commission's success. But that success also brought changes which, in the long run, undercut the College's commitment to *virtuoso* ideals of learning.

Several reforms were made which transformed both pedagogy and the curriculum. The Visitors abolished the regenting system and created fixed chairs, which meant that after 1727 specialist professors taught only one branch of learning. This probably improved instruction, but it ultimately worked to break up the *virtuosi*'s sense of the unity of knowledge. Students might come away with metaphysical, logical or moral beliefs no longer entirely consonant with the rest of the philosophy they had learned. Secondly, the end of regenting meant that many students ceased to take degrees, for there was no longer a regent urging them to complete their course because he would be rewarded with their graduation fees. More boys studied only what suited them or their parents. The new teaching system placed a premium upon useful learning quickly attained. For many, this meant no Greek, a minimum of Latin, no history and perhaps no natural or moral philosophy. The Visitors required that a course of experiments be conducted annually by the new Professor of Natural Philosophy.[60] This certainly gave a boost to the study of experimental philosophy, and to methods and empirical attitudes not much touched upon in traditional logic courses. But the emphasis on experiments and experimental methods, combined with the introduction of fixed chairs, may also have encouraged the belief that the sciences could be developed and elaborated independently of one another.

This shift away from the *virtuoso* ideal of the unity of knowledge can be seen most clearly in the teaching of William Cullen.[61] In his lectures, Cullen portrayed chemistry as a useful, autonomous science that was not reducible to physics, and that provided a set of probable claims about the specifically chemical properties of matter. Of course, such probabilities were very likely to change and, as Hume argued, they revealed more about man's experience and patterns of thought than about nature and God's real design of it. Pursuing chemistry or any other field might be a rational, polite or genteel amusement, but chemistry was regarded as a distinct discipline that depended upon its own methods, concepts and subject matter. Moreover, it had ends which were not necessarily or primarily related to piety. As John Christie has remarked, Cullen 'was not a chemical theist'; rather Cullen's 'philosophical' chemistry was intrinsically secular in character.[62] Neither Cullen, nor Black, nor most of the other Glasgow chemists later in the century worked with the *virtuosi*'s definition of natural knowledge.[63] After 1727, then, the *virtuoso* tradition should be seen as being in slow decline in Glasgow. More emphasis was gradually placed on a narrowly utilitarian conception of knowledge, and the stress on practical utility was accompanied by a growing sense of human nescience which undercut the *virtuosi*'s confidence in the intelligibility of nature. This did not happen immediately, but the shift had begun and can be traced in the courses taught by the professors of natural philosophy and other branches of the natural sciences.

The first of the specialist professors of natural philosophy, Robert Dick I, replaced John Law as regent in 1714. While he was a regent, Dick was involved in the maintenance of the stock of experimental hardware owned by the University and, in the period 1726 to 1728, serious efforts were made by Dick and his colleagues to upgrade the collection of scientific and mathematical instruments and to improve the room in the college used for experimental demonstrations.[64] By 1730 at the latest, the University engaged Henry Drew (a Glasgow clock, watch and instrument maker) as a demonstrator.[65] Following his appointment as Professor, Dick organised his teaching in the light of the regulations laid down by the Visitation Commission in 1727. In addition to his normal course of experimental philosophy given during the day, Dick included two evening lectures per week on the subject, and all of his classes were open to artisans and interested outsiders from the town.[66] The masters of the University had originally proposed to allow townsmen to attend experimental philosophy lectures in 1710 and, although they can thus be seen as enlarging the audience for natural knowledge, they also tried to control the presentation of such knowledge to the public. In 1724 they blocked an application to the Provost of Glasgow by two itinerant lecturers to offer a course in the town, but by the 1740s such courses were occasionally given, with or without the sanction of the University.[67]

Inventories of the instrumental paraphernalia which Dick I employed in his classes have survived, and we therefore have some idea of what he discussed in his lectures.[68] Given the instruments listed, Dick probably offered a fairly standard course of experimental philosophy covering mechanics, pneumatics, hydrostatics, magnetism, optics, astronomy and, after 1749, electricity.[69] He probably spent somewhat less time in relating his science to the other branches of human learning and to religion, and more time showing the uses of the knowledge he purveyed.[70] With Dick's courses, a step had therefore been taken toward both the democratisation of learning and the insertion of natural knowledge into a more secular utilitarian context; such a step at Edinburgh and Aberdeen would come only in the latter part of the century.[71] Dick's son, Robert Dick II, went even further in this direction in his natural philosophy classes, as did John Anderson.[72] For the two of them, the pressures to address the needs of their students who could not make careers in Scotland fostered an emphasis on topics of a very practical nature, notably those skills especially needed by soldiers and sailors.

Scottish science had always had some relation to the needs of navigators and military men. Navigation had long been taught both extramurally and in the colleges by regents and professors of mathematics. So too had surveying, which was incorporated into geometry. The wars of the late seventeenth and eighteenth centuries attracted many Scottish mercenaries for whom gunnery and fortification were useful subjects. At Glasgow, George Sinclair, Robert Simson and Robert Dick I dealt with these topics as did Dick II during his brief tenure of the chair of natural philosophy from 1751 until 1757.[73] There were particularly good reasons to teach such subjects after 1725 when the 2nd Duke of Argyll became Master-General of Ordnance. He used his patronage to place disproportionate numbers of Scots as engineers or as artillery officers. Many of them came from the North and West of Scotland, areas from which many Glasgow students also hailed. By the 1770s and 1780s John Anderson was not only teaching topics related to military activities, he was also designing cannon and shot with an eye to his personal profit and the country's benefit.[74] While his general outlook was that of a *virtuoso*, the knowledge he sought to disseminate had a largely utilitarian cast and, by the end of his career, he believed that women, as well as working men, ought to form part of the audience for natural philosophy.[75] Anderson was thus no longer tied to an elitist ideal of politeness, even though he had earlier espoused it.[76] Instead, he articulated a vision of academe similar to that of the Popular Party pamphleteer, the Rev. William Thom, which challenged the pedagogical ideals shared by the university men with whom Anderson had spent his entire career.[77] By the 1790s, then, a serious challenge to the University's virtual hegemony over natural knowledge emerged. But this did not come from Anderson alone. Throughout the eighteenth century, natural knowledge circulated in an extramural realm which now and again sought institutional form. Central to this development was the world of the Glasgow medical men who worked both inside and outside the University.

Glasgow's medical school was the realisation of hopes initially voiced in the 1630s and again with increasing frequency between 1690 and 1720. The apprentices of local physicians and surgeons needed and wanted someone to teach them botany. That demand was partially satisfied in 1704 when John Marshall began to teach in the new College 'physick garden', and the Faculty of Physicians and Surgeons later subsidised a second extramural instructor, John Wodrow.[78] There was less clamour for an anatomist, although John Gordon is thought to have taught the subject extramurally, beginning around 1714. It is not certain that he did so as a professorial deputy, and it would seem that his audience was small because he did not offer classes annually.[79] The newly elected Professor of Medicine, Dr. John Johnstoun, also had few students and did little teaching. In 1718, a new chair in botany and anatomy was founded by the King and, once Marshall died in 1719, the College moved to fill the position. Dr. Thomas Brisbane was eventually presented to the chair in February 1720. Like Johnstoun, Brisbane was not a dedicated pedagogue, and he failed to provide any anatomy demonstrations even though he did lecture on botany. Despite the warnings and admonitions of Visitors in 1725 and 1727, neither Brisbane nor Johnstoun could be made fully effective.[80] The local demand for anatomical teaching was satisfied by a series of men lecturing either inside or outside the College but with its approval. The first of these was probably John Paisley, who gave classes from roughly 1730 to 1740, and who certainly served as Brisbane's deputy around 1740. He was followed by John Love in 1740 and then by John Crawford, who in 1741–42 seems to have taught anatomy and surgery extramurally in conjunction with Dr. Robert Hamilton.[81] Crawford was at this point affiliated with the Glasgow Town's Hospital, where teaching was done in the infirmary, beginning in 1740, often by physicians and surgeons who were also attached to the University.[82] In 1744 William Cullen joined this group of extramural instructors but his lectures on medicine could not compete with those on botany and anatomy given by Robert Hamilton, who had succeeded to Brisbane's chair in 1742.[83] Beginning in the 1746–47 session, Cullen served as Johnstoun's deputy and the Glasgow medical school was effectively established. Its future was assured in 1751 when Johnstoun resigned in favour of Cullen.[84]

This story has several important features. First, academic medicine was, with the exception of Johnstoun and Paisley, taught by men who were not conspicuous for their *virtuoso* interests. This would remain the case after 1751. Cullen and Hamilton were first and foremost medical men. They were trained surgeons interested in anatomy, physiology, botany, chemistry and medical theory, but not in the wider range of polite topics which appealed to Johnstoun or expatriates like William Hunter. Specialisation and a degree of professionalisation had thus come early to this part of the scientific community.

Secondly, extramural teachers found no suitable homes outside the university in Glasgow as they did in Edinburgh, where they often taught at Surgeons' Hall. The market in Glasgow was simply too small. While there were a number of men teaching midwifery and anatomy, until the 1780s Glasgow seems really to have had only two other extramural lecturers on medical subjects, John Gibson (midwifery and history of medicine) and Andrew Morris (medicine).[85] Both probably lectured in the hall of the Faculty of Physicians and Surgeons in 1764. What is more interesting, they seemingly taught subjects for which they presumed some demand but which were not then covered by the university teachers. The University did not create a chair of midwifery until 1791, but in January 1747 it did respond to pressure from the marketplace by establishing a lectureship in chemistry. Cullen was given the post, and the classes he began in early 1747 continued under Black and John Robison. When Robison was appointed in 1766, the faculty created a lectureship in materia medica for William Irvine, who subsequently combined the two lectureships after Robison's departure in 1769.[86] These positions were the last permanent ones to be created before John Anderson died in 1796. Calls were made, however, for chairs of surgery, midwifery and natural history.[87] By 1795 extramural lecturers were teaching midwifery, anatomy and surgery.[88] Anderson himself thought that Glasgow should have a teacher of natural history, a topic which both he and Dick II before him covered in their courses.[89] Lastly, the demands for industrial chemistry clearly exceeded the University's ability to meet them. The appearance of the Scrutons' laboratory in the 1790s and the proliferation of classes given by extramural lecturers point to the inability of the College to cater to Glasgow's commercial and industrial needs.[90] The city now clearly had qualified men who knew they would not get university chairs; it had effective demand for subjects not adequately taught at the College or its medical school; and there were students eager to acquire practical training but uninterested in, or unable to afford, a comprehensive, polite and liberal education. Such was the context for the birth of Anderson's Institution.

'For the good of Mankind and the Improvement of Science': Anderson's Institution

Professor John Anderson had quarrelled for years with most of his Glasgow colleagues.[91] He was long a thorn in the side of Principal William Leechman and, around the time of Leechman's death in 1785, Anderson made himself even more unpopular by lobbying for a visitation of the College by a Royal Commission. He also differed politically with all but three of his colleagues (Reid, John Millar and George Jardine), being himself an extreme Foxite Whig in an increasingly Tory institution.[92] He was initially sympathetic to the

French Revolution and, in 1791, he travelled to Paris to present the Convention with a cannon of his own design. He seems to have had little regard for those closely affiliated with the Edinburgh Moderates, and he cultivated friendships with evangelical clerics who belonged to the Popular Party or to dissenting sects.[93] His closest associates were in the city's industrial and mercantile community, and it was their needs that he intended to address in 1796 when he left 'the whole of my other property, of every sort, to the Public, for the good of Mankind and the Improvement of Science, in an Institution to be denominated "Anderson's University"'.[94] Over the course of his career, Anderson had given some thought to pedagogical issues. In late 1761 he was one of the professors involved in the promotion of 'a scheme for An Academy of dancing, fencing and riding to be established at Glasgow under the direction of the University' and, in an attempt to encourage competition among the arts students, he established prizes for both elocution and the best essay on a topic in natural philosophy in 1770. A few years later he spoke to the Glasgow Literary Society on the question 'What would be the consequence to Literature and to the State (in so far as Literature is connected with the Interest of the State) if the salaries of all the Professors in the Universities were five hundred pounds? And what would be the consequence to Literature and the State if all the Universities in Scotland were abolished and no such Institutions made in their place[?]'. Then, in 1792, he was consulted by the Edinburgh Professor of Natural History, John Walker, about improving the instruction given to divinity students.[95] The plan for a new university outlined in his will thus registers not only Anderson's frustration with what he regarded as the corrupt practices of his professorial colleagues but also his long-standing interest in educational reform.

Through his bequest, Anderson created, or rather hoped to create, a degree-granting university with colleges of arts, medicine, law and theology. Each of the colleges was to have nine professors, of whom one was to serve as President or Dean. These were larger faculties than any Scottish university then possessed. It is instructive to notice the new chairs Anderson hoped to see established. Arts was to get specialists in civil history, chemistry and an extra Latin teacher. Medicine was to have chairs in natural history, botany, practical surgery and obstetrics. Whereas Glasgow College had but one professor of law, Anderson stipulated that the faculty of law in his university was to have teachers of Roman law, Scots law, English law, the law of nature and of nations, ecclesiastical law, commercial law, 'Practice in the Scottish Courts' and 'Roman Antiquities'. In divinity, he allotted a theology professor to each of the prominent dissenting groups and went so far as to assign a chair to a representative of the local Gaelic community.[96] In his allocation of chairs in each of the faculties he tried to satisfy needs that he thought were otherwise not met by the Scottish universities and he sought to appoint proper men to address these needs. While the structure of his projected university to some extent embodies the *virtuoso* outlook, the governance of his institution was vested in a body of eighty-one trustees that was to function as a sort of extramural senate and that included artisans as well as more politely-educated gentlemen. This was an institution whose practical bent was assured by those whom he named to hold office within it.

Twenty-three of those whom Anderson designated as trustees or professors were members of the Glasgow scientific community. Only two of these men were M.D. – his personal physician, Peter Wright, whom he named Dean of the College of Medicine, and John Balmanno, whose family had a commercial herb garden and pharmaceutical business.[97] Like Anderson, Balmanno would have been something of a chemist. Nine surgeons were also named. Two had taught extramurally prior to Anderson's death: John Burns had lectured on anatomy while James Monteath had lectured on midwifery, which was in fact the specialty of both men. A third surgeon, William Anderson, was named Professor of Obstetrics. The Professor of Chemistry, William Couper, was an investor in chemical concerns controlled by Charles Tennant and Charles Macintosh.[98] Macintosh was also to serve as a trustee. Taking chemistry out of the medical faculty and appointing Couper as professor was a clear sign of what Anderson expected the chair to do. The man Anderson picked to teach clinical medicine, John Scruton, ran a (pharmaceutical?) laboratory in Glasgow with his brother William.[99]

Several others also knew some chemistry. Robert Cowan, Anderson's botanist, would have had training in pharmacy similar to that of Balmanno, who was assigned the materia medica chair.[100] The surgeon John Cross, whom Anderson ambiguously decribed as 'my late Operator', also performed chemical experiments.[101] Another trustee, James Crichton, supplied instruments and equipment to manufacturers.[102] Finally, John Wilson of Hurlet was an owner of an alum works and was also a Macintosh partner.[103] At least ten of the twenty-three trustees or professors are known to have had interests in chemistry; in six cases, these interests tied them to industries flourishing in the region. Most of the others had some practical involvement with the community that had recommended them to Anderson. It did not matter to him that his Professor of Natural History, David Ure, was a self-taught weaver. What mattered was that Ure had managed to learn enough to conduct surveys of Roxburghshire, Dumfriesshire, Kinross-shire and Dumbartonshire for Sir John Sinclair. Perhaps it also mattered to Anderson that Ure was given a church living in 1796 by the 11th Earl of Buchan whose Whiggish views Anderson shared.[104]

The other seven trustees are an odd job lot. They include Anderson's protégé, William Meikleham, who was named Professor of Astronomy, an office he later held in the University of Glasgow.[105] Robert Wright, the brother of Anderson's physician, was an artillery officer who shared Anderson's interest in explosives, shot and cannon. The remaining trustees included a mathematics teacher, a merchant and two men who are as yet unidentified.

These professors and trustees would have worked within an institution far more committed than the University of Glasgow to serving the needs of an expanding industrial city dependent upon chemistry. The medicine which was to have been taught at Anderson's university included surgery and midwifery, increasingly urgent concerns in a city with ever more industrial accidents and higher infant mortality rates. Surgery was also boosted by wartime needs. When it was formally established in 1796, Anderson's Institution represented the perceived need for, and the possibility of, a focus upon utilitarian science and general medicine and surgery that Glasgow University did not provide.[106] Anderson's far-sighted provisions for lectures for women and working men evoked derision even from liberal faculty members. Although his Institution had no chairs in subjects like engineering that by the 1820s were urgently needed in the city, it was admirably contrived to meet the real educational requirements of 1796. Given its complicated management structure, it is likely that had it blossomed as its founder had hoped, it would have adapted quickly to changing local conditions. Why, then, did Anderson's Institution not challenge Glasgow University and become a second institutional focus for the interests of Glasgow scientists?

Part of the answer is clearly found in the inadequate funding provided by Anderson's estate. Anderson's Institution was left a sizeable library, an extensive collection of natural history specimens and an ample stock of excellent experimental hardware that was far superior to what remained at the University.[107] It also carried a heavy political burden. This was a Whig foundation created in wartime and during a period in which radicals were being persecuted.[108] Those who intended to succeed in life cultivated Henry Dundas and his local manager, the 3rd Duke of Montrose, who was the University's Chancellor.[109] By 1799 Meikleham held a professorship given by the Tory Duke. By c.1810 six of the twenty-three trustees and professors belonged to Tory-sponsored volunteer regiments, and three others held patronage posts granted by Tories. By 1815 another was teaching at Glasgow University's medical school. Anderson's Institution could not flourish in the face of the financial and political pressures upon it. In the end, its usefulness as a centre for research and teaching was limited. Initially its lecturers reached artisans but not the creative members of the scientific community except in the case of anatomy. Moreover, its most successful early lecturers, Thomas Garnett and George Birkbeck, were from England. Only after 1815, when political tensions were relaxed, did Anderson's medical school take off. Even then this venture did not float the second university for which he had hoped. Ultimately, Anderson's Institution merely provided a home for the extramural teaching which inevitably thrived in a city the size of nineteenth-century Glasgow. It catered to the needs of a university-oriented clientele unwilling to take impractical arts courses or to those for whom a full university education was unthinkable because of costs. In neither case could it really develop a distinctive ethos or identity or sustain research programmes that differed from those pursued at the University. At best, Anderson's Institution merely showed the premium that Glaswegians placed upon useful knowledge.

Institutional Sites: Clubs, Societies and Voluntary Institutions

If the University was not displaced as *the* structuring institution for the pursuit of science in Glasgow during the long eighteenth century, a number of other, largely complementary sites for the production and dissemination of natural knowledge emerged prior to the opening of Anderson's Institution. In addition to the Glasgow Town's Hospital, clubs and societies increasingly provided extramural *foci* for men of science, although it was only after 1800 that these groupings really moved out of the precincts of the University. As noted above, among the more significant of the early societies which cultivated natural knowledge were Robert Simson's various clubs and the Considerable Club. Simson not only conversed about a wide range of subjects with his associates on Saturdays in the village of Anderston, but he also discussed matters of science with his friends while dining in the evenings at a tavern across from the College.[110] Less is known about the Considerable Club, to which both John Anderson and the distinguished natural philosopher Thomas Melvill belonged.[111] A letter from Anderson to Gilbert Lang dating from 1750 indicates that the Club heard discourses written by its members, and that they debated such topics as the veracity of miracle reports. The fact that many of the members eventually became ministers suggests that the Club may have focused on theological issues, but given the involvement of Anderson and Melvill, the Club probably discussed scientific questions as well.[112] It is likely too that the religious implications of natural knowledge were explored in the various theological societies at the University which existed from at least the 1740s onwards.[113]

The most important eighteenth-century coterie involving men of science was the Glasgow Literary Society, which was one of the longest-lasting voluntary clubs founded in the Scottish Enlightenment.[114] Like the Aberdeen Philosophical Society (1758–1773), the Literary Society was a generalist body which preserved something of the outlook of the *virtuosi* but, unlike the Wise Club, it was not explicitly committed to a scientistic agenda. Following the standard eighteenth-century pattern, the Literary Society heard book reviews, research papers and discussions based on questions proposed by individual members. Given the disparate interests of those involved in the Glasgow group, it is hardly surprising that a wide variety of subjects were canvassed in the meetings, ranging from *belles lettres* to political economy and religion. Unfortunately, the record of the Society's proceedings is fragmentary, which means that we cannot know exactly what the balance of topics addressed was.[115] Thus while Richard Sher is undoubtedly right to see a connection between the Society's discussions and the cultural priorities of the Hutchesonian Enlightenment, his claim about the marginality of natural knowledge in the Society ignores the fact that we lack minutes for those periods (especially the years 1752 to 1764) in which the Society had its largest number of notable men of science, and he does not take into account the evidence provided by the minutes for meetings held from 1794 to 1800, which indicate that the natural sciences and medicine were by no means peripheral to the concerns of the Society's members.[116] What the extant minutes show is that there was an inevitable ebb and flow of interests within the Society, which reflected the varying levels of participation of key individuals and constituencies within the membership. After 1764 the astronomer Alexander Wilson had little to do with the Society. By 1794 John Anderson had withdrawn entirely from the meetings, and Anderson's successor, James Brown, played as negligible a role in the Society as he did in the College more generally. While in the early years the College's medical men may have participated in the life of the Society to only a limited extent, in the later period they were active members reading papers relevant to their teaching and research. Natural knowledge and medicine were of more importance to the institutional life of the Society in some periods than in others. Consequently, Sher's account does little justice to the nuances to be found in the surviving sources for the history of the Society.

The questions debated and discourses read to the Literary Society on scientific and medical topics register all of the main areas of research cultivated by Glasgow's men of science during the second half of the eighteenth century. The various branches of medicine collectively account for the highest number of known papers given to the Society, with the bulk of these presented in the 1790s. As already noted, prior to the closing decade of the century, the professors of medicine were conspicuous in the Society by their absence once William Cullen and Joseph Black had left Glasgow. Following Black's departure in 1766, his successor Alexander Stevenson gave only one discourse on 'How far the Mind ought to be an Object of attention in the Cure of Diseases', while the Professor of Anatomy and Botany, Thomas Hamilton, was inactive.[117] The life sciences were not, however, entirely neglected in the 1760s and 1770s. William Irvine contributed papers on the nerves, the comparative physiology of plants and animals and 'On the Degrees of Heat necessary for supporting the Lives of Animals and Vegetables', a subject which dovetailed with his chemical researches.[118] Physiology was also later the focus of essays read by Thomas Reid. As a coda to his protracted examination of Joseph Priestley's materialist system, Reid explored the role of the 'principle of life' in two papers from the late 1780s or early 1790s, and in 1795 he discussed muscular motion.[119]

Reid evidently had a good sense of his audience, for by 1795 the medics had turned the Society into an important venue for the presentation of their research. In sharp contrast with their predecessors, Thomas Charles Hope, James Jeffray, Richard Millar, Robert Cleghorn and Robert Freer were all active participants in the Society who used it to further their intellectual and professional interests. Before leaving for Edinburgh, Hope discoursed in March 1795 on the application of pneumatic chemistry to medicine, probably in response to recently published work by Thomas Beddoes, whom Hope knew from their student days in Edinburgh.[120] Reflecting his position as the materia medica lecturer, Millar spoke on the use and effects of drugs as well as on medicine and materia medica in the medieval period; in 1800, however, he turned his attention to the subject of obesity. Jeffray took up various topics in anatomy which were also relevant to midwifery, while Freer delivered a series of discourses on fevers, and Cleghorn discussed diabetes and quack medicines.[121] Within the context of the Society, therefore, the medical men were able to converse with their colleagues about issues rooted in the local context such as midwifery, along with matters of broader topical concern like Beddoes's innovative therapeutic techniques.

If medicine took on greater importance in the Literary Society during the latter part of the eighteenth century, the one major area of research that was of continuous interest was chemistry. Initially at least, the papers on chemical topics read before the Society focused on the science of heat and reflected the research tradition in that science fostered in Glasgow by Cullen and Black.[122] Cullen apparently gave two papers summarising his investigations on heat and mixture in the period 1752 to 1755, while Black detailed his experiments on the use of thermometers to measure heat on 28 March 1760 and outlined his theory of latent heat on 23 April 1762.[123] As is well known, Black's ideas were taken up in Glasgow by a number of his pupils and associates, including William Ogilvie, James Watt, John Robison, William Irvine, William Trail and Thomas Reid.[124] What is distinctive about the work done by this circle is its quantitative character, a feature already present in Cullen's researches. Robison, whose own scientific style involved rigorous quantification, later commented to James Watt that Irvine 'had a true mathematical taste, and delighted in reducing every thing to measure by means of Equations. He hunted for opportunities of doing this, and was very quick sighted in discovering the means of procedure[.] Willy Traill and he were almost constantly occupied in this way'.[125] A number of Irvine's discourses to the Society exemplify his quantitative approach and his engagement with Black's theories, especially those on the heat produced by mixture, the heat produced by combustion and on the quantity of matter in bodies.[126] But Irvine's contributions were not entirely restricted to the science of heat, for he also gave papers on such topics as the chemical properties of various waters, evaporation, fermentation and cements.[127] After Irvine's death in 1787, Thomas Charles Hope was appointed lecturer in chemistry and he discoursed to the Society on the composition of water and on strontium.[128] Hope transferred to the chair of the practice of medicine in 1791, and his successor as chemistry lecturer, Robert Cleghorn, read two papers on fire and a further two on combustion.[129] With Hope and Cleghorn, however, the tradition of chemical inquiry initiated by Cullen and Black was transformed, because Hope was an early convert to the system of Lavoisier, and it is likely that Cleghorn also taught the doctrines of the French chemists.[130]

Although the University did not have a chair of natural history until 1807, this branch of knowledge attracted a significant degree of interest among the members of the Society, for we know of eleven discourses delivered that were related to the subject. John Anderson, who followed the lead of Dick II in lecturing on natural history as part of his natural philosophy course, addressed the debate between naturalists like Buffon and Linnaeus over the merits of natural as opposed to artificial classification systems and also considered the question of the varieties of humankind in 1774. Four years later, Anderson returned to the topic of the natural history of man when he twice explored the question of which mental powers and physical characteristics distinguish humankind from the animal kingdom, a question made topical in Scotland in the 1770s through the publication of works by Lords Kames and Monboddo as well as the appearance of successive editions of John Gregory's *A Comparative View of the State and Faculties of Man with those of the Animal World*.[131] Accounts of voyages and travels had long served as empirical grist to the mill of natural historians of humankind, and it may be that the reliability of such accounts was touched on in Patrick Graham's 'The idea of that Species of Writing called Voyages and Travels', delivered on 13 February 1795.[132] A related genre was the geographical and natural historical survey, and the Society heard an example of one such survey in April 1770, when John Walker's 'A Discourse . . . concerning the natural History of Inch Columkill [Iona]' was read over the course of two meetings.[133] The vegetable and mineral kingdoms also attracted some attention from William Irvine, who talked about the roots and seeds of plants and the properties of diamonds, and from Lockhart Muirhead, who discoursed on 'some indigenous plants observed on the coast of Fife' in 1799 and again in 1800.[134] In addition, geology (which in the eighteenth century was regarded as a branch of natural history) was the subject of two discourses by the Glasgow Professor of Divinity (1761–75), Robert Trail, who theorised in 1768 and 1770 about the 'ancient state of our Globe' on the basis of evidence drawn from fossils and the stratigraphy of mountains.[135] Buffon may well have been the target of Trail's papers; later, another great Enlightenment theorist, James Hutton, prompted a discourse in 1789 or 1790 by Thomas Charles Hope, 'Observations on the Theory of [the] Earth by Dr. Hutton and Answers to some of Mr. De [Luc]'s remarks upon it'.[136]

Closely related to natural history (and to chemistry) in the constellation of knowledge of the Enlightenment was agricultural improvement. Because few of them came from landowning families and the local audience for natural knowledge was not dominated by landowners, Glasgow's men of science were less interested in agricultural improvement than their counterparts in Aberdeen and Edinburgh. Nevertheless, those who had landed properties to improve like William Dunlop, William Cullen, John Glassford and Lord Dundonald all gave some thought to the subject, while Cullen did his best to focus attention on the application of natural knowledge to agricultural practice by giving lectures on agriculture to his chemistry students in 1748.[137] Within the Literary Society, one of its earliest members, the printer Robert Foulis, discoursed on 'the Improvement of Agriculture and at the same time diminishing the Expence' in November 1771, and Irvine followed suit with a paper 'On the Fertility of Soils' in December 1778.[138] We know of no further treatment of issues related to agricultural improvement until the 1790s, when

Archibald Arthur discussed the effects of frost and snow on agriculture, and the manner in which lime served to ameliorate the soil. Cleghorn contributed a discourse on mosses, while Alexander Rankine considered the causes of blight in grain and turned to history to survey 'the State of Agriculture among the Gauls upon the 9th Century'.[139]

Meteorology was a field of inquiry which came into its own during the long eighteenth century in Britain, not least because of the creation of new instruments like the barometer.[140] Glasgow men of science shared the more general enthusiasm for this field and its apparatus. In particular, Alexander Wilson (in collaboration with Thomas Melvill) began to use paper kites in order to measure the temperature of the atmosphere at great heights in 1749, and he later used these kites to investigate atmospheric electricity in conjunction with his son Patrick.[141] Both Wilsons also recorded changes in the weather, and published papers in the *Philosophical Transactions* on severe cold spells which affected Glasgow.[142] One of the Wilsons' collaborators, John Anderson, was likewise interested in meteorology and kept meticulous records of rainfall in Glasgow from the 1760s to the 1790s, using rain gauges built to his specifications.[143] Based on this work, Anderson presented a discourse on rain gauges to the Literary Society in 1780.[144] William Irvine was another Glasgow man of science who collaborated with the Wilsons and who engaged in meteorological investigations. In May 1779, he introduced the question 'What is the cause that the falling of Mercury in the Barometer is commonly followed by rain?', and in the period 1784 to 1787 he apparently read an essay to the Society in which he addressed James Hutton's theory of rain.[145] Meteorological concerns again emerge in the last years of the century with Alexander Rankine's 'Inquiry into the Causes & Appearances of Changes in the Atmosphere', two discourses by Patrick Graham on 'The condition of the Atmosphere in reference to the state of the weather' and Meikleham's consideration of barometrical measurement.[146]

Alexander and Patrick Wilson also played a prominent role in the cultivation of astronomy in Glasgow during the second half of the eighteenth century. Robert Simson and Dick II were closely involved in the acquisition of astronomical instruments and efforts to establish an observatory in 1754, but it was not until the alumnus Alexander Macfarlane bequeathed his collection of instruments to the University in 1756 that the observatory scheme made significant headway.[147] By the time the new facility opened in 1760, the College had made two appointments which directly contributed to the flourishing of the sciences in Glasgow, James Watt and Alexander Wilson. As the University's mathematical instrument maker, Watt was able to provide the technical and manual expertise necessary to maintain the collection of instruments owned by the College, and his shop became a site for scientific discussion and debate.[148] Wilson's selection by the Crown, on the recommendation of the 3rd Duke of Argyll, to become the new Professor of Practical Astronomy in January 1760 gave the University another man of science with practical gifts as an instrument maker, and his connections with astronomers in

London helped to strengthen the scientific links between Glasgow and the metropolis. Moreover, Wilson occasionally brought together a group of men of science at the College to carry out astronomical observations, and then reported on their findings to astronomers elsewhere.[149] Although Wilson was not, as noted above, very active in the Literary Society after 1764, he did seek permission to deliver a 'Discourse on the Solar Spots' in December 1769 which in a revised version was subsequently published in the *Philosophical Transactions*.[150] Patrick Wilson later took up issues at the interface of optics and observational astronomy before the Society. In November 1770, he was given permission to read 'a Dissertation concerning the Velocity of Light shewing that this as deduced from Dr. Bradley's Theory of the Aberration of the fixed Starts is erroneous and a new Principle which affects that pointed out and examined', and in January 1777 he discussed the refraction and refrangibility of light, a topic which harked back to the optical researches of Thomas Melvill.[151] Apart from the Wilsons, only James Williamson and James Millar contributed papers on astronomy, the former on the history of Copernicanism and on Newtonian astronomy and the latter on 'the desiderata in the present state of practical Astronomy'.[152]

Under the heading of natural philosophy, six somewhat miscellaneous discourses were delivered to the Society, of which three were given by the Wilsons. In 1757–58, Alexander Wilson discussed the use of the 'hydrostatical glass-bubbles' he had recently invented to measure the specific gravity of fluids, and in 1758 he displayed an improved clock he had designed on the basis of his analysis of the motion of pendulums.[153] Instrumentation was again the subject of a discourse by John Walker in March 1765 on the use of barometers to measure heights, while in March 1775 John Anderson outlined the implications of Nevil Maskelyne's measurements taken on Schiehallion in 1774 for Newton's theory of gravitation.[154] In March 1795, Patrick Wilson read an essay with the curious title, 'Some Account of Certain Motions Acquired by small Lighted Wicks when made to swim in a Bason of Oil; with Observations upon the phenomena tending to explain the principles upon which such motions depend', and in February 1800 James Millar gave one of the last known discourses on a natural philosophical topic on weights and measures.[155]

Despite the fact that Robert Simson was among the earliest members of the Society, mathematics did not figure prominently in the Society's proceedings, presumably because the intricacies of mathematical reasoning did not lend themselves to collegial discussion amongst a diverse membership. Simson himself is not known to have read a paper related to his own work to the Society, and only four discourses appear to have been given on mathematical subjects in the latter part of the eighteenth century. One of Simson's protégés, John Robison, 'gave some observations on the study of Mathematics tending to shew the Excellence of that branch of Science', in February 1766,[156] but it was not until the 1790s that other members of the Society ventured to speak about mathematical subjects. At some point in the early 1790s Thomas Reid gave a critical analysis of the definitions and axioms of Euclid's *Elements*, with particular reference to the vexed question of Euclid's treatment of parallel lines.[157] James Millar, who was James Williamson's assistant from 1789 before succeeding Williamson as the Glasgow Professor of Mathematics in 1796, introduced his fellow members of the Society to the study of fluxions and subsequently paid homage to Simson in 'Observations on the Life and Writings of the late Dr Robert Simson'.[158]

The final area of interest to Glasgow men of science in the Literary Society was what we can call, following Michael Barfoot, 'scientific metaphysics'.[159] Thomas Reid read no fewer than five discourses in the 1780s and 1790s in response to the materialism of Joseph Priestley, but he was not the only member of the Society who engaged with such heterodox ideas, for in March 1795 James Couper reflected on the materialist speculations of Priestley's fellow Unitarian, Thomas Cooper.[160] Earlier that year, George Jardine had explored the question, 'What are the proper boundaries of that branch of Science which we denominate Metaphysics?', which would inevitably have involved him in mapping the boundaries between metaphysics and the natural sciences.[161] The pursuit of natural knowledge and the practical applications of such knowledge also figured in a further twelve discourses on miscellaneous topics, ranging from the epistemological dimensions of scientific inquiry, to the history of medicine and the sciences, gunnery and the writings of Maupertuis.[162] In sum, the known contributions to the Glasgow Literary Society show that it functioned much like the Aberdeen Philosophical Society, the Manchester Literary and Philosophical Society or the provincial academies of France as an institution which fostered the discussion of the natural sciences and medicine. While recognising that the pursuit of natural knowledge did not dominate the Society's proceedings, it is important to acknowledge that the Society served as a venue for both the announcement of important scientific discoveries and the display of significant innovations in instrumental design, with Joseph Black, William Cullen, Alexander Wilson and Thomas Charles Hope all presenting papers that were of considerable moment in the development of the natural sciences during the Enlightenment. Furthermore, the Society provided a forum in which its members could discuss the latest theories and experimental or therapeutic techniques preoccupying men of science throughout the Atlantic world. Like the University to which it was so closely attached, therefore, the Literary Society served as one of the primary sites for the cultivation of natural knowledge in the Glasgow Enlightenment.

The virtual hegemony over learned and literary culture enjoyed by the Glasgow Literary Society from its inception was, however, increasingly challenged in the latter decades of the long eighteenth century. Late in the 1770s and then again around 1785 the University doctors and students along with some outsiders formed a chemical society. There is little extant information about this group, so that we do not know, for example, how long it survived.[163] By 1787 there was an agricultural society in Glasgow, and by 1802 the Philosophical Society had come into existence.[164] Both of them attracted local medical practitioners and industrial chemists as members. A student Medico-Chirurgical Society was formed at the University in 1802 but there was not a genuinely professional medical body meeting for more than conviviality until 1814.[165] A small society for chemists met briefly in 1800–1, which had its own room and laboratory in Graeme Street.[166] The amateur astronomers organised the Glasgow Society for Promoting Astronomical Science and opened an observatory on Garnethill in 1808.[167] Of greater moment was the founding of the Literary and Commercial Society of Glasgow, which appears to have taken place on 27 March 1805.[168] This seems to have been a splinter group of the Literary Society which considered questions concerning trade, social policy, the poor and politics; these topics were of periodic concern to the parent society, but they presumably became of marginal interest after 1805. The Literary and Commercial Society was attended by city merchants, the evangelical Thomas Chalmers and reforming members of the Town Council who seem to have led discussions by the 1820s. Inevitably, all of these groupings would have had some impact upon the Literary Society. The new societies again show the effect of specialisation upon the generalist culture of the *virtuosi* and the enlightened *literati*, and they point directly to the growth of the city which could now support diverse specialised institutions. This development was also manifest in other initiatives.

Glasgow was slower in building a broad range of voluntary institutions than some cities. As we have seen, the Glasgow Town's Hospital opened its doors in 1733, but it was only in the 1790s that Glasgow witnessed the beginnings of the creation of the kind of voluntary institutions that had existed elsewhere for much of the eighteenth century. In 1792 James Towers organised a lying-in ward for poor women funded by the University, although it was not until 1834 that Glasgow had adequate facilities for natal care.[169] The Royal Infirmary received its charter in 1791 and admitted its first patients in 1794.[170] Ten years later the idea of an asylum for the insane was mooted, but it took another ten years before the Royal Lunatic Asylum on Dobbie's Loan was finished.[171] Libraries appeared at roughly the same rate. The Stirling Library opened in 1791; there was then an hiatus until after 1814, when other large subscription libraries were instituted.[172] The city had to wait for an adequate botanical garden until 1816; two years later the Royal Botanic Institute was founded.[173] Glasgow also lagged behind many provincial towns in Scotland and England in the construction of Assembly Rooms. Whereas Edinburgh's opened in 1787, it was not until 1796 that Glasgow had such a facility, which was occasionally used for public lectures on scientific subjects.[174] While the various societies and institutions thus offered venues for intellectual pursuits to some men not in the University, these bodies were still mostly dominated by academics. The Glasgow Professor of Logic and Rhetoric, George Jardine, for example, played a significant role in the foundation of the Royal Infirmary, while Jardine and a number of his colleagues were members of the General Committee struck to establish the Lunatic Asylum.[175] In 1802, therefore,

Glasgow continued to lack the institutional bases to sustain scientific work that differed significantly from that carried out in the College. Only the applied research of the city's chemists stood apart from what was done in the University.

The Challenge of the Industrial Chemists

We have seen that a fascination with chemistry had come to Glasgow by the time William Cullen arrived in 1744. From the start chemical knowledge was of interest to estate owners, businessmen and entrepreneurs like John Glassford. The study of chemistry was also encouraged by the Duke of Argyll. As the city industrialised, chemical knowledge became more widely valued, not least because manufacturers needed to produce goods to make the tobacco trade more profitable and then to lessen the shock of the disruption of this trade in the 1770s and 1780s. Thanks to Whitney's cotton gin, the 1790s saw a great increase in cotton imports, which also accelerated the search for bleaches, dyes, mordants and better equipment, as had the earlier expansion of the linen trade.[176] Although some of the University faculty were in touch with these developments, outside the College there had grown up a sizeable group of medical practitioners and a number of men directly interested in industrial chemistry. The medics may not have been effectively organised until the 1810s but the chemists apparently were.

Glasgow's industrial chemists (perhaps a fifth of the men of science in 1800) were connected by investments and to some extent by ties of kinship. They also benefited from the efforts of businessmen to organise in order to influence government economic policy. Glasgow's profitable American tobacco trade depended upon favourable legislation. Its disruption led merchants to press the government for special consideration in 1778 and 1779, and again in the mid- and late 1780s over the settlement of American debts. Proposals for 'economic reform' in the early 1780s, for the deregulation of Irish trade in the early 1780s and discussions of the Eden treaty with France in 1786 all forced Glasgow businessmen to consider their interests. The Glasgow Chamber of Commerce was formed in 1782–83 to do just that. Other English chambers followed in the next few years, linking Glaswegians to businessmen in the larger British cities. Active in the Glasgow Chamber were four members of the scientific community: James MacGregor, his son-in-law James Watt and George and Charles Macintosh. These men of science dealt with Josiah Wedgwood, Joseph Henry and other figures elsewhere who were likewise engaged in chemistry, commerce and industry.[177] The drawing rooms and cabinets, factories and counting houses of industrialists like John Glassford, James MacGregor, the Macintoshes, Charles Tennant, James Knox, John Finlay and their partners and kin were undoubtedly *fora* in which natural knowledge was discussed. As the city grew, such men became more numerous, confident and independent of the University's intellectuals. They, not Anderson's Institution, posed the real challenge to the traditional institutional structure of Glasgow science. For such men, science could not be a mere polite amusement because its successful pursuit was a prerequisite for their economic success. Their educations had left most of them without the skills to fit into a *virtuoso*'s world which, in any case, must have seemed full of useless interests. What mattered to them were profits and the need to keep afloat enterprises that now sometimes employed several hundred people. This was a social group which sustained its own research projects and, by the late 1780s, defined its own priorities.

Conclusion

We began this chapter by asking if the place of natural knowledge in the cultures of the three largest Scottish burghs was the same. It clearly was not. At Glasgow, the *virtuoso* ideal began to decline by the 1740s; elsewhere it lasted longer among more of the cultural leaders. Indeed, even in 1800 the Royal Society of Edinburgh and the Society of Antiquaries of Scotland were in some disarray because both were formally wedded to this ideal, even though it no longer reflected the realities of intellectual life.[178] Science in Glasgow was not wholly university-based as it more or less was in Aberdeen. The difference between Glasgow and Aberdeen is illustrated by the fact that new bleaching processes reached the Gordons of Aberdeen through Patrick Copland, the Professor of Natural Philosophy at Marischal College. At Glasgow, the transmission of the information was from James Watt to his father-in-law, James MacGregor, and then on to the Macintoshes. The University played no mediating role whatever.[179] In Edinburgh, physicians rather than industrialists and chemists tended to be the key players in the scientific community, which in 1800 was not split as it was in Glasgow. The men of science affiliated with Edinburgh University and the Royal Society dominated the industrial scene in the Scottish capital as their Glasgow counterparts could not do. Moreover, Edinburgh was almost exclusively a Tory town, whereas Glasgow was not. While Anderson's Institution may have met with limited success in Glasgow, it is hard even to imagine a comparable institution in Edinburgh.

If the cities were all a bit different, does that mean that their Enlightenments differed? To some extent it does. All three Enlightenments originated in the views of the *virtuosi* at the turn of the eighteenth century who accepted the new science, taught it and institutionalised it with the countenance of powerful politicians. However, the Glasgow Enlightenment had its own distinctive character that was formed by a variety of local factors. The growth of the city, the secessions from the Kirk, the dynamism of the Evangelicals and their implacable hostility to those affiliated with the Moderates, and the increased importance of a merchant oligarchy that was not assimilable into the landed aristocracy all helped to shape the nature of the Glasgow Enlightenment. So too did the complement of institutions within the city. Glasgow had fewer lawyers, politicians, generals and resident aristocrats than did Edinburgh. Hence it is not surprising that Glasgow's cultural leaders were more often men like John Anderson or Charles Macintosh than figures such as the Rev. William Robertson, David Dalrymple (Lord Hailes), Henry Mackenzie, General Sir James Oughton or the learned canal managers Dr. James Hutton and Sir George Clerk-Maxwell. The particularities of the local context were important.[180] In Glasgow, by the end of the eighteenth century business made chemistry rather than medicine *the* science. It placed a premium on utility and showed less regard for speculative sciences. Polite culture found a home within the University, but it was not well nourished in the city. Enlightenment was present and bound up with the cultivation of natural knowledge, but in both the University and the town it was somewhat less tolerant, more shrill and more closely tied ideologically to middle-class traders than to the genteel and landed aristocracy. Because of the dominance of industry and commerce, the Glasgow Enlightenment was different. The differences between Glasgow and Edinburgh are perhaps best epitomised in the career of Patrick Colquhoun. A tobacco merchant who became an industrialist and then Provost in 1782, Colquhoun eventually migrated to London and became the first important Scottish convert to Benthamism. For their part, the Benthamites in Edinburgh ran the *Edinburgh Review* and theorised instead of managing mills and becoming stipendiary magistrates. Common sense philosophy may have dominated the curriculum at Glasgow University following the appointment of Thomas Reid in 1764, but the town and its men differed in outlook, although not necessarily in the ways that Richard Sher, Ned Landsman and Robert Kent Donovan have described. For us, the history of science in eighteenth-century Glasgow is not a relatively insignificant narrative of the development of ideas and institutions in a minor provincial town. Rather, the story is one of natural knowledge in a dynamic urban context being captured by industry and made subservient to new conceptions of utility. We have shown that Glasgow's scientific community was not made up of 'marginal men' as characterised by Arnold Thackray, and indicated that the cultivation of natural knowledge was a central feature of the Glasgow Enlightenment. Our story thus challenges influential accounts of both provincial science in Britain and the Scottish Enlightenment, and it is therefore of more than merely parochial interest.

Notes

The authors would like to thank: Matthew Eddy, Fiona Macdonald and Charles Withers for references and for comments on earlier drafts of this chapter; and Rosemary Barlow for her assistance with the Tables.

1. For example, there are only two indexed entries on Glasgow in Ian Inkster and Jack Morrell (eds.), *Metropolis and Province: Science in British Culture, 1780–1850* (London, 1983). Notwithstanding the geographical/national focus of the article, Edinburgh is mentioned in Roy Porter, 'Science, Provincial Culture and Public Opinion in Enlightenment England', *British Journal for Eighteenth-Cen-*

tury Studies 3 (1980), 20–46; Glasgow is not. Glasgow also does not figure as a provincial centre for the natural sciences in Porter's recent *Enlightenment: Britain and the Creation of the Modern World* (London, 2000). Recognition of, and evidence for, Glasgow's significance in relation to Manchester is found in Arnold Thackray, 'Natural Knowledge in Cultural Context: The Manchester Model', *American Historical Review* 79 (1974), 672–709, but Thackray does not elaborate on the connection.

2. Recent works containing materials on Glasgow include: Andrew Gibb, *Glasgow: The Making of a City* (London, 1983); T. M. Devine and Rosalind Mitchison (eds.), *People and Society in Scotland, Volume I: 1760–1830* (Edinburgh, 1988); Richard B. Sher and Jeffrey R. Smitten (eds.), *Scotland and America in the Age of the Enlightenment* (Edinburgh, 1990); Andrew Hook and Richard B. Sher (eds.), *The Glasgow Enlightenment* (East Linton, UK, 1995); T. M. Devine and Gordon Jackson (eds.), *Glasgow, Volume I: Beginnings to 1830* (Manchester and New York, 1995). See also T. M. Devine, *The Scottish Nation* (London and New York, 2000).

3. Robert Kent Donovan, 'The Popular Party of the Church of Scotland and the American Revolution', in Sher and Smitten, *Scotland and America*, 81–99; *idem*, 'Evangelical Civic Humanism in Glasgow: The American War Sermons of William Thom', in Hook and Sher, *The Glasgow Enlightenment*, 227–45; Ned C. Landsman, 'Presbyterians and Provincial Society: The Evangelical Enlightenment in the West of Scotland', in John Dwyer and Richard B. Sher (eds.), *Sociability and Society in Eighteenth-Century Scotland* (Edinburgh, 1993), 194–209; *idem*, 'Witherspoon and the Problem of Provincial Identity in Scottish Evangelical Culture', in Sher and Smitten, *Scotland and America*, 29–45; *idem*, 'Liberty, Piety and Patronage: The Social Context of Contested Clerical Calls in Eighteenth-Century Glasgow', in Hook and Sher, *The Glasgow Enlightenment*, 214–26.

4. Richard B. Sher, 'Commerce, Religion and the Enlightenment in Eighteenth-Century Glasgow', in Devine and Jackson, *Glasgow*, 312–59. Sher's view of the development of academic moral philosophy in Glasgow is criticised in Paul Wood, '"The Fittest Man in the Kingdom": Thomas Reid and the Glasgow Chair of Moral Philosophy', *Hume Studies* 23 (1997), 277–313.

5. James Moore, 'Hutcheson's Theodicy: The Argument and the Contexts of *A System of Moral Philosophy*', in Paul Wood (ed.), *The Scottish Enlightenment: Essays in Reinterpretation* (Rochester, NY, 2000), 239–66; M. A. Stewart, 'John Smith and the Molesworth Circle', *Eighteenth-Century Ireland* 2 (1987), 89–102; *idem*, 'Rational Dissent in Early Eighteenth-Century Ireland', in Knud Haakonssen (ed.), *Enlightenment and Religion: Rational Dissent in Eighteenth-Century Britain* (Cambridge, 1996), 42–63.

6. Roger L. Emerson, 'Medical Men, Politicians and the Medical Schools at Glasgow and Edinburgh, 1685–1805', in A. Doig, J. P. S. Ferguson, I. A. Milne and R. Passmore (eds.), *William Cullen and the Eighteenth-Century Medical World* (Edinburgh, 1993), 186–215.

7. Roger L. Emerson, *Professors, Patronage and Politics: The Aberdeen Universities in the Eighteenth Century* (Aberdeen, 1992); P. B. Wood, 'Science and the Pursuit of Virtue in the Aberdeen Enlightenment', in M. A. Stewart (ed.), *Studies in the Philosophy of the Scottish Enlightenment* (Oxford, 1990), 127–149; *idem*, 'Science and the Aberdeen Enlightenment', in Peter Jones (ed.), *Philosophy and Science in the Scottish Enlightenment* (Edinburgh, 1988), 39–66. The story is also sketched in Roger L. Emerson, 'Science and the Origins and Concerns of the Scottish

Enlightenment', *History of Science* 26 (1988), 333–66. For recent views of Scottish culture and thought *c.* 1700, see also: David Allan, *Virtue, Learning and the Scottish Enlightenment: Ideas of Scholarship in Early Modern History* (Edinburgh, 1993); Colin Kidd, *Subverting Scotland's Past: Scottish Whig Historians and the Creation of an Anglo-British Identity, 1689–1830* (Cambridge, 1993); John Robertson (ed.), *A Union for Empire: Political Thought and the Union of 1707* (Cambridge, 1995).

8. Our interpretation of the Scottish Enlightenment is attacked in Richard B. Sher, 'Science and Medicine in the Scottish Enlightenment: The Lessons of Book History', in Wood, *The Scottish Enlightenment*, 99–156.
9. Our terminology reflects the fact that the culture of science in Glasgow during our period was dominated by men.
10. Similar changes were also taking place in Aberdeen and Edinburgh. The adoption of the professorial system at King's College, Aberdeen in 1800 was a belated recognition of academic specialisation; see Paul B. Wood, *The Aberdeen Enlightenment: The Arts Curriculum in the Eighteenth Century* (Aberdeen, 1993), 160. Moreover, the Leslie Affair of 1805 was symptomatic of the dissolution of the social and cultural formation which had sustained the Edinburgh Enlightenment since the 1750s. On the process of differentiation and specialisation in Scotland more generally, see: Roger L. Emerson, 'The Scottish Enlightenment and the End of the Philosophical Society of Edinburgh', *The British Journal for the History of Science* 21 (1988), 33–66; Paul Wood, ' "Jolly Jack Phosphorus" in the Venice of the North; or, Who was John Anderson?', in Hook and Sher, *Glasgow Enlightenment*, 111–32.
11. For Kuhn's definition of a 'scientific community', see Thomas S. Kuhn, *The Structure of Scientific Revolutions*, 2nd edn (Chicago and London, 1970), 176–81. Our understanding of the term 'community' is closer to that found in John R. R. Christie, 'The Origins and Development of the Scottish Scientific Community, 1680–1760', *History of Science* 12 (1974), 122–41, and Steven Shapin and Arnold Thrackray, 'Prosopography as a Research Tool in History of Science: The British Scientific Community, 1700–1900', *History of Science* 12 (1974), 1–28.
12. Anderson's will is transcribed in James Muir, *John Anderson: Pioneer of Technical Education and the College He Founded*, ed. James M. Macaulay (Glasgow, 1950), 129–62.
13. The sources for our list and the statistical information discussed below are too numerous to mention but they include, *inter alia*, directories, institutional histories, university manuscripts and lists, local histories, genealogies and biographies and histories of chemistry, medicine and the industries of Glasgow.
14. Two of our men of science, George and Charles Macintosh, were leading figures in the Glasgow Gaelic community, and employed a Gaelic workforce at their Dunchattan factory; see D. W. F. Hardie, 'The Macintoshes and the Origins of the Chemical Industry', in A. E. Musson (ed.), *Science, Technology and Economic Growth in the Eighteenth Century* (London, 1972), 168–94, esp. p. 176, and Charles W. J. Withers, *Urban Highlanders: Highland-Lowland Migration and Urban Gaelic Culture, 1700–1900* (East Linton, UK, 1998), 136–37, 165, 189.
15. Robert Smart, 'Some Observations on the Provinces of the Scottish Universities, 1560–1850', in G. W. S. Barrow (ed.), *The Scottish Tradition: Essays in Honour of Ronald Gordon Cant* (Edinburgh, 1974), 91–106, esp. pp. 97, 102–3.
16. Roughly 25% of Glasgow's matriculants in the period 1750 to 1799 came from abroad: Ireland (15%), England (6%), and other (3%). See W. M. Matthew, 'The

Origins and Occupations of Glasgow Students, 1740–1839', *Past and Present* 33 (1966), 74–94.
17. How much more 'middle class' and 'well-off' these men were than the average Glasgow undergraduate student can be seen by comparing these figures with the recruitment figures given by Smart, 'Some Observations', 104, and by Matthew, 'Origins and Occupations'.
18. Few Scottish ministers made £75 a year or more. The average living was probably closer to £50 *per annum* plus housing, a glebe and a few perquisites.
19. This assumes that all of the instrument makers were apprenticed and that the surgeons were all qualified even if they were not known to have attended a university. Those whose educations are unknown are mostly teachers.
20. This assumes that those who had M.D.s from an unknown university had gone to at least one university.
21. S. G. Checkland, *Scottish Banking: A History, 1695–1973* (Glasgow and London, 1975), 202–3.
22. Roger L. Emerson, 'Politics and the Glasgow Professors, 1690–1800', in Hook and Sher, *The Glasgow Enlightenment*, 21–39.
23. Emerson, 'Medical Men, Politicians and Medical Schools', 193–203. One reminder of Argyll's close links with Glasgow is the splendid Allan Ramsay fulllength portrait of the Duke commissioned by the Town Council in 1749, for which see Alistair Smart, *Allan Ramsay: A Complete Catalogue of His Paintings*, ed. John Ingamells (New Haven and London, 1999), entry 14 and fig. 303. Another reminder is the tribute penned by Carlo Denina, who praised not only Francis Hutcheson but also the Duke as men who 'seem to have been particularly destined by heaven to raise, and bring to maturity, in the cold regions of the north, what had heretofore been foolishly supposed incapable of taking root but in the warmer climes of ASIA MINOR, GREECE, and ITALY . . . [Argyll] patronized the ingenious with a bounty worthy of himself, and paid particular attention to the university of GLASGOW, which has since become one of the most renowned in EUROPE'; Carlo Denina, *An Essay on the Revolutions of Literature*, trans. John Murdoch (London, 1763), 276–77.
24. See below, pp. 102–7.
25. See below, pp. 101.
26. Thomas F. Donald, *The Hodge Podge Club, 1752–1900* (Glasgow, 1900), esp. pp. 59–60 for a list of the members in our period; John Strang, *Glasgow and Its Clubs; or Glimpses of the Condition, Manners, Characteristics and Oddities of the City, During the Past and Present Century* (Glasgow, 1856), 43–66. Something of the spirit of the Club can be found in the 'Verses On the Hotch Potch Club at Glasgow By the late Dr Moore', National Library of Scotland, MS 5003, fol. 107; Strang (pp. 48–51) prints an expanded and somewhat different version of this poem.
27. For the history of the Philosophical Society, as well as lists of members, see: Roger L. Emerson, 'The Philosophical Society of Edinburgh, 1737–1747', *The British Journal for the History of Science* 12 (1979), 154–91; idem, 'The Philosophical Society of Edinburgh, 1748–1768', *The British Journal for the History of Science* 14 (1981), 133–76; idem, 'The Philosophical Society of Edinburgh, 1768–1783', *The British Journal for the History of Science* 18 (1985), 255–303; and Emerson, 'The Scottish Enlightenment and the End of the Philosophical Society of Edinburgh'. For the Fellows of the Royal Society of Edinburgh we have used a typescript list compiled by the late Professor Eric Forbes.
28. The Fellows were: Robert Houston (1725), John Anderson (1759), Matthew

Stewart (1764), David Steuart Erskine, 11th Earl of Buchan (1765), William Cullen (1777), James Watt (1785), Adair Crawford (1786), James Adair (1788), John Finlay (1788), James Makittrick Adair (1788), Thomas Brisbane (1810), Thomas Charles Hope (1810), Thomas Thomson (1812), Andrew Ure (1821) and Charles Mackintosh (1824).

29. On the Glasgow Faculty of Physicians and Surgeons, see Johanna Geyer-Kordesch and Fiona Macdonald, *Physicians and Surgeons in Glasgow: The History of the Royal College of Physicians and Surgeons of Glasgow, 1599–1858* (London and Rio Grande, 1999), 154–57.

30. The Faculty of Physicians and Surgeons tried to improve its library in the 1730s and again in 1745. Significantly, the acquisition of books in 1745 was overseen by Robert Hamilton, who was also buying books for the University. Geyer-Kordesch and Macdonald suggest that the Faculty and the College were in fact trying to co-ordinate their holdings; Geyer-Kordesch and Macdonald, *Physicians and Surgeons*, 229–30.

31. A number of gardens were maintained in Glasgow during the eighteenth century, but they were typically private (and often commercial) ventures which provided medicinal plants. Two of the most important of these gardens were those maintained by the Balmannos and by Dr. John Wodrow. Wodrow's garden was in operation from sometime before 1760 until 1769 and was subsidised by the Faculty of Physicians and Surgeons; see Geyer-Kordesch and Macdonald, *Physicians and Surgeons*, 199–204. The Balmannos' garden was open for many years and the property was sold to the burgh *c.* 1794: Robert Renwick (ed.), *Extracts from the Records of the Burgh of Glasgow, 1781–1795* ([Glasgow], 1913), 576, 687.

32. Fiona A. Macdonald, 'The Infirmary of the Glasgow Town's Hospital, 1733–1800: A Case for Voluntarism?', *Bulletin of the History of Medicine* 73 (1999), 64–105; *idem*, 'The Infirmary of the Glasgow Town's Hospital: Patient Care, 1733–1800', in Wood, *The Scottish Enlightenment*, 199–238.

33. On the University and the creation of a public sphere in eighteenth-century Glasgow, see: Paul Wood, 'Science, the Universities and the Public Sphere in Eighteenth-Century Scotland', *History of Universities* 13 (1994), 99–135, esp. pp. 111–15; Charles W. J. Withers, 'Towards a History of Geography in the Public Sphere', *History of Science* 37 (1999), 45–78.

34. For an account of the early teaching of the Greek professors, see M. A. Stewart, 'The Origin of the Scottish Greek Chairs', in E. M. Craik (ed.), *'Owls to Athens': Essays on Classical Subjects Presented to Sir Kenneth Dover* (Oxford, 1990), 391–400.

35. Professorships in mathematics had been founded in the seventeenth century at Marischal College, Aberdeen (1626) and Edinburgh (1620), while St. Andrews had one dating back to the late sixteenth century which was re-established as a Regius Chair in 1668. During the eighteenth century, King's College, Aberdeen failed to create a permanent and funded chair, although it tried to do so in 1703, 1718 and *c.*1730–32. The teaching of mathematics thus fell to the regents.

36. How a gentleman interested in natural knowledge reacted to the Glasgow and Leyden curricula may be seen in John M. Gray (ed.), *Memoirs of the Life of Sir John Clerk of Penicuik, Baronet* (Edinburgh, 1892), 12–19. For a sketch of the Glasgow curriculum prior to the end of regenting in 1727, see John Loudoun's undated memorandum, 'Methods of Teaching in the Philosophy Classes', Glasgow University Archives, MS GUA 43228. As this memorandum shows, Lou-

doun used experimental demonstrations in his teaching and, like his counterparts elsewhere in Scotland, he was preoccupied with countering the threat of atheism and infidelity in his lectures.

37. There is no literature on the teaching of history at Glasgow, but see: James Coutts, *A History of the University of Glasgow* (Glasgow, 1909), 170, 192; Andrew Browning, 'History', in J. B. Neilson (ed.), *Fortuna Domus: A Series of Lectures Delivered in the University of Glasgow in Commemoration of the Fifth Centenary of Its Foundation* (Glasgow, 1952), 41–57, esp. pp. 41–43; Bruce P. Lenman, 'The Teaching of History in the Scottish Universities', *Scottish Historical Review* 53 (1973), 165–90; Allan, *Virtue, Learning and the Scottish Enlightenment*, 36.
38. Andrew Ross, 'The Method in which Humanity is taught in the University of Glasgow', Glasgow University Library MS Gen. 357. Ross's short report was perhaps done for the Visitation of 1717.
39. Matthew Brisbane to Sir Robert Sibbald, in 'Ane Essay Relating to the Natural History of Scotland by way of Supplement to the Podromus Naturalis Historae Scotae published anno 1684', National Library of Scotland, Advocates MS 33.5.19, 242.
40. William Dunlop, 'Description of Renfrew Shire', in William Hamilton of Wishaw, *Descriptions of the Sheriffdoms of Lanark and Renfrew, Compiled about MDCCX with Illustrative Notes and Appendices* (Glasgow, 1831), 142–48. See also, Denys Hay, 'The Historiographers Royal in England and Scotland', *Scottish Historical Review* 30 (1951), 15–29.
41. Sibbald, 'Ane Essay', 59, 300.
42. Sibbald, 'Ane Essay', 125, 342, 486.
43. There is no good extended biographical account of Sinclair but see Paul Wood's entry on him in Andrew Pyle (ed.), *The Dictionary of Seventeenth-Century British Philosophers*, 2 vols. (Bristol, 2000), 2:750–51.
44. Sibbald, 'Ane Essay', 110, 330. Robert Sinclair is identified as the son of George by Ian Sneddan in his biographical entry for Robert Simson in the *DSB*.
45. Wodrow's activities as a *virtuoso* have been discussed in Roger L. Emerson, 'Sir Robert Sibbald, Kt, the Royal Society of Scotland and the Origins of the Scottish Enlightenment', *Annals of Science* 45 (1988), 41–72.
46. Catalogues of this collection survive; see 'A Catalogue of minerall and figur'd stones found in Scotland', National Library of Scotland, Advocates MS 13.2.8, and 'Eastwood Museum: List of Materials for a Natural History 1703', Andersonian Library, University of Strathclyde, MS 11. Some items from Wodrow's cabinet are undoubtedly preserved but unidentified as such in the Hunterian Museum at the University of Glasgow.
47. On Wodrow as a field naturalist, see Charles W. J. Withers, 'Reporting, Mapping, Trusting: Making Geographical Knowledge in the Late Seventeenth Century', *Isis* 90 (1999), 497–521, esp. pp. 504–5. On Wodrow's connections with the *virtuosi* in England, see also Michael Hunter (ed.), *The Occult Laboratory: Magic, Science and Second Sight in Late Seventeenth-Century Scotland* (Woodbridge, UK, 2001), 205–15.
48. Robert Wodrow to James Sutherland, 29 December 1701, in L. W. Sharp (ed.), *Early Letters of Robert Wodrow, 1698–1709* (Edinburgh, 1937), 184–85; see also pp. xxx-xxxi. On this episode, see also A. D. Boney, '"For the student's improvement in the skill of Bottany": Botanists and Botany in the University of Glasgow, July 1704–July 1994' (unpublished typescript, 2000, on deposit in Glasgow University Archives), 4.
49. These statements are based on a reading of letters to him and from him held in the

National Library of Scotland. Much of Wodrow's correspondence has been published but much has not. Letters to him are in Edinburgh and Glasgow University Libraries, as well as in the National Library of Scotland. His nineteenth-century editor tended not to publish Wodrow's secular correspondence.

50. Christine M. K. Shepherd, 'Philosophy and Science in the Arts Curriculum of the Scottish Universities in the Seventeenth Century' (unpublished Ph.D. thesis, Edinburgh University, 1974), 77–78, 240–41; *idem*, 'Newtonianism in the Scottish Universities in the Seventeenth Century', in R. H. Campbell and A. S. Skinner (eds.), *The Origins and Nature of the Scottish Enlightenment* (Edinburgh, 1982), 65–85, esp. p. 76. Law is described as a teacher of elementary mathematics, content to repeat what he himself had been taught, in J. D. Mackie, *The University of Glasgow, 1451 to 1951* (Glasgow, 1954), 161. In the 1710s Law continued to teach in a traditional way using dictates and disputations; W. J. Duncan (ed.), *Notices and Documents Illustrative of the Literary History of Glasgow, During the Greater Part of Last Century* (1831; reprinted Glasgow, 1886), 124–25.

51. David Murray, *Memories of the Old College of Glasgow* (Glasgow, 1927), 260. Murray states that Law's 1698 theses 'relate largely to Physics and Astronomy and refer to a new form of Sun-dial devised by him'.

52. On Carmichael, see, Shepherd, 'Philosophy and Science', 78–80, 241; Murray, *Memories*, 507n; C. A. Campbell, 'Philosophy', in Neilson, *Fortuna Domus*, 103; James Moore and Michael Silverthorne, 'Gershom Carmichael and the Natural Jurisprudence Tradition in Eighteenth-Century Scotland', in Istvan Hont and Michael Ignatieff (eds.), *Wealth and Virtue: The Shaping of Political Economy in the Scottish Enlightenment* (Cambridge, 1985), 73–87.

53. Coutts, *History*, 482–83; Geyer-Kordesch and Macdonald, *Physicians and Surgeons*, 194–96.

54. Murray, *Memories*, 172; Colin Maclaurin to John Johnstoun, 9 June 1737, inaccurately transcribed in Stella Mills (ed.), *The Collected Letters of Colin MacLaurin* (Nantwich, 1982), 72–74; W. S. Craig, *History of the Royal College of Physicians of Edinburgh* (Oxford, London, Edinburgh and Melbourne, 1976), 1072.

55. Rev. William Trail, *Account of the Life and Writings of Robert Simson, M.D. Late Professor of Mathematics in the University of Glasgow* (Bath and London, 1812); [John Robison], 'Simson, Dr Robert', in *Encyclopaedia Britannica*, 3rd edn, 18 vols. (Edinburgh, 1797), 17:504–9. See also Ian Sneddan's entry on Simson in the *DSB*.

56. Glasgow University Library still has his books shelved together as a collection, and the Department of Special Collections houses his papers and many letters to and from him. A further substantial holding of Simson's correspondence is to be found in the University Archives, and there are also letters in Edinburgh University Library, the Royal Society of London and the Saltoun Correspondence in the National Library of Scotland.

57. Argyll's scientific interests are detailed in Roger L. Emerson, 'The Scientific Interests of Archibald Campbell, 1st Earl of Ilay and 3rd Duke of Argyll, 1682–1761', *Annals of Science* 59 (2002), 21–56. His library is discussed in *idem*, '*Catalogus Librorum A.C.D.A.*, or the Library of Archibald Campbell, Third Duke of Argyll, 1682–1761', in Paul Wood (ed.), *The Culture of the Book in the Scottish Enlightenment* (Toronto, 2000), 12–39; *idem*, 'The Library of Archibald Campbell, Third Duke of Argyll', *Bibliotheck* 28 (2003), forthcoming.

58. King's College, Aberdeen granted Argyll an honorary M.D. in 1708: P. J.

Anderson (ed.), *Officers and Graduates of University and King's College Aberdeen, 1495–1860* (Aberdeen, 1893), 125.

59. Emerson, 'Politics and the Glasgow Professors', 29–31.
60. The 'Act regulating the University of Glasgow', dated 18 September 1727, is included in *Munimenta Alme Universitatis Glasguensis*, 4 vols. (Glasgow, 1854), 2:569–81, p. 577.
61. Our discussion of Cullen's teaching at Glasgow is indebted to: A. L. Donovan, *Philosophical Chemistry in the Scottish Enlightenment: The Doctrines and Discoveries of William Cullen and Joseph Black* (Edinburgh, 1975); J. R. R. Christie, 'Ether and the Science of Chemistry: 1740–90', in G. N. Cantor and M. J. S. Hodge (eds.), *Conceptions of Ether: Studies in the History of Ether Theories, 1740–1900* (Cambridge, 1981), 85–110; idem, 'William Cullen and the Practice of Chemistry' in Doig, Ferguson, Milne and Passmore, *William Cullen*, 98–109; Jan Golinski, *Science as Public Culture: Chemistry and Enlightenment in Britain, 1760–1820* (Cambridge, 1992), Ch. 2. The basic source for Cullen's teaching at Glasgow remains John Thomson's *An Account of the Life, Lectures and Writings, of William Cullen, M.D. Professor of Medicine and General Pathology in the University of Edinburgh*, 2 vols. (1832–59; reprinted with an introduction by Michael Barfoot, Bristol, 1997), 1:24–26, 29–31, 35, 44–45.
62. Christie, 'Ether and Chemistry', 93.
63. For his part, Black distinguished sharply between the spheres of the natural philosopher, the natural historian and the chemist; see Joseph Black, *Lectures on the Elements of Chemistry, Delivered in the University of Edinburgh*, ed. John Robison, 2 vols. (Edinburgh and London, 1803), 1:337–39.
64. Coutts, *History*, 195–96; Peter Swinbank, 'Experimental Science in the University of Glasgow at the Time of Joseph Black', in A. D. C. Simpson (ed.), *Joseph Black, 1728–1799: A Commemorative Symposium* (Edinburgh, 1982), 23–35, esp. pp. 23–24, 31–32. Dick was in contact with the Edinburgh Professor of Natural Philosophy, Robert Steuart, since he is listed as a donor to Steuart's Physiological Library; see Murray, *Memories*, 166, and *The Physiological Library. Begun by Mr. Steuart, and Some of the Students of Natural Philosophy in the University of Edinburgh April 2 1724: And augmented by some Gentlemen; and the Students of Natural Philosophy, December 1724* ([Edinburgh], 1725), 6.
65. Mackie, *University*, 218; Swinbank, 'Experimental Science', 30–32. One of Drew's accounts, dated 2 November 1735, survives; see 'Harry Drews Acct for Experiment room 1736 [sic]', Glasgow University Archives, MS GUA 5194.
66. *Munimenta*, 2:578; Swinbank, 'Experimental Science', 24–25.
67. Swinbank, 'Experimental Science', 31; Coutts, *History*, 196. These lecturers were almost certainly John Thorold and Martin Triewald; see Larry Stewart, *The Rise of Public Science: Rhetoric, Technology and Natural Philosophy in Newtonian Britain, 1660–1750* (Cambridge, 1992), 281, 361–66. Thorold and Triewald also had contacts with William Adam and Sir John Clerk; see William Adam to Sir John Clerk, 20 March, 24 March, 12 April, 16 April 1725, National Archives of Scotland, Clerk of Penicuik MSS, MS 4728/7, 8, 11, 12. On some of the early public lectures in Glasgow, see John A. Cable, 'The Early History of Scottish Popular Science', *Studies in Adult Education* 4 (1972), 34–45.
68. 'Inventory of Instruments for Philosophical Experiments and Observations belonging to the University of Glasgow, As they were delivered by Mr Lowdoun to Mr. Dick February 13 1727', Glasgow University Archives, MS GUA 5191; this manuscript includes another list headed 'Instruments Sent from London by Mr Hauksbee April 8 1727'.

69. Dick acquired electrical apparatus in 1749: Coutts, *History*, 196. The electrical equipment is described in 'A List of Instruments belonging to the University of Glasgow', Glasgow University Archives, MS GUA 5210, fol. 1r. This list, dated 26 June 1760, was drawn up by John Anderson.
70. P. R. Dee, 'Natural Philosophy', in Neilson, *Fortuna Domus*, 314–15. The elder Dick seems to have been something of an evangelical since he supported the candidacy of the evangelical minister John Maclaurin for the vacant chair of divinity in 1740 and subsequently opposed the appointment of William Leechman in 1743. See Neil Campbell, John Johnstoun, Robert Dick and George Ross to ?, 13 February 1740, Saltoun Papers, National Library of Scotland, MS 16582, fol. 51.
71. Wood, 'Science and the Public Sphere', 106–8, 117–18.
72. For Dick II's course, see *A Course of Experiments and Lectures on Mechanics, Hydrostatics, Pneumatics and Optics* (Glasgow, 1753). Dick II's course outline signals his interest in the practical applications of natural knowledge, but it is silent on the religious implications of the experimental philosophy. Dick II was a highly respected lecturer; see John Robison, 'Preface', in Black, *Lectures*, 1:vii, xx–xxi. Anderson's teaching is discussed in Wood, 'Jolly Jack Phosphorus', 115, 123–24.
73. For Sinclair's interest in navigation, see his *The Principles of Astronomy and Navigation* (Edinburgh, 1688). Although we do not know the precise details of what Simson taught in his classes, it is likely that he covered the basics of navigation, especially given his close association with the mathematican Humphrey Ditton, who was Mathematical Master at Christ's Hospital; see Traill, *Life of Simson*, 3, 5. Later in the century the practical concerns of the Glasgow Professor of Mathematics, James Millar, were manifest in his extensive lectures on geography; see 'Notes of Lectures on Geography. Delivered in the University of Glasgow. By Mr. James Millar, Professor of Mathematics. In the Session of 1802-3', National Library of Scotland, MS 2801. The instruments at Dick I's disposal suggest that he may have taught navigation and perhaps some surveying. Dick II dealt with gunnery in the section of his course devoted to projectiles; Dick, *Course*, 11–12.
74. See Wood, 'Jolly Jack Phosphorus', 114, 117; Muir, *John Anderson*, 40–61. For Anderson's unsuccessful dealings in 1789 with the Master General of the Ordnance, the Duke of Richmond, see 'Copy of the Correspondence with the Duke of Richmond concerning Professor Anderson's Field Piece', Andersonian Library, University of Strathclyde, MS 7.
75. Wood, 'Science and the Public Sphere', 114. Women had, of course, formed part of the audience for natural philosophy courses given by itinerant lecturers throughout the eighteenth century, and Colin Maclaurin and at least one of his students lectured to mixed audiences in extramural courses given in Edinburgh during the 1740s: Wood, 'Public Sphere', 105. Women were also targeted by lecturers on geography in the period; see Withers, 'Geography in the Public Sphere'.
76. Wood, 'Jolly Jack Phosphorus', 117.
77. See especially Thom's *Letter to J—— M——, Esq. On the Defects of an University Education, and Its Unsuitableness to a Commercial People; with the Expediency and Necessity of Erecting at Glasgow an Academy for the Instruction of Youth*, reprinted in *The Works of the Rev. William Thom, Late Minister of Govan* (Glasgow, 1799), 263–301. For another relevant attack on the pedagogical practices of the universities, inspired by the ideas of Rousseau, see [Patrick Clason], *Essays. Viz. I.*

On the Origin of Colleges, or Universities. II. On the Origin of the Custom of Lecturing in Latin. III. On the Impropriety of this Custom, at Present (Glasgow, 1769).

78. Geyer-Kordesch and Macdonald, *Physicians and Surgeons*, 198–99; A. D. Boney, *The Lost Gardens of Glasgow University* (London, 1988), 27–47. The minutes of the Faculty record payments to John Wodrow of £10 annually and then of an annuity to his widow and unmarried daughters.
79. Coutts, *History*, 484; Geyer-Kordesch and Macdonald, *Physicians and Surgeons*, 215.
80. Brisbane's teaching responsibilities were spelled out in detail by the Visitors in 1727; see *Munimenta*, 2:579–80. For a more sympathetic view of Brisbane and Johnstoun, see Geyer-Kordesch and Macdonald, 195–96. Their view, however, does not take into account the 1727 regulations.
81. Boney, *Lost Gardens*, 69–70; Mackie, *University*, 168–70.
82. Macdonald, 'Infirmary of the Glasgow Town's Hospital', 88–94. Macdonald emphasises the close connections between the Town's Hospital and the University.
83. Thomson, *Life of Cullen*, 1:19–26; Donovan, *Philosophical Chemistry*, 49–50.
84. Thomson, *Life of Cullen*, 1:70; Donovan, *Philosophical Chemistry*, 69–70.
85. Gibson seems to have lectured c.1764, as did Andrew Morris. Both had M.D.s (Leyden and Rheims). Morris's short teaching career was probably an effort by an incapacitated physician to make a bit of income. By 1763 he had become a pensioner of the Glasgow Faculty of Physicians and Surgeons: 'Transcript of the Minutes of the Faculty of Physicians and Surgeons, 11 November 1763 to 3 November 1788', Royal College of Physicians and Surgeons of Glasgow. Morris's name is replaced by that of his daughter in 1788. The Faculty accounts note a payment to Morris of £20 'pr Sundry orders & receipts'. This was twice his usual pension and may represent a payment for lectures previously given between November 1765 and August 1767; 'Transcript of the Minutes', 126.
86. On the chemical lectures of Cullen, Black, Robison and Irvine, see: Thomson, *Life of Cullen*, 1:25, 35, 44–45; Donovan, *Philosophical Chemistry*, 64–65; Robison, 'Preface', in Black, *Lectures*, 1: xxii; and the relevant chapters in Andrew Kent (ed.), *An Eighteenth Century Lectureship in Chemistry* (Glasgow, 1950). On the creation of the lectureship in materia medica for Irvine, see Alexander Duncan, *Memorials of the Faculty of Physicians and Surgeons of Glasgow, 1599–1850* (Glasgow, 1896), 131–32, and the anonymous life of Irvine in *Medical Commentaries* 2 (1787), 455–61, esp. p. 456.
87. John Anderson to William Mure of Caldwell, 8 January 1763; John Graham to Mure 10 January 1763, in *The Caldwell Papers*, 3 vols. (Paisley, 1883), 1:163–66; the original of the Anderson letter is now in the National Library of Scotland, MS 2524, fols. 5–6. John Towers had annual appointments from the College as Lecturer in Midwifery from 1790 until 1815 when he became the first Professor of Surgery and Midwifery; Geyer-Kordesch and Macdonald, *Physicians and Surgeons*, 270–71, 281. Some topics in midwifery were also taught after 1790 by the Professor of Botany and Anatomy, James Jeffray.
88. By around 1792 John Burns was teaching anatomy, surgery and midwifery in the city, to be joined later by his brother Allan (for whom see the *DNB* entry). Earlier (c.1778) James Monteath had taught midwifery; see Geyer-Kordesch and Macdonald, *Physicians and Surgeons*, 271–73, 280–81; Duncan, *Memorials*, 261.
89. Wood, 'Jolly Jack Phosphorus', 115; [John Anderson], *A Compend of Experimental Philosophy; Containing Propositions Proved by a Course of Experiments in*

Natural Philosophy, and the General Heads of Lectures which Accompany Them (Glasgow, 1760), 53–56. The copy of Anderson's *Compend* in Glasgow University Library is unfortunately incomplete and therefore does not provide a complete picture of what Anderson discussed in his natural history lectures at this early point in his career.

90. John and William Scruton seem to have maintained a chemistry laboratory toward the end of the period on King Street; Duncan, *Memorials*, 263.
91. Anderson's career and his struggles with his colleagues are admirably summarised in John Butt, *John Anderson's Legacy: The University of Strathclyde and Its Antecedents, 1796–1996* (East Linton, UK, 1996), Ch. 1.
92. James Millar and Robert Cleghorn later joined this small group of Whigs; on Cleghorn, see below, Ch. 9.
93. Several of Anderson's clerical friends were named in his will to posts in the Faculty of Theology in 'Anderson's University'.
94. Muir, *John Anderson*, 131.
95. William Robert Scott, *Adam Smith as Student and Professor* (1937; reprinted New York, 1965), 149; *Regulations concerning the Premiums given to the Masters of Arts by Mr. Anderson, Professor of Natural Philosophy in the University of Glasgow* ([Glasgow], [1770]), a copy of which is in Glasgow University Archives, GUA 27330; 'Laws of the Literary Society in Glasgow College', Glasgow University Library, MS Murray 505, 52; John Anderson to John Walker, 3 May 1792, Edinburgh University Library, MS La.III.352.1.
96. Muir, *John Anderson*, 142. Anderson named the Rev. Angus McIntosh of the Ingram Street Gaelic Chapel to teach in Gaelic. On this chapel, see Withers, *Urban Highlanders*, 165–67.
97. On Wright and Balmanno, see: Duncan, *Memorials*, 259–60, 268; Geyer-Kordesch and Macdonald, *Physicians and Surgeons*, 202–3, 342.
98. E. W. D. Tennant, *A Short Account of the Tennant Companies, 1797–1922* (London, 1922), 4.
99. Duncan, *Memorials*, 263.
100. Cowan's family had breweries in Glasgow: W. Innes Addison (ed.), *The Matriculation Albums of the University of Glasgow From 1728 to 1858* (Glasgow, 1913), 134; Duncan, *Memorials*, 265.
101. Muir, *John Anderson*, 134. Cross practised in Hamilton.
102. Murray, *Memories*, 115; James Cleland, *Enumeration of the Inhabitants of the City of Glasgow and the County of Lanark, for the Government Census of 1831*, 2nd edn (Glasgow, 1832), 104; David Bryden, *Scottish Scientific Instrument Makers, 1600–1900* (Edinburgh, 1972), 46.
103. The original partners of the Hurlet Alum Works included Wilson, Charles Macintosh, James Knox and Major John Finlay. The latter was Inspector of the Powder Mills and Major Commanding at the Faversham Arsenal. He was for a time secretary to the 3rd Duke of Richmond, Master of the Ordnance, to whom Anderson addressed his memoranda concerning his improved field cannon. All but Knox had belonged to the Chemical Society discussed below, p. 107; see George Macintosh, *Biographical Memoir of the Late Charles Macintosh, F.R.S. of Campsie and Dunchattan* (Glasgow, 1847), 6, 53n; Hardie, 'Macintoshes', 180–81.
104. On Ure, see the *DNB*.
105. Murray, *Memories*, 263.
106. The minute books from the meetings of the Managers and Trustees of the Institution survive in Strathclyde University Archives; see especially MSS B/1/1–2. The early years of Anderson's Institution are discussed in Butt, *John*

Anderson's Legacy, Ch. 2. Geyer-Kordesch and Macdonald observe that 'with Anderson's College beginning to operate in 1799/1800, in particular with the appointment of John Burns to teach anatomy, surgery and midwifery, the first serious challenge to university hegemony in teaching requirements for the *surgical* licence [from the Faculty of Physicians and Surgeons] was mounted'; Geyer-Kordesch and Macdonald, *Physicians and Surgeons*, 239.

107. Anderson's collection of instruments is described by the Institution's first Professor of Natural Philosophy, Thomas Garnett, in his *Observations on a Tour through the Highlands and part of the Western Isles of Scotland*, 2 vols., new edn. (London, 1811), 2:195–96; Garnett notes (p. 196) that the Trustees of the Institution had considerably enlarged the stock of chemical apparatus. He further remarks that a public subscription had to be established because 'the funds left by Professor Anderson were deemed inadequate to carry on the institution with confidence' (p. 194). Catalogues of Anderson's library, as well as lists of his instruments and natural history collections, are to be found in the Andersonian Library, University of Strathclyde, MSS 46 and 48. MS 46 contains lists of books, instruments and 'fossils' dating from around 1799. See also *A Catalogue of Minerals in the Museum of Anderson's Institution, Glasgow* (Glasgow, 1798). In the years following Anderson's death, the University had to invest heavily in the improvement of its own collection of instruments for the natural philosophy class, for it had fallen into serious disrepair. A committee consisting of Patrick Wilson, James Jeffray and James Millar reported in October 1797 that 'before some Intelligent Artist or workman, well acquainted with such Apparatus, should go over all the parts and clean them, and put them severally into good working order, it might be hazardous in any Person however well prepared in other respects to undertake a Course of lectures with this Apparatus, this present Session'; 'Report of the Committee on the Experimental Apparatus', Glasgow University Archives, MS GUA 5204, fol. 1v.

108. For a sense of the reactionary political climate in Glasgow, see the anonymous pamphlet *Asmodeus; or, Strictures on the Glasgow Democrats* (Glasgow, 1793), esp. p. 24, where Anderson is mentioned. The atmosphere of the 1790s is evoked in Henry W. Meikle, *Scotland and the French Revolution* (Glasgow, 1912) and Christine Bewley, *Muir of Huntershill* (Oxford, 1981). Even more evocative is Henry Cockburn, *Memorials of His Time*, ed. Karl Miller (Chicago and London, 1974), 40–41, 73–76; *idem, An Examination of the Trials for Sedition in Scotland*, 2 vols. (1888; New York, 1970).

109. Appointed as Chancellor in 1781, Montrose was also an amateur chemist interested in agricultural improvement and a friend of Charles Macintosh: Macintosh, *Biographical Memoir*, 75.

110. There is no complete listing of the members of Simson's Friday and Anderston Clubs. In 1743 Simson seemingly met with: Robert Dick I (Professor of Natural Philosophy); Hercules Lindsay (Lecturer in Law and later Professor of Law); James Moor (Librarian and later Professor of Greek); James Purdie (Rector of Glasgow Grammar School); Matthew Stewart (Divinity student and later Edinburgh Professor of Mathematics); Alexander Carlyle (Divinity student and later minister of Inveresk). In 1752 we know that Simson met with Thomas Melvill (tutor); John Millar (later Professor of Law); Adam Smith (Professor of Moral Philosophy); Joseph Black (later Professor of Botany and Anatomy, Medicine and Lecturer in Chemistry); William Cullen (Professor of Medicine); and James Watt (later instrument maker to the University). At a later date Simson's circle included: Thomas Hamilton (Professor of Botany and Anatomy); George Ross

(Professor of Humanity); Andrew Foulis (printer); Robert Foulis (printer); William Ruat (Professor of Ecclesiastical History); John Anderson (Professor of Natural Philosophy). Information on Simson's clubs is to be found in: Alexander Carlyle, *Anecdotes and Characters of the Times*, ed. James Kinsley (Oxford, 1973), 41–42; Strang, *Glasgow and Its Clubs*, 19–33. Most of these men had three things in common: competence in Greek, in Latin and interests in mathematics or natural philosophy. Although Alexander Carlyle suggested that Simson did not discuss mathematical problems with his dining companions, this is contradicted by an anecdote about a conversation between Simson and Alexander Wilson retold in James Millar to the Earl of Buchan, 9 February 1809, James Marshall and Marie-Louise Osborn Collection, Beinecke Rare Book and Manuscript Library, Yale University, Osborn Files 10274.

111. Melvill had at this point already published 'An Essay on Dreaming; or, the State of Mind in Sleep', *The Scots Magazine* 11 (1749), 442–46. By 1753 he was corresponding with the Astronomer Royal, James Bradley. See 'A Letter from Mr. T. Melvil to the Rev. James Bradley, D.D. F.R.S. With a Discourse Concerning the Cause of the Different Refrangibility of the Rays of Light', *Philosophical Transactions* 48 (1753), 261–70; and Melvill to Bradley, 2 June 1753, in S. P. Rigaud (ed.), *Miscellaneous Works and Correspondence of the Rev. James Bradley,. D.D. F.R.S. Astronomer Royal, Savilian Professor of Astronomy in the University of Oxford* (Oxford, 1832), 483–87. Melvill also remained in touch with William Cullen; see his letters to Cullen, 18 July 1750 and 21 February 1752, in Thomson, *Life of Cullen*, 1:587–89. Perhaps his most important paper, which was read to the Philosophical Society of Edinburgh in early 1752, was his 'Observations on Light and Colours', *Essays and Observations, Physical and Literary* 2 (1756), 12–90.

112. John Anderson to Gilbert Lang, 13 February 1750 and Anderson to Lang, 27 December 1750, Strathclyde University Archives, MSS A/2/1 and A/3. The members of the Considerable Club included: John Cunninghame, President (later minister of Dalmellington and Monkton); John Anderson (later Professor of Oriental Languages and Natural Philosophy); Gilbert Lang (later minister of Largs); Frederick Hamilton (son of Lord Archibald Hamilton); John Leslie (tutor and later Professor of Greek, King's College, Aberdeen); Henry Maxwell (MA 1754); William Patoun (Keeper of Glasgow University Library, 1747–50, and later MD); Frederick Greer (identified as a preacher); Andrew McVey (later minister of Dreghorn); Alexander Grahame (preacher); William Lindsay (preacher); James Frame (later minister of Dalziel and Alloa); John Melvill; Thomas Melvill; James Wodrow (Keeper of Glasgow University Library, 1750–55, and later minister of Dunlop and Stevenston); 'Hunter from Galloway' (unidentified); Daniel Noble (unidentified); and Andrew Ross (later minister of Loudon and Inch).

113. Alexander Carlyle mentions joining a theological as well as a literary club in Glasgow in 1743: Carlyle, *Anecdotes and Characters*, 40. Archibald Arthur was active in a theological society at the University, most probably in the 1760s, and some of his discourses to that society were subsequently published in his posthumous *Discourses on Theological and Literary Subjects*, ed. William Richardson (Glasgow, 1803). The first of these discourses deals with the design argument.

114. On the general British context of the Glasgow Literary Society, see Peter Clark, *British Clubs and Societies, 1580–1800: The Origins of an Associational World* (Oxford, 2000).

115. Roughly two-thirds of the minutes of the Literary Society have not survived. We lack minutes for: 10 January 1752 to May 1764, 22 November 1771 to January

1773, 11 February 1773 to May 1776, and May 1779 to April/May 1794. The minutes which survive are contained in the manuscript in the Murray Collection of Glasgow University Library cited above in note 95 (which is a transcript of the original manuscript currently held in the Library of the Royal Faculty of Procurators of Glasgow) and in 'Minutes of the College Literary Society, 1790[sic]-1799', Glasgow University Library, MS Gen.4. The extant minutes cover the periods: 1 November 1764 to 22 November 1771, 4 February 1773, 1 November 1776 to 14 May 1779, and 7 November 1794 to 26 April 1800. Further information about the Society (derived from notes taken by William Richardson from minute books which are now lost) is contained in Duncan, *Notices and Documents*, 132–35.

116. Sher, 'Commerce, Religion and Enlightenment', 336–38. Sher's analysis of the Society is akin to that found in D. D. McElroy, *Scotland's Age of Improvement: A Survey of Eighteenth-Century Literary Clubs and Societies* (Pullman, WA, 1969), insofar as they both emphasise the 'literary' character of the Society at the expense of the scientific and medical aspects of the Society's proceedings.

117. Stevenson did discuss questions on the American colonies and on the state of the fine arts: 'Laws of the Literary Society', 17, 33, 53.

118. 'Laws of the Literary Society', 26, 30, 65. Irvine's discourse 'On the Effects of Heat and Cold on Animal Bodies' was later published by his son in William Irvine I and William Irvine II, *Essays Chiefly on Chemical Subjects* (London, 1805), 191–205. The topic of animal heat was taken up by Irvine's student Adair Crawford in *Experiments and Observations on Animal Heat, and the Inflammation of Combustible Bodies; Being an Attempt to Resolve these Phenomena into a General Law of Nature*, 2nd edn (London, 1788). This work was first published in 1779 and was based on experiments carried out in Glasgow in 1777.

119. 'Minutes of the College Literary Society', 34. For the texts of these discourses and commentary, see Paul Wood (ed.), *Thomas Reid on the Animate Creation: Papers Relating to the Life Sciences* (Edinburgh, 1995), 28–30, 48–52, 103–24, 217–41.

120. 'Minutes of the College Literary Society', 23; James Kendall, 'Thomas Charles Hope, M. D.', in Kent, *Eighteenth-Century Lectureship*, 162. By 1795 Beddoes had published *Observations on the Nature and Cure of Calculus, Sea Scurvy, Consumption, Catarrh and Fever: Together with Conjectures upon Several other Subjects of Physiology and Pathology* (1793), *A Letter to Erasmus Darwin, M.D., on a New Method of Treating Pulmonary Consumption and Some Other Diseases Hitherto Found Incurable* (1793) and *Considerations on the Medicinal Use of Factitious Airs and on the Manner of Obtaining them in Large Quantities* (1794), the latter co-authored with James Watt. On Beddoes' pneumatic medicine and the Beddoes-Watt collaboration, see Dorothy A. Stansfield, *Thomas Beddoes M.D., 1760–1808: Chemist, Physician, Democrat* (Dordrecht, 1984), 145–58. Watt presented one of his medical pneumatic devices to the Glasgow Royal Infirmary, although it is unclear exactly when he did so. Watt's correspondence with Robert Cleghorn indicates that Cleghorn was using such a device in January 1796, and he probably did so at the Infirmary. Significantly, a letter from Thomas Garnett to Watt dated 24 January 1797 shows that Watt had sent one of the pneumatical devices for use at Anderson's Institution. We thank Fiona Tait, Archivist at the Archives of Soho Project, Birmingham City Archives, for this information.

121. 'Minutes of the College Literary Society', 26, 27, 47, 48, 67, 73, 74, 94, 95, 96, 118, 119, 137, 141, 142, 143.

122. In addition to the works cited above in note 61, see the useful overview of the tradition in Arthur L. Donovan, 'William Cullen and the Research Tradition of

Eighteenth-Century Scottish Chemistry', in Campbell and Skinner, *Scottish Enlightenment*, 98–114.
123. Thomson, *Life of Cullen*, 1:53–54; Black, *Lectures*, 1:xxxviii, xxxix.
124. Thomas Reid first learned of Black's theory of latent heat from William Ogilvie, who was then in Glasgow; see Reid to [Ogilvie], [1763], Aberdeen University Library, MS 2131/3/III/13. Reid's engagement with chemistry is detailed in Paul Wood, 'Thomas Reid, Natural Philosopher: A Study of Science and Philosophy in the Scottish Enlightenment' (unpublished Ph.D. thesis, University of Leeds, 1984), 270–84. Ogilvie (who later taught at King's College) and Trail (who became Professor of Mathematics at Marischal College in 1766) both helped to spread Black's theories to Aberdeen.
125. Robison to Watt [October 1800], in Eric Robinson and Douglas McKie (eds.), *Partners in Science: Letters of James Watt and Joseph Black* (London, 1970), 359–60.
126. 'Laws of the Literary Society', 39, 42, 59. Irvine's discourses 'On Heat Produced by Mixture' and 'On the Quantity of Matter of Bodies' were published in Irvine and Irvine, *Essays*, 163–89 and 407–21. In the latter discourse Irvine discusses Black's theory of the negative weight of phlogiston, for which see Carleton E. Perrin, 'Joseph Black and the Absolute Levity of Phlogiston', *Annals of Science* 40 (1983), 109–37.
127. Irvine and Irvine, *Essays*, 207–25, 307–26, 327–43, 345–66, 367–82.
128. Murray, *Memories*, 190. Hope published his analysis of strontium in 'An Account of a Mineral from Strontian, and of a Peculiar Species of Earth which It Contains', *Transactions of the Royal Society of Edinburgh* 4 (1798), 3–39.
129. 'Minutes of the College Literary Society', 19, 42, 89, 113.
130. Kendall, 'Thomas Charles Hope', 157–58; George Thomson, 'Robert Cleghorn, M.D.', in Kent, *Eighteenth-Century Lectureship*, 166–67.
131. Wood, 'Jolly Jack Phosphorus', 115, 118–24; 'Laws of the Literary Society', 64, 74; 'Discourses Of Natural and Artificial Systems in Natural History and on The Varieties in the Human Kind Read to the Literary Society in Glasgow College by Mr Anderson, MDCCLXXIV', Andersonian Library, University of Strathclyde, MS 9; see also MS 20 for another treatment of this topic. First published in 1765, Gregory's *Comparative View* reached a seventh edition in 1777.
132. 'Minutes of the College Literary Society', 18. In 1797 and 1798 Lockhart Muirhead read discourses based on his travels in France which probably contained a certain amount of natural historical information: 'Minutes of the College Literary Society', 75, 98.
133. 'Laws of the Literary Society', 35. Among Walker's surviving manuscripts is to be found 'A Natural History of the Island of Icolumbkil', Edinburgh University Library, MS La.III.575. A published account of Iona appeared in John Walker, *Essays on Natural History and Rural Economy* (Edinburgh, 1808), 111–217. Walker was given an honorary M.D. by Glasgow in 1765, and was elected a member of the Literary Society before 1764. Walker's Glasgow discourse was largely based on his trip to the Hebrides in 1764, and he incorporated some of his materials in his 'Report on the Hebrides' submitted to the Board of Annexed Estates in 1771, for which see Margaret M. McKay (ed.), *The Rev. Dr. John Walker's Report on the Hebrides of 1764 and 1771* (Edinburgh, 1980). On the importance of the geographical and natural historical survey in eighteenth-century Scotland, see Charles W. J. Withers, 'Toward a Historical Geography of Enlightenment in Scotland', in Wood, *Scottish Enlightenment*, 77–79.
134. 'Minutes of the College Literary Society', 121, 145; Irvine and Irvine, *Essays*, 273–88, 289–306, 383–401.

135. 'Laws of the Literary Society', 27, 34. Earlier, Trail had discussed using the transit of Venus of 1761 to measure the parallax of the sun in the Aberdeen Philosophical Society; see H. Lewis Ulman (ed.), *The Minutes of the Aberdeen Philosophical Society, 1758–1773* (Aberdeen, 1990), 189. On geology as a branch of natural history in the Scottish context, see James Ritchie, 'Natural History and the Emergence of Geology in the Scottish Universities', *Transactions of the Edinburgh Geological Society* 15 (1952), 297–316.
136. Murray, *Memories*, 190, n.1. Murray seems to have misread 'De Sae' for 'De Luc', but if so, this complicates the dating of Hope's discourse, for De Luc's criticisms of Hutton only began to appear in the *Monthly Review* in 1790. Another Glasgow man of science interested in Hutton's system was James Millar, who briefly discussed Hutton when lecturing on geography; see Millar, 'Notes of Lectures on Geography', 154–56.
137. Charles W. J. Withers, 'William Cullen's Agricultural Lectures and Writings and the Development of Agricultural Science in Eighteenth-Century Scotland', *The Agricultural History Review* 37 (1989), 144–56, esp. pp. 149–50, 151–52.
138. 'Laws of the Literary Society', 39, 75; Irvine and Irvine, *Essays*, 255–70. Irvine may have been encouraged to turn his mind to the subject of his discourse by Lord Kames, who in 1779 issued a second edition of *The Gentleman Farmer*, which was first published in 1776. Kames had earlier corresponded with Thomas Reid on agricultural matters; see Reid to Kames, 1 October 1775, National Archives of Scotland, MS GD24/1/569/13–14.
139. 'Minutes of the Literary Society', 17, 51, 52, 69, 138.
140. For an illuminating discussion of barometers and the place of meteorology in Enlightenment culture, see Jan Golinski, 'Barometers of Change: Meteorological Instruments as Machines of Enlightenment', in William Clark, Jan Golinski and Simon Schaffer (eds.), *The Sciences in Enlightened Europe* (Chicago and London, 1999), 69–93. On eighteenth-century meteorology, see also Theodore S. Feldman, 'Late Enlightenment Meteorology', in Tore Frängsmyr, J. L. Heilbron and Robin E. Rider (eds.), *The Quantifying Spirit in the Eighteenth Century* (Berkeley, Los Angeles and Oxford, 1990), 143–77.
141. Patrick Wilson, 'Biographical Account of Alexander Wilson, M.D. Late Professor of Practical Astronomy in Glasgow', *The Edinburgh Journal of Science* 10 (1829), 1–17, esp. pp. 5–8; Alexander Wilson, 'A Method of Raising Artificial Kites to any Great Heights; and a Proposal of Using Wire in Place of a Common Line; with a View of Rendering Them More Fit for Experiments upon the Electricity of the Atmosphere', Royal Society of London, Letters and Papers VII.29.
142. Alexander Wilson, 'An Account of the Remarkable Cold Observed at *Glasgow*, in the Month of *January, 1768*; in a Letter from Mr. *Alexander Wilson*, Professor of Astronomy at *Glasgow*, to the Rev. Mr. *Nevil Maskeline*, B.D. F.R.S. and Astronomer Royal', *Philosophical Transactions* 61 (1771–72), 326–31; Patrick Wilson, 'An Account of a Most Extraordinary Degree of Cold at Glasgow in January Last; together with some New Experiments and Observations on the Comparative Temperature of Hoar-frost and the Air near to It, Made at the *Macfarlane* Observatory Belonging to College. In a Letter from *Patrick Wilson*, M.A. to the Rev. *Nevil* Maskelyne, D.D.F.R.S. and Astronomer Royal', *Philosophical Transactions* 70 (1780), 451–73; idem, 'Farther Experiments on Cold, Made at the *Macfarlane* Observatory Belonging to *Glasgow College*. In a Letter from *Patrick Wilson*, M.A. to the Rev. *Nevil Maskelyne, D.D.F.R.S.* and Astronomer Royal', *Philosophical Transactions* 71 (1781), 386–94; Wilson to Henry Maty, 8 November 1781, Royal Society of London, MS RS.1.47. In Patrick Wilson's 1780 paper he

mentions taking measurements with John Anderson, William Irvine and Adair Crawford, and says that he discussed their observations with Thomas Reid.
143. Anderson's work is detailed in C. G. Wood, 'John Anderson's Rain Gauge', *The Philosophical Journal: Transactions of the Royal Philosophical Society of Glasgow* 5 (1968), 138–50.
144. 'Of Rain Gauges. A Discourse Read to the Literary Society in Glasgow College November 1780', Andersonian Library, University of Strathclyde, MS 15^2.
145. 'Laws of the Literary Society', 86; Irvine and Irvine, *Essays*, 227–52. Hutton outlined his theory before the Royal Society of Edinburgh on 2 February and 12 April 1784.
146. 'Minutes of the Literary Society', 91, 112, 139, 150. Rankine appears to have discussed meteorology with Anderson, for in the codicil to Anderson's will he asks his executors to 'get from the Revd. Mr. Rankine my Essay on Rain Gauges and Rainy Climates'; Muir, *John Anderson*, 157.
147. Robert Dick to Robert Simson, 8 August 1754, 27 August 1754, 6 September 1754 and 14 September 1754, Glasgow University Archives, MSS GUA 26303, 26305a/b, 26306 and 26308. An inventory of Macfarlane's instruments (in French) survives in Glasgow University Archives, MS GUA 27403. A protracted correspondence about instruments between Simson and William Ruat dating from the period 1754 to 1756 also survives in Glasgow University Archives. On the foundation of the Macfarlane Observatory, see Coutts, *History*, 229, and Murray, *Memories*, 260–62. For a helpful survey of astronomical research in eighteenth-century Glasgow, see David M. Gavine, 'Astronomy in Scotland, 1745–1900', 2 vols. (unpublished Ph.D. thesis, Open University, 1981), 1:55–65.
148. As noted above in Chapter 2, the nature of Watt's appointment at the University is unclear. His arrangement with the College began in 1757, but we do not know when it terminated. On the importance of Watt's shop, see, for example, John Robison's account of his first meeting Watt in Robinson and McKie, *Partners in Science*, 256–57. Watt initially set up a workshop in the College, but by the end of 1759 he also had a shop in the town; see D. J. Bryden, 'James Watt, Merchant: The Glasgow Years, 1754–1774', in Denis Smith (ed.), *Perceptions of Great Engineers: Fact and Fantasy* (London, 1994), 9–21, esp. pp. 10–11.
149. This group included James Williamson, Thomas Reid, William Irvine, John Anderson and Wilson's son Patrick; see Alexander Wilson, 'Observations on the Transit of *Venus* Over the Sun, Contained in a Letter to the Reverend *Nevil Maskeleyne*, Astronomer Royal, from *Dr. Alexander Wilson*, Professor of Astronomy in the University of *Glasgow*', *Philosophical Transactions* 59 (1769), 333–38.
150. 'Laws of the Literary Society', 33; Alexander Wilson, 'Observations on the *Solar Spots*', *Philosophical Transactions* 64 (1774), 1–30. Wilson mentions in his article (p. 6) that he had made his initial observations of sunspots in November 1769. His discourse to the Society must therefore have contained his first thoughts on their nature.
151. 'Laws of the Literary Society', 36, 45–46. Patrick Wilson later published 'An Experiment Proposed for Determining by the Aberration of the Fixed Stars, Whether the Rays of Light, in Pervading Different Media, Change their Velocity According to the Law which Results from Sir Isaac Newton's Ideas Concerning the Cause of Refraction; and for Ascertaining their Velocity in Every Medium whose Refractive Density is Known', *Philosophical Transactions* 72 (1782), 58–70. Wilson's interest in aberration was shared by Thomas Reid; see Wood, 'Thomas Reid', 248–56, and the references cited there.
152. 'Laws of the Literary Society', 12–13, 16; 'Minutes of the Literary Society', 116.
153. Wilson, 'Biographical Account', 8–9. As his son notes, Alexander Wilson also

designed an improved thermometer in the late 1750s which was much in demand by Scottish men of science. Wilson first began making thermometers in the mid-1730s, when he collaborated with George Martine in St. Andrews.

154. 'Laws of the Literary Society', 13. What appears to be a copy of Walker's paper survives among the manuscripts of John Anderson; see 'Experiments Upon the mensuration of heights by the Barometer', Andersonian Library, University of Strathclyde, MS 26. For Anderson's account of Maskelyne's expedition, see 'Question For the Literary Society in Glasgow College March 17th 1775. For what reason did the Royal Society send Mr Maskeline in Summer last to Sheehallion in Perthshire?', Andersonian Library, University of Strathclyde, MS 18. While he was up Schiehallion, Maskelyne was visited by Anderson, Thomas Reid and the Wilsons; Derek Howse, *Nevil Maskelyne: The Seaman's Astronomer* (Cambridge, 1989), 137.
155. 'Minutes of the Literary Society', 21, 140.
156. 'Laws of the Literary Society', 17.
157. Aberdeen University Library, MS 3061/11. See also Paul Wood, 'Reid, Parallel Lines and the Geometry of Visibles', *Reid Studies: An International Review of Scottish Philosophy* 2 (1998), 27–41.
158. 'Minutes of the Literary Society', 48, 71. Millar read his discourse on Simson in March 1797 and may have been prompted to write about his predecessor by the appearance of John Robison's account of Simson in the *Encyclopedia Britannica* cited above in note 55. However, the wording of the title of Millar's discourse suggests that he may have been commenting on a draft of Traill's biography, which was not published until 1812.
159. Michael Barfoot, 'James Gregory (1753–1821) and Scottish Scientific Metaphysics, 1750–1800' (unpublished Ph.D. thesis, University of Edinburgh, 1983). This thesis is a brilliant exploration of the interconnections of medicine, natural philosophy and metaphysics in the second half of the eighteenth century.
160. 'Minutes of the Literary Society', 24. Couper was commenting on Thomas Cooper's *Tracts, Ethical, Theological and Political* (Warrington, 1787), in which Cooper outlined his materialist metaphysics.
161. 'Minutes of the Literary Society', 13.
162. Duncan, *Notes and Documents*, 132; 'Laws of the Literary Society', 24, 26, 50–51, 84–85; 'Minutes of the Literary Society', 33, 99, 125, 144; 'A Dissertation Read before the Literary Society at Glasgow. 1 April 1763', Glasgow University Library, MS.Gen.915; John Anderson, 'Of War as a Scientific Art', and 'Of the Moisture in Houses that are situated on prominent Rocks', Andersonian Library, University of Strathclyde, MSS 23 and 40. Arthur's discourse 'On the Importance of Natural Philosophy' from November 1795 was later published in his *Discourses*, 407–40.
163. The existence of this Society was first noted in Murray, *Memories*, 188. See also J. A. V. Butler, 'John Maclean, Charles Macintosh and an Early Chemical Society in Glasgow', *Journal of Chemical Education* 19 (1941), 43–44. The two basic sources for information on it are: Macintosh, *Biographical Memoir*, 6; John Maclean, *A Memoir of John Maclean, M.D., the First Professor of Chemistry in the College of New Jersey*, 2nd edn ([Princeton], 1885), 8–10, 13. The Chemical Society was apparently founded by William Irvine around 1785 (or perhaps slightly earlier), and it may be that it did not last much beyond his death in 1787. The members included: Mr. Archer; James Candlish (or M'Candlish); William Couper (surgeon); Adair Crawford; Stewart Cruikshanks (matriculated 1781); Major John Finlay (Royal Engineers); Charles Macintosh; John Maclean (MD); Joseph Monroe (matriculated

1786); Dr. Alexander Tilloch or Tulloch (printer); John Wilson of Hurlet. Some of Charles Macintosh's essays read before it are summarised in Macintosh, *Biographical Memoir*, 7–9. The titles are as follows: 'On the Application of the Blue Colouring Matter of Vegetable Bodies'; 'On Alcohol' (dated 1787); 'On Crystallization'; 'On Alum'. Maclean's contributions are likewise noted in Maclean, *Memoir*, 9–10. Maclean's son states that his father read at least seven papers to the Society, and identifies the topics of three: on respiration, fermentation and alkalis.

164. James Cleland, *Annals of Glasgow, Comprising an Account of the Public Buildings, Charities and the Rise and Progress of the City*, 2 vols. (Glasgow, 1816), 2:146, 457; Strang, *Glasgow and Its Clubs*, 378. According to Cleland, the agricultural society sponsored 'an annual plowing match, and [gave] premiums to three of the competitors whose specimens [were] adjudged the best', while Strang says that the society united 'the men of science with the working mechanic and chemist, [and] formed, ere long, a most effective nucleus for mutual encouragement and advancement'.

165. Strang, *Glasgow and Its Clubs*, 293–94, 303–4; Murray, *Memories*, 520; Duncan, *Memorials*, 187–89; Jacqueline Jenkinson, *Scottish Medical Societies, 1731–1939: Their History and Records* (Edinburgh, 1993), 159, 160–61, 213–14.

166. Jenkinson, *Scottish Medical Societies*, 129–30. See also the 'Minute Book of the Chemical Society of Glasgow, 1800–1', Strathclyde University Archives, MS T/MIN/2.

167. Murray, *Memories*, 264–66; Gavine, 'Astronomy in Scotland', 1:64–65, 2:315–20.

168. Cleland, *Annals of Glasgow*, 2:448; see also Thomas Atkinson, *Sketch of the Origin and Progress of the Literary and Commercial Society of Glasgow* (Glasgow, 1831); Anon., *List of Essays Read by the Members of the Literary and Commercial Society of Glasgow* (Glasgow, 1831). The published list, which was probably compiled by Thomas Atkinson, covers the period 27 November 1806 to 22 April 1830. This list shows that the Society occasionally heard papers on scientific and medical topics. Summaries of essays which Charles Mackintosh may have read to the Society are provided in Mackintosh, *Biographical Memoir*, 9–16, although one of these papers dates from 1792.

169. Coutts, *History*, 501; Geyer-Kordesch and Macdonald, *Physicians and Surgeons*, 271, 274–92. In fact, in 1834 two rival lying-in hospitals were founded in Glasgow by the University and the Faculty of Physicians and Surgeons, who were then locked in a power struggle.

170. M. S. Buchanan, *History of the Glasgow Royal Infirmary, from Its Commencement in 1787, to the Present Time* (Glasgow, Edinburgh and London, 1832); Jacqueline Jenkinson, Michael Moss and Iain Russell, *The Royal: The History of the Glasgow Royal Infirmary, 1794–1994* (Glasgow, 1994), Ch. 1; Geyer-Kordesch and Macdonald, 318–37.

171. Jonathan Andrews and Iain Smith (eds.), *'Let There be Light Again': A History of Gartnell Royal Hospital from Its Beginnings to the Present Day* (Glasgow, 1993), esp. Ch. 1.

172. Cleland, *Annals of Glasgow*, 2:435–47.

173. Boney, *Lost Gardens*, 237–64.

174. John M. Leighton, *Select Views of Glasgow and Its Environs* (Glasgow, 1828), 82 (we thank Fiona Macdonald for this reference).

175. Jenkinson, Moss and Russell, *The Royal*, 1–2, 5; Andrews and Smith, *History of the Gartnell Royal Hospital*, 2, 4, 12. Another Glasgow professor who was extremely active in the foundation of the Infirmary was Thomas Reid; see Wood, 'Thomas Reid', 188–89.

176. There may be no better guides to these developments than: Henry Hamilton, *The Industrial Revolution in Scotland*, 2nd edn (London, 1966); Archibald Clow and Nan L. Clow, *The Chemical Revolution: A Contribution to Social Technology* (1952; Freeport, NY, 1970); R. H. Campbell, 'The Making of the Industrial City', in Jackson and Devine, *Glasgow*, 184–213.
177. For such connections, see, for example, Macintosh, *Biographical Memoir*, 5–6, 24, 27, 28; Robinson and McKie, *Partners in Science*; A. E. Musson and Eric Robinson, *Science and Technology in the Industrial Revolution* (Manchester and Toronto, 1969), esp. Chs. 7 and 8. On the growing links between Glasgow and the industrial towns of the north of England see Hamilton, *Economic History*, 270–79; John Money, *Experience and Identity: Birmingham and the West Midlands, 1760–1800* (Montreal, 1977), 35–47.
178. On the decline of the Literary Class within the Royal Society of Edinburgh, see especially Steven Shapin, 'The Royal Society of Edinburgh: A Study of the Social Context of Hanoverian Science' (unpublished Ph.D. thesis, University of Pennsylvania, 1971), 224–35. Shapin's thesis also contains much useful material on the Edinburgh scientific community at the turn of the nineteenth century.
179. Clow and Clow, *Chemical Revolution*, 168–98.
180. On this point, see Roger L. Emerson, 'The Enlightenment and Social Structures', in David Williams and Paul Fritz (eds.), *City and Society in the 18th Century* (Toronto, 1973), 99–129.

Table 4.1: The Growth of Glasgow and Its Resident Scientific Community

Dates	Interval totals	Number in selected years	*Dictionary of National Biography*	*Dictionary of Scientific Biography*	Glasgow's population*
1690		5	2		*c.* 15,000
1709	12	9	2		*c.* 13,000
1710		11	2		*c.* 12,000
1729	20	14	2	1	*c.* 15,000
1730		14	3	1	
1749	38	35	6	5	17,000
1750		38	6	4	*c.* 22,000
1769	58	34	6	1	*c.* 27,000
1770		37	6	1	*c.* 33,000
1789	96	57	10	1	42,000
1790		64	11	3	*c.* 65,000
1802	104	66	12	0	*c.* 82,000+
Total	168		42	9	

* Sources: James Cleland, *Annals of Glasgow*, 2 vols. (Glasgow, 1816), 2:516; James Playfair, *A Geographical and Statistical Description of Scotland*, 2 vols. (Edinburgh, 1819), 2:224; George MacGregor, *The History of Glasgow* (Glasgow, 1881), 315; J. Cunison and J. B. S. Gilfillan (eds.), *The Third Statistical Account of Glasgow* (Glasgow, 1958), 58; Henry Hamilton, *An Economic History of Scotland in the Eighteenth Century* (Oxford, 1963), 18; T. M. Devine and Gordon Jackson (eds.), *Glasgow, Volume I: Beginnings to 1830* (Manchester and New York, 1995), 10; T. M. Devine, *The Scottish Nation: A History, 1700–2000* (New York, 1999), xx.

Table 4.2: Glasgow's Men of Science (by birth cohorts)

Key: + = *DNB*; * = *DSB*; b = by; a = after; GLS = member of the Glasgow Literary Society; APS = member of the Aberdeen Philosophical Society; PSE = member of the Edinburgh Philosophical Society; FRSE = Fellow of the Royal Society of Edinburgh; FRS = Fellow of the Royal Society of London; FPSG = Fellow of the Faculty of Physicians and Surgeons of Glasgow. We have distinguished between positions in Anderson's University (which were designated only) and those in Anderson's Institution (which were actually held).

Born before 1650 (3)

1.	Matthew Brisbane	M.D., Dean of Faculty and Rector (Glasgow), Town Physician, FPSG	c.1638–1699
2.	James Corss(e)	Mathematics Teacher	?–1679/80
3.	George Sinclair +	Regent and Professor of Mathematics (Glasgow), Engineer, Extramural Lecturer and Mathematics Teacher (Edinburgh)	c.1630–1696

Born 1650–1669 (5)

4.	William Dunlop +	Tutor, Minister, Major, Principal of Glasgow University, virtuoso	1654–1700
5.	William Jameson +	Lecturer on Civil and Ecclesiastical History (Glasgow), *virtuoso*	c.1660–1720
6.	John Law	Regent (Glasgow)	c.1660–1718
7.	John Marshall	Surgeon, Lecturer on Botany (Glasgow), Botanist, FPSG	c.1655–1719
8.	Robert Sinclair	M.D., Professor of Mathematics, Lecturer on Hebrew, Extraordinary Professor of Medicine (Glasgow)	b1665– a1716

Born 1670–1689 (8)

9.	Thomas Brisbane	M.D., Professor of Anatomy and Botany (Glasgow), FPSG	1684–1742
10.	Archibald Campbell, 3rd Duke of Argyll +	Honorary M.D., Mathematician, Chemist, Improver	1682–1761
11.	Gershom Carmichael	Regent (St. Andrews), Regent and Professor of Moral Philosophy (Glasgow)	1672–1729
12.	Robert Houston	M.D., Surgeon, FPSG, FRS	b1690–1734
13.	John Johnstoun	M.D., Professor of Medicine (Glasgow), FPSG, PSE	1685–1762
14.	Robert Simson +*	Professor of Mathematics (Glasgow), Botanist, GLS	1687–1768
15.	George Thomson	M.D., Physician to Workhouse, FPSG	1674–c.1736
16.	Robert Wodrow +	University Librarian (Glasgow), Minister, *virtuoso*	1679–1734

Born 1690–1709 (7)

17.	Robert Dick I	Tutor, Regent and Professor of Natural Philosophy (Glasgow)	c.1690–1751
18.	Henry Drew	Watch and Mathematical Instrument Maker	c.1700–1763
19.	John Gordon	M.D., Surgeon, Manufacturer, FPSG	b1700–1770
20.	George Montgomerie II	M.D., Extramural Medical Lecturer, Physician (Glasgow Town's Hospital), FPSG	c.1705–1778
21.	John Paisley	Surgeon, Deputy Professor of Anatomy (Glasgow), Town Surgeon, FPSG	c.1698–1740
22.	William Stirling	Surgeon, Manufacturer, FPSG	c.1690–1757
23.	John Wodrow	M.D., Extramural Teacher of Botany, FPSG	1695–1769

Born 1710–1729 (35)

24.	James Makittrick Adair +	M.D., Chemist, Manager of Chemical Works, FRS	1728–1802
25.	John Anderson +	Professor of Oriental Languages and Natural Philosophy (Glasgow), GLS, FRSE, FRS	1726–1796
26.	Joseph Black +*	M.D., Professor of Medicine and Lecturer in Chemistry (Glasgow), Physician (Glasgow Town's Hospital), Professor of Chemistry (Edinburgh), FPSG, GLS, PSE, FRSE	1728–1799
27.	John Brisbane	M.D., Physician (Glasgow Town's Hospital), FPSG (Hon), GLS	b1730–1776
28.	James Buchanan	Professor of Oriental Languages (Glasgow), Botanist, Mathematician, GLS	1727–1761
29.	Patrick Carmichael	M.D., Surgeon, GLS	b1720–1792
30.	John Carrick	Surgeon, Assistant Lecturer in Chemistry, Assistant to Professor of Anatomy (Glasgow), Surgeon (Glasgow Town's Hospital), FPSG	1724–1750
31.	Andrew Craig	Surgeon (Glasgow Town's Hospital), FPSG	b1725–1782
32.	John Crawford	Deputy Professor of Anatomy (Glasgow), Extramural Lecturer in Anatomy and Surgery, Surgeon (Glasgow Town's Hospital), FPSG	b1720–a1755
33.	William Cullen +*	Extramural Lecturer on Chemistry and Professor of Medicine (Glasgow), Physician (Glasgow Town's Hospital), Professor of Chemistry and Medicine (Edinburgh), FPSG, GLS, PSE, FRSE, FRS	1710–1790
34.	Sir John Dalrymple, Bt.	Advocate, Baron of Exchequer, Chemist, Man of Letters, FRSE	1726–1810
35.	Robert Dick II	M.D., Professor of Natural Philosophy (Glasgow), Physician (Glasgow Town's Hospital), FPSG, GLS	c.1722–1757
36.	John Gibson	M.D., Extramural Lecturer on Midwifery, FPSG	1718–a1769
37.	John Glassford	Merchant, Manufacturer	1715–1783
38.	Robert Hamilton	M.D., Professor of Anatomy and Botany (Glasgow), Physician (Glasgow Town's Hospital), FPSG, GLS	1714–1756
39.	Thomas Hamilton	Surgeon, Professor of Anatomy and Botany (Glasgow), FPSG, GLS	1728–1782
40.	Ninian Hill	Surgeon (Glasgow Town's Hospital), FPSG	b1730–a1787
41.	George Jamieson	Instrument Maker	fl.1729–1755
42.	George Jardine	Smith, Watch and Mathematical Instrument Maker	fl.1737–1770
43.	John Love	Surgeon, Deputy Professor of Anatomy (Glasgow), FPSG	fl. 1740
44.	Hector McLean	Surgeon (Glasgow Town's Hospital), FPSG	b1720–1782
45.	Thomas Melvill +*	Tutor, Natural Philosopher, GLS, PSE	c.1726–1753
46.	James Moor +	Tutor, Schoolmaster, Professor of Greek (Glasgow), Mathematician, GLS	1712–1779
47.	John Moore +	M.D., Surgeon (Glasgow Town's Hospital), FPSG	1729–1802
48.	Andrew Morris	M.D., Army Surgeon, Extramural Medical Lecturer, Apothecary (Glasgow Town's Hospital), FPSG	1717–1788
49.	James Muir	Surgeon, Extramural Lecturer on Midwifery, FPSG	b1720–a1787
50.	William Ralston	Surgeon, Physician (Glasgow Town's Hospital), FPSG	b1730–a1769

51.	Thomas Reid +	Minister, Regent (King's College Aberdeen), Professor of Moral Philosophy (Glasgow), Mathematician and Natural Philosopher, GLS, APS, FRSE	1710–1796
52.	James Russel	M.D., Surgeon (Glasgow Town's Hospital)	c.1722–?
53.	Adam Smith +	Professor of Logic and Moral Philosophy (Glasgow), GLS, PSE, FRSE, FRS	1723–1790
54.	Alexander Stevenson	M.D., Professor of Medicine (Glasgow), Physician (Glasgow Town's Hospital), Improver, FPSG, GLS, FRSE	1725–1790
55.	Matthew Stewart +*	Minister, Professor of Mathematics (Edinburgh) FRSE, FRS	1717–1785
56.	Robert Traill	Minister, Professor of Divinity (Glasgow), GLS, APS	1720–1775
57.	James Williamson	Minister, Professor of Mathematics (Glasgow), GLS	c.1725–1795
58.	Alexander Wilson +*	Surgeon, Instrument Maker, Typefounder, Professor of Astronomy (Glasgow), GLS, PSE, FRSE	1714–1786

Born 1730–1749 (23)

59.	Archibald Arthur +	Assistant and Professor of Moral Philosophy (Glasgow), GLS, FRSE	1744–1797
60.	Samuel Charters (Charteris)	Minister, Natural Philosopher, GLS	1742–1825
61.	Archibald Cochrane 9th Earl of Dundonald +	Army, Navy, Manufacturer, Chemist, FRSE	1749–1831
62.	Adair Crawford +	M.D., Professor of Chemistry (Woolwich)	1748–1795
63.	William Dalrymple	Army Officer, Manufacturer	c.1730–1791
64.	Alexander Dunlop	Surgeon (Glasgow Town's Hospital), FPSG, GLS	c.1739–1815
65.	Erskine, David 11th Earl of Buchan +	Man of Science and Letters, Agricultural Improver, GLS, PSE, FRS	1742–1829
66.	Robert Freer	M.D., Surgeon, Army Officer, Professor of Medicine (Glasgow), Professor of Anatomy and Theory of Surgery Designate (Anderson's University), FPSG, GLS, FRSE	1745–1827
67.	John Gardner	Land Surveyor, Instrument Maker	c.1730–c.1791
68.	John Gillies	Tutor, Historian, Man of Letters	1747–1836
69.	James Gordon	Manufacturer, Chemist?	1746–1811
70.	Gilbert Hamilton	Merchant, Manufacturer, GLS	1744–1808
71.	William Irvine I	M.D., Lecturer in Materia Medica and Chemistry (Glasgow), FPSG, GLS, FRSE	1743–1787
72.	George Mackintosh +	Manufacturer	1739–1807
73.	Robert Marshall	M.D., FPSG, GLS	c.1730–1788
74.	Patrick Millar +	Banker, M.P., Manufacturer	1731–1815
75.	William Ogilvie +	Regent and Humanist (King's College Aberdeen), *virtuoso*, GLS, APS, FRSE	1736–1819
76.	James Parlane	Surgeon (Glasgow Town's Hospital), FPSG	b1745–1805
77.	John Robison +	Lecturer in Chemistry (Glasgow), Teacher of Mathematics (Cronstadt), Professor of Natural Philosophy (Edinburgh), GLS, PSE, FRSE	1739–1805
78.	James Watt +	Instrument Maker, Engineer, Chemist, Manufacturer, FRSE, FRS	1736–1819
79.	Charles Wilson(e)	Surgeon (Royal Infirmary), FPSG	b1750–1820
80.	Patrick Wilson	Typefounder, Professor of Astronomy (Glasgow), GLS, FRSE	1743–1811
81.	Peter Wright	M.D., First President of Anderson's Institution, Physician (Glasgow Town's Hospital), Professor of the Institutes of Medicine Designate (Anderson's University), FPSG	c.1740–1819

Born 1750–1769 (38)

82.	James Alexander	Surgeon, Extramural Lecturer in Materia Medica, FPSG	b1770–1817
83.	William Anderson	Surgeon, Professor of Obstetrics (Anderson's University), FPSG	1766–1819
84.	James 'Bubbly' Brown	Instrument Maker	fl.1773–1789
85.	James Brown	Teacher, Minister, Deputy Professor of Mathematics (St. Andrews), Professor of Natural Philosophy (Glasgow), GLS	1763–1836
86.	James Candlish (M'Candlish)	Chemist, Extramural Lecturer on Medicine (Edinburgh)?	c.1760–1806
87.	Robert Cleghorn	M.D., Lecturer in Materia Medica and Chemistry (Glasgow), Physician (Glasgow Town's Hospital), FPSG, GLS, FRSE	1755–1821
88.	James Couper	Minister, Professor of Astronomy (Glasgow), GLS	1752–1836
89.	William Couper (Cooper)	Surgeon, Professor of Chemistry Designate (Anderson's University), Manufacturer, FRSE	1757–1843
90.	Robert Cowan	Surgeon, Chemist, Professor of Botany Designate (Anderson's University), Surgeon (Glasgow Town's Hospital), FPSG	1768–1808
91.	James Crichton	Instrument Maker	fl.1785–1835
92.	John Cross	Surgeon	fl.1790
93.	? Cruickshanks	Medic?, Chemist	fl.1790
94.	John Finlay	Army Officer, Engineer, Manufacturer, Naturalist	c.1770–1802
95.	Thomas Garnett +*	M.D., Professor of Natural Philosophy (Anderson's Institution and Royal Institution, London), FPSG	1766–1802
96.	William Hamilton +	M.D., Professor of Anatomy and Botany (Glasgow), GLS, FRSE	1758–1790
97.	Robert Hastie	Surgeon, Natural Philosopher	?1766 -1827
98.	James Headrick	Natural Philosopher, Chemist, Minister	1757–1841
99.	William Henderson	M.D., FPSG	c.1760–1806
100.	Thomas Charles Hope +*	M.D., Lecturer in Materia Medica and Chemistry and Professor of Medicine (Glasgow), Professor of Chemistry (Edinburgh), FPSG, GLS, FRSE, FRS	1766–1844
101.	James Jeffray	M.D., Professor of Anatomy and Botany (Glasgow), FPSG, GLS	1759–1848
102.	Charles Macintosh +	Manufacturer, Agriculturist, FRS	1766–1843
103.	James Millar	Banker, Professor of Mathematics (Glasgow), GLS	1762–1838
104.	Richard Millar	M.D., Lecturer and Professor of Materia Medica (Glasgow), FPSG, GLS	b1770–1833
105.	James Monteath (Monteith)	Surgeon, Extramural Lecturer on Midwifery, President Anderson's Institution, Professor of the Practice of Medicine Designate (Anderson's University), FPSG	? -1834
106.	Lockhart Muirhead	Lecturer and Professor of Natural History (Glasgow), GLS	1766–1829
107.	William Ramsay	Chemist, FPSG	fl.1790
108.	Alexander Rankin + (Ranken)	Minister, Naturalist, GLS	1755–1827
109.	William Reid	Printer, Bookseller	1764–1831
110.	Peter Rolland	Surgeon, Professor of Practical Surgery Designate (Anderson's University), FPSG	b1770–1825
111.	John Scruton	Surgeon, Chemist, Professor of Clinical Cases Designate (Anderson's University), FPSG	c.1764–1833

in Glasgow, 1690–1802 135

112.	William Scruton	Surgeon, FPSG	c.1760–1803
113.	William Symington +	Engineer	1763–1831
114.	Charles Tennant +	Manufacturer	1768–1838
115.	Alexander Tilloch (Tulloch) +	Publisher, Naturalist	1754–1825
116.	James Tower	Surgeon, Lecturer and Professor of Midwifery (Glasgow)	b1770–1820
117.	David Ure +	Minister, Natural Historian, Professor of Natural History Designate (Anderson's University)	c.1758–1798
118.	John Wilson of Hurlet	Chemist, Merchant, Manufacturer	1761–1847

Born 1770–1799 (17)

119.	John Balmanno	M.D., Botanist, Professor of Materia Medica Designate (Anderson's University)	c.1775–1840
120.	George Birkbeck +	M.D., Professor of Natural Philosophy (Anderson's Institution)	1776–1860
121.	Sir Thomas Brisbane Bt. +	Army Officer, Astronomer, Natural Philosopher, FRSE, FRS	1773–1860
122.	Thomas Brown	M.D., Surgeon, Deputy Professor of Botany (Glasgow), FPSG, GLS	1774–1853
123.	Allan Burns	Surgeon, Physician, Extramural Lecturer on Anatomy	1781–1813
124.	John Burns +	Surgeon, Extramural Lecturer on Anatomy, Professor of Anatomy (Anderson's Institution), Professor of Surgery Glasgow FPSG	1774–1850
125.	William Dunlop	Surgeon, Clinical Lecturer (Glasgow), FPSG	c.1770–1809
126.	William Irvine II	M.D., Chemist, FPSG, FRSE	1776–1811
127.	Thomas Jackson	Deputy Professor of Natural Philosophy (Glasgow), Professor of Natural Philosophy (St.Andrews), GLS, FRSE	1773–1837
128.	John Maclean	M.D., Extramural Lecturer on Chemistry, Professor of Chemistry (Princeton), Professor of Mathematics (College of William and Mary), FPSG	1771–1814
129.	William Meikleham (McIlquham) +	Teacher, Assistant and Professor of Natural Philosophy and Professor of Astronomy (Glasgow), Professor of Physics Designate (Anderson's University), GLS	1771–1846
130.	David Mushet +	Chemist, Manufacturer	1772–1847
131.	John Warroch Parsell (Pursall)	M.D., Assistant to John Anderson	c.1770–1834
132.	Thomas Thomson +*	M.D., Professor of Chemistry (Glasgow), FPSG, FRSE, FRS	1773–1852
133.	Andrew Ure +	M.D., Surgeon, Professor of Natural Philosophy and Lecturer on Chemistry (Anderson's Institution), Consulting Chemist, FPSG, FRS	1777–1857
134.	James Watt	M.D., Surgeon, Minister	1763–1821
135.	Robert Watt +	M.D., Surgeon, Extramural Lecturer on Medicine, Bibliographer, FPSG	1774–1819

Men With Unknown Birthdates (27)

136.	George? Montgomerie I	M.D., Assessor of Medical Degrees (Glasgow), FPSG	fl.1712–c.1743
137.	John Campbell	Surgeon	fl.1715
138.	James Mariot	Instrument Maker	fl.1742
139.	James Hutchison	Assistant to the Professor of Natural Philosophy (Glasgow)	fl.1748
140.	Robert Dobson	Mathematics Teacher	fl.1750s

141.	James Stirling	Geography and Navigation Teacher	fl.1750–60
142.	James Laurie	Instrument Maker	b1760–a1810
143.	James Galbreath	Geography Teacher	fl.1760s
144.	John Holden	Mathematics Teacher	fl.1760s
145.	Alexander Jack	Mathematics Teacher	fl.1760s
146.	William Gordon	Mathematics Teacher	fl.1760s
147.	James McGrigor	Merchant, Manufacturer	fl.1760–90
148.	John McArthur	Geography Teacher	fl.1770s
149.	Robert Irvine	M.D., Botanist	fl.1779
150.	Samuel Birnie	M.D., Industrial Chemist	fl.1780s
151.	[John or Nicholas] Archer	M.D., Surgeon, Chemist	fl.1780s
152.	Ronald Crawford	Chemist	fl.1780s
153.	Joseph? Monroe	Chemist	fl.1780s
154.	Robert Nichol	Geography and Navigation Teacher	fl.1780s
155.	John Risso	Thermometer and Barometer Maker	fl.1783
156.	Murray Osborne	Mathematical Instrument Maker	fl.1783–1817
157.	James Crichton II	Instrument Maker and Locksmith	fl.1789–1835
158.	John Fullerton	Surgeon, Diplomat, Anderson Trustee	?–1849
159.	John Geddes	Chemist, Manager Verreville Glass Works	fl.1790s–a1820
160.	George Cuthbert Gordon	Merchant, Manufacturer	?–a1791
161.	James Knox	Manufacturer	fl.1790s
162.	John Laurie	Merchant, Anderson Trustee	fl.1790s
163.	Robert Lothian	Instrument Maker, Navigation Teacher, Professor of Mathematics Designate (Anderson's University), Mathematics and Geography Lecturer (Anderson's Institution)	fl.1790s
164.	James Spruel	Surgeon?, Anderson Trustee	fl.1790s
165.	James Sym	Instrument Maker	fl.1790s–1846?
166.	William Twaddell	Instrument Maker	fl.1770s–1839?
167.	George Wright	Anderson Trustee	fl.1790s
168.	Robert Wright	Army, Engineer, Anderson Trustee	fl.1790s

Note to Tables 4.3–4.9. Because the individual percentages in Tables 4.3–4.9 have been rounded off, the totals do not always equal 100%.

Note to Tables 4.4–4.5. To take account of the multiple occupations of the fathers and sons (e.g. those who were both professors and clerics, or clerics who were also landowners), we have double counted such persons. The totals of the three columns do not, therefore, directly correspond to the real number of the cohort.

Table 4.3: Places of Origin of Glasgow Men of Science

	Born or presumed born before 1730		Born or presumed born before 1730–1790		Other		Totals	
England	2	3%	2	3%			4	2%
Other	1	2	3	3			4	2
Glasgow	11	17	21	22			32	19
Edinburgh			4	2			4	2
Aberdeen								
St. Andrews	1	2					1	1
Lanarkshire	5	8	4	2			9	5
Renfrewshire	7	11	6	7			13	8
Dunbartonshire	2	3	1	1			3	2
Stirlingshire	2	3	3	4			5	3
Ayrshire	2	3	6	7			8	5
Wigtonshire								
Kirkcudbrightshire								
Dumfriesshire	1	2					1	1
Roxburghshire								
Selkirkshire								
Berwickshire								
Peeblesshire								
Midlothian	1	2	1	1			2	2
Haddington	1	2					1	1
W. Lothian			1	1			1	1
Fife	2	3	4	2			6	4
Perth			3	2			3	2
Clackmannan								
Kinross								
Angus			1	1			1	1
Kincardine	2	3					2	1
Aberdeenshire								
Banff								
Moray								
Elgin	1	2	1	1			2	1
Nairn								
Argyll								
Bute	1	2					1	1
Inverness	1	2					1	1
Ross & Cromarty			1	1			1	1
Sutherland								
Caithness								
Orkney								
Unknown	20	31	29	31	14	100	63	38
Totals	63	100%	91	100%	14	100%	168	100%

Table 4.4: Fathers' known Occupation

Occupations	Born or presumed born before 1730		Born or presumed born after 1730		Totals	
Clerics	17	27%	10	10%	27	16%
Professors	3	2	5	5	8	5
Teachers & Tutors	1	1	3	2	4	2
Advocates			1	1	1	1
Writers, Factors	2	3			2	1
Civil Office	2	3			2	1
Navy						
Army						
Physicians & M.D.s	4	6	4	3	8	5
Surgeons	2	3	4	3	6	4
Merchants	6	10	14	13	20	12
Manufacturers			1	1	1	1
Artisans			5	5	5	3
Farmers	1	1	7	7	8	5
Landed	8	13	13	13	21	13
Titled	2	3	4	3	6	4
Unknown	23	37	44	43	67	28
Totals	64	100%	104	100%	168	100%

Table 4.5: Initial Occupations of Glasgow Men of Science

Occupations	Born and presumed born before 1730		Born and presumed born after 1730		Totals	
Clerics	4	6%	4	4%	8	5%
Professors	3	5			3	2
Teachers & Tutors	12	19	12	12	24	14
Advocates	1	2			1	1
Writers, Factors			1	1	1	1
Civil Office						
Navy			1	1	1	1
Army	1	2	6	6	7	4
Physicians & M.D.s	18	28	22	21	40	24
Surgeons	18	28	19	18	37	22
Merchants & Bankers	1	2	3	3	4	2
Manufacturers			10	10	10	6
Artisans	4	6	14	13	18	11
Surveyor, Engineer			3	3	3	2
Farmers						
Landed	2	3			2	1
Landed at death	10	16	21	20	31	18
Titled	1	2	1	1	2	1
Unknown	1	1	3	3	4	2
Totals	64	100%	104	100%	168	100%

Table 4.6: University Attendance

	Born or presumed born before 1730 (64)		Born or presumed born 1730 and after (104)		Totals (168)	
1 University	33	51%	50	48%	83	49%
2 Universities	20	31	15	14	35	21
3 Universities	7	11	3	3	10	6
4 Universities	1	2			1	1
Unknown number	8	13	5	4	13	8
Total	49	77%	69	66%	118	70%
Glasgow	36	56	53	51	89	53
Edinburgh	12	19	18	17	30	18
St. Andrews	3	5	2	2	5	3
King's College			3	3	3	2
Marischal College	1	2			1	1
Unknown Scots College	3	5	5	5	8	5
Total Scots	45	70%	65	63%	110	65%
Leyden	12	19	1	1	13	8
Utrecht	5	8			5	3
Other or Unknown Dutch	1	2			1	1
Total Dutch	14	22%	1	1%	15	9%
French Universities	3	5			3	2
Oxford	1	2	1	1	2	1
Cambridge	1	2	1	2	2	1
Total at some university	49	77%	69	66%	118	70%
Other English Schools	3	5	1	1	4	2
Hunter's Medical School	2	3	8	8	10	6
Partially educated in England but not at schools	1	2	9	9	10	6
Total	5	8%	15	14%	20	12%
Apprenticed	7	11%	14	13%	21	13%
Total partially educated outside Scotland	31	48%	12	12%	43	26%
Unknown	3	5%	22	21%	25	15%

Recipients of MDs from St. Andrews and the Aberdeen universities have been counted as attending a university but not those universities. Surgeons have not been included among those counted as apprentices although most served apprenticeships. Those educated outside Scotland generally studied in hospitals or with chemists.

140 Science and Enlightenment

Table 4.7: Characteristics of Glasgow Men of Science

		-1650	1650-69	1670-89	1690-1709	1710-1729	1730-1749	1750-1769	1770-1789
Age of Entry to Profession	Ave	?	28	22	22	24	23	24	22
	Med	?	31	22	22	23	22/23	23/24	22/23
	No	?	1	1	4	22	12	18	12
	Set	3	5	8	7	35	23	38	17
Age of Entry to Scientific Community	Ave	?	28	22	22	22	22	24	21
	Med	?	32	22	22	23	23	21/22	22/24
	No	?	3	1	1	22	11	17	10
	Set	3	5	8	7	35	23	38	17
Married		3	3	5	4	18	16	21	7
Single		0	1	1	0	4	1	3	2
Unknown		0	1	2	3	13	6	13	8
Age at 1st Marriage	Ave	?	?	?	?	34	44	29	33
	Med	?	?	?	?	27/31	44	27/31	30
	No	?	?	?	?	4	5	6	3
	Set	3	5	8	7	35	23	38	17
Known Time in the Scientific Community	Ave	34	32	42	42	35	45	38	37
	Med	38	27	43/45	48/50	40	44/45	42/44	40/42
	No	3	3	8	6	29	16	28	14
	Set	3	5	8	7	35	23	38	17
Known Time in Glasgow	Ave	22	24	40	40	23	26	30	25
	Med	22	17/21	43	48/50	22/25	25	33/35	13/20
	No	3	4	7	4	20	21	28	14
	Set	3	5	8	7	35	23	38	17
Age at Death	Ave	66	57	63	64	61	72	68	63
	Med	66	58/60	72	67/73	67	75	68/70	65
	No	1	4	3	4	24	19	20	13
	Set	3	5	8	7	35	23	38	17

Key: Ave – average for the number known; Med – median for the number known; No – number in the set for which data exist; Set – number in the birth cohort.

1) All ages given as e.g. 'c.1660' have used 1660 as if it were certain. All dates are assumed to be New Style and all averages have been rounded off to the nearest whole number.
2) The age of entry to a profession is calculated from the time of entry to a guild, the receipt of a medical or legal qualification, the end of apprenticeship, the acceptance of a first teaching position, ordination, or commissions in the army or navy.
3) Age of entry to the scientific community is based on the date at which an adult interest in science is clear.
4) The time in the scientific community includes the time spent in Glasgow but extends beyond it usually to their death. For some Glasgow men of science their time in the scientific community began while they were students.
5) Insufficient data exist for the 32 unknown men of science and they have therefore been excluded from this table.

Table 4.8: Principal Occupation of Glasgow Men of Science by Birth Cohort

	1630–49		1650–69		1670–89		1690–1709		1710–29		1730–49		1750–69		1770–89		Unknown before 1730		Unknown after 1730		Total Unknown		Totals	
Clerics	1	33%	1	20%	1	13%					1	4%	4	11%	1	6%							9	5%
Professors or Equivalents			2	40	2	25	1	14	14	40%	5	22	8	21	8	47			7	25%	7	21%	40	23
Tutors & Teachers	1	33							1	3													9	5
Advocates & Judges									1	3													1	1
W.S. & Factors											1		3										1	1
Civil Officers																							1	1
Naval Officers											1		1		1				1		3		4	2
Army Officers	1	33	1	20	3	38	2	29	4	11	3	13	1	3	5	29	2	50%	3	11	5	16	25	15
Physicians & M.D.s			1	20	1	13	3	43	11	31	3	13	10	26	1	6			1	4	3	3	31	18
Surgeons									2		2	8	2	5	1				1	4	1	3	3	2
Merchants & Bankers							2		6		4	17	1	3	6				4	14	4	13	13	8
Manufacturers											1		3						1	4	1	3	3	2
Managers & Engineers							1	14	2	6			4	11			1	25	6	21	7	22	14	8
Instrument Makers											2		5										2	1
Printers & Publishers																							1	1
Man of Letters					1	13					1	4											2	1
Titled											1	4											2	1
Unknown													4	11			1	25	4	14	5	16	9	5
Totals	3	100%	5	100%	8	100%	7	100%	35	100%	23	100%	38	100%	17	100%	4	100%	28	100%	32	100%	168	100%

This table lists the profession which was most important in the working lives of individual men of science and not all those in which they qualified or with which they were involved while members of the Glasgow scientific community.

142   *Science and Enlightenment*

Table 4.9: The Changing Occupational Composition of the Glasgow Scientific Community by Ten-Year Intervals

	1690		1700		1710		1720		1730		1740		1750		1760		1770		1780		1790		1802	
Clerics	1	20%	1	13%	1	9%	1	8%	1	7%	3	13%	7	19%	9	26%	3	9%	5	13%	7	11%	4	6%
Professors & Equivalents	2	40	5	64	5	45	5	38	4	29	3	13	7	19	6	17	10	29	8	21	12	19	10	16
Tutors & Teachers			1	13	1						1	4	3	8	6	17	1	3	2	5	2	3	1	2
Extramural Lecturers											2	9			1	3								
Students	1		1	13	1	9			1	7			4	11							1	2		
Advocates & Judges																			1		1	2	1	2
Writers, W.S., Factors																								
Civil Officers									1		1		1	3							1	2		
Naval Officers									1				3		3				1	3	2	3	1	2
Army Officers											3	1							2		3		1	2
Physicians & M.D.s	2	40	1	13	4	36	5	38	6	43	5	22	9	25	6	17	7	20	6	15	11	17	11	18
Surgeons	1	20	1	13	1	9	3	23	3	21	9	39	12	33	10	29	11	31	9	23	15	24	17	27
Merchants													1	3	2	6	1	3	3	8	3	5	2	3
Manufacturers											2	9	5	14	4	11	5	14	9	23	13	21	15	24
Managers & Engineers													1	3	1	3	3	9	1	3	2	3		3
Artisans & Instrument Makers							1	8	1	7	4	17	6	17	6	17	5	14	6	15	10	10	7	11
Printers/Publishers																			1	3	1	2	2	3
Landed			1	13	2	18			2	14	3	13	4	11	3	9	2	6	3	8	6	10	5	8
Titled									1	7	1	4	1	3	1	3					2	3	2	3
Unknown													1	3					1	3	4	6	5	8
Number in Set	5	100%	8	100%	11	100%	13	100%	14	100%	23	100%	36	100%	35	100%	35	100%	39	100%	63	100%	62	100%

The principal roles of the Glasgow men of science have been noted in this table, which has resulted in the ascription of more than one role for many members of each set in the given year (where the set number represents the actual number of individuals in the community). Alexander Wilson, for example, was a manufacturer type, a university professor, an instrument maker and began life as a surgeon apothecary. Placing him in all of these categories means that the total percentages will exceed 100%. We think this method of classification gives a more adequate description of the interests and range of knowledge of the Glasgow men of science in our period.

5

Maclaurin and Newton: The Newtonian Style and the Authority of Mathematics

JUDITH V. GRABINER

Sir Isaac Newton was a mythic figure even in his lifetime, as the well-known lines of Alexander Pope illustrate:

> Nature and Nature's Laws lay hid in Night.
> GOD said, *Let Newton be!* and all was Light.

As Newton's disciple, Colin Maclaurin, who lived from 1698 to 1746, basked in some of that reflected light.[1] Arguably the most significant Scottish mathematician and physicist of the eighteenth century, Maclaurin achieved recognition not only in Britain but also on the Continent. Besides his original scientific contributions, he was a Professor of Mathematics at Marischal College, Aberdeen, and then at the University of Edinburgh, a premier expositor of Newtonian science, a winner of two prizes from the Académie des Sciences in Paris, a defender of Newtonian calculus, one of the moving spirits in founding the Philosophical Society of Edinburgh, an accomplished 'improver' and a champion of natural theology. He moved in key intellectual and social circles of the early Scottish Enlightenment, from the Rankenian Club to the orbit of the Earl of Ilay.[2]

Maclaurin's influence has been widely acknowledged, but less is known about the ideas that motivated his work. Collections like the present volume make clear that students of the relationship between science, thought and society no longer treat the content of science as a black box that somehow generates 'impacts' on society, nor do they characterise a scientist's ideas by focusing on the scientist's rhetoric. In recent work on the history of eighteenth-century science, scholars like Theodore Porter, Steven Shapin, Simon Schaffer and Lorraine Daston have grappled with the specifics of scientific work in enough detail to understand that work on its own terms at the same time as they treat it as a social construct and situate it in its cultural context. This is my aim also. In the present chapter, my chief concern is to explore how Colin Maclaurin applied what has come to be called 'the Newtonian style' to areas ranging from the actuarial evaluation of annuities to the shape of the earth. Indeed, Maclaurin thought that using this approach could guarantee success in any scientific endeavour. I shall also briefly examine how Maclaurin's career was promoted by Newton and by other patrons, and how his scientific successes helped to validate his philosophical and religious views.

Maclaurin was not the first to bring Newtonian ideas to Scotland, but Maclaurin's Newtonianism seemed both the cause and the outward sign of his scientific success. Maclaurin's career also illustrates and embodies the way mathematics and mathematicians, building on the historical prestige of geometry and the success of Newtonianism, were understood to exemplify certainty and objectivity during the Scottish Enlightenment.

The Young Newtonian: The Newtonian Style and Its Applications

Maclaurin mastered a good deal of Newtonian science while still a student in Glasgow. This meant not only learning Newton's mechanics and theory of gravity, but also appreciating how mathematics can drive scientific understanding. Much has been written about Newtonian methodological precepts, especially about his dictum *'hypotheses non fingo'*.[3] Besides the theoretical physics of Newton's *Principia*, there was an important experimental tradition based on Newton's *Opticks*.[4] But Maclaurin's key concern was the pursuit of Newtonian mechanics, for which the usual capsule characterisations of Newtonian method are far from adequate. It is the 'mathematical' in the full title of Newton's *Principia*, 'Mathematical Principles of Natural Philosophy', that was most important for Maclaurin.

Maclaurin said much in praise of Newton's methods. But Maclaurin not only adopted what Newton said; he internalised what Newton did. The way Newton did theoretical physics has best been described by I. Bernard Cohen, who has dubbed the approach 'the Newtonian style'.[5] In the *Principia*, according to Cohen, Newton first separated problems into their mathematical and physical aspects. A simplified or idealised set of physical assumptions was then treated entirely as a mathematical system, and the consequences of these assumptions were deduced by applying mathematical techniques. But since the mathematical system was chosen to duplicate the idealised physical system, the propositions deduced in the mathematical system could now be compared to the data of experiment and observation. The test against experience often required modifying the original system. Further mathematical deductions and comparisons with nature would then ensue, in what G. E. Smith has called 'a sequence of successive approximations'.[6] Success finally comes 'when the system seems to conform to (or at least to duplicate) *all* the major conditions of the external world'.[7] It is sophisticated mathematics, not only a series of experiments or observations, that links a mathematically describable law to a set of causal conditions.

A good example of how the Newtonian style works is the way Newton treated planetary motion in Book I of the *Principia*. He began with one body, a point mass, moving in a central-force field. He then introduced Kepler's laws, then a two-body system with the bodies acting mutually on each other, then many bodies, then bodies which are no longer mass points. The force which explained this entire system was universal gravitation, a force which Newton argued 'really exists'.[8]

Maclaurin knew the Newtonian style intimately, and treated a number of such examples in his own exposition of Newton's work.[9] Maclaurin beautifully expressed his deep understanding of the Newtonian style when he wrote: '[E]xperiments and observations ... could not alone have carried [Newton] far in tracing the causes from their effects, and explaining the effects from their causes: a sublime geometry ... is the instrument, by which alone the machinery of a work [the universe], made with so much art, could be unfolded'.[10] Throughout his career, this is the methodology Maclaurin followed.

Mathematics and Morality

A sort of trial run of Maclaurin's use of the Newtonian style was his youthful attempt to build a calculus-based mathematical model for ethics. In an unpublished essay, 'De viribus mentium Bonipetis',[11] Maclaurin analysed mathematically the forces by which our minds are attracted to different morally good things. This essay was written in 1714 when Maclaurin was no older than sixteen.[12] It is not the subject to which a modern mathematician would expect to see Maclaurin applying the calculus, but it is nonetheless Maclaurin's first original application of the Newtonian style.

In the 'De viribus' Maclaurin postulated that the 'forces with which our minds are carried towards different good things are, other things being equal, proportional to the quantity of good in these good things'.[13] Also, the force of a good one hour in the future would exceed that of the same good several hours in the future. Maclaurin represented the total quantity of good as the area under a curve whose x-coordinate gives the duration and the y-coordinate the intensity of the good at a particular instant. He added that one could find the maximum and minimum intensities of any good or evil using the calculus; to do this, he used fluxional notation, and referred to the discussion of limits in Newton's *Principia*. He graphed the total attraction of a good as a function of how far that good is from the mind, and, by integration, derived the equation of that attraction function. Good men, he concluded, need not complain 'about the miseries of this life' since 'their whole future happiness taken together' will be greater; this is because as the 'intensity is increased ... gratitude will be greater and very many virtues will be exercised'.[14] Maclaurin has tested his mathematical model against Christian doctrine, and has found that the results fit.[15]

Since Maclaurin never published this essay, we may conclude that he did not long believe that he could use mathematics to buttress morality in this way. However, his having sent the 'De viribus' to Colin Campbell suggests that at one time he did take this use of mathematics seriously. And, at least for Maclaurin himself, it was an instructive exercise in ingeniously applying mathematical technique. He had shown, by computing the areas under curves, that the Christian doctrine of salvation maximises the future happiness of good men. 'Maximising' and 'minimising' are important problems solved by the calculus. Applications of this technique abound in eighteenth-century physics, from curves of quickest descent to the principle of least action. Maclaurin systematically treated finding maxima and minima in Chapter XIII of his *Treatise of Fluxions*, applying the technique in many novel situations and then comparing his results to the best data. He described the same subject in a more popular fashion in his *Account of Sir Isaac Newton's Philosophy*.[16] He also wrote a paper on how bees, in constructing their honeycombs, use the shape that, for a given volume of honey, minimises the amount of wax used to make the cell.[17]

The 'De viribus' itself resembles some familiar ideas of Maclaurin's contemporary at Glasgow, Francis Hutcheson, and Maclaurin himself seems to have recognised a kinship between his views and Hutcheson's.[18] Mary Poovey has observed that Hutcheson seems to have believed that 'mathematics provided the analytic method by which one could move beyond observed particulars to the design that informed them, because mathematics demonstrated general principles or regularities'.[19] Like Maclaurin, Hutcheson used what he appeared to think were mathematical principles both to describe and to demonstrate his laws of virtue: 'In computing the Quantities of Good and Evil . . . when the Durations are equal, the Moment is as the Intenseness . . . when the Intenseness of Pleasure is the same, or equal, the Moment is as the Duration'.[20] In a chapter entitled 'The manner of computing the morality of actions', Hutcheson wrote, 'the Virtue is in proportion to the Number of Persons to whom the Happiness shall extend . . . and in equal Numbers, the Virtue is as the Quantity of the Happiness, or Natural Good . . . so That Action is best, which procures the greatest Happiness for the greatest Numbers'.[21] What makes Hutcheson's seem more than just a conventional appropriation of mathematics of a kind going back to the Middle Ages is the appeal to mathematical optimisation, a technique used by Maclaurin and others in questions ranging from the construction of honeycombs to the justification of the laws of mechanics on the basis of the principle of least action, to explain why things should be as they are.

A 'most clear & mathematical proof of the existence of a god'[22]

Another Newtonian theme in Maclaurin's work is the way the order of the universe demonstrates the existence and nature of God. Maclaurin first expressed this in his graduation thesis, defended at Glasgow when he was fifteen. In his thesis, Maclaurin explained Newton's quantitative argument for identifying terrestrial gravitation with the force that holds the moon in its orbit, showed how Newton's law of gravitation explained both lunar and planetary motions, argued that the grandeur and fitness of the creation showed the omnipotence and goodness of the Creator, and concluded that the cause of gravity 'is the will of some incorporeal and intelligent cause uniformly exerting its power according to a fixed general law . . . the Author of gravity fully deserves to be recognized as the lord of the earth and the preserver of men'.[23]

The thesis is marked also by opposition to Cartesianism, and the way Maclaurin dismisses what he calls Descartes' hypotheses in favour of Newton's emphasis on observation – all for a theological purpose – is akin to the views of the Glasgow Professor of Moral Philosophy, Gershom Carmichael, better known as Hutcheson's teacher. Carmichael moved from a Cartesian to a more Newtonian viewpoint, writing in 1713 that 'knowledge of natural law . . . is to be derived from the nature of things and their uninterrupted course, and a proper use of reason'.[24] Carmichael saw human duties as following from our recognition of the divine goodness. Carmichael's ideas must have helped Maclaurin find a more harmonious relationship between his science and his Church than Newton had had with Anglican orthodoxy. Maclaurin's linking of God's providence with Newtonian physics was appreciated by Moderates in the Church of Scotland, and helped reinforce the alliance between science, religion and the Scottish universities.

'Electus Ipso Newtono suadente':[25] *Maclaurin's Patrons and His Career*

In 1731, Maclaurin described his career strategy in a letter to Martin Folkes, writing that 'I . . . had it always in my View to make some Interest amongst those that are in Power to see if by their favour I could get myself made easier'.[26] Indeed, Maclaurin seems always to have had his eye on possible patrons, studying the possibilities open to him as society changed. For instance, he owed his first position, at Marischal College, Aberdeen, from 1717 to 1722, to Squadrone politicians. Marischal's faculty owed more to patronage than to nepotism, and its quality could serve the interests of these patrons.[27] Maclaurin, following his ambitions, left Aberdeen without leave in mid-1722 to go to the Continent as tutor to the eldest son of Lord Polwarth, Patrick Hume, returning only in 1724 and soon afterwards being appointed at Edinburgh.

Maclaurin's first significant personal patron was Newton himself. 'A paper of mine about the mensuration of some series of curves fell into the hands of Dr Halley', Maclaurin related to Colin Campbell, and Halley wrote back asking Maclaurin to 'communicate any things else I had of that kind'. So he sent, and Halley published, an abstract of Maclaurin's *Geometria organica*. After having been thanked by the Royal Society, Maclaurin went to London in 1719 'to see and converse', bringing along the complete manuscript of his *Geometria organica*.[28]

Fortunately for Maclaurin, the aging Newton had become interested in helping the cause of promising young men of science.[29] So, as Maclaurin told Campbell, 'I received the greatest civility . . . and particularly from the great Sir Isaac Newton with whom I was very often. I was admitted a member into the Royal Society & Sir Isaac encouraged me to publish the Theorems I had discovered'.[30] Maclaurin dedicated the published version of his *Geometria organica* to Newton, who sponsored the publication.

In 1725, Newton supported Maclaurin's appointment at Edinburgh – apparently to the extent of offering to supplement Maclaurin's salary. Newton did this, he told Maclaurin, 'not only because you are my friend, but principally because of your abilities, you being acquainted as well with the new improvements of Mathematicks as with the former state of those sciences'.[31] Edinburgh Town Council evidently agreed, and it appears that Newton's support, as well as that of others, made a difference. In advocating Maclaurin's appointment, the Council's Register states that Maclaurin 'had made surprising appearances in that part of learning . . . had a very favourable character bestowed on him by very great men, and even by Sir Isaac Newton himself . . . it was impossible for us to hope for any opportunity of doing a thing more honourable and advantageous for the city, that could contribute more to the reputation of the university, and advance the interest of learning in this country, than the giving Mr M'Laurin suitable encouragement to settle among us'.[32]

With Newton's encouragement, then, Maclaurin became the chief spokesman in Scotland for Newtonian science. At Edinburgh, Maclaurin gave a three–year course in mathematics and its applications, ranging from arithmetic to Newton's *Principia* and fluxions.[33] Much of his teaching became the basis for later published work, some posthumous: the elementary *Treatise of Algebra* (1748), the decidedly non-elementary *Treatise of Fluxions* (1742), and the *Account* (1748).

Maclaurin enjoyed the patronage of the Argyll family throughout his stay in Edinburgh. From 1725 to 1761 Archibald Campbell, the Earl of Ilay, brother of the 2nd Duke of Argyll and after 1743 the 3rd Duke, was the dominant figure in Scottish affairs.[34] His family represented the largest and most influential aristocratic interest in Scotland, and in return for patronage, Ilay and Argyll kept most Scots Members of Parliament behind Walpole.[35] Ilay had a finger in every pie, from being Governor of the Royal Bank of Scotland (where the Trustees for Manufactures, an Argathelian fief, duly deposited their funds),[36] to the appointment of university professors. Men like Ilay wanted their appointees to be highly competent. Argathelian patronage was exercised not only on behalf of mathematicians like Maclaurin and the Glasgow Professor Robert Simson, but, by 1761, on behalf of the majority of the professors of Scotland.[37] That Maclaurin sought, and benefited from, Ilay's patronage is abundantly clear from the Maclaurin correspondence.[38] The 2nd Duke was another patron, and it was to him that Maclaurin dedicated the *Treatise of Fluxions*. When the 2nd Duke died in 1743, Maclaurin included a fulsome eulogy in a letter to a friend.[39]

Patronage, of course, is designed to benefit the patrons. One cause of Ilay's political success was that his protégés could deliver when needed, and Maclaurin, as I show below, did so. For instance, to help quiet political dissatisfaction with excise taxes on molasses at Glasgow, in 1735 he wrote a *Memorial to the Honourable Commissioners of Excise* on measuring the amount of molasses in barrels. This aided the powerful Excise service, an important battleground in the struggle between the Argathelians and their rivals, the Squadrone.[40]

Despite enjoying Argathelian patronage, Maclaurin found his work at the University of Edinburgh arduous. In 1731 he enlisted the Secretary of the Royal Society, Martin Folkes, as a possible patron by flatteringly saying that after Newton's passing he regarded Folkes as 'in his place'. Maclaurin complained that he had been 'drudging here these six years in an intolerable way', teaching six hours a day, and mostly elementary science rather than 'the sublimer parts of it', and he asked Folkes to speak to Walpole on his behalf to get him a job in the Exchequer, which, Maclaurin said, 'would make me easy for Life'.[41] Perhaps the enormous effort he expended on his *Memorial to the Honourable Commissioners of Excise* was not merely to serve his Argathelian patrons and his country, but to impress the Excise Commission and their associates enough to secure such a position.[42] If so, the ploy did not work.

Although Maclaurin had no immediate successor of remotely comparable brilliance, some notable figures among his students at Edinburgh rated him highly. For instance, the Rev. Alexander Carlyle said that Maclaurin 'made mathematics a fashionable study'.[43] The great geologist James Hutton 'enjoyed MacLaurin's lectures particularly'.[44] Hutton's uniformitarian geology treated the earth's changing cycles of erosion and uplift of rock as a self-renewing machine, much like the eighteenth-century view of the solar system as 'the Newtonian world-machine', and Hutton closely followed Maclaurin's language in inferring God's wisdom and foresight from nature.[45] Matthew Stewart, the father of Dugald Stewart, studied higher mathematics with Maclaurin and eventually succeeded him in the Mathematics Chair in Edinburgh. The other side of Maclaurin's complaint about his large classes is the fact that he taught mathematics and science to a large number of prospective doctors, engineers, merchants and clergy. His influence extended outside the university, since he gave public courses of lectures, on one occasion even reluctantly admitting women.[46]

The course of Maclaurin's career, and his relationship with various patrons and universities, exemplifies what Larry Stewart has written about other disciples of Newton, like John Theophilus Desaguliers, John Harris, Humphry Ditton, Stephen Hales, Henry Beighton, Mårtin Triewald, John Horseley and Edmond Halley. These men found their way 'into the industrial and commercial transformation of Britain', where, by trading on their expertise, they could find both patronage and opportunity.[47] Throughout his life, Maclaurin continued to advance Newtonian science and to bring Newtonian science and its applications to bear on the 'improvement' of Scotland.

Maclaurin as Defender of the Fluxional Faith: The Treatise of Fluxions

Part of the authority of mathematics and science in the eighteenth century stemmed, as Maclaurin said in the preface to his *Treatise of Fluxions*, from the belief that mathematical demonstration left 'no place for doubt or cavil'.[48]

Maclaurin and Newton held that the mathematically based natural philosophy of the *Principia* demonstrated the existence of God, and, in a more secular vein, that their mathematically based methods underlay any serious investigation of the nature of the world. Essential to Newtonian natural philosophy were the ideas of absolute space and time, ideas also essential to the velocity-based explanations Newton gave for his calculus.

But the point of view just described was not shared by George Berkeley. He saw Newtonian science as encouraging Deism and undermining the authority of Scripture. Besides criticising the Newtonian philosophy, in *The Analyst, or a Discourse Addressed to an Infidel Mathematician* (1734), Berkeley attacked the logical validity of the calculus itself. Berkeley held that 'we can form no clear conception of abstract terms', like infinite quantities, infinitesimals, or absolute space, time and motion.[49] Furthermore, and using well-chosen examples, Berkeley argued that, on the evidence of the way the calculus was actually being explained, mathematicians reasoned worse than theologians.[50]

Mathematicians were not the only ones who wanted Berkeley's *Analyst* to be definitively refuted in print. In 1737 Francis Hutcheson warned Maclaurin to hurry up and publish 'your Fluxions' for fear that Berkeley 'should have some silly answer ready before yours'.[51] Maclaurin had indeed been delaying his response; the *Treatise of Fluxions* did not appear in final form until 1742, although parts of it were circulated earlier. But Maclaurin had greater ambitions for his own exposition of the calculus than a simple answer to Berkeley.[52]

Maclaurin did begin his *Treatise of Fluxions* by replying to Berkeley's objections. Unfortunately, since he said he was answering Berkeley by using geometric rigour, the *Treatise* has often been described as a step backwards, as a reduction of modern calculus to ancient geometry. Maclaurin said that he intended to deduce the basic laws of fluxions 'after the manner of the Antients, from a few unexceptionable principles, by Demonstrations of the strictest form'.[53] Nevertheless, the key idea here is rigour of method, which could be applied to the far more fruitful modern mathematics – that is, algebra and calculus – as well. In particular, to defend the calculus, Maclaurin linked the authority of geometry with the authority of Newton.

'In explaining the Notion of a Fluxion', wrote Maclaurin, 'I have followed Sir Isaac Newton . . . imagining that there can be no difficulty in conceiving Velocity wherever there is Motion.'[54] Maclaurin gave four axioms which presuppose the notions of space, time, and velocity, and a definition of fluxion as a velocity. Together, these allowed him to prove the basic results of the calculus.[55]

Contrary to the views of Leibniz and the accusations of Berkeley, Maclaurin made it clear that infinitesimals and infinite quantities are not needed in the calculus.[56] Maclaurin used the same sorts of arguments against infinitesimals that Newtonians had used against Cartesian ideas – infinitesimals are fictions, they have no basis in nature, the idea of infinitesimals is not clear and plain – and explicitly characterised the ideas of infinity and infinitesimals as Cartesian and as related to vortices.[57] Having identified his own views with the Newtonian repudiation of Descartes, Maclaurin eloquently concluded, 'Geometry is best established on clear and plain principles . . . Geometricians cannot be too scrupulous in admitting or treating of infinites, of which our ideas are so imperfect . . . an absurd philosophy is the natural product of a vitiated geometry'.[58] Fluxions (unlike infinitesimals) are real, because they are based on time and motion, and they are logically sound because the proofs about them meet the same standards of rigour as do proofs in ancient geometry. Nor need the calculus be geometric; in Book II of the *Treatise* Maclaurin showed that the same rigour can be achieved in algebraic formulations of the calculus.[59] In fact, Maclaurin's improved, algebraic, and inequality-based understanding of the limit concept played an important role in the eventual rigorisation of the calculus.[60]

Besides refuting Berkeley's challenge, Maclaurin's defence of the calculus continues the themes of the errors of Cartesianism, the correctness of the Newtonian method and the authority, based on universal agreement, of mathematics. Maclaurin's *Treatise of Fluxions* was received as the major work of a major mathematician. Although its 754 pages originated as a polemic, the book contains a wealth of important results both in mathematics and in physics.

'A new theorem or problem is still one of my greatest ragouts'[61]: Rotating Bodies and the Shape of the Earth

This chapter is not the place for an account of all the mathematical and physical achievements of the *Treatise of Fluxions*.[62] But we should at least describe one of Maclaurin's major claims to fame as a mathematical physicist, namely his treatment of the shape of the earth, for it exemplifies his use of the Newtonian style, especially the sophisticated use of mathematical models, to solve problems of great importance for physics and astronomy.

The earth is not a sphere. As Newton had predicted in the *Principia*, its rotation brings about real forces which cause it to be flattened at the poles and to bulge at the equator. Pierre de Maupertuis had empirically verified the general shape in the 1730s. Maclaurin carried the theory of this subject substantially farther.

In 1740, Maclaurin produced the first rigorously exact, geometrical theory of homogeneous figures shaped like ellipsoids of revolution whose parts attract according to the inverse-square law. His method was geometric, based on his deep knowledge of the conic sections, and worked for figures whose ellipticities were finite as well as infinitesimal. In this work, Maclaurin exploited new techniques, such as the method of balancing columns and the principle of the plumb line.[63] He showed how to determine the exact value of the attraction at arbitrary points on the surface of a homogeneous ellipsoid by reducing the problem to finding the exact value of the attraction at the poles and equators. Then he demonstrated that such a homogeneous figure, revolving around its axis of symmetry, is indeed a possible figure of equilibrium.[64] In addition,

Maclaurin treated stratified figures of equilibrium (that is, made of shells of different density), both in his 1740 paper on the tides and, more fully, in the *Treatise of Fluxions*.[65] Besides developing the mathematical theory of rotating bodies, their shapes, and their attractions, Maclaurin took a keen interest in getting the latest news from the expeditions to various parts of the world, undertaken in the 1730s, to measure the earth's shape, as his correspondence abundantly testifies.[66] Maclaurin compared the predictions of his theory to a table 'of all the properties of the diameters of the Earth that result from the Observations made' by his countryman George Graham, whom he regarded as the most accurate of all such observers, and by the various French academicians.[67] In the *Treatise of Fluxions*, all of this work is brought together. Maclaurin achieved a sophisticated mathematical theory of rotating ellipsoids whose parts gravitate, developed precise predictions about the earth's shape, and tested these predictions against the latest observations – an impressive and successful application of the Newtonian style.[68]

'[W]hen men, having got into the right path, prosecuted useful knowledge'[69]: Maclaurin, Mainsails and Molasses

Employing fashionable Baconian rhetoric, Maclaurin rejoiced that advances in science could produce useful knowledge, and enthusiastically participated in that production. Summarising Maclaurin's career in 1748, Patrick Murdoch said that Maclaurin

> often had the pleasure to serve his friends and country by his superior skill. Whatever difficulty occurred concerning the construction or perfecting of machines, the working of mines, the improvement of manufactures, the conveying of water, or the execution of any other public work, Mr. *Maclaurin* was at hand to resolve it.[70]

Many examples of Maclaurin's interest in applying Newtonian physics and his serving of patrons can be seen in his correspondence. For instance, in 1739, writing to advise the Earl of Morton about a proposed expedition to map Shetland, Maclaurin told those on the ship to keep a journal, and he detailed which observations they should make, which clocks and telescopes to use and how to use them, how to observe variation of the compass, tides, meteors, barometric pressures and which items of natural history, from herrings to shrubs, to describe.[71] In other letters, he discussed the motions of engines and ships, the sails of windmills, the flow of rivers, the syphon for draining a pond at Lord Minto's estate, the search for a Northern passage to the East Indies 'for the benefit of poor Scotland', and remapping the north coast of Britain, important 'especially in the event of a french warr'.[72] Even in the *Treatise of Fluxions*, Maclaurin used physics and calculus to discuss water-driven and air-driven machines, and the best position of a sail for making the wind move a ship in the desired direction.[73] Murdoch's biography tells us further that

Maclaurin, before his untimely death in 1746, 'had resolved . . . to compose a course of practical mathematics, and to rescue several useful branches of the science, from the bad treatment they often meet with in less skilful hands'.[74] Perhaps as a first step toward this task, in 1745 Maclaurin published a manuscript by David Gregory on finding the volumes of barrels, or 'gauging', supplemented by comments of his own.[75]

Gauging, as already mentioned, had been the subject of one of Maclaurin's most detailed contributions to applied science. In 1735, he wrote an extensive memoir for the Scottish Excise Commission, explaining the most accurate way to find the volume of molasses in the barrels in the port of Glasgow.[76] This memoir reflects the importance of the molasses trade to the prosperity of the Glasgow of the 1730s, the crucial role played by excise taxes in financing the Scottish government, and the need for the Argyll-dominated Excise Commission, and Scottish government in general, to stifle any cause for further protest about taxes. This may be seen as a case study in the use of mathematical authority to achieve consensus in a problem – taxation – clearly rife with disagreement between parties with vastly different interests.[77] In the present context, Maclaurin's solution of the problem is another example of his use of the Newtonian style.

Maclaurin's memoir provided a set of clear rules about where the excise officer was to place his dipstick to find the amount of molasses in the barrel, and which calculations he was to make afterwards. To derive these rules, Maclaurin began by treating the case where the barrels had precise mathematical shapes. The mathematics itself involved the properties of barrels in the shape of solids generated by revolving conic sections on their axes, and he used the fluxional calculus to prove the results which underlay his practical rules. His use of fluxions, rather than the classical geometry he would use to present the same results in the *Treatise of Fluxions*, seems designed more to enhance the authority of his memoir than to be read easily by excise officers.[78] Once the mathematical theory was completed, he gave a thorough analysis of the way the actual barrels deviated from those shapes, and, crucially, showed how new calculations could deal with those deviations.[79] Once again, Maclaurin began with a mathematical model and corrected it according to observation to produce an authoritative solution.

Nor was this Maclaurin's only major accomplishment for Scottish society. He recognised that the progress of science and its applications was a social matter. He wrote that in the age of experimental philosophy 'not private men only, but societies of men . . . prosecuted their enquiries into the secrets of nature'.[80] This statement reflects Maclaurin's leading role in organising the Philosophical Society of Edinburgh in 1737. The members included enlightened aristocrats and merchants as well as scientists, with the aim of 'improving arts and sciences and particularly natural knowledge'; one example of the Society's engagement with improvement is the surveys of the north coast and the islands sponsored by Lord Morton and undertaken by several of Maclaurin's students in the 1740s. The eighteenth-century ideal of consensus appears clear in the Society's proposals, as it had earlier in the publications of the Royal Society of London: 'Religious or Political Disputes' were proscribed for the Philosophical Society of Edinburgh, and members were warned that 'in their Conversations, any Warmth that might be offensive or improper for Philosophical Enquiries is to be avoided'.[81]

The Society's members served as consultants to Scottish development agencies like the Board of Trustees for Arts and Manufactures, the Annexed Estates Commission, the British Linen Company and the Royal Bank of Scotland, as well as being involved in mapping Scotland and her coasts, making astronomical observations that could assist navigation and determining longitude, pumping water out of mines, examining various medical questions, designing or building canals, investigating the distribution of minerals in Scotland and measuring the force of winds at sea as a function of latitude. The applications of science pursued by the Society reinforce the picture drawn here of Maclaurin's interests.[82] These activities, just like his work on the volumes of barrels, embody the ideal of 'improvement' prevalent in eighteenth-century Scotland: the idea that science and technical expertise could, and should, be applied to solve problems of economic and social importance in this new and yet-to-be-developed North Britain.

Maclaurin and his circle – aristocrats and school teachers, public officials and physicians, scientists and craftsmen – fit closely the broader features of eighteenth-century public science in Europe.[83] Popularisers of science (in whose number Maclaurin should be included) helped make science into public knowledge.[84] So many of Newton's disciples were engaged in projects of various kinds that, as Larry Stewart has observed, the role of the natural philosopher underwent a profound change.[85] Scientists were 'much sought after by investors and improvers throughout Britain and parts of the Continent'.[86] The science to be applied did not need to be the latest: all that was required was a degree of expertise, an evident need and a patron who knew enough of the science to appreciate the possibilities – a Lord Ilay, for instance.

In Maclaurin's rhetoric, as in that of the Philosophical Society of Edinburgh, one mark of the validity of the science to be applied came from the fact that everybody could agree about it. This, many Enlightenment thinkers asserted, gave eighteenth-century science its authority, and this universal agreement was, for Maclaurin, the achievement of Newton and Newton's 'right path'. Newton, Maclaurin claimed, 'had a particular aversion to disputes' (a hagiographical comment indeed) and 'weighed the reasons of things impartially and coolly'. And, just as Maclaurin had used the universal agreement, beyond 'doubt or cavil', deserved by mathematics to motivate his refutation of Berkeley, so he used his idealised picture of Newton to formulate his own version of the achievement of Enlightenment science:

> We are now arrived at the happy æra of experimental philosophy; when men, having got into the right path, prosecuted useful knowledge; when their views of nature did honour to them, and the arts received daily

improvements; when not private men only, but societies of men, with united zeal, ingenuity and industry, prosecuted their enquiries into the secrets of nature, *devoted to no sect or system*.[87]

'[T]he authority of his name was of great use'[88]: Maclaurin and the Scottish Ministers' Widows' Fund

We now consider an episode combining illustrations of Maclaurin's authority as a mathematician, his service to Scottish institutions and the Newtonian style: his actuarial work for the Scottish Ministers' Widows' Fund in 1743. Robert Wallace, Moderator of the General Assembly of the Church of Scotland, wrote to Maclaurin that, when objections to the soundness of the pension scheme were raised in Parliament, 'I answered them that the Calculations had been revised by you . . . this entirely satisfied them'.[89] Or, as Patrick Murdoch put it, 'the authority of [Maclaurin's] name was of great use'. The calculations as revised by Maclaurin were indeed satisfactory; the fund was, in 1992, still providing 'for the widows and fatherless children of ministers of the Church of Scotland, the Free Church of Scotland and some of the professors of the four old Scottish Universities'.[90]

The goal of the Church in 1743 was to keep its widows and orphans out of poverty by providing them with an annuity. Synods had tried to do this before, but the funds tended to run out of money. Everyone involved in the 1743 scheme recognised that making it work required finding data and creating a mathematical model based on those data. As it turned out, the task also involved checking and refining data, and then revising the model to fit those data accurately. The men who did the first two steps were Alexander Webster and Robert Wallace. By carrying out the two last steps, Maclaurin improved their scheme and made it work. He thus used the Newtonian style once again, this time to put the annuity scheme into its successful form, and also to become a pioneer of actuarial science.

To start planning for the Fund, Webster, Minister at the Tolbooth Kirk, Edinburgh, collected the relevant data by sending questionnaires to every parish of the Church, asking how many ministers there were, how many widows there were and how many remained unmarried, how many orphans there were, of what ages, and so on.[91] Here Webster drew on an earlier British quantitative tradition exemplified by William Petty.[92] Wallace, Minister of New North, Edinburgh, and later Moderator of the General Assembly, was himself a competent mathematician who had assisted James Gregory, and who had helped found the Rankenian Club and the Philosophical Society. Wallace made the first mathematical model predicting the changes in the widow and orphan populations for the Fund. Both Webster and Wallace applied quantitative methods to study populations in broader contexts as well. For instance, Webster used parish survey methods in his *Account of the Number of People in Scotland*, begun in 1743 and published in 1755, a work whose focus on the 'political arithmetic' of the nation makes it an important step toward the modern census,[93] while Wallace constructed mathematical models to study human populations in works often recognised as forerunners of the work of Malthus: *A Dissertation on the Numbers of Mankind* (1753) and *Various Prospects of Mankind, Nature and Providence* (1761).

Webster and Wallace stated that their purpose for the Widows' Fund was to provide 'an annuity to the widow . . . and a stock to the children of such as should leave no widow, founded on an annual tax payable out of their benefices and a capital arising from the surpluses of these taxes during the earlier years of the scheme when there would be no great burden on the fund'.[94] This formulation was eventually accepted with the addition that the provisions for families of the ministers dying soonest would be just as great as for later ones. Maclaurin, perhaps as a representative of the Edinburgh professors, was sent a copy of the scheme by Wallace. Using Wallace's model, Maclaurin prepared tables to predict the future of the scheme 'according to the doctrine of chances' (an eighteenth-century term for what we now call 'probability theory').[95] He showed that the fund would run out of money unless it reduced the children's benefits.[96] This bankruptcy would occur because Wallace had used a mathematically simple, but arbitrary, way of estimating the number of widows, assuming that one in eighteen of the widows would die each year, ignoring the varying age-distribution of the group.

Maclaurin sought empirical data to check Wallace's assumption. He found the data he wanted by studying more carefully than Wallace had done the mortality tables from Breslau published by Edmond Halley in 1673. According to the tables, said Maclaurin, the average annual mortality rate of the widows would at first be lower. More precisely, Maclaurin assumed the average age of the predicted eighteen new widows each year was fifty, and that their mortality would follow Halley's table, rather than having one in eighteen of them die each year. Even in the second year, the predicted number of widows differs in the two models. For Wallace, the eighteen in the first year will become, with a mortality rate of one in eighteen, seventeen, and adding eighteen new ones gives thirty-five; Maclaurin's figure was 35.42, with its empirical basis taken from Halley. These differences become compounded over time. For instance, after thirty years, Maclaurin's model predicts 306.60 widows, as contrasted with Wallace's 257.11.[97] Maclaurin's recalculations were 'almost certainly the earliest actuarially-correct fund calculations ever carried out' because he had used 'a realistic and accurate life table'. And, most important for the Fund itself, what actually happened corresponded remarkably well with the calculated figures.[98]

This was a very influential piece of work. In 1761, the first real life insurance plan in the United States was undertaken by the Rev. Francis Alison of the First Presbyterian Church of Philadelphia for the relief of Presbyterian ministers there. Alison said that his plan was 'in imitation of the laudable example of the Church of Scotland'.[99] The example of the Widows' Fund was likewise followed by the first Scottish life assurance company to be established, as that company's name, 'Scottish Widows', indicates.[100] And, although the ideas behind the first real insurance company, the Equitable of London in 1762, seem to have been independent of the Widows' Fund, it is interesting to observe that Richard Price, consulting actuary to the Equitable in 1771, criticised the Church of Scotland's scheme in his *Observations on Reversionary Payments* (1769); Webster, who continued to direct and improve the Fund, replied at length, evidently persuading Price, so that later editions of Price's book, which became a standard actuarial textbook, praised the soundness of the Widows' Fund plan and commended Webster, treating him as its founder.[101]

The story of the Widows' Fund is also of interest in the light of the work of several modern historians. For instance, Mary Poovey has asked when 'gestural mathematics' gave way to 'literal counting'; that is, when one first needed to have real data, numbers that claim actually to refer to particulars that had been counted, rather than numbers used as illustrations. Poovey sees Malthus as a key figure in the change. She asks whether the lack of an infrastructure sufficiently developed or centralised to collect real data may have delayed such 'literal counting', observing that the parish system in Scotland may have impeded national information-gathering, even though this system was 'successfully used in the 1790s to implement the first such project', which, she writes, 'culminated in the publication of Sir John Sinclair's twenty-one-volume *Statistical Account of Scotland*'.[102] But there had already been other such projects,[103] and, in particular, the Widows' Fund was an earlier and successful example not only of collecting real data through surveys of each parish, but of exploiting those data in long-term projections. Wallace's later work on population links the ideas behind the Widows' Fund to the work of Malthus, just as Webster's is related to Sinclair's. Furthermore, Theodore Porter has pointed out the importance of actuarial and insurance work in the nineteenth century in promoting quantitative methods in the 'objective' study of society.[104] Maclaurin's contribution was much earlier, and it was trusted, with his mathematical authority 'removing any doubts . . . concerning the sufficiency of the proposed fund, or the due proportion of the sums and annuities'.[105]

Observing that probability-based insurance schemes took a long time to develop, Lorraine Daston has asked why they did not appear before the early nineteenth century. She identifies two inhibiting factors: the assumption that mortality statistics are regular, and the perception of insurance as akin to gambling. She saw a third factor as helping to promote such schemes: a change of values toward prizing frugality and thus commending the purchase of insurance.[106] Maclaurin and his Newtonian style saved the Widows' Fund from the false assumption of too-regular mortality. Because Maclaurin followed the work of men like Jakob Bernoulli and Abraham De Moivre and saw probability theory as a part of applied mathematics concerning the frequencies of events, rather than identifying it with its origins in gambling, he could apply 'the doctrine of chances' to strengthen an insurance scheme. As for the third factor, the Church of Scotland seems to have had the appropriate values. The success of the Widows' Fund stemmed from the way Maclaurin joined mathematical modelling and empirical data, this time in the social realm, to solve an economic problem for the church and the universities in Scotland.

'[A]s in mathematics, so in natural philosophy'[107]: Maclaurin Explains It All to Everyone

Maclaurin's most influential public explanation of Newtonianism was his *Account of Sir Isaac Newton's Philosophical Discoveries*. This book was apparently begun around 1730 at the urging of John Conduitt, worked on throughout Maclaurin's university career, but published only after his death. The book boasted an impressive list of subscribers – a list which testifies to his place in his society and to the wide diffusion of his ideas.[108]

As L. L. Laudan has observed, Maclaurin set out to explain both the mathematics and the physics of the *Principia* 'without compromising its profundity by the ... oversimplifications to which most Newtonian popularizers resorted'.[109] Maclaurin, however, sought not only to explain, but also to defend, the Newtonian position. Like the anti-Berkeleian polemic of the *Treatise of Fluxions*, the *Account* is an apologetic work. Maclaurin's goal was to consolidate Newtonianism and establish its authority beyond all doubt: 'In order to proceed with perfect security, and to put an end forever to disputes'.[110] To do this, Maclaurin argued for the metaphysical foundations of Newtonian physics: absolute space and time are needed, forces are not fictional but real. Maclaurin wanted to answer those thinkers (including Berkeley, Hume, Daniel Bernoulli, Leibniz, Maupertuis, Huygens and Spinoza) who disagreed with some of the religious, metaphysical or methodological implications of the *Principia*, such as the status of absolute space and time. Accordingly, Book I of the *Account* begins with a chapter on scientific method done Newton's way, and then describes other scientific 'methods' to show their deficiencies. Book II introduces elementary classical mechanics. Having laid the foundation, Books III and IV proceed to address the Newtonian system of the world. Book III covers 'Gravity demonstrated by analysis', while Book IV deals with 'The effects of the general power of gravity deduced synthetically'. Maclaurin arranged all of this to justify the Newtonian method and the Newtonian style. But some explanation is required of the terms 'analysis' and 'synthesis' to which Maclaurin gave such prominence.

A reader might think that 'analysis' – the breaking-up of complex things into their component parts – is the method of all natural science, including the Cartesian system. But Newton justified his own version of 'analysis' in natural science by referring to the Greek usage, where problem-solving in mathematics is done by 'analysis'. In Query 31 of the *Opticks*, Newton wrote, 'As in Mathematicks, so in Natural Philosophy, the Investigation of difficult Things by the Method of Analysis, ought ever to precede the method of Composition'.[111] In Greek geometry, 'analysis' (or 'solution backward') meant assuming a problem solved, then working backwards from that assumed 'solution' until reaching something already known; once this had been done, the 'synthesis' could proceed, by beginning from the known thing and demonstrating the problem's solution.[112] Newton said that, by analogy, in natural philosophy, 'This Analysis consists in making Experiments and Observations, and in drawing general Conclusions from them by Induction, and admitting of no Objections against the Conclusions, but such as are taken from Experiments, or other certain Truths. For Hypotheses are not to be regarded in experimental Philosophy'. As for synthesis in natural philosophy, it 'consists in assuming the Causes discover'd, and establish'd as Principles, and by them explaining the Phaenomena proceeding from them, and proving the Explanations'.[113] But here, too, analysis comes first.

This passage from the *Opticks* is one of the most quoted statements of Newton's methodology. The disdain for hypotheses he expressed suggests that he is contrasting the mathematical method, which requires analysis first, to the premature synthesis of Cartesian physics. Newton had similarly contrasted analysis and synthesis in science in an unpublished preface he had prepared for the second edition of the *Principia*, implying that he had discovered the truths of the *Principia* by 'analysis' (calculus) and only later chose to present them by 'synthesis' (geometry) – presumably to help establish his priority over Leibniz in the dispute over inventing the calculus.[114] But whatever Newton's reasons were for wording the passage as he did, Maclaurin gave it a different slant.

In his own gloss on this passage, Maclaurin emphasised the certainty that was achieved by obeying Newton's precepts. And to demonstrate this certainty, he identified the success of Newtonian physics with its following the methods of mathematics (with a slap at the Cartesians on the way). It was 'in order to proceed with perfect security, and to put an end for ever to disputes', wrote Maclaurin, that Newton had told us

> in our enquiries into nature, the methods of analysis and synthesis should be both employed in a proper order ... It is evident that, as in mathematics, so in natural philosophy, the investigation of difficult things by the method of analysis ought ever to precede the method of composition, or the synthesis. For in any other way, we can never be sure that we assume the principles *which really obtain in nature*; and that our system, after we have composed it with great labour, is not mere dream and illusion.[115]

According to Maclaurin, Newton did not claim what he could not prove; he 'had a particular aversion to disputes [and] weighed the reasons of things impartially and coolly'.[116] Newton's scientific virtues were modesty, rather than pride; impartiality, rather than disputation; mathematics, rather than hypotheses; and, above all, success in finding the laws of nature by mathematical means, or, as Maclaurin put it, 'by so skilful an application of geometry to nature'.[117]

This is not the end of the story. The very last chapter of the *Account* covers natural theology and is entitled 'Of the Supreme Author and Governor of the Universe, the True and Living God'. Maclaurin chose to conclude by linking science and theology, not, as some have suggested, because of his religious background and theological training, and certainly not because of the imminence of his death when he wrote the *Account*,[118] since natural theology had been a major part of his Glasgow thesis. Rather, Maclaurin's conclusion can be seen as sharing the goals of both his reply to Berkeley and his youthful thesis, namely to refute the 'false schemes of natural philosophy [which] may lead to atheism, or suggest opinions, concerning the Deity and the universe, of most dangerous consequence to mankind'.[119] It is an example of what Larry Stewart has called 'the appropriation of providence by Newton's natural philosophy'.[120]

Maclaurin's natural theology eloquently argues from the nature of the solar system, not to the pale deity of the Deists, but to a God with Christian attributes:

> The admirable and beautiful structure of things for final causes, exalt our idea of the Contriver: the unity of design shews him to be One. The great motions in the system, performed with the same facility as the least, suggest his Almighty Power . . . the subtility of the motions and actions in the internal parts of bodies, shews that his influence penetrates the inmost recesses of things, and that He is equally active and present everywhere. The simplicity of the laws . . . the beauty . . . suggest his consummate Wisdom. The usefulness of the whole scheme, so well contrived for the intelligent beings that enjoy it . . . shew his unbounded Goodness . . . The Deity's acting and interposing in the universe shew that he governs it as well as formed it . . . He is not the object of sense; his essence, and indeed that of all other substances, is beyond the reach of all our discoveries, but his attributes clearly appear in his admirable works . . . Matter is not infinite or necessary, but he created as much of it as he thought proper . . . he formed the planets of such a number, and disposed them at various distances from the sun, as he pleased . . . we plainly perceive the vestiges of a wise agent, but acting freely and with perfect liberty.[121]

In this concluding chapter, Maclaurin intended to address the key concerns of his contemporaries in eighteenth-century Scotland. He had already stated that natural philosophy's chief value was to lay 'a sure foundation for natural religion and moral philosophy'.[122] But a naive Baconianism would not do. All we can obtain from experiments and observations alone are natural history and description; this does not take us from phaenomena 'to the powers or causes that produce them'. For that, we need something more.[123]

Let me end this section, then, with Maclaurin's answer to the question raised by men like Hobbes and Hume as to how a set of correlated observations could produce any understanding of causal connexions – natural laws capable of supporting science, technology, moral philosophy and theology. 'Experiments and observations', wrote Maclaurin, 'could not alone have carried [Newton] far in tracing the causes from their effects and explaining the effects from their causes: a sublime geometry was his guide in this nice and difficult enquiry'.[124] Maclaurin's Newtonianism, and anti-Cartesianism, did not make him oppose the use of mechanical principles in explanations: 'mechanism has its share in carrying on the great scheme of nature'.[125] Nevertheless, the 'sublime geometry' was, to Maclaurin, 'the instrument, by which alone the machinery of a work, made with so much art, could be unfolded'.[126] The Newtonian style alone could reveal the true nature of the universe that God had made.

Conclusion

The 'Newtonian style' seemed to Maclaurin to guarantee success, not only in science, but in areas ranging from theology to astronomy, from gauging to insurance. Maclaurin's successes and reputation helped others expect comparable achievements from approaches which were or which pretended to be Newtonian, from the quantified ethics of Hutcheson and the analysis-synthesis ideas of George Turnbull, to the geological world-machine of Maclaurin's student James Hutton and the economic systems of Adam Smith.[127]

Given Maclaurin's significance, it is remarkable how much research remains to be done on his mathematics and physics and on how they relate to one another, and on the way his scientific work informed his other endeavours. There is much yet to be done on the social roles he and his associates played, on the interaction between his ideas and those of others, on the role of his students in popularising Newtonianism among both men and women[128] and on the construction of audiences for popular science in Scotland in general. I hope that this chapter can help direct such work. Meanwhile, it is now clearer how Maclaurin's work embodied Newtonianism's mathematical prowess, its use of mathematical models to solve problems, its links with particular social classes and social forces to advance the development of society, its harmony with revealed religion, and, above all, its authority. The mathematical core of the Newtonian style helped inspire Maclaurin's quest to gain, for Newtonian natural philosophy, all the authority of mathematics. Colin Maclaurin's influence on scientists, craftsmen, theologians, philosophers and patrons provides a bridge from the doctrines of Isaac Newton to all aspects of Scottish thought and society.

Notes

1. Alexander Pope, *Epitaphs*, in *Poetical Works*, ed. Herbert Davis (Oxford and New York, 1985), 651. A classical ode in the manner of Lucretius, praising Newton's mathematically based superiority over all predecessors, was published in the *Principia* by Edmond Halley. The last line of Halley's 'Ode to Newton' reads, 'No

closer to the gods can any mortal rise'. See: Isaac Newton, *The Principia: Mathematical Principles of Natural Philosophy*, ed. I. Bernard Cohen, trans. I. Bernard Cohen, Anne Whitman and Julia Budenz (Berkeley and Los Angeles, 1999), 380; W. R. Albury, 'Halley's Ode on the *Principia* of Newton and the Epicurean Revival in England', *Journal of the History of Ideas* 39 (1978), 24-43. Maclaurin's epitaph in Greyfriars Churchyard (which was written by his son) was inspired by some lines of Virgil, and reads (in part): 'Electus Ipso Newtono suadente./ . . . / Hujus enim scripta evolve / Mentemque tantarum rerum capacem / Corpori caduco superstitem crede'. In Erik Sageng's translation: 'Elected at the urging of Newton himself/ . . . /Study the writings of this man, and believe that the mind capable of such great things/ Survives the fallen body'; Erik L. Sageng, 'Colin MacLaurin and the Foundations of the Method of Fluxions' (unpublished Ph.D. thesis, Princeton University, 1989), 2. See also Charles Tweedie, 'Samuel Johnson's Contribution to the Epitaph of MacLaurin', *The Mathematical Gazette* 9 (1919), 305; I thank Erik Sageng for this information.

2. On Maclaurin's scientific career and influence, see: Sageng, 'Colin MacLaurin'; Judith V. Grabiner, 'Was Newton's Calculus a Dead End? The Continental Influence of Maclaurin's *Treatise of Fluxions*', *American Mathematical Monthly* 104 (1997), 393-410; L. L. Laudan, introduction to Colin Maclaurin, *Account of Sir Isaac Newton's Philosophical Discoveries* (1748; New York and London, 1968), ix-xxv; John L. Greenberg, *The Problem of the Earth's Shape from Newton to Clairaut: The Rise of Mathematical Science in Eighteenth-Century Paris and the Fall of 'Normal' Science* (Cambridge, 1995); J. F. Scott, 'Maclaurin', in Charles C. Gillespie (ed.), *Dictionary of Scientific Biography*, 18 vols. (New York, 1970-90), 8:609-12; Charles Tweedie, 'A Study of the Life and Writings of Colin Maclaurin', *Mathematical Gazette* 8 (1915), 132-51; Herbert Westren Turnbull, *Bicentenary of the Death of Colin Maclaurin (1698-1746): Mathematician and Philosopher, Professor of Mathematics in Marischal College, Aberdeen (1717-1725)* (Aberdeen, 1951); Niccolò Guicciardini, *The Development of Newtonian Calculus in Britain, 1700-1800* (Cambridge, 1989); George Elder Davie, *The Democratic Intellect: Scotland and Her Universities in the Nineteenth Century* (Edinburgh, 1961), esp. pp. 137-48. I thank Erik Sageng for letting me draw on his yet-unpublished study.

3. 'I do not feign hypotheses'; 'General Scholium', Newton, *Principia*, 943 and n. See also: L. L. Laudan, 'Thomas Reid and the Newtonian Turn of British Methodological Thought', in Robert E. Butts and John W. Davis (eds.), *The Methodological Heritage of Newton* (Toronto, 1970), 103-31, esp. p. 104; Simon Schaffer, 'Newtonianism', in Robert C. Olby, G. N. Cantor, J. R. R. Christie and M. J. S. Hodge (eds.), *Companion to the History of Modern Science* (London and New York, 1990), 610-62; I. Bernard Cohen and Richard S. Westfall (eds.), *Newton: Texts, Background, Commentaries* (New York, 1995); Jed Z. Buchwald and I. Bernard Cohen (eds.), *Isaac Newton's Natural Philosophy* (Cambridge, MA, 2000).

4. I. Bernard Cohen, *Franklin and Newton: An Inquiry into Speculative Newtonian Experimental Science and Franklin's Work in Electricity as an Example Thereof* (Philadelphia, 1956).

5. I. Bernard Cohen, 'Newton's Method and Newton's Style', in Cohen and Westfall, *Newton*, 126-43, 132; I. Bernard Cohen, *The Newtonian Revolution: With Illustrations of the Transformation of Scientific Ideas* (Cambridge, 1980), Ch. 3.

6. G. E. Smith, 'The Newtonian Style in Book II of the *Principia*', in Buchwald and Cohen, *Newton's Natural Philosophy*, 249–313, at p. 249.
7. Cohen, 'Newton's Method and Style', 139 (emphasis added). Compare Cohen, *Newtonian Revolution*, 62–63. Cohen shows that Newton used this style by examining, not Newton's rhetoric, but his work. Still, it is worth quoting Newton's formulation: 'Mathematics requires an investigation of those quantities of forces and their proportions that follow from *any* conditions that may be supposed. Then, coming down to physics, these properties must be compared with the phenomena, so that it may be found out which conditions [or laws] of forces apply to each kind of attracting body. And then, finally, it will be possible to argue more securely concerning the physical species, physical causes, and physical proportions of these forces:' Newton, *Principia*, 588–89 (emphasis added).
8. Cohen, *Newtonian Revolution*, 66. Newton's best-known affirmation of the reality of the gravitational force is in the 'General Scholium' to the *Principia*: '[G]ravity really exists and acts according to the laws that we have set forth and is sufficient to explain all the motions of the heavenly bodies and of our sea'; Newton, *Principia*, 943. Another instructive example of Newton's use of the Newtonian style is the way he proved that if bodies move in circular orbits, the times taken vary according to the nth powers of the radii, that is, as R^n, if and only if the centripetal force varies inversely as R^{2n-1}. One consequence of this general mathematical relation is that the periodic times vary with the 3/2 power of the radii (Kepler's Third Law) if and only if the force varies inversely with the square of the radii. Thus the test of the general mathematical law against Kepler's observation establishes that the inverse-square force law holds for the solar system: Newton, *Principia*, Bk. I, Theorem 4 and Scholium, 449–52. Indeed, the very structure of the *Principia* exemplifies the Newtonian style. The first two Books give what Newton calls the mathematical principles of his physics, and only the third book is devoted to the actual 'system of the world'; see Newton's introduction to the third Book, *Principia*, 793.
9. For instance, Maclaurin discussed the consequences of an inverse-cube law and other laws for gravitation, and explained how Newton treated the motion of the apsides of a planetary orbit 'in *any* law of gravity' (emphasis added): Maclaurin, *Account*, 279, 284–85, 318.
10. Maclaurin, *Account*, 8.
11. Colin Maclaurin, 'De viribus mentium Bonipetis [On the Good-Seeking Forces of Mind]', Colin Campbell Collection, University of Edinburgh Library, MS 3099.15.6.
12. Maclaurin wrote the Rev. Colin Campbell that 'since I have mentioned the use of Mathematics I shall beg your pardon for troubling you with some thoughts I have relating to their use in morality ... under the title of De viribus mentium Bonipetis': Maclaurin to Campbell, 12 September 1714, in Stella Mills (ed.), *The Collected Letters of Colin Maclaurin* (Nantwich, UK, 1982), 161.
13. Maclaurin, 'De viribus', 1; Sageng, 'Colin Maclaurin', 124. The Newtonian character and language of this discussion of attractive forces and to what they are proportional should be apparent.
14. Maclaurin, 'De viribus', 4–5.
15. The idea for Maclaurin's essay may have been suggested by John Craige's *Theologiae Christianae Principia Mathematica* (1699), in which Craige graphs the intensity of pleasures as various functions of time, calculates the total pleasure by integration and concludes, using arguments so Newtonian that nineteenth-

century readers saw his work as a parody of Newton's *Principia*, that one should eschew the finite pleasures of this world for the infinite pleasures of the world to come. Craige was a Scot who, as early as 1688, was closely associated with many of the people important in Maclaurin's career, including David Gregory, Archibald Pitcairne, Colin Campbell, Edmond Halley and Newton. Maclaurin and Craige were also eventually acquainted. See Richard Nash, *John Craige's Mathematical Principles of Christian Theology* (Carbondale and Edwardsville, IL, 1991).

16. For instance, in the study of 'mechanical engines', Maclaurin laid new emphasis on the importance of finding 'what ought to be the proportion of the power and weight to each other . . . that it may produce *the greatest effect possible, in a given time*'. Maclaurin, *Account*, 149 (emphasis in the original). Elsewhere in the *Account* he determined the best design for water wheels (pp. 172–75), windmills (pp. 175–76) and the placement of the sails of a ship while sailing 'with a side wind' (pp. 177–78).

17. The pyramidal base of a cell of a honeycomb, whose sides are hexagons, is bounded by three rhombuses. Maclaurin demonstrated geometrically that the minimal amount of wax for a given volume is obtained when the angle of the rhombus is $2 \arctan \sqrt{2}$, about $109° 28'$; Maclaurin then cited the recent observations by G. F. Maraldi and Réamur of $110°$. This confirmed the honeycomb's 'Regularity and Beauty, connected of Necessity with its Frugality'; Sageng, 'Colin Maclaurin', 133, quoting from Colin Maclaurin, 'Of the Bases of Cells wherein Bees Deposit Their Honey', *Philosophical Transactions* 42 (1743), 565–71. Compare Maclaurin, *Account*, 176. For a draft of the paper on bees, see Maclaurin to Martin Folkes, 30 June 1743, in Mills, *Correspondence*, 386–91. See also Maclaurin to Folkes, 31 January 1744, in Mills, *Correspondence*, 397–400, where Maclaurin applies the same principles to the most economical way to build a barn.

18. In 1728, Maclaurin wrote to Hutcheson praising the latter's *An Inquiry into the Original of our Ideas of Beauty and Virtue* (1725), expressing admiration for Hutcheson's 'principles' and 'moral sense': Maclaurin to Hutcheson, 22 October 1728, in Mills, *Correspondence*, 25–27.

19. Mary Poovey, *A History of the Modern Fact: Problems of Knowledge in the Sciences of Wealth and Society* (Chicago and London, 1998), 185.

20. Francis Hutcheson, *An Essay on the Nature and Conduct of the Passions and Affections with Illustrations on the Moral Sense* (1729; Gainesville, FL, 1969), 40; this passage is quoted in Poovey, *Modern Fact*, 186.

21. Francis Hutcheson, *An Inquiry into the Original of Our Ideas of Beauty and Virtue*, 2nd edn (1726; Charlottesville, VA, 1993), 117; quoted in Poovey, *Modern Fact*, 186.

22. Maclaurin to Colin Campbell, 12 September 1714, in Mills, *Correspondence*, 159–60.

23. Colin Maclaurin, *Dissertatio philosophica inauguralis, de gravitate, aliisque viribus naturalibus, quam cum annexis corollariis* (Edinburgh, 1713), vi (my translation).

24. Roger L. Emerson, '"The affair" at Edinburgh and the "project" at Glasgow: The Politics of Hume's Attempts to Become a Professor', in M. A. Stewart and John P. Wright (eds.), *Hume and Hume's Connections* (Edinburgh, 1995), 1–22, p. 19.

25. 'Chosen at the urging of Newton himself' – from Maclaurin's tombstone in Greyfriars Church, Edinburgh.

26. Maclaurin to Folkes, 25 March 1731, in Mills, *Correspondence*, 32.

27. Roger L. Emerson, 'Aberdeen Professors 1690–1800: Two Structures, Two

Professoriates, Two Careers', in J. J. Carter and J. H. Pittock, *Aberdeen and the Enlightenment* (Aberdeen, 1987), 155–67 , esp. pp. 159, 160, 164.

28. Maclaurin to Campbell, 6 July 1720, in Mills, *Correspondence*, 162. The *Geometria organica* generalises the Newtonian 'organic' generation of conics – that is, the theoretical construction of curves by means of idealised instruments – to all orders of curves, and classifies curves on the model of Newton's *Enumeratio linearum tertii ordinis*: Sageng, 'Colin Maclaurin', 13–16. While in London Maclaurin bought scientific instruments for the collection at Marischal College: Paul B. Wood, *The Aberdeen Enlightenment: The Arts Curriculum in the Eighteenth Century* (Aberdeen, 1993), 15.
29. Richard S. Westfall, *Never At Rest: A Biography of Isaac Newton* (Cambridge, 1980), 832. On Newton, Halley, and patronage patterns during this period, see also Richard S. Westfall and G. Funk, 'Newton, Halley and the System of Patronage', in Norman Thrower (ed.), *Standing on the Shoulders of Giants: A Longer View of Newton and Halley* (Berkeley, 1990), 3–13.
30. Maclaurin to Campbell, 6 July 1720, in Mills, *Correspondence*, 163. A. Rupert Hall has written that Newton and Maclaurin 'never met', but clearly they did: A. Rupert Hall, *Isaac Newton: Adventurer in Thought* (Cambridge, 1992), 367.
31. Newton to Maclaurin, 21 August 1725, in Mills, *Correspondence*, 171–72; see also Maclaurin to Folkes, 25 March 1731, in Mills, *Correspondence*, 32.
32. Edinburgh Council Register, 51:15, quoted in Sageng, 'Colin Maclaurin', 29–30. The other 'great men' whose influence the Council felt could well have included Lord Ilay and the 2nd Duke of Argyll (I thank a referee for this suggestion), but the references to reputation and learning would seem to follow from the scientific qualifications praised by Newton himself.
33. Maclaurin's courses were described at length in the *Scots Magazine*: First 'College': arithmetic, decimals, Euclid Books I-6, plane trigonometry, how to use logarithmic and sine tables, surveying, fortification, 'other practical parts' of applied mathematics, the elements of algebra, and geographical lectures; Second: review and advancement of algebra, solid geometry, spherical trigonometry, navigation, conic sections, gunnery, astronomy, optics; Third: perspective, more astronomy and optics, Newton's *Principia*, and the direct and inverse method of fluxions. Maclaurin also gave a course in experimental philosophy, including astronomical observation with telescopes; see 'A Short Account of the University of Edinburgh, the Present Professors in It, and the Several Parts of Learning Taught by Them', *Scots Magazine* 3 (1741), 371–74, as cited in Sageng, 'Colin Maclaurin', 31.
34. For a detailed and thoroughly documented account of Ilay's scientific interests, political career and role as a patron, which convincingly argues for his key role in modernising eighteenth-century Scottish institutions, see Roger L. Emerson, 'The Scientific Interests of Archibald Campbell, 1st Earl of Ilay and 3rd Duke of Argyll (1682–1761)', *Annals of Science* 59 (2002), 21–56.
35. Alexander Murdoch, *The People Above: Politics and Administration in Mid-Eighteenth Century Scotland* (Edinburgh, 1980), 7.
36. John Stuart Shaw, *The Management of Scottish Society, 1707–1764* (Edinburgh, 1983), 130.
37. Emerson, 'Hume's Attempt', 2.
38. See, for example, Maclaurin to Folkes, 25 March 1731, John Conduitt to Maclaurin, 10 March 1732/3 and 24 August 1734, and Maclaurin to Archibald Campbell of Knockbuy, 9 April 1725 and 4 October 1743, in Mills, *Correspondence*, 32, 46, 60, 111–12, 171.

39. Maclaurin to Archibald Campbell of Knockbuy, 4 October 1743, in Mills, *Correspondence*, 111.
40. R. Scott, 'The Politics and Administration of Scotland, 1725–1748' (unpublished Ph. D. Thesis, University of Edinburgh, 1981), 87. This episode is discussed more fully below. See also Judith V. Grabiner, '"Some Disputes of Consequence": Maclaurin among the Molasses Barrels', *Social Studies of Science* 28 (1998), 139–68.
41. Maclaurin to Folkes, 25 March 1731, in Mills, *Correspondence*, 31–33.
42. I thank J. R. R. Christie for suggesting this possibility.
43. Alexander Carlyle, *Anecdotes and Characters of the Times*, ed. James Kinsley (London, 1973), 17.
44. V. A. Eyles, 'Hutton', *Dictionary of Scientific Biography*, 6:577–89, at p. 578.
45. D. B. McIntyre, 'James Hutton's Edinburgh: The Historical and Political Background', *Earth Sciences History* 16 (1997), 100–157, at pp. 138–39.
46. Sageng, 'Colin Maclaurin', 38–39; Maclaurin to Archibald Campbell, 21 February 1745, in Mills, *Correspondence*, 123–24.
47. Larry Stewart, *The Rise of Public Science: Rhetoric, Technology and Natural Philosophy in Newtonian Britain, 1660–1750* (Cambridge, 1992), 392.
48. Colin Maclaurin, *A Treatise of Fluxions in Two Books*, 2 vols. (Edinburgh, 1742), 1:1.
49. Geoffrey Cantor, 'Berkeley's *The Analyst* Revisited', *Isis* 75 (1984), 668–83; *idem*, 'Anti-Newton', in John Fauvel, Raymond Flood, Michael Shortland and Robin Wilson (eds.), *Let Newton Be!* (Oxford, 1988), 211–13.
50. This point is made clearly in the subtitle of *The Analyst*: *Wherein It is Examined Whether the Object, Principles and Inferences of the Modern Analysis Are More Distinctly Conceived, or More Evidently Deduced, than Religious Mysteries and Points of Faith*; see George Berkeley, *The Analyst: Or, a Discourse Addressed to an Infidel Mathematician*, in A. A. Luce and T. E. Jessop (eds.), *The Works of George Berkeley, Bishop of Cloyne*, 8 vols. (London, 1951), 4:65–102. For a summary of Berkeley's attack, see Carl B. Boyer, *The History of the Calculus and Its Conceptual Development* (New York, 1959), 224–27.
51. Hutcheson to Maclaurin, 21 April 1737, in Mills, *Correspondence*, 274.
52. As I have argued at length elsewhere, Maclaurin contributed a great deal to the eventual algebraic rigorisation of the calculus in the nineteenth century; Grabiner, 'Was Newton's Calculus a Dead End?', esp. 398–404.
53. Maclaurin, *Treatise of Fluxions*, 1:vii-viii.
54. Maclaurin, *Treatise of Fluxions*, 1:x.
55. The proofs are by double *reductio ad absurdum*; such a proof requires inequalities, and Maclaurin's axioms provide them: Maclaurin, *Treatise of Fluxions*, 1:59.
56. Newton had briefly flirted with infinitesimals, but then denounced them.
57. Maclaurin, *Account*, 87, and *Treatise of Fluxions*, 1:39; see also Sageng, 'Colin Maclaurin', 155–58.
58. Maclaurin, *Treatise of Fluxions*, 1:47; compare Maclaurin, *Account*, 87.
59. Maclaurin, *Treatise of Fluxions*, 2:576; see also Grabiner, 'Was Newton's Calculus a Dead End?', 394–95. Maclaurin, like Newton, was both a consummate geometer and an accomplished and enthusiastic practitioner of algorithmic mathematics (contrary to George Elder Davie's characterisation of Maclaurin as a mathematical Hellenist like Simson; see Davie, *Democratic Intellect*, esp. pp. 111–12, 128–48, 151–55). On Maclaurin's willingness to use the power of algebra when appropriate, and the influence of this point of view on Thomas Reid and John Playfair, see Paul B. Wood, 'Thomas Reid, Natural Philosopher: A Study of

Science and Philosophy in the Scottish Enlightenment' (unpublished Ph. D. thesis, University of Leeds, 1984), 225–26, 229–31.
60. Grabiner, 'Was Newton's Calculus a Dead End?', 398–401; Sageng, 'Colin Maclaurin', Chs. III-IV. For Britain, see Guicciardini, *Newtonian Calculus*, 47–51; Judith V. Grabiner, *The Origins of Cauchy's Rigorous Calculus* (Cambridge, MA, 1981), 80–87.
61. Maclaurin to Sir Andrew Mitchell, 18 November 1735, in Mills, *Correspondence*, 68.
62. Nor does such an account exist. But see Greenberg, *Problem of the Earth's Shape*, 412–25, 585–601; Grabiner, 'Was Newton's Calculus a Dead End?', 397–404; Turnbull, *Maclaurin Bicentenary*, 17–19; Tweedie, *Life and Writings of Maclaurin*, 146–49; and Sageng, 'Colin Maclaurin', 47–76, 190–233. See also Erik Sageng, 'Maclaurin: Treatise on Fluxions', in I. Grattan-Guinness (ed.), *Landmark Writings in Western Mathematics, 1640–1940* (Amsterdam and London, forthcoming).
63. That is, assuming that the columns from the centre to the surface of the spheroid all balance, and treating the effective gravity at any point on the surface of the figure as perpendicular to the surface at that point.
64. Greenberg, *Problem of the Earth's Shape*, 172. Newton had not proved this. For Newton's treatment of the topic, see Newton, *Principia*, Book III, Prop. 19, and 347–50.
65. Maclaurin, *Treatise of Fluxions*, 2:551–66. Maclaurin remarked that the observations made by various observers at various latitudes (summarised in the *Treatise of Fluxions*, 2:550) gave reason to believe that the variation of the density of the earth was 'considerable' (2:551), and for that reason, the mathematics of it needed to be pursued. Maclaurin's work on this topic, extended by eighteenth-century physicists like Clairaut, Legendre and Laplace, has continued to be important to physics and astronomy. Nobel Laureate in physics Subrahmanyan Chandrasekhar devoted an entire chapter to what are now called 'Maclaurin spheroids' – the figures that arise when homogeneous bodies rotate with a uniform angular velocity – in his classic, *Ellipsoidal Figures of Equilibrium* (1969; New York, 1987), 77–100. Chandrasekhar also found it essential to begin his modern theoretical discussion by proving a number of Maclaurin's results: Chandrasekhar, *Ellipsoidal Figures*, 38, 49.
66. See Maclaurin to John Johnstoun, 9 May 1738, Stirling to Maclaurin, 26 October 1738, Maclaurin to Folkes, March 2 1738/9 and 6 March 1738/9, and Maclaurin to the Earl of Morton, 26 February 1740, in Mills, *Correspondence*, 292–93, 304–6, 308–13, 345–46. John L. Heilbron, *The Sun in the Church: Cathedrals as Solar Observatories* (Cambridge, MA, 1999), 166–75, tells how such measurements were carried out by Cassini, Boscovich and others.
67. Maclaurin to Folkes, March 2 1738/9, in Mills, *Correspondence*, 308–9. On the role of Graham's observations in determining the earth's shape, see also Richard Sorrenson, 'George Graham, Visible Technician', *British Journal for the History of Science* 32 (1999), 203–21.
68. Maclaurin, *Treatise of Fluxions*, 2:513–66; compare Greenberg, *Problem of the Earth's Shape*, 590–97.
69. Maclaurin, *Account*, 62.
70. Patrick Murdoch, 'An Account of the Life and Writings of the Author', in Maclaurin, *Account*, xix.
71. Maclaurin to the Earl of Morton [1739], in Mills, *Correspondence*, 441–45.
72. Mills, *Correspondence*, esp. pp. 79, 87, 90, 96, 97, 105–110, 166, 172, 252–53, 255–56, 308–13, 314–21, 327–33, 338–39, 340–42, 370–72, 394–95, 404–9.

73. Maclaurin, *Treatise of Fluxions*, 2:727–28, 733–35, 735–42. Maclaurin worked in all of the principal areas of applied mathematics in the age of Newton: hydraulic engineering, fortification, navigation, map-making, political arithmetic, mortality tables and gauging. On these areas, see the editors' introduction to Buchwald and Cohen, *Newton's Natural Philosophy*, xvii; Richard S. Westfall, 'Background to the Mathematization of Nature', in Buchwald and Cohen, *Newton's Natural Philosophy*, 321–39, esp. 321–27; Grabiner, 'Disputes of Consequence', 154–56; and Ch. 3 above.
74. Murdoch, 'Account', xix.
75. David Gregory, *A Treatise of Practical Geometry in Three Parts*, ed. Colin Maclaurin (Edinburgh, 1745).
76. Judith V. Grabiner, 'A Mathematician Among the Molasses Barrels: MacLaurin's Unpublished Memoir on Volumes', *Proceedings of the Edinburgh Mathematical Society* 39 (1996), 193–240.
77. Grabiner, 'Disputes of Consequence', 152–53.
78. Grabiner, 'Unpublished Memoir', 239–40. See Maclaurin, *Treatise of Fluxions*, 1:24–26, for a proof by classical geometry. Excise officers did have a fair amount of mathematical training, and the British Excise Service in the eighteenth century embodied clear, rationalised practice and quantitatively based rules: John Brewer, *The Sinews of Power: War, Money and the English State, 1688–1783* (London, 1989), esp. pp. 94–114. On the parallel between the rationalised Excise Department and the standard gauging manuals' rational analysis of idealised barrels, see Miles Ogborn, 'The Capacities of the State: Charles Davenant and the Management of the Excise', *Journal of Historical Geography* 24 (1998), 289–312, esp. pp. 289, 301, 306.
79. Grabiner, 'Disputes of Consequence', 140–43; see Grabiner, 'Unpublished Memoir', 201–22, for the full details. Ogborn argues that the mathematical idealisation which treats physical barrels as solids of revolution is clearly false, limiting mathematical authority and allowing for negotiation: Ogborn, 'Capacities of the State', 301, 303. But in the present case, Maclaurin's use of the Newtonian style to handle the deviations from the ideal form extended his authority beyond that of the ordinary mathematically trained gauger.
80. Maclaurin, *Account*, 62.
81. Roger L. Emerson, 'The Philosophical Society of Edinburgh, 1737–1747', *British Journal of the History of Science* 12 (1979), 154–191, esp. pp. 162, 164, 178. On the rhetoric of consensus and of avoiding political and religious controversy, and this rhetoric's situation in the Royal Society of London, see Steven Shapin, *The Scientific Revolution* (Chicago, 1996), 105–108, 112, 134–35.
82. Discussions of matters of this sort may be found throughout Maclaurin's correspondence. A good sampling may be obtained using the entries under the Edinburgh Philosophical Society in the index to Mills, *Correspondence*.
83. A wide 'Maclaurin circle' can be identified by looking at the overlapping subscription lists to works by Abraham de Moivre, Henry Pemberton, Nicholas Saunderson, Robert Smith, James Stirling and Maclaurin's *Account*; see P. J. Wallis, 'The MacLaurin "Circle": The Evidence of Subscription Lists', *The Bibliotheck* 11 (1982), 38–54. On the importance of subscription publication for Newtonian science in general, and the way it interacted with different forms of patronage, see Stewart, *Public Science*, 151–60. Maclaurin was also a member of the philosophically minded Rankenian Club.
84. Stewart, *Public Science*, 119–23. J. T. Desaguliers was an especially influential populariser. On this topic, see also Roger Cooter and Stephen Pumfrey, 'Separate

Spheres and Public Places: Reflections on the History of Science Popularization and Science in Popular Culture', *History of Science* 32 (1994), 237–67.
85. Stewart, *Public Science*, 391.
86. Stewart, *Public Science*, 361.
87. Maclaurin, *Account*, 13, 62 (emphasis added).
88. Murdoch, 'Account', xix.
89. Wallace to Maclaurin, 26 February 1744, in Anon., 'The Second Paper of Memoirs of Mr MacLaurin beginning 1725', Glasgow University Library, MS Gen 1378/2. The Wallace letter is copied in an eighteenth-century hand; the Library catalogue says of this manuscript volume that it 'Contains material not used by Patrick Murdoch in his account of Maclaurin's life'.
90. See the editor's introduction in A. Ian Dunlop (ed.), *The Scottish Ministers' Widows' Fund, 1743–1993* (Edinburgh, 1992), xii.
91. J. B. Dow, 'Early Actuarial Work in Eighteenth-Century Scotland', in Dunlop, *Widows' Fund*, 27.
92. Andrea A. Rusnock, 'Biopolitics: Political Arithmetic in the Enlightenment', in William Clark, Jan Golinski and Simon Schaffer (eds.), *The Sciences in Enlightened Europe* (Chicago and London, 1999), 55–57.
93. Michael W. Flinn (ed.), *Scottish Population History from the Seventeenth Century to the 1930s* (Cambridge, 1977), 58–64, 251–52; Charles W. J. Withers, 'How Scotland Came to Know Itself: Geography, National Identity and the Making of a Nation, 1680–1790', *Journal of Historical Geography* 21 (1995), 378.
94. A. Ian Dunlop, 'Provision for Ministers' Widows in Scotland – Eighteenth Century', in Dunlop, *Widows' Fund*, 9.
95. Dow, 'Early Actuarial Work', 53; Maclaurin to Robert Wallace, 23 May 1743, in Mills, *Correspondence*, 108.
96. Maclaurin to Sir Andrew Mitchell, 24 May 1743, in Mills, *Correspondence*, 109.
97. Maclaurin to Robert Wallace, 23 May 1743, in Mills, *Correspondence*, 108. Compare D. J. P. Hare and W. F. Scott, 'The Scottish Ministers' Widows' Fund of 1744', in Dunlop, *Widows' Fund*, 63, and Dow, 'Early Actuarial Work', 53.
98. Hare and Scott have called the Widows' Fund of 1744 'the earliest actuarially-based fund in the world'. What they mean by 'actuarially based' is that the mathematical model must include the mortality rates of the lives involved, the rate of interest likely to be earned on the fund's investments, the expenses of management, and – Maclaurin's contribution – the use of an accurate and realistic life table: Hare and Scott, 'Widow's Fund of 1744', 57, 68.
99. Dunlop, 'Provisions for Widows', 19–20.
100. Dow, 'Early Actuarial Work', 24.
101. Dow, 'Early Actuarial Work', 41–42. Price revised his views in the third edition, published in 1773. Later editions include the preface from the third edition; see, for example, Richard Price, *Observations on Reversionary Payments*, 7th edn. (London, 1812), xxi-xxvii; for Price's acceptance of Webster's corrections, see pp. xxi, 117–18.
102. Poovey, *Modern Fact*, 280–81.
103. Withers, 'How Scotland Came to Know Itself'.
104. Theodore M. Porter, 'Precision and Trust: Early Victorian Insurance and the Politics of Calculation', in M. Norton Wise (ed.), *The Values of Precision* (Princeton, NJ, 1995), 173–97; Theodore M. Porter, *Trust in Numbers: The Pursuit of Objectivity in Science and Public Life* (Princeton, NJ, 1995). See also the emerging literature on the question of trust and science, notably Steven Shapin, *A*

Social History of Truth: Civility and Science in Seventeenth-Century England (Chicago and London, 1994).

105. Murdoch, 'Account', xix. Other difficulties faced by the scheme in 1742–43 included the unpopularity of Walpole's government, anxiety about possible rebellion in Scotland, and the circumstance that the Widows' Fund was not only the type of scheme that often proved unreliable but also that it was exclusively Scottish: Dow, 'Early Actuarial Work', 32. The threat of rejection by Parliament was real: even the Equitable's first application for a charter was 'rejected by our Law-officers' because life insurance was 'so little understood'. W. Morgan, 'Introduction to the Seventh Edition', in Price, *Observations on Reversionary Payments*, viii.
106. Lorraine J. Daston, 'The Domestication of Risk: Mathematical Probability and Insurance, 1650–1830', in Lorenz Krüger, Lorraine J. Daston and Michael Heidelberger (eds.), *The Probabilistic Revolution, Volume 1: Ideas in History* (Cambridge, MA, 1987), 237, 253–55; Lorraine J. Daston, *Classical Probability in the Enlightenment* (Princeton, NJ, 1988), 171–87.
107. Maclaurin, *Account*, 9; compare Isaac Newton, *Opticks, or a Treatise of the Reflections, Refractions, Inflections and Colours of Light*, 4th edn (1730; New York, 1952), 404.
108. Wallis, 'Maclaurin Circle'.
109. See Laudan's editorial introduction to Maclaurin, *Account*, x.
110. Maclaurin, *Account*, 8–9.
111. Newton, *Opticks*, 404–5.
112. Carl B. Boyer, 'Analysis: Notes on the Evolution of a Subject and a Name', *Mathematics Teacher* 47 (1954), 450–62. In addition, in Newton's time, the term 'analysis' often indicated the particular problem-solving methods of algebraic symbolism or the algorithms of fluxions; see I. Bernard Cohen's editorial commentary in Newton, *Principia*, 122–23.
113. Newton, *Opticks*, 404–5.
114. See Cohen's editorial commentary in Newton, *Principia*, 49, 122–23. Newton's claim is doubtful. D. T. Whiteside, the leading scholar of Newton's mathematics, says emphatically that all the evidence indicates that Newton composed the *Principia* exactly as it is published. It is, of course, widely recognised that there are many calculus-based ideas in the *Principia*; see Cohen's editorial commentary in Newton, *Principia*, 123–27.
115. Maclaurin, *Account*, 8–9 (emphasis added). Compare Maclaurin's statement in the preface to the *Treatise on Fluxions* that mathematical demonstration left 'no place for doubt or cavil'(1:1).
116. Maclaurin, *Account*, 13.
117. Maclaurin, *Account*, 11.
118. See Laudan's editorial introduction in Maclaurin, *Account*, xviii.
119. Maclaurin, *Account*, 4.
120. Stewart, *Public Science*, 59.
121. Maclaurin, *Account*, 381–83.
122. Maclaurin, *Account*, 3.
123. Maclaurin, *Account*, 221.
124. Maclaurin, *Account*, 221; compare p. 8.
125. Maclaurin, *Account*, 388.
126. Maclaurin, *Account*, 8.
127. P. B. Wood, 'Science and the Pursuit of Virtue in the Aberdeen Enlightenment', in M. A. Stewart (ed.), *Studies in the Philosophy of the Scottish Enlightenment*

(Oxford, 1990), 127–49, esp. 130–33; McIntyre, 'Hutton's Edinburgh', 138, which should be compared with Maclaurin, *Account*, 390; Richard Olson, *Science Deified and Science Defied: The Historical Significance of Science in Western Culture*, 2 vols. (Berkeley, Los Angeles and London, 1990), 2:219–22.

128. Writing to Sir John Clerk, Maclaurin said that 'One of my scholars gives a short course . . . and several Ladies of quality attend. This I hope will incite the Young Gentlemen to study, that they might be able to answer the questions of the Ladies': Maclaurin to Clerk, 23 March 1743, in Mills, *Correspondence*, 87.

6

The Burden of Procreation: Women and Preformation in the Works of George Garden and George Cheyne

ANITA GUERRINI

In the *Philosophical Transactions* of the Royal Society of London for February 1691 a paper appeared entitled 'A Discourse concerning the Modern Theory of Generation', written by the Aberdeen *virtuoso*, Dr. George Garden. Apart from the early use of the term 'modern' in the title, the paper offered a spirited defence of the animalculist version of preformation theory, which declared that the preformed embryo was contained in the semen of the father. Garden stated his premises clearly: '1. That Animals are *ex animalculo*. 2. That these Animalcles [sic] are originally *in semine Marium & non in Foeminis*. 3. That they can never come forward, nor be formed into Animals of the respective kind, without the *Ova in foeminis*'.[1]

Fifty years later, in his *Natural Method of Cureing the Diseases of the Body* (1742), Garden's former pupil, the physician George Cheyne, reiterated these ideas and placed them in the context of his discussion of female illnesses and the role of women in procreation. In this chapter, I will examine Garden's and Cheyne's ideas in the context of the philosophical and religious beliefs that they developed in Aberdeen. Cheyne further developed Garden's particular fusion of natural philosophy and theology in the light of Garden's own subsequent theological journey along the path of the 'mystics of the northeast', a journey which Cheyne repeated. In his popular medical works of the 1720s and 1730s Cheyne translated the language of mysticism into prescriptions for everyday behaviour. His work added a spiritual dimension to preformation theory which emphasised the importance of the role of women, although he continued to believe in the animalculist interpretation. In Garden's formulation, animalculism gave women little role in reproduction, but Cheyne's positive portrayal of the role of women reflected his concern for the health of his many female patients.

Together, the work of Garden and Cheyne gives a broader background to the development of Enlightenment thought in Scotland. The role of natural knowledge in the Scottish Enlightenment is still much debated among historians.[2] Most historiography of the European Enlightenment, and especially of Enlightenment science (with the notable exception of Margaret C. Jacob), has minimised the role of religion in its formulation.[3] 'Religion' and 'Enlightenment' indeed have been thought of as antithetical and, even in

Jacob's work, the emphasis is on the movement towards deism and irreligion rather than the survival of personal belief. Recent work on the history of eighteenth-century Britain, in contrast, recognises in various ways the fundamental role played by religion.[4] Similarly, historians of science, beginning with Richard S. Westfall's magisterial biography of Isaac Newton, have begun to recognise that personal religion in its many forms cannot simply be written out of the story of eighteenth-century science.[5]

In the historiography of the Enlightenment in Scotland, both religion and local contexts have gained increasing attention. Aberdeen, in particular, has been the focus of much study in the past decade.[6] Its unique nexus of religion, politics and natural philosophy makes it an especially useful case study to challenge the traditional characterisations of the Scottish Enlightenment. As a crossroads not only of Episcopalianism and Presbyterianism but also with the addition of a strong Catholic presence and various sects such as the Quakers, Aberdeen, unlike other cultural centres in Scotland, was forced to grapple with the implications of a multiplicity of Christian beliefs.

George Garden and Preformation Theory

George Garden, who lived from 1649 to 1733, was a doctor of divinity, not of medicine. Although Elizabeth Gasking in her *Investigations into Generation* identified him as a bishop, he in fact never attained that status.[7] He was a Regent at King's College, Aberdeen, in 1673, and taught there throughout the 1670s and 1680s. The son of an Episcopalian clergyman, Garden had particularly close ties with the most prominent Scottish Episcopalians of his era. He was ordained in 1677 by Patrick Scougal, Bishop of Aberdeen, and was a close friend of Scougal's son Henry. Henry Scougal was Professor of Divinity at King's, and his *Life of God in the Soul of Man* (1677) was celebrated as a treatise on practical mysticism. Scougal advocated a form of personal religion to counteract the religious factionalism that had divided Scotland throughout the seventeenth century; Roger L. Emerson has called him 'the most influential Scottish episcopalian writer of the period'.[8] Scougal placed himself in a Scottish Episcopalian tradition of personal religion dating back to the 1620s that included such figures as John Forbes of Corse and Bishop Robert Leighton. 'True Religion', wrote Scougal, 'is an Union of the Soul with God'.[9] It had little to do with the quarrelling churches of his day. Both Scougal and, before him, Leighton, had declared that control of the body and its passions was a step toward achieving grace. Among other topics, Scougal lectured on ethics at King's College, and his colleagues continued to deliver these lectures for many years after his death. The ethics lectures, like his book, emphasised that the love of God was the basis for true morality, since reason could never induce virtue. He mentioned Descartes and Boyle, and was clearly influenced by the Cambridge Platonists, especially John Smith and Henry More. In a funeral sermon for Scougal, Garden claimed he was the first man in Aberdeen 'who taught the youth that philosophy which has now the preference by all the knowing world', that is, the mechanical philosophy.[10] Scougal was succeeded in the divinity chair by Garden's brother, James.

Scougal and the Gardens established King's College as a centre of Scottish Episcopalian thought during the period of the second episcopacy between the Restoration and the Glorious Revolution. As a student at Marischal in the late 1680s, George Cheyne probably heard much discussion of theology and its relationship to moral philosophy. He would also have learned something of the mechanical philosophy of nature. In the late 1680s, Descartes was prominent in natural philosophy lectures at King's and Marischal, and the experiments of Boyle and others earned mention. By the early 1690s, Newton's ideas also entered the curriculum.[11] Cheyne's education in Aberdeen combined religion and natural philosophy in ways that mutually informed each.

George Garden had a well-established reputation as a *virtuoso* in natural philosophy. He had already published several articles in the *Philosophical Transactions* on an eclectic collection of topics – including the weather, bees and caterpillars – prior to the appearance of his 1691 article on generation. One of his essays, in the form of a letter addressed to Scougal, indicated that they had had regular conversations on topics in natural philosophy.[12] But by 1691, his circumstances had changed considerably from the time of his first publication in 1685. The Glorious Revolution had permanently altered the position of Scottish Episcopalians. The Presbyterian Kirk was re-established as the Church of Scotland, and Episcopalians in the ministry and the universities were required to take oaths to William and Mary and to the established (Presbyterian) church to retain their positions. In 1692, George Garden was removed from his post as Minister at the church of St Nicholas, the town parish of Aberdeen, for refusing to pray for William and Mary. His brother James was deprived of the King's College professorship of divinity for refusing to sign the Westminster Confession in 1696.

When his article on preformation appeared, therefore, George Garden was in the midst of a personal and professional crisis. His motivation for writing the article remains mysterious. Did the practice of natural philosophy provide a relief from this state of crisis, as it did for some during the English Civil War, or was it an expression of his crisis? Was Garden trying to establish a professional identity outside Scotland, should he be compelled to leave? If so, he did not follow this up. The Council of the Royal Society approved his election as a Fellow in 1695, but for unknown reasons he was never elected.[13] Nor did he leave Scotland, although many others in similar positions did. Garden's article was in the form of a letter to William Musgrave, a physician and Fellow of the Royal Society. As the editor of the *Philosophical Transactions* in the mid-1680s, Musgrave had published some of Garden's first essays.

Garden's article summarised several decades of speculation about the nature of reproduction. William Harvey's *De generatione* (1651), with its motto of '*ex ovo omnia*', argued strongly both for the primacy of the egg in generation and for a theory of epigenesis which saw foetal development as a gradual process of growth and differentiation from an undifferentiated embryo. This theory was soon challenged by a number of competing theories which together centred on the idea that the embryo somehow contained all of its parts at the outset; foetal development, then, was a process of growth, not of differentiation. Nothing was created anew. While some believed, with Harvey, that the foetus did not exist in any sense at the moment of generation, they diverged from him in claiming that the foetus came into being all at once, with all its parts, immediately after generation. The remaining process of parturition was simply one of growth. This notion of metamorphosis differed from preformation, namely the idea that the foetus was a miniature adult formed before conception in the body of its parent. Opinions differed over whether this parent was female or male (ovism or spermatozooism, also known as animalculism). While the Dutch naturalist Jan Swammerdam had argued for ovism in the 1670s, the majority of thinkers favoured animalculism after the Dutch microscopist Anthony van Leeuwenhoeck published his observations of spermatozoa in the *Philosophical Transactions* in 1679. Preformation must in turn be differentiated from pre-existence, which stated that the preformed embryo was formed, not by the parent, but by God at the Creation, and had therefore existed for all time. The ovist version of this theory thus claimed that Eve held the eggs for all of humanity.[14]

Descartes asserted in the 1630s that all aspects of the animal, including generation, could be explained in mechanical terms and, for much of the seventeenth century, mechanical philosophers endeavoured to elaborate these theories. Descartes himself opted for a theory of epigenesis, which, he explained, occurred mechanically by means of a combination of fermentation and condensation.[15] Garden followed most of his contemporaries in contemptuously dismissing Descartes on this matter: 'And indeed all the Laws of Motion which are as yet discovered, can give but a very lame account of the forming of a Plant or Animal. We see how wretchedly *Des Cartes* came off when he began to apply them to this Subject'.[16]

Shirley Roe has argued that preformation theories responded to two problems which arose from a mechanistic theory of epigenesis. As Garden pointed out, it was difficult to see how mechanism could explain how living beings came about. In addition, mechanistic epigenesis posed a significant theological problem, since if matter could organise itself into life, what role remained for God the Creator?[17] Although the teaching of Descartes dominated the natural philosophy curriculum at the Aberdeen universities, Garden's dismissal of Descartes is thus not as surprising as it seems at first glance. The mediation of the Cambridge Platonists, who viewed natural knowledge as a means to know God, is evident here. Natural philosophy by itself had its limits:

> they [i.e. plants and animals] are form'd by Laws yet unknown to Mankind, and it seems most probable that the *Stamina* of all the Plants and Animals that have been, or ever shall be in the World, have been formed *ab Origine Mundi* by the Almighty Creator within the first of each respective kind.[18]

Garden therefore not only supported preformation against epigenesis, he also supported pre-existence. Quite apart from the theological question, Garden marshalled a number of scientific reasons for his support of preformation and pre-existence, displaying an impressive range of knowledge. His interlinked scientific and religious arguments emerged from the Aberdeen curriculum which conjoined morals, religion and natural philosophy. The theological tensions evident in Aberdeen by the early 1690s made the task of reconciling natural philosophy with theology all the more important.

Garden claimed that visual evidence supported the contention that animals are preformed. The Italian physician and natural philosopher Marcello Malpighi saw 'the Rudiments of an Animal in the shape of a Tadpole' in the *cicatricula* or embryo-forming part of the chicken egg. Garden explained that what Harvey had seen as merely the *'punctum saliens'* – the pulsating point of blood from which the foetus developed – was seen more accurately by Malpighi under the microscope as a preformed being, visible within thirty hours of the start of incubation. Garden explained that it was more probable that these parts merely grew from a smaller size, upon receiving the nutrition provided by the remainder of the egg, than that they were formed anew in such a short time.[19] Further evidence included Swammerdam's observations of the folded-up parts of the butterfly in the pupa. Swammerdam had first enunciated a theory of preformation in 1669, and he viewed metamorphosis as both the model for preformation and conclusive evidence of God's creative power. Spontaneous generation, which many had used to explain metamorphosis, struck him as profoundly irreligious. In a 1686 letter on caterpillars (not published until 1698), Garden also argued against spontaneous generation, although he did not give a theological argument. He argued similarly in 1691, mainly from observational data. While the individual facts, said Garden, might not be wholly convincing, taken together the evidence was strong.[20]

In his *Recherche de la vérité* (1674), the French Cartesian philosopher Nicholas Malebranche had argued that the tiny organisms revealed by the microscope might be only the most visible of the subset of smallness, which extended to the infinitely small.[21] Garden accepted the idea that visible nature may indeed contain an infinity of tiny parts.[22] Yet the true structures of things, like the laws governing reproduction, remained hidden. While Garden implied that we might some day have this knowledge, he thought that for now it must be accepted on trust, a trust that God would not deceive us.

Garden went on to demonstrate that the pre-existing individual could only be contained in the male seed and not in the female. Here he relied on Leeuwenhoek's observations of spermatozoa, which established their existence in the semen as well as their active nature which seemed to imply some kind of vital role in generation. Since, said Garden, the observed rudiments of the foetus appeared only in fertilised eggs, it seemed obvious that the pre-existing being did not exist in the egg, but in the sperm. Malpighi's tadpole-like early embryo, which took the shape of the spermatozoon, provided additional confirmation.[23]

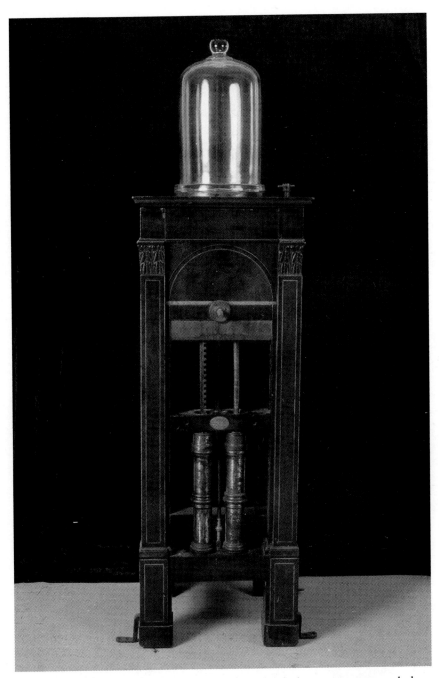

1. Air pumps were one of the standard instruments used by experimental philosophers in the Enlightenment. This air pump was made by the Glasgow instrument maker James Crichton between 1786 and 1796.

2. Although more common that the air pump, barometers could be expensive precision instruments. This barometer (with, inset, details of the brass scale and the categories of weather) was manufactured by James Watt in about 1770 with an eye to portability and was aimed at the cheaper end of the market.

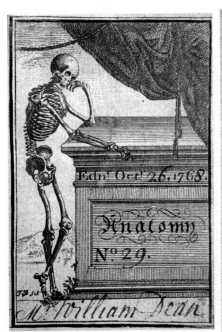 

3. Class cards were issued to students who had paid their fees to their instructors, although cards were sometimes given *gratis* to less well-off students by their professors. These cards – one (left) for the anatomy class of Alexander Monro *secundus* and the other for private anatomical demonstrations by John Innes – belonged to William Dean, who matriculated at the University of Edinburgh in 1768.

THE
ART
OF
LAND-MEASURING
EXPLAINED.

IN FIVE PARTS, VIZ.

I. TAKING DIMENSIONS.
II. FINDING CONTENTS.
III. LAYING OUT GROUND.
IV. DIVIDING.
AND
V. PLANNING.

WITH AN APPENDIX CONCERNING INSTRUMENTS.

BY JOHN GRAY,
TEACHER OF MATHEMATICS IN GREENOCK,
AND LAND-MEASURER.

GLASGOW:
PRINTED BY ROBERT AND ANDREW FOULIS FOR THE AUTHOR.
SOLD BY D. WILSON AND J. DURHAM, IN THE STRAND, LONDON;
G. HAMILTON AND J. BALFOUR, EDINBURGH;
AND R. AND A. FOULIS, GLASGOW.
M.DCC.LVII.

4. The title page of John Gray's *The Art of Land Measuring Explained* (1757). The book was printed by the Foulis Press of Glasgow, using a typeface cast by Alexander Wilson, who later became the Professor of Practical Astronomy at Glasgow University. The Foulis Press set new standards in scientific and other publishing in the print culture of Enlightenment Scotland.

5. This portrait of William Cullen (1710–1790) was done after William Cochrane. Cullen's writings in chemistry and medicine were read throughout the Atlantic world. Like other men of science in his day, Cullen was actively interested in other subjects, including agricultural improvement.

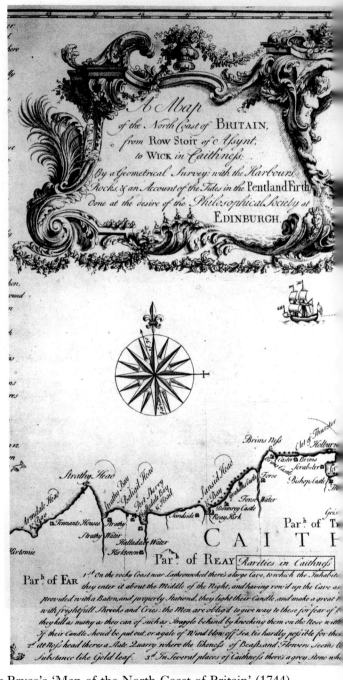

6. Part of Alexander Bryce's 'Map of the North Coast of Britain' (1744). This map is of interest for its combination of geometrically derived survey information with written descriptions of local customs, geology and natural history.

Part of ORKNEY

Whiteford Hill Kirkwall
Scapa Flow

Hoy Head
Hoy Hill
Red Head
Kir Hope
Hoccey Head
South Ronaldsha
Cantock Head
Swithia
Barrock Hill
Swana
Stroma Isld
Lother

PENTLAND FIRTH

Windy knap or Dunnet Head
Ham Bay Freskerry
Brugh
Kotter Mey Castle Gills
Murkle Bay
Dunncioft
Thurristoft
Clerick Kirk Garth
Par.h of Dunt
Par.h of Cannesbay
Cannesbay Kirk Ishon Grouts
Stroma Swithie
Mell's Ebb
Cannesbay
Dungsbay Head
Stacks of Dungsbay
Middle Skerry
Little Skerry

Par.h of Olerick
Tester
Treswick
Treswick Bay
Ochingill
Keiss

NESS
Waster
Sinclairs Bay
Acrigil Noss Head
Carnigo
Storehouse Staxigo
Wick Papigo
Old Wick Castle Wick Bay
Hempriss

Boat in the month of October, to kill Seals or Selchs
can, they oce a Shore, each person being well
brings down the Selchs in a confused Body,
own by them, but when this throng is over
ns these they trail to their Boats & carry home
their lives.
n upon the Stones, with some Yellow
s bright as a Candle, without having its

7. This painting from the early 1740s by an unknown artist, 'Sir Archibald Grant of Monymusk and Family', illustrates the way in which scientific instruments functioned as cultural artefacts and indicates that natural knowledge was the subject of family conversation in a domestic context.

8. Science in the field involved careful observation and direct contact less with Nature. Fieldwork was crucial to the development of James Hutton's ideas on erosion, uplift and repair and to the concept of 'deep time'. Hutton and John Clerk, accompanied by Hutton's dog, are examining an exposure of sandstone and overlying sill of teschenite.

Some natural philosophers worried about the theological implications of the loss of so many pre-existing individuals in unused spermatozoa. Garden did not mention this problem, and indeed found in the numbers of male seed an explanation for multiple births. This, in turn, he wrote, 'gives a new light as it were to the first Prophecy concerning the *Messiah*, that the Seed of the Woman shall bruise the Head of the Serpent, all the rest of Mankind being thus most properly and truly the Seed of the Man'.[24] Thus despite his emphasis on observed facts, the final affirmation of preformation theory was for Garden not philosophical, but theological.

While, by this argument, Adam rather than Eve held humankind in his loins, it could not be denied that the female played an important role in reproduction since both the egg and the seed were necessary for the process to occur. The female egg provided both nest and nourishment for the foetus. Garden argued that the egg, and not simply the uterus, was the proper receptacle for the foetus, by means of numerous examples of extra-uterine conceptions from the pages of the *Philosophical Transactions* itself. In addition, Garden claimed that the foetus did not immediately attach itself to the uterus, further proof that the egg and not the uterus was the *nidus*. Eve Keller has argued that preformation theory, even in the ovist version, assumed a passive role for the egg. The animalculist version, of course, makes this passivity more obvious by giving agency to the male sperm. Garden's language, she further argues, endowed the preformed foetus with an identity and a will, for it consciously sent out feelers to attach to the uterus.[25] Finally, Garden cited de Graaf, Malpighi, Leeuwenhoek and Harvey to support his claim that the female eggs were contained in the '*testes foeminae*' or ovaries, and asserted that conception must take place either in the ovary or in what we call the Fallopian tubes.[26]

As Elizabeth Gasking points out, Garden was the first to give 'a physical account of conception itself which was in accord with both Harvey's and Leeuwenhoek's discoveries'.[27] But in addition, and perhaps more importantly for his purposes, Garden placed the entire argument in a specific theological context. To him, natural philosophy served the aims of Biblical exegesis. Over the course of the 1690s, as Garden's political and professional situation worsened, he retreated from public life into the personal religion advocated by Scougal. In about 1695 he encountered the works of the Flemish mystic Antoinette Bourignon (whom Swammerdam had known), who also argued that the goal of personal religion was to attain purity of spirit through the conduct of daily life. But far more than the Scottish Episcopalians, she urged discipline of the body as a means of attaining that goal. While rejection of the things of the world was a standard Christian message, Bourignon went farther than most mystics in emphasising the mortification of the body, along with spiritual passivity, as the most important means of attaining a state of grace. She was widely considered to be theologically suspect, and Garden's wholehearted adoption of her ideas – he translated several of her works into English, wrote a long commentary, and even led a Bourignonist commune for a time –

put him even farther outside the political and religious mainstream than he had already been. In 1701 he was ejected from the ministry by the General Assembly of the Church of Scotland, although he continued to preach and hold services surreptitiously for his circle of followers.[28]

Cheyne, Garden and Preformation: The Philosophical Principles of 1705

Cheyne did not encounter Garden again after he left Aberdeen in 1689 until around the end of 1705.[29] Earlier that year, Cheyne published his *Philosophical Principles of Natural Religion*. Written in the style of the well-known Boyle Lectures, this book, as its title implied, gave proofs of 'natural religion' from natural philosophy. Natural religion emphasised reason rather than revelation, and Cheyne intended his substantial text to place him at the forefront of that throng of Newtonian apologists who argued that Newtonian theory proved the existence of God. Cheyne's book failed resoundingly to bring him professional success, but his account of generation reveals how much he borrowed from Garden, and how he fitted that account into his larger Newtonian framework.

Cheyne's God, like Newton's, is all-powerful and can suspend the laws of nature at will. Mechanism alone cannot adequately explain creation without the active intervention of God. 'The Production of Animals,' he declared, 'is altogether inconsistent with the Laws of *Mechanism*.'[30] A 'casual concourse of *Atoms*' could never produce a living creature, and

> very few who have considered the Matter but own that ev'ry Animal proceeds from a pre-existent *Animalcul*, and that the Parents conduce nothing but a convenient Habitation and suitable Nourishments to it, till it is fit to be trusted with the Light, and capable of receiving the Benefit of the Air.[31]

Although the animal was itself a machine, a machine could not reproduce itself. Like Garden, Cheyne contemptuously dismissed Descartes on this issue, and made it clear that the most compelling argument for his claims was their consilience with Christian theology.[32] Pre-existence was one of a web of concepts which together proved the existence of God, one among Cheyne's many proofs against Epicureanism and the notion that the world is eternal and uncreated.

Yet in this work Cheyne never made entirely clear *which* parent carried the pre-existent animalcule. Presumably it was the female who provided 'habitation and nourishments', but Cheyne wrote 'parents': what was the role of the father? In his typically prolix manner, Cheyne wandered from a discussion of generation to a broader discourse on divine design as exhibited by the animal body, and did not return to the question of generation until nearly forty pages later. Here he asserted his belief in pre-existence: 'that ev'ry generated *Animal* is produc'd from a preexistent *Animalcul* of the same *Species*'. Since he had disproved the Cartesian – 'our Adversary's' – theory that matter and motion alone could produce life, then plants and animals 'must of necessity have been from all Eternity. And consequently that all the *Animals* and *Vegetables* that have existed or shall exist, have actually been all included in the first of ev'ry *Species*'. Since Adam was the first of the human species and not Eve, this would confirm Garden's assertion that Adam contained within himself the seed for all of humankind. But Cheyne never actually stated this. Instead, he went on to argue that pre-existence precludes the existence of the world from eternity, because it would imply an infinitely large number of plants and animals in the world, which we know is not the case.[33]

Was Cheyne's ambiguity deliberate? Perhaps not. He would have assumed that his readers knew the ideas of preformation, and the most common version at that time was animalculism. Nonetheless, the reference to 'parents' and the omission of specific references to gender is intriguing, and a point to which I will return.

Discouraged by his lack of success in London, at the end of 1705 Cheyne experienced a personal crisis and returned to Aberdeen.[34] His attempts to enter the inner circle of Newton's apostles and to establish a fashionable medical practice had both failed, and after four years of strenuous effort to settle himself in London he had little to show. He became seriously depressed, and his weight, always difficult to keep under control, ballooned to over thirty stones. The combination of physical and mental symptoms led in turn to a crisis of faith, and he began to doubt that the 'natural theology' of his *Philosophical Principles* was adequate for salvation. When he re-encountered Garden at this time, he was open to Garden's mystical religious ideas. 'Natural religion' was no longer spiritually satisfying to Cheyne, just as the London culture of natural philosophy which produced it was not professionally fulfilling. Cheyne's retreat to Scotland and especially to Aberdeen, and not to the Edinburgh of his mentor in natural philosophy, Archibald Pitcairne, is therefore highly significant. He believed that his crisis was spiritual, and turned to a spiritual mentor. The very successful medical and scientific ideas he developed after 1705 were deeply influenced by this specific Aberdeen context, and this is especially apparent in his views on reproduction.

At the end of 1705 Cheyne was depressed, unemployed and overweight. Garden's message of personal religion, and Bourignon's emphasis on the rejection of the body, held particular meaning for him at this moment. Over a period of several months, Cheyne underwent a spiritual conversion, lost weight, and regained his self-confidence. He abandoned 'natural religion' and embraced Garden's peculiarly Scottish blend of mysticism and personal religion. His subsequent works reflected this profound change in sensibility.

In 1715 he published a new edition of the *Philosophical Principles of Natural Religion*. Now entitled *The Philosophical Principles of Religion, Natural and Revealed*, the 1715 edition illustrated how far Cheyne had travelled in the previous decade. His previous confidence in natural philosophy and natural religion were now replaced with an acknowledgement of the contingent nature of natural knowledge, and the overriding role of the spirit in human affairs.[35]

However, his comments on generation remained unchanged from the 1705 edition.

The 'women's doctor' and the Burden of Procreation

Following the appearance of the 1715 edition of the *Philosophical Principles*, Cheyne devoted himself to his steadily growing medical practice in Bath, where he resided permanently after 1718. His success owed much to his appeal to women patients, and when he returned to the question of preformation in the late 1730s he could draw on a wealth of clinical experience. In Bath, he specialised in nervous disorders, and in a series of bestselling popular works on medicine – *An Essay on the Gout* (1720), *An Essay of Health and Long Life* (1724), *The English Malady* (1733) – he helped to define the fashionable disorder of 'the vapours'. While 'the vapours' or 'the spleen' afflicted both men (in whom it was often known as hypochondria or 'the hyp') and women, women's weaker constitutions were generally believed to be more susceptible to it. In addition, the spleen affected those primarily of a more refined and intellectual character, while the working classes hardly ever experienced it.[36]

In Dr. Cheyne, women found an especially sympathetic and trustworthy caregiver. Here he followed the example of his fellow Scot, David Hamilton, whose care of Queen Anne focused less on her obstetrical problems than on her mental state. Hamilton's sympathy and conversation earned him the Queen's gratitude and trust, as well as a knighthood and considerable wealth.[37] Cheyne himself continued to struggle both with his weight and with melancholy, experiencing another personal crisis in the 1720s. At one point his weight rose to thirty-three stones and he could barely walk. The fact that his own symptoms centred on his excessive weight may have made his therapeutic emphasis on diet inevitable. But his own experience connected therapy to the religious ideas of Garden and Bourignon, who argued that self-denial in food and drink was a means toward achieving grace. In his popular works, he urged his audiences to reform both their diet and their souls. Moderation in diet could keep unruly passions in check and at the same time relieve many physical symptoms. He demonstrated this in his own case. By the time he described 'the case of the author' in *The English Malady*, he had lost weight with a vegetarian diet and had resolved his spiritual crisis by reading mystical literature.

One of Cheyne's female patients was the young Catherine Walpole, daughter of the Whig politician. She suffered from a combination of 'hysteria', lack of appetite and fainting. Her hysterical fits often lasted for hours. While a modern physician might diagnose her as a typical anorexic teenager, her symptoms do not precisely correspond with any modern category. Some contemporary physicians labelled her disorder 'green sickness' from the extreme pallor of the skin, which looked greenish. Green sickness was thought to be a particular ailment of unmarried young women, often cured by marriage. Cheyne did not give Catherine Walpole's illness a name, and he treated her with the usual remedies of emetics and cathartics, and later prescribed both bathing in and drinking Bath waters. Catherine nonetheless died at the age of nineteen in 1723. Cheyne's therapies reveal no startling innovations, albeit with one exception. In an age when many physicians prescribed without seeing a patient, and avoided touching them, Cheyne took another tack. He visited Catherine often, talking to her sometimes for hours. Conversation became an important part of his therapy, as it had been for Hamilton.[38]

By the time Cheyne met the Countess of Huntingdon a decade later, he had written two bestselling books and refined his therapeutic techniques. The Countess, the former Selina Shirley, married Theophilus Hastings, the young Earl of Huntingdon, in 1728. The Countess very soon began to fulfil one of the chief obligations of an aristocratic wife, namely to produce an heir. Amanda Vickery has noted that 'genteel women were positively immersed in the business of child-bearing and child-rearing'. Since Theophilus was the only surviving male among four sisters, securing the inheritance was especially important. He was raised by his half-sister Lady Elizabeth Hastings (usually known as Lady Betty), who never married.[39]

The Countess had her first child about a year after her marriage. By the time she met Cheyne in 1732, she was twenty-four and the mother of three; she had four more children before 1740. This was a standard pattern among aristocratic women. Commenting on a slightly later period, Judith Schneid Lewis, who surveyed childbirth among the aristocracy between 1760 and 1860, noted that most women became pregnant within a year of marriage and bore on average eight children. For aristocratic women, Lewis comments, 'adult life [was] nearly synonymous with pregnancy'. Cheyne knew Lady Betty and had corresponded with the Countess, beginning about 1730.[40]

Lady Huntingdon did not have fits like Catherine Walpole, but she too suffered from 'lowness of spirits', stomach pains and 'heats'. She also seems to have had some sort of skin inflammation from time to time. In a letter of 1730 Cheyne detected bladder stones, that curse of the upper classes, but reassured the Countess that the stone 'was never known dangerous in your sex'.[41] Cheyne sympathised with her 'pains and sufferings' and described Lady Huntingdon's case as a 'distracting, sinking nervous complaint'. But her symptoms, he said, while in part hereditary, were largely due to her almost continuous state of pregnancy, for she gave birth in 1729, 1730, 1731, 1732, 1735, 1737 and 1739.[42] Cheyne wrote in 1735, 'I greatly hope the worst is over with your ladyship if you do not breed again, for had it not been for that you must have been well ere now'.[43] Many of Catherine Walpole's symptoms, particularly lack of appetite and of menstruation, made no appearance.

Cheyne told a fellow physician that Lady Huntingdon's ailment was an 'acrid, hot humour of the juices'.[44] A cooling regimen was thus in order. While he recommended the standard practices of bleeding, vomiting and purging to lessen the heat of the body and to get rid of the excessive hot humour, he also recommended a 'cool and scanty diet' consisting mainly of milk and vegetables, and liberal use of mineral water. He was tempted, he wrote her, to urge her to follow a completely vegetarian diet, which would effect a complete cure. But he hesitated: 'it is particular and inconvenient in the world, and all man and womenkind will be up in arms against me, and your Ladyship will often be told you are killing yourself by Dr Cheyne's whims'.[45]

During her pregnancies, he prescribed abstinence from alcohol and the mildest of purgatives, avoiding the stronger metallic concoctions. A child's diet, he said, would suit both mother and child the best. Cheyne frequently praised the Countess for following his regime, while at other times he 'earnestly beseech[ed]' her to continue on her course. Although a 'low' diet had long been recommended for pregnant women, not everyone agreed, and Lady Huntingdon heard much opposing advice.[46] 'Keep to your diet inviolably,' urged Cheyne, 'let who will say to the contrary.' But she did not take his word for any of his assertions, and asked him to give his opinion of her case to another doctor.[47] Each physician had his own favourite medicines, and the Countess seems to have tried them all. Her relatives had their own favourite cures as well. Lady Betty sent a recipe for 'snail-watter' in 1731, noting that 'Phisicians will seldome approve what they don't recomend'.[48] In addition, Lady Huntingdon prescribed for herself. She wrote her husband in 1735, 'I was so much out of order . . . I was forced to take a puke last night'.[49] While Cheyne often recommended vomits, it is not clear that he would have approved of such self-diagnosis.[50] He reprimanded Lady Huntingdon in 1734: 'you must give over all medicines. I fear you have suffered by taking so many without a full notion of your case'. The Countess did not reveal all of her symptoms to her physicians either, because, said Cheyne, of her natural 'modesty and reservedness'.[51]

Lady Huntingdon, unlike many of her contemporaries, disliked Bath, which she thought 'stupid'. As her letters home attest, she begrudged being away from her family on her trips to the spa. However, she continued to have children, and her symptoms do not seem to have lessened. She therefore persisted in seeking relief at Bath as well as at other spas such as Bristol, Buxton and Scarborough, and in consulting Cheyne and other doctors.[52] Her sisters-in-law, the Ladies Elizabeth, Margaret, Ann and Frances Hastings, also followed the doctor's prescriptions intermittently. Lady Betty commented to her brother that Lady Ann's problem was that 'she eates & drinks as she use'd to do'.[53] Lady Betty was famous for her hospitality, although she herself lived frugally, and a luncheon menu from the 1720s illustrates the usual meat-heavy diet of the aristocracy. The only vegetables served were salad and cabbage.[54] Presumably this was what Cheyne advised Lady Huntingdon against.

Preformation and the Preservation of the Aristocracy

Cheyne's final two books presented his mature views on gender, preformation and childbirth. The *Essay on Regimen*, published in 1740, summarised his natural philosophy, while his last book, *The Natural Method of Cureing the Diseases of the Body and the Disorders of the Mind Depending on the Body*, summarised his medical practice. In the *Essay on Regimen* he returned to the theme of preformation, reiterating ideas which dated back to 1705. But he now added a new spiritual dimension. In the 1715 edition of the *Philosophical Principles*, he had argued in opposition to Cartesian dualism for a Platonic continuum between matter and spirit. In the *Essay on Regimen*, he combined this with preformation theory. In this view, the animal body proceeded to grow from an original spiritual body:

> [the] spiritual animal *Body*, at first divinely organised, may be rolled up, folded together and contracted in this present State of its Duration, into an infinitely small *Punctum Saliens* . . . lodg'd in the Loins of the Male of all Animals (for it is highly probable the Female was but a *secondary* Intention, or a *Buttress* to a falling Edifice) and proceeding in a *diverging Series*, and progressive Gradation, that in due Time it may be fit to be nourished, and increased by the Juices of the proper Female, and thereby enabled to bear the Coarseness and Injuries of this *ruinous Globe*, and gross Element, to which it is to be Condemned for a certain Period.[55]

Garden's view that preformation allows us to understand Biblical prophecy is fully elaborated here, in a manner which also solves the problem of the quantity of physical spermatozoa. Adam's seeds were not physical seeds but spiritual seeds. In keeping with Cheyne's earlier reference to the role of both parents in nourishment, Adam also provided the initial nutrition and embodiment which then allowed the spermatozoa to assume physical form and allow conception to occur. Eve's contribution was to allow the physical body to exist on Earth. This role was no minor task, and Cheyne noted that 'imprudent and improper Nourishment' by mothers and nurses led to 'suffering and misery' of the body. Although a certain amount of suffering was a necessary part of being human, Cheyne believed that most human suffering was both unnecessary and undesirable.[56] Traditional thinking, following Aristotle, had asserted that the male provides mind while the female provides body. Cheyne offered a specific theological context for this theme and a partial reinterpretation of the relative roles of male and female.

Cheyne dedicated the *Essay on Regimen* to the Earl of Huntingdon, noting that 'Good *Lady* HUNTINGDON, I believe, has benefited by some of the Rules for Health laid down in this Treatise'.[57] But the *Natural Method*, not the *Essay on Regimen*, contained his final word on medical care. In the first part of the *Natural Method*, he reiterated his views on preformation and elaborated his ideas on gender. Cheyne believed that the sexes formed a continuum rather than a dichotomy, and that the notion of placing 'two *equal* Human Souls' in two differently formed bodies only occurred to God after he realised that Adam by himself was inadequate.[58] Cheyne had told his friend, the writer and publisher Samuel Richardson, that 'I know no Difference between the Sexes but in their Configuration. They are both of the same Species and differ only in Number as in Number two is after one'.[59] He had previously only minimally distinguished between male and female ailments, viewing hypochondria and hysteria as almost identical.

However, Cheyne's association with Lady Huntingdon and his reconsideration of generation in the *Essay on Regimen* now led him also to reconsider female illnesses as specific to the sex. Women's reproductive capacity, he now believed, played a central role in the state of their health, and *vice versa*. He confined the female role in generation to providing nourishment for the embryo pre-formed in the male seed. Yet because diet was at the centre of Cheyne's ideas of sickness and health, the female role in generation was absolutely critical:

> There is no Nation in Europe, perhaps, where great and opulent Families sooner become extinct, or change Lineage so quickly, as they do in *England*; or where such Devastation of the Female Sex, especially among those of Rank and Condition, is made by Abortions [i.e. miscarriages] and dangerous Child-bearing.[60]

The Earl of Huntingdon owed his title to the early death of his older half-brother. Queen Anne had endured seventeen pregnancies, and her only child to survive beyond early childhood died at the age of eleven. Many other examples of early deaths and forfeited titles could be found. In his study of early modern marriage, Lawrence Stone pointed out that 'in three out of four cases of all first marriages among the squirearchy that were broken by death within ten years, the cause was the death of the wife'.[61] The health of women was essential to the health of the English aristocracy.

The English aristocracy were victims of their own economic good fortune. The poor and the middle classes, said Cheyne, suffered no such reproductive misfortunes, 'and no-where is there a finer or more numerous Posterity, than among the *Highlanders* of *Scotland*, or the *native Irish*', whose offspring he contrasted to the 'deformed, diseased, stunted and short-lived' English.[62] In view of the famine of the 1690s in Scotland, which Cheyne presumably witnessed, and ongoing famines in Ireland in the 1720s and again in 1740–41, this claim seems ironic, if not disingenuous. However, there is some evidence that Irish military recruits, at least, were healthier than their English counterparts.[63]

Cheyne attributed both difficult pregnancies and the early deaths of adults and children to a luxurious adult lifestyle which included too much rich food and drink. The increase in difficult births in his time, which, he said, made '*Man-Midwifery* so necessary and profitable a profession', was directly attributable to bad regimen. David Hamilton's early success was as a man-midwife to the aristocracy, and the number of man-midwives was increasing. Man-midwives had originally been called to assist female midwives in difficult or unusual births. Their growth in numbers signified to Cheyne that difficult births were increasing.[64]

Physicians agreed that women were naturally weaker in body than men, in part, argued Cheyne, from the stresses of childbirth. Although women were naturally more *'temperat, abstemious,* and *modest'* than men, the coarser, stronger male body was better able to withstand the effects of 'their Licentiousness and Luxury'.⁶⁵ Women were naturally more susceptible, he further argued, to the effects of lifestyle. In particular, he complained to Richardson, women were more susceptible to luxury.⁶⁶ They preferred the urban life of the London season to a dull, if virtuous, winter on a country estate. But *'London Houses',* Cheyne wrote, 'are made to last Fifty Years, *Country Houses* for Five hundred'.⁶⁷ The 'childlike' diet recommended by Cheyne was also a plain country diet, as opposed to the more elaborate food of the city; 'such a *Diet* makes *Children, Farmers* and *Country Persons,* who can afford no other *Diet,* look the freshest, most healthy and lively'.⁶⁸ For women, especially childbearing women, the combination of luxury and their natural temperament could be disastrous. Lady Huntingdon, for example, was continuously ill during her decade of childbearing, and she lost at least one child shortly after birth. Others fared worse.⁶⁹

The renunciation of luxury was to Cheyne not only medically advisable but a moral obligation. He argued against both those who felt that life without luxury was not worth living and those who, already ill, felt that self-denial was too dear a price to prolong their lives. Such 'immoral and murderous *Indulgence'* was a form of suicide (and a form of murder for physicians who allowed it), and therefore a most serious sin. While his methods might not lead to a perfect cure, they would ameliorate most conditions, 'under the *Order* of Providence over them, and the benign *Influence* of the *Sun* [sic] of *Righteousness, who has Healing under his Wings'.* In other words, God would care for those who led a thoughtful and abstemious life.⁷⁰ As he had learned from Garden and others, regimen was both physical and spiritual. Restraint in diet was, as Garden and Bourignon had argued, one means of attaining grace.

Cheyne had attempted to resolve Lady Huntingdon's symptoms with physical regimen, but management of pregnancy could not in itself solve the deeper problem of her overall health, which had been undermined at an early age, even perhaps in the womb. Young virgins such as Catherine Walpole had demonstrated to Cheyne that the vapours and other symptoms began long before marriage. While most contemporary physicians followed popular wisdom in declaring that green sickness could be cured by marriage and childbirth, Cheyne emphatically disagreed, calling the conventional wisdom a 'common profane *Joke'.*⁷¹ If anything, childbearing would make things worse. He told Lady Huntingdon that she would only find relief if she stopped having children. She should have followed the proper regimen from an early age. Catherine Walpole presumably died because the doctor had been called in too late. A proper regimen would prevent symptoms from appearing in the first place.⁷²

Cheyne advised women, therefore, to take control of their bodies by maintaining their own health. He viewed himself as a guide and adviser to his patients, who themselves had the free will to make good and bad choices. Medicine was just part of what was necessary to live a complete life. For childbearing women, maintenance of their bodies also took on a spiritual significance, since self-restraint was one means of attaining grace. In addition, the significance of nutrition in procreation meant that the health and happiness of the human race, and its continuation, rested squarely on women.

Conclusion

If we are to see Garden and Cheyne as participating in the Scottish Enlightenment, and not merely by chronological coincidence, we must broaden our view of that movement to include both natural knowledge and religion. In addition, we must extend our definition of religion in this era to include elements previously considered opposed to enlightened principles, even going as far as to include mysticism. It is nearly seventy years since G. D. Henderson first described the group he called the 'mystics of the northeast', and his study has yet to be surpassed in its thoroughness. The movement we might call the 'Episcopalian Enlightenment' of Aberdeen reminds us of the heterogeneous nature of the Scots intellectual cosmos. I have endeavoured to show just how sharply it contrasts with the more familiar intellectual movements in Edinburgh and (as discussed in Chapter 4) Glasgow.

It could be argued that both Garden and Cheyne can be viewed as rather conventional Enlightenment figures in their use of reason. Cheyne's late works in particular expressed basic principles of the Enlightenment: the use of reason to determine moral behaviour, the lawfulness of nature which made human bodies essentially alike, the responsibility of the individual and, above all, the goal of the good life. But to see Cheyne and Garden merely as men of reason is to dismiss their own motivations for doing natural philosophy at all. They were 'enlightened' both in their use of reason and in their spiritual enlightenment by revelation. In the Aberdeen context, both reason and revelation were necessary to make sense of conflicting religious and philosophical points of view. While Garden described preformation primarily in terms of its logical and observational proofs, the final proof of its truthfulness as a theory could only come from its agreement with theology. Later, Cheyne resolved the tension between sensibility and reason by allowing a space for the spiritual. His immense influence on the notion of sensibility, that central concept of the Enlightenment, means that we cannot ignore the role he gave to spirit. In so doing, he demonstrates how much we need to re-evaluate our notions of 'Enlightenment'.

Notes

An earlier version of this chapter was delivered at the Gender and History Seminar, History Department, University of California, Santa Barbara. I am grateful to Erika Rappaport and other participants for their comments, as well as to the editors and to Michael A. Osborne.

1. George Garden, 'A Discourse Concerning the Modern Theory of Generation, by Dr. George Garden of Aberdeen, being Part of a Letter to Dr. William Musgrave, L.L.D. Reg. Soc. S. and by him Communicated to the Royal Society', *Philosophical Transactions* 16 (1691), 474–83, at p. 475.
2. Paul Wood, 'Introduction: Dugald Stewart and the Invention of "The Scottish Enlightenment"', in Paul Wood (ed.), *The Scottish Enlightenment: Essays in Reinterpretation* (Rochester, NY, 2000), 19–20. This article provides an excellent overview of the historiography of the Scottish Enlightenment.
3. See, for example, William Clark, Jan Golinski and Simon Schaffer (eds.), *The Sciences in Enlightened Europe* (Chicago and London, 1999), which mentions religion only in passing. Compare Margaret C. Jacob's *The Newtonians and the English Revolution* (Ithaca, NY, 1976), and *The Radical Enlightenment: Pantheists, Freemasons and Republicans* (London, 1981).
4. See, for example, Paul Langford, *A Polite and Commercial People: England, 1727–1783* (Oxford, 1992); Roy Porter, *English Society in the Eighteenth Century*, 2nd edn. (London, 1990); Linda Colley, *Britons: Forging the Nation, 1707–1837* (New Haven, CT, 1992); and Frank O'Gorman, *The Long Eighteenth Century: British Political and Social History, 1688–1832* (London, 1997); only the latter two treat Britain rather than England.
5. R. S. Westfall, *Never at Rest: A Biography of Isaac Newton* (Cambridge, 1980); Roger L. Emerson, 'Natural Philosophy and the Problem of the Scottish Enlightenment', *Studies on Voltaire and the Eighteenth Century* 242 (1986), 243–291; *idem*, 'Science and Moral Philosophy in the Scottish Enlightenment', in M. A. Stewart (ed.), *Studies in the Philosophy of the Scottish Enlightenment* (Oxford, 1990), 11–36.
6. J. J. Carter and J. H. Pittock (eds.), *Aberdeen and the Enlightenment* (Aberdeen, 1987); Roger L. Emerson, *Professors, Patronage and Politics: The Aberdeen Universities in the Eighteenth Century* (Aberdeen, 1992); Paul Wood, *The Aberdeen Enlightenment: The Arts Curriculum in the Eighteenth Century* (Aberdeen, 1993). On religion in Aberdeen, see G. D. Henderson, *Mystics of the North-East* (Aberdeen, 1934); *idem*, 'The Aberdeen Doctors', in *The Burning Bush: Studies in Scottish Church History* (Edinburgh, 1957), 75–93; Gordon DesBrisay, 'Quakers and the University: The Aberdeen Debate of 1675', *History of Universities* 13 (1994), 87–98; *idem*, 'Catholics, Quakers and Religious Persecution in Restoration Aberdeen', *Innes Review* 47 (1996), 136–68.
7. Elizabeth Gasking, *Investigations into Generation, 1651–1828* (Baltimore, 1967), 56.
8. Emerson, 'Science and Moral Philosophy', 13. Emerson gives a full explanation of Scougal's intellectual background; see also G. D. Henderson, 'Henry Scougal', in *The Burning Bush*, 94–104; Anita Guerrini, *Obesity and Depression in the Enlightenment: The Life and Times of George Cheyne* (Norman, OK, 2000), 13–15, where I also give further details of Garden's career.
9. Henry Scougal, *The Life of God in the Soul of Man* (London, 1677), 7.
10. Quoted in Emerson, 'Science and Moral Philosophy', 15.

11. Christine King Shepherd, 'The Arts Curriculum at Aberdeen at the Beginning of the Eighteenth Century', in Carter and Pittock, *Aberdeen and the Enlightenment*, 148–50.
12. George Garden, 'A Discourse concerning Weather, etc. By the Learned Dr Garden of Aberdene. Written by Way of Letter, to Mr Henry Scougall, Professor of Theology in That University; And Communicated to the Phil. Soc. Of Oxford, by the Reverend Dr Middleton, Provost of the Kings College in Aberdeen', *Philosophical Transactions* 15 (1685), 991–1001.
13. Michael Hunter, *The Royal Society and its Fellows, 1660–1700: The Morphology of an Early Scientific Institution*, 2nd edn (Oxford, 1994), 60. Hunter identifies 'Dr Garden' as James Garden, but George seems a more likely candidate.
14. Jacques Roger, *The Life Sciences in Eighteenth-Century French Thought*, ed. K. R. Benson, trans. Robert Ellrich (Stanford, CA, 1997), Part II; Gasking, *Investigations into Generation*, Chs. 1–4; Peter Bowler, 'Preformation and Pre-existence in the Seventeenth Century', *Journal of the History of Biology* 4 (1971), 221–44; A. J. Pyle, 'Animal Generation and the Mechanical Philosophy: Some Light on the Role of Biology in the Scientific Revolution', *History and Philosophy of the Life Sciences* 9 (1987), 224–54.
15. Roger, *Life Sciences*, 113–23; Pyle, 'Animal Generation', 230–32.
16. Garden, 'Discourse', 476.
17. Shirley Roe, *Matter, Life and Generation: Eighteenth-Century Embryology and the Haller-Wolff Debate* (Cambridge, 1981), 7–8.
18. Garden, 'Discourse', 476–77; Emerson, 'Science and Moral Philosophy', 15–17.
19. Garden, 'Discourse', 476–77.
20. Garden, 'Discourse', 477; Edward G. Ruestow, 'Piety and the Defense of the Natural Order: Swammerdam on Generation', in Margaret J. Osler and Paul L. Farber (eds.), *Religion, Science and Worldview* (Cambridge, 1985), 217–41, at pp. 224–31; George Garden, 'A Letter from Dr Geo Garden, Dated Aberdeen, Dec. 8, 1686. Concerning Caterpillars that Destroy Fruit', *Philosophical Transactions* 20 (1698), 54–55.
21. Gasking, *Investigations into Generation*, 43, 46; Roe, *Matter, Life and Generation*, 84.
22. Garden, 'Discourse', 477.
23. Garden, 'Discourse', 477.
24. Garden, 'Discourse', 478.
25. Garden, 'Discourse', 478–80; Eve Keller, 'Embryonic Individuals: The Rhetoric of Seventeenth-Century Embryology and the Construction of Early-Modern Identity', *Eighteenth-Century Studies* 33 (2000) 340–41.
26. Garden, 'Discourse', 481–83.
27. Gasking, *Investigations into Generation*, 59.
28. On Bourignon and Garden, see Guerrini, *Obesity and Depression*, Ch. 1; on Swammerdam, see G. A. Lindeboom, 'Antoinette Bourignon's First Letter to Jan Swammerdam: A Contribution to His Biography', *Janus* 61 (1974), 183–99.
29. For biographical information on Cheyne, see Guerrini, *Obesity and Depression*.
30. George Cheyne, *The Philosophical Principles of Natural Religion* (London, 1705), Part II, 17.
31. Cheyne, *Philosophical Principles*, Part II, 24.
32. Cheyne, *Philosophical Principles*, Part II, 12.
33. Cheyne, *Philosophical Principles*, Part II, 60–63.
34. The following is summarised from Guerrini, *Obesity and Depression*, 4–12.
35. George Cheyne, *Philosophical Principles of Religion: Natural and Revealed* (Lon-

don, 1715). For a detailed analysis of the two editions, see Guerrini, *Obesity and Depression*, Ch. 4.
36. See Guerrini, *Obesity and Depression*, Chs. 5 and 6 and the references cited there.
37. Philip Roberts (ed.), *The Diary of Sir David Hamilton, 1709–1714* (Oxford, 1975).
38. On the case of Catherine Walpole, see Guerrini, *Obesity and Depression*, Ch. 5. On green sickness, see Karl Figlio, 'Chlorosis and Chronic Disease in Nineteenth-Century Britain', *Social History* 3 (1978), 167–97; Joan Jacobs Brumberg, *Fasting Girls* (New York, 1989).
39. Edwin Welch, *Spiritual Pilgrim: A Reassessment of the Life of the Countess of Huntingdon* (Cardiff, 1995); Amanda Vickery, *The Gentleman's Daughter: Women's Lives in Georgian England* (New Haven, CT, 1998), 96.
40. Judith S. Lewis, *In the Family Way: Childbearing in the British Aristocracy, 1760–1860* (New Brunswick, NJ, 1986), 122–24; Welch, *Spiritual Pilgrim*, 20. Although Lewis discusses a slightly later period, her conclusions seems applicable to the 1730s. Vickery similarly notes that 'the average' mother in this period bore six to seven live children': *Gentleman's Daughter*, 97.
41. Cheyne to Countess of Huntingdon, undated (c. 1730), in C. F. Mullett (ed.), *The Letters of Dr. George Cheyne to the Countess of Huntingdon* (San Marino, CA, 1940), 2.
42. Cheyne to Countess of Huntingdon, 12 August 1732, *Letters to Countess of Huntingdon*, 9; I have inferred Lady Huntingdon's pregnancies from an examination of the Hastings MSS, Huntington Library; see also Welch, *Spiritual Pilgrim*, 22. Only one child survived her.
43. Cheyne to Countess of Huntingdon, 3 November 1735, *Letters to Countess of Huntingdon*, 53.
44. Cheyne to Dr. Harding, 3 March 1732/3, *Letters to Countess of Huntingdon*, 18.
45. Cheyne to Countess of Huntingdon, 19 November 1733, *Letters to Countess of Huntingdon*, 28.
46. Cheyne to Countess of Huntingdon, 4 September 1733, *Letters to Countess of Huntingdon*, 23; Melitta Weiss-Amer, 'Medieval Women's Guides to Food during Pregnancy: Origins, Texts and Traditions', *Canadian Bulletin of Medical History* 10 (1993), 5–23; Lewis, *In the Family Way*, 131–32, 147–48.
47. Cheyne to Countess of Huntingdon, 28 May 1733, and Cheyne to Dr. Harding, 3 March 1732/3, *Letters to Countess of Huntingdon*, 18, 21.
48. Lady Elizabeth Hastings to Countess of Huntingdon, 17 December [1731], Huntington Library, Hastings MSS HA 4741.
49. Countess of Huntingdon to Lord Huntingdon, undated (c. 1735), Huntington Library, Hastings MSS HA 5846.
50. Cheyne prescribed a monthly vomit 'a week before the critical time', presumably before menstruation: Cheyne to Countess of Huntingdon, 7 January 1733/4, *Letters to Countess of Huntingdon*, 33.
51. Cheyne to Countess of Huntingdon, 7 January 1733[/4], *Letters to Countess of Huntingdon*, 32–33.
52. 'This place is as stupid as usual': Countess of Huntingdon to Lord Huntingdon, 20 December, c. 1743, Huntington Library, Hastings MSS HA 5856; Welch, *Spiritual Pilgrim*, 24.
53. Lady Elizabeth Hastings to Lord Huntingdon, 21 June 1738, Huntington Library, Hastings MSS HA 4758.
54. Ruth Perry, *The Celebrated Mary Astell: An Early English Feminist* (Chicago, 1986), Appendix E, 456. For more detail on diet see Anita Guerrini, 'A Diet for a

Sensitive Soul: Vegetarianism in Eighteenth-Century Britain', *Eighteenth-Century Life* 23 (1999), 34–42.
55. George Cheyne, *An Essay on Regimen* (London, 1740), 7–8.
56. Cheyne, *Essay on Regimen*, 23.
57. Cheyne, *Essay on Regimen*, unpaginated Dedication.
58. George Cheyne, *The Natural Method of Cureing the Diseases of the Body, and the Disorders of the Mind Depending on the Body* (London, 1742), 4–5. Cheyne does not address the theological inconsistency of an omniscient God who makes mistakes.
59. Cheyne to Richardson, 24 August 1741, in C. F. Mullett (ed.), *Letters of Doctor George Cheyne to Samuel Richardson (1733–1743)* (Columbia, MO, 1943), 69.
60. Cheyne, *Natural Method*, 276.
61. Lawrence Stone, *The Family, Sex and Marriage in England, 1500–1800* (London and New York, 1977), 80; Stone gives many examples on pp. 73–81.
62. Cheyne, *Natural Method*, 277.
63. See Cormac Ó Gráda, *Ireland: A New Economic History* (Oxford, 1994). I am grateful to the many members of the H-Albion list who responded to my query on this topic. On the argument for the healthiness of Highlanders and others, see also Guerrini, 'A Diet for a Sensitive Soul', 40–42.
64. Cheyne, *Natural Method*, 276–77; see Adrian Wilson, *The Making of Man-Midwifery: Childbirth in England, 1660–1770* (Cambridge, MA, 1995).
65. Cheyne, *Natural Method*, 281–82.
66. Cheyne to Richardson, 2 February 1742, *Letters to Richardson*, 82; according to Cheyne, women 'would rather renounce Life than Luxury'.
67. Cheyne, *Natural Method*, 228.
68. Cheyne, *Natural Method*, 228, 236.
69. *Letters to Countess of Huntingdon*. Lady Huntingdon's second daughter Selina, born in 1735, was dead by 1736. For other examples, see Lewis, *In the Family Way*.
70. Cheyne, *Natural Method*, 241–42.
71. Cheyne, *Natural Method*, 283.
72. Cheyne, *Natural Method*, 276–83. See Patricia Crawford, 'Attitudes to Menstruation in Seventeenth-Century England', *Past and Present* 91 (1981), 47–73.

7

William Smellie and Natural History: Dissent and Dissemination

STEPHEN W. BROWN

The study of the book trade in eighteenth-century Scotland has been commonly neglected by scholars. Book culture was seldom discussed as an integral aspect of the Scottish Enlightenment before Warren McDougall's seminal work on the publishing firm of Hamilton, Balfour and Neill in 1974.[1] While bibliographers, book historians and reader response theorists like David Foxon, James McLaverty, Kathleen Lynch, Michael Treadwell and Robert Darnton demonstrated the compelling relationship among printers, publishers, authors and readers in the major works of John Dryden, Alexander Pope, Jonathan Swift and Jean-Jacques Rousseau,[2] the sites for intellectual production in eighteenth-century Scotland were commonly thought to be situated in the universities and learned societies of the period, not the printing houses and book shops.[3] This discrepancy may be explained in part by the fact that the French and the English produced and read more poetry and novels than the Scots did during the eighteenth century.[4] Edinburgh, Glasgow and Aberdeen were centres of distinction in the natural sciences, medicine, moral philosophy and history, not imaginative literature. Moreover, Pope, Swift, Diderot and Rousseau were all aware of the characteristics of the book as a material artefact and took advantage of their medium to manipulate their readers; in creative fiction the medium can be the message, as Diderot and Sterne show respectively in *Le Neveu de Rameau* (begun in 1762) and *Tristram Shandy* (1759–67).[5] We are more inclined to think that the texts of natural science, medicine, philosophy and history can easily be distinguished from the books that deliver their message, that the ideas, facts and theories expressed in such disciplines are separable from the material 'thing' that conveys them in print.[6]

In some sense the culture of the Scottish Enlightenment remained an oral one, even in print, drawing as it did on lectures, sermons, debate and conversation. Certainly its chief creative artist, Robert Burns, was demonstrably more interested in the singing voice than the printed page, unlike Alexander Pope or Thomas Gray. Because the primary spaces for the exchange of ideas in Scotland were lecture halls, masonic lodges, churches, taverns and dinner tables, the printed page, in all its curious production, has seemed less crucial to our understanding of eighteenth-century Scotland than it has been to the appreciation of eighteenth-century France and England. But now that the work of Roger L. Emerson, Steven Shapin and Charles Waterston, among others, has helped us to negotiate the politics of such spaces as the Scottish universities, the Royal Society of Edinburgh, the Society of Antiquaries of Scotland and the Royal College of Physicians of Edinburgh, it may be an appropriate time to explore the role of the book in Enlightenment Scotland. Just as the moderate culture of the University of Edinburgh encouraged a certain kind of historical and political writing,[7] and as the medical faculty generated a particular form of preferred clinical theories,[8] so too the book as learned institution in its own right exerted a peculiar influence on the authors and texts it processed, both creating a print culture that was distinctively Scottish – even when the publishing firms were based in London or Philadelphia – and promoting Scots books as international commodities.[9]

William Zachs's study of the working relationship between the first John Murray, a Scots publisher in London, and the Whig historian and journalist Gilbert Stuart identified a productive paradigm for understanding how the book trade made certain aspects of the Scottish Enlightenment possible.[10] Zachs shows that a sort of symbiosis obtained between Murray and Stuart as the one provided the ideas and words, the other the format and readers for Stuart's idiosyncratic Scottish narrative history. Murray and Stuart also encouraged one another's interests in journalism with Murray finding international markets for Stuart's review writing. We see Murray learning from Stuart's journalistic misadventures in Edinburgh how not to run a magazine, lessons that served him well when he later established the successful *English Review* in 1783.[11] In a later study of the first John Murray's London publishing firm, Zachs decisively establishes the importance of a publisher's self-promotion in the contemporary periodical literature in creating a readership for his authors and a profit for himself. Zachs indicates that 'fifteen to thirty percent of the total cost of production might be put toward advertising', quoting Murray's assertion that 'when the advertising is at an end so is the sale'.[12] The growth of newspapers and magazines was thus demonstrably an extension of the publishers' interests in the book trade, ventures often made financially justifiable by the use of such periodical literatures to promote the booksellers' titles and determine the marketplace for their authors. Certainly book advertisements made up the greater part of all advertising in Edinburgh newspapers in the mid- to late eighteenth century.[13] At the same time, the expansion of the periodical press in Edinburgh alone was staggering, increasing from three identifiable newspapers, magazines or reviews in the 1740s to twenty-one in the 1770s, with continued variation and growth throughout the 1780s and 1790s.[14] It was also common practice for publishers to insist that magazine covers be entirely given over to advertisements for current and forthcoming titles from their own presses. This was particularly the case with specialised medical and science journals, as emphasised by John Murray in a letter to Andrew Duncan, editor of Murray's prized and financially successful periodical, *Medical and Philosophical Commentaries*: 'I have always conceived that the cover of any periodical publication was the Bookseller's, and all persons of good sense will esteem it so'.[15]

Taking his lead from Zachs, Richard Sher has delineated other symbiotic pairings of authors and publishers with regard to the publication history of William Buchan's *Domestic Medicine* and Andrew Millar's repackaging of David Hume,[16] while Warren McDougall has investigated Charles Elliot's interactions with the medical profession, especially his efforts in securing the copyright to William Cullen's *First Lines of the Practice of Physic* and Benjamin Bell's *A System of Surgery*, as well as Elliot's continuing interest in the dissemination of basic natural knowledge as revealed in his reconceptualisation of the audience for the *Encyclopaedia Britannica* when publishing its third edition in 1788.[17] Still, the scholarly examination of the Scottish book trade during the Enlightenment is hardly exhausted. John Balfour, Alexander Donaldson, Alexander Kincaid, John Bell, James Sibbald, William Creech and Peter Hill – to name only the stars in the Edinburgh trade – have yet to be studied from this perspective.[18] And with the recent discovery of several decades of ledgers from the firm of Bell and Bradfute, combined with the materials at the National Archives of Scotland, and the extensive archives now available for John Murray, Charles Elliot, William Strahan, William Smellie and William Creech, the opportunity for research in this field is significant.[19] Very little has been done to examine the book trade in the natural sciences beyond medical publications, with natural history particularly neglected. In this context, two figures show promise for scholarly investigation: William Smellie and James Tytler, respectively the editors of the first and the second editions of the *Encyclopaedia Britannica*, and both notable men of science, Edinburgh printers and journalists.[20]

In what follows, I examine William Smellie's career as Edinburgh's most celebrated printer from the perspective of his role in the publication of both academic and popular works in the natural sciences, and especially in natural history. Smellie edited, translated, reviewed, printed and wrote extensively in medicine and natural history, while his work as a journalist led him early to a belief that the wide dissemination of natural knowledge in texts and language that were sensitive to differences in readership was essential to the democratisation of learning, something evident in his attempts to reach the poor and women readers. As we shall see, Smellie's Whig instincts moved him toward increasingly radical postures as a journalist while he came, without apparent conscious intent, to equate dissemination with dissent in the later stages of his career as a printer. Paradoxically, Smellie began his intellectual life in natural history as a conservative thinker and emerged only gradually as a controversialist, in large part not because of his philosophical position as an author of natural science but because of his practices and experiences as a printer. Smellie is thus an ideal case study for the integral involvement of ideas with business in the book trade of Enlightenment Scotland.

A Learned Printer, 1740–1771

Although William Smellie would become Scotland's most successful printer of scientific works,[21] his nineteenth-century biographer, Robert Kerr, reports that Smellie's father had originally intended him to become a stay maker.[22] The family had been for several generations stonemasons, with both Smellie's father and grandfather becoming master builders. They were also confirmed dissenters and members of one of the more extreme Covenanting sects, the Cameronians. When Smellie was born in 1740 in the Pleasance, a suburb close to the original wall of the Old Town of Edinburgh, his father still regularly walked into the Pentlands for religious meetings that had been long suppressed by the government. Smellie's earliest memories, told to his son Alexander, are recollections of being carried on his father's shoulders to those secluded gatherings. His was a severe Presbyterian upbringing, which inculcated respect for the pursuit of useful learning together with an austere sense of God and love. In correspondence with the Rev. Dr. Samuel Charteris of Wilton, whom he met while pursuing the curriculum in theology at the University of Edinburgh, Smellie recalls a childhood in which silence pervaded the home, his father seldom speaking and entertaining no guests. The elder Smellie (who together with Smellie's grandfather had designed the Martyrs Monument in Greyfriars kirkyard) spent his time away from his trade studying Latin and taking a leading role in the Mary's Chapel Masonic Lodge. Smellie went on to study at Edinburgh High School before joining the firm of Hamilton, Balfour and Neill in 1754 as an apprentice. Although he would work over the next forty years with the leading publishers of medical and science books (including John Balfour, Charles Elliot, John Murray, William and Andrew Strahan, the Cadells and Bell and Bradfute) and obtain an international reputation as an editor, translator and author of works on natural history, Smellie would retain his sense of individualism, self-sufficiency and suspicion of authority that he acquired from his upbringing in the Cameronians – a 'movement which had dangerously republican tendencies' – and from his involvement with Freemasonry with its invocation 'to polish and adorn', especially in the study of the 'wonderful properties of nature' and the pursuit of 'moral truths enriched with useful knowledge'.[23] In Smellie's earliest extant manuscript writings on natural history, he asserts that for 'the mind [to] have full liberty' in examining the 'oeconomy of Nature', suspicion of all theory and authority is an essential posture.[24] This equation of liberty with suspicion, perhaps rooted in Smellie's Cameronian childhood, certainly became the credo of his adult practices both in business and in the natural sciences to such a high degree that it became, as we shall see, a fatal flaw. During his apprenticeship, Smellie was granted the exceptional freedom of absenting himself from the printing house to attend classes at Edinburgh University. He began his studies by pursuing the curriculum which normally led to preparation for the ministry, although his correspondence with Charteris indicates that Smellie had doubts about his vocation from the outset, and was particularly reluctant to join the Church of Scotland. Still, Smellie was an outstanding scholar of ancient languages and his training in Latin was to serve him well as a printer of medical and natural science books.[25]

Smellie, however, shifted the focus of his university studies from the arts to the natural sciences late in 1759 when he also took up a position as the editor of the *Scots Magazine*. By this time, through his membership in the Newtonian Society at Edinburgh University, Smellie had become a close friend of Dr. William Buchan, who advised Smellie on his selection of courses. In their correspondence between 1761 and 1766, while Buchan was a physician, first in Ackworth, Yorkshire and then in Sheffield, we see Smellie wrestling with the moral significance of different sorts of knowledge and the need for learning to be valued for its usefulness in improving the general way of life among all classes. In a letter to Samuel Charteris, Smellie indicated that he was moving from theology to medicine, because he could not take seriously the saving of human souls while so many lived in physical destitution.[26] He began to see medicine as an alternative ministry, dedicated to relieving the human condition through 'the study of Nature in all her various operations'.[27] Smellie and Buchan exchanged opinions of the various professors in the medical school, with Smellie readily accepting Buchan's view that William Cullen had 'his head full of Theory and vague hypothesis' whereas Robert Whytt was 'ingenious' but nonetheless 'solid'.[28] Smellie also shared Buchan's deep admiration of John Gregory and may, in fact, have been the conduit for Buchan's familiarity with Gregory's famous phrase about 'laying medicine open', which he would have heard while attending Gregory's lectures in 1766 and 1767, long before their publication in 1772. Certainly Smellie accepted Gregory's criticism of those in the medical profession who, 'stimulated by vanity . . . have had recourse to various mean and unworthy arts to raise their importance among the ignorant . . . some of these arts have been an affectation of knowledge inscrutable to all, except adepts at the science'.[29] Smellie also developed lasting relationships with the Professor of Botany, John Hope, and the Professor of Natural History, John Ramsay, printing all of Hope's pamphlets and catalogues and helping him to prepare Dr. Charles Alston's lectures for publication in 1770.[30] Although Smellie eventually abandoned his aspirations to become a physician because of the demands of his growing family and his financial dependence on his printing trade, he retained his connections with medicine, printing many of the period's major titles beginning with Buchan's *Domestic Medicine* in 1769.

While attending the courses in medicine, Smellie had begun to allow his interest in the natural sciences to influence his career in the book trade. As the editor of the *Scots Magazine* from 1760 through early 1766, he effectively increased the articles on medicine and natural history. In 1759, the year before Smellie arrived, the *Scots Magazine* made no serious attempt to list, let alone to extract or review, books on medicine or the natural sciences, but by the end of Smellie's tenure the magazine had regular sections devoted to such works.[31] While with the *Scots Magazine*, Smellie came to appreciate the importance of periodical publications for the wide dissemination of learning. He wrote regularly to William Tod, a printer in London who had been a fellow apprentice, extolling the virtues of journalism and fantasising about writing and publishing his own magazine. He praised the *London Chronicle*, a publication often on the wrong side of the government,[32] as 'the only newspaper worth reading', and lamented the failure of Gavin Hamilton's ambitious and intelligent *Edinburgh Chronicle*, which he had helped to print at the end of his apprenticeship.[33] But Smellie's most formative experiences in combining his interests in natural science, his nascent sense of democratic access to knowledge and his now reaffirmed commitment to a career in publishing came with his work on Buchan's *Domestic Medicine* and his editorship of the first edition of the *Encyclopaedia Britannica* (1768–71).

By 1766 Smellie had moved from the *Scots Magazine* into a partnership with William Auld and John Balfour which brought with it two advantages: he became with Balfour the official printer of all the theses for Edinburgh University, with his proficiency in Latin a definite asset; and he joined Auld as co-publisher of the *Edinburgh Weekly Journal*, fulfilling briefly his desire to have his own periodical. At the same time, Buchan wrote from Sheffield asking Smellie to edit and print his *Domestic Medicine*,[34] while Andrew Bell and Colin Macfarquhar approached him to compile and edit the first edition of the *Britannica*. These ventures combined to put to a demanding test Smellie's ideals about the dissemination of useful knowledge. Buchan's project took precedence over the *Encyclopaedia*, and Smellie probably started revising the text in January 1766. He certainly began the crucial business of advertising by the summer of 1766, when he printed an abridged version of Buchan's 'Proposal for Subscriptions'. Here is the first evidence of Smellie's editorial hand. He adeptly reduced Buchan's overly long sixteen-page text (printed at Sheffield in 1765) to an essential two pages, listing the conditions for subscribers and the 'General Heads of the Work'. This he printed in the *Edinburgh Weekly Journal* no later than 20 August 1766.[35] Subsequently, Smellie completely rewrote the text of the proposal, altering its style and tone substantially, and printing it again in the *Edinburgh Weekly Journal* for 3 June 1767. On this date, Smellie indicated that the work was 'in the Press and will be published with all convenient speed'. In this case, 'convenient speed' meant two years. After an initial flurry of advertisements in 1766, all mention of the book ceased until May 1769, when it was finally available. No manuscript exists, and it is, therefore, difficult to estimate how much editing Smellie did for Buchan. Kerr claims it was substantial, while Richard Sher suggests that it might not have been.[36] But judging from the rewrites of the 'Proposal' which Smellie printed in his *Edinburgh Weekly Journal*, and the similarities between these and the advertisement-cum-preface of the Edinburgh edition in 1769, Smellie's hand would have been more than modestly evident in the final product. Typical among the identifiable changes is Smellie's modification of Buchan's Part VI of the 'General Heads', which in their first guise, read:

> Will contain friendly Cautions and Advice to all Persons who are so foolish as to trust their Lives in the Hands of Quacks, Conjurors, Mountebanks, and such like: Wherein the low cunning, gross ignorance, and consummate Villainy of these Pretenders to Physic are evidently

pointed out and the prodigious Hazard every Person runs of losing his Life who trust to their Medicines plainly demonstrated.

These were altered to the far simpler 'Cautions against trusting to Quacks, Mountebanks, etc. on using their medicines' in the *Edinburgh Weekly Journal* of 3 June 1767. And where Buchan begins his 'Account of the Work' by saying that 'This attempt is not intended to encourage quackery or put ignorant people upon the use of medicines which they do not understand; but to shew how far it is in the power of every man to preserve his own health by a proper conduct, or restore it when lost', Smellie begins his by observing, 'It may seem strange, at a time when Physic is universally cultivated as a liberal science, to say that sundry of its most useful parts are almost entirely neglected: But whoever attentively considers the matter will find it to be true'.[37] In the prefatory advertisement to the 1769 edition of *Domestic Medicine*, we encounter a sentiment that expresses succinctly Smellie's position on the popularisation of learning:

> Those who follow the beaten tract of a teacher seldom make any useful discoveries. Accordingly we find that most of the real improvements in medicine have either been the effect of chance, or have been made by persons not bred to physic. Men who think and reason for themselves, who are not fettered by theories nor wrapt by hypotheses, bid the fairest for improving any art.[38]

In the introduction to his translation of Buffon, in his many criticisms of Linnaeus and throughout his reviews of works of natural history, Smellie returned to this same point: that authorities, systems and theories obscure learning and impede creativity. When he joined the Earl of Buchan in founding the Society of Antiquaries of Scotland, Smellie was looking for a democratic learned society that did not defer to titles whether social or academic, and one in which men could 'think and reason for themselves'.

In William Buchan, Smellie may have had a verbose and aggressive literary ally – he found another such in Gilbert Stuart – but at least they shared the belief that the affectation of mystery in any science was fatal, and that all knowledge should be available and accessible to those who needed and sought it. While Smellie was editing Buchan's book between 1766 and 1768, he was struggling to bring his knowledge of the natural sciences to the widest readership through compiling and writing articles for the first edition of the *Encyclopaedia Britannica*. In this instance his employers were less patient and less egalitarian than Buchan had been.[39] Smellie seems to have disagreed fundamentally with Bell and Macfarquhar on everything from the format (it was originally proposed as an octavo in six volumes)[40] to the length and specialisation of the articles (Smellie significantly delayed the publication by writing far longer and more detailed entries on agriculture, algebra and anatomy than the publishers desired). Smellie's sustained and testy defences of his editorial work appeared in the *Edinburgh Evening Courant*, the *Edinburgh Advertiser*, the *Caledonian Mercury* and the *Edinburgh Weekly Journal* throughout July 1769, asserting the need for subscribers to recognise that the intention behind the elaborate articles was to allow readers to 'become the master of any science', which they could not do from 'Mr Chambers or any other dictionary'.[41] Smellie also drew attention to the *Britannica*'s 'particular improvements such as the Scots Law, [and] the method observed in Natural History' in the *Edinburgh Evening Courant* for 5 July 1769.[42] As the publication of subsequent parts of the *Encyclopaedia* proceeded, Smellie relented and reduced the extent of the coverage, although he continued to emphasise natural science and included a revision of a paper attacking Linnaeus on the sexuality of plants which he had first read to the Newtonian Society in 1761 and later submitted to John Hope, for which he received the gold medal in botany.[43] By 1771, when his work was complete on the *Britannica*, Smellie had dissolved his partnership with William Auld, losing his interest in the *Edinburgh Weekly Journal* but retaining a connection with John Balfour to print the Edinburgh theses.[44] His work as an editor for the *Scots Magazine*, Buchan's *Domestic Medicine* and the first edition of the *Britannica* had persuaded him that 'the best method of promoting useful science is to communicate it with perspicuity, and in a language as void of technical terms as the nature of the subject will permit'. Smellie now became increasingly concerned with using the press to extend public education, especially in providing 'genuine descriptions of Nature in any of her forms', and exposing what he called 'contempt of the Vulgar . . . [that] very fashionable and very universal affectation. This modish species of cruelty is practised by almost every person who is able to purchase a genteel suit of cloaths . . . If the vulgar are poor or illiterate, we should . . . use our utmost endeavours to dispel their ignorance'.[45] In the next two decades, Smellie would bring to his own writing and printing of natural history more of a dissenter's zeal, insisting that 'men of knowledge' also be men 'of generous and ardent minds' and that 'the art of printing . . . diffuse light and learning'.[46]

Adventures in the Book Trade: Natural History, 1772–1791

By 1773, Smellie was engaged in two innovative periodical publications, *The Medical and Philosophical Commentaries*, edited by Andrew Duncan, which he was contracted to print for John Murray,[47] and the *Edinburgh Magazine and Review*, of which he owned a one-third share and was printer and co-editor with the firebrand Gilbert Stuart. The *Commentaries* have been described by Roy Porter as bringing 'a more active, vocal journalistic persona' into the discourse of the learned societies because Duncan 'cultivated a certain editorial presence', striving to be both inclusive and accessible in ways that elicited 'the occasional individual apologia'.[48] Smellie's friendship with Duncan continued a pattern – initiated with Gregory and Buchan and completed with John Brown – of close relationships with physicians who were no allies of William

Cullen, something Cullen would remember during the Society of Antiquaries charter crisis in the early 1780s. In these relationships, Smellie sought out men of science who were outside the establishment, and it is a testimony to his great talents as a printer of the natural sciences and medicine that he was typically the first choice to print texts, even for Cullen, despite his opposition role in Edinburgh's academic politics. With his training in medicine and his lifelong study of natural history, Smellie was capable, unlike most other printers, of fully understanding the scientific works he printed, which greatly improved the chances that the final text would be superior. Such was his skill that John Murray and Charles Elliot both endured considerable inconvenience in order to keep Smellie as their printer. Another Scots printer in London, William Strahan, tried unsuccessfully to lure Smellie to London to print exclusively for his firm.[49]

In late 1772, Smellie began to make notes about a proposed new Scottish review that he and Gilbert Stuart were to undertake. Two manuscripts entitled respectively 'Hints for Reviewing Scotch Books' and 'Hints for Magazine' reveal a definite intent on Smellie's part to give over significant space in the publication to the natural sciences, with particular heads for mathematics, medicine, chemistry, natural philosophy, natural history and medical theses, which Smellie intended to 'have a touch at' twice a year.[50] As these preliminary notions evolved into the *Edinburgh Magazine and Review* (which ran from November 1773 to August 1776),[51] not all that Smellie had envisioned came into being. But the publication was radical for three reasons that pertain significantly to the current discussion: first, it placed an emphasis upon reviewing current scientific titles from the point of view of an informed but non-professional readership; second, it stressed Scottish authors; and third, it did not hold back in frankly criticising authors for obscurity, conservatism or inappropriate speculation and theory.[52]

Smellie's reviews were always direct, often blunt and at times scathing and caustic. His dismissal of Edward Harwood was not untypical: 'The vanity and petulance of the writer are in proportion to his incapacity'. But there are recurring themes in the reviews as a whole which emphasise Smellie's concerns about the writing of natural science and natural history and define his still emerging methodology. He was troubled by what might best be called revelationists, theologians practising a sort of pseudo-science whom he felt 'created a greater number of infidels than the whole group of deistical writers'. He demanded a clear and accessible style, disdaining academic writing for its 'affectation of superior abilities industriously wrapped up in obscure hints, and more obscure language'. He decried systems, declaring with reference to Buffon that 'the spirit of system [will] draw the most penetrating writers' into 'wildness and absurdity'. He lauded and promoted the utility of natural science, displaying a remarkable knowledge of agrarian arts and calling for 'liberal encouragement' of those engaged in forest plantations in Scotland. And he asserted that a chief attention of natural history should be the study of the 'instincts of brute creation' because such an acquaintance was essential to 'minutely investigating the principles of human nature'.[53] The *Edinburgh Magazine and Review* demonstrably looks forward to the *Edinburgh Review* (1802) and *Blackwood's Magazine* (1817) in its intellectual ferocity, its combination of politics with literary reviewing and its modern sense that literature shapes society. Furthermore, Smellie's own editorial stance on the natural sciences anticipates *Putnam's Monthly Magazine* (1852), an American periodical which insisted on discussing natural science in controversial and often political contexts.[54] We also find Smellie returning to his equation of suspicion with liberty, suggesting at various points in his harsh criticisms of Thomas Hardy, Robert Henry and William Worthington that doubt, both religious and secular, plays an essential part in the advancement of natural knowledge and the consequent betterment of the human condition. Everywhere Smellie sees the intellectual dependence upon systems – something that he finds in common with theology and certain kinds of science – as an impediment to the pursuit of truth. And he suggests time and again that the love of reputation sustains systems in the Church and the academy, criticising both theologians and scientists for this shared fault:

> These remarks, it is possible will give offence to those who are attached to respectable names. But is not truth more respectable than any name whatever? And what is it that hurts so effectually its investigation, as the gaping wonder with which the mass of men look towards a system of reputation, and as the idle censure which they are often so assiduous to pour out against those who are disposed to doubt and reason?[55]

Interestingly, Smellie disliked systems because they did not depend upon narration and, accordingly, dismisses one natural historian abruptly because 'nothing of the history of manners is narrated'. In Kames, Smellie found the inspiration for the sort of natural science he wanted to write, one that used moral philosophy to create contexts within which to explore the natural world, with an appreciation of human psychology as its final outcome. In his essays on Kames's *Sketches*, Smellie observes that 'we mean not to depreciate system but we are sorry to see it subverting utility and philosophy'. That subversion of utility by systems was also becoming equated for Smellie with systems of political authority in the academy and in local and national government. Slowly Smellie was linking the dissemination of learning not just with the improvement of the physical conditions of humankind but also the intellectual and, by extension, the political lot of the 'common man'. In the *Edinburgh Magazine and Review*, Smellie reflects upon the need for 'an intelligent Farrier' and begins to speculate about laying open agrarian science in the way Gregory and Buchan had proposed for medicine.[56] And, by the late 1770s, Smellie commonly wrote and spoke about the republic of letters, joining the Earl of Buchan in viewing the Society of Antiquaries as an ideal for Edinburgh.

When the *Edinburgh Magazine and Review* folded in 1776 under the pressure of lawsuits and cancelled subscriptions arising from the periodical's confrontations with various Edinburgh establishments, from the town government to the intellectual elite, Smellie turned first to the *Thesaurus Medicus* (1778–85), and then to translating Buffon's *Histoire Naturelle* (1780), before moving on in the 1780s to the decade-long project of his *Philosophy of Natural History*. Along the way he was caught up pivotally in the battle between the Antiquaries and the Royal Society of Edinburgh, suffering a defeat that put him on the political fringes for the rest of his life. When Smellie's original intent to write twice a year in the *Magazine and Review* about Edinburgh medical theses did not come about, he began to look at other ways to take advantage of his unique access to this material as University Printer. Smellie's commitment to the dissemination of scientific writing and his desire to see the spread of learning by the lesser known practitioners of medicine, combined with what he was observing in the success of Duncan's *Commentaries*, seem to have suggested to him a compilation of the abstracts of all of the theses written in Edinburgh since the founding of the medical school. He brought out the first two volumes in 1778, after which Charles Elliot took up the project, following up with the third and fourth volumes in 1785.[57] But, where the *Thesaurus* was simply an extension of Smellie's belief in the distribution of knowledge, the Buffon translation was more philosophical. Smellie had been urged by many, among them Kames, Thomas Pennant and eventually Buffon himself, to undertake the work and to dedicate himself to natural history. However, Smellie was ultimately aroused to translate Buffon by the success of Goldsmith's *History of the Earth*, which he considered an incompetent derivation from the *Histoire Naturelle*, copying 'the French naturalist so servilely, that [it] transcribes his blunders, his contradictions, and, we are sorry to add, even many of his atheistical principles'.[58] Thus Smellie undertook the more ambitious plan to adapt and edit Buffon, adding many of his own observations and notes, and making a loose translation that was well received and which continued to be reprinted well into the nineteenth century.[59] This publication established Smellie internationally as a natural historian and began his reputation as a 'natural theologian' as a consequence of his insistence that final causes stimulated rather than silenced inquiry since 'they only tend to reconcile us to the circumstances in which we are placed', providing consolation not answers.[60]

When the Earl of Buchan was assembling the group of men he hoped would become the Society of Antiquaries of Scotland, he no doubt included Smellie for four reasons: Smellie was a respected member of both the Newtonian Society and the Philosophical Society, and a keen public speaker; Smellie was now a noted naturalist and Buchan wanted to include the most current natural knowledge; Smellie was the leading academic printer and a proven editor, and Buchan had from the outset aspirations to publish transactions; and, most importantly, Smellie was intellectually disenfranchised, having increasingly aligned himself with the town's growing Whig faction since his failed attempt at the chair in natural history at Edinburgh University. But if Buchan was attracted by Smellie's reputation for intellectual dissent, he could not have anticipated the degree of his acolyte's resentment of the establishment. Smellie entered the Antiquaries as its Superintendent of Natural History with the intention of delivering public lectures on his subject, a plan which was rightly seen as a direct affront to his nemesis, the Rev. Dr. John Walker who had received the university appointment for which Smellie had vigorously campaigned.[61] Smellie's lectures would have been free to the general public and would have focused on the materials that became the *Philosophy of Natural History*, especially emphasising dream and instinct with regard to human nature, subjects far from Walker's less speculative interests.[62] The planned lectures became a chief point of contention between the Antiquaries and the University establishment, which now pushed its own learned agenda with a proposal to dissolve the Philosophical Society into the Royal Society of Edinburgh. Both groups sought royal charters. Contrary to Roger L. Emerson, there is good evidence to believe that Smellie promoted the lectures on his own initiative, far beyond what Buchan encouraged and was willing to support. The Antiquaries' Minute Book for January 1782 shows Buchan acceding to a motion from the floor to pre-empt Smellie's lectures. Subsequently, on 15 July 1782, Smellie circulated a motion calling for a procedure to prevent motions from the floor of the sort that had undermined his proposal to lecture. Indeed, Smellie so persisted in defying Buchan that in order to make peace with the University, Buchan actually compelled Smellie to write a letter of reassurance to Walker, sending what he called 'hints' and what were in fact clear directions on precisely what such a letter should say. Smellie did what was asked of him but Buchan's distrust was such that he required Smellie to send the letter through him rather than directly to Walker, and kept a copy of the letter as surety.[63] Despite this betrayal by Buchan, Smellie remained loyal to the Society – although he abstained from defending Buchan when the latter was forced to resign in 1790 – taking pleasure in its egalitarian composition and finding, in what Horace Walpole called Buchan's 'congregation of Wiseacres at Edinburgh',[64] an opportunity to communicate with thinkers outside Scotland about natural knowledge and to have some acknowledged role on the local scene.

Walpole may have questioned – as did Boswell – Buchan's desire to associate with an intellectual rabble, but Smellie was very much a democrat in the world of letters. Throughout the late 1780s and early 1790s, he laboured to bring into print works of natural science by and for the common people. Perhaps because of his necessary exchanges with the many students whose theses he printed and his lifelong intellectual curiosity, Smellie always seemed eager to help new writers on medicine and the natural sciences get first works into print, regardless of their backgrounds. We have already seen the extraordinary assistance he lent to William Buchan, and he was just as generous with others, including Hugo Arnot and the young James Gregory whom Smellie encouraged to develop his *Conspectus medicinae theoreticae* in 1780, printing it himself for the publisher William Creech. Later in the decade, Smellie would delay the completion of his *Philosophy of Natural History* for over two years while dedicating his very limited free time to helping Malcolm McCoig, the head gardener at the Botanical Gardens, to complete his *Flora Edinburgensis* (1788), and then setting about to edit substantially and write an introduction and numerous advertisements for Dr. John Taylor's *A Medical Treatise on the Virtues of St Bernard's Well* (1790). Charles Elliot had been placing advertisements as early as 3 March 1788 in the *Edinburgh Evening Courant* and elsewhere, declaring the *Philosophy of Natural History* to be 'in the press', while in his correspondence with various booksellers in Europe Elliot praised Smellie's work as the equal of both Cullen's *First Lines* and the third edition of the *Encyclopaedia Britannica*.[65] But despite his desperate financial need for the balance of the 1000 guineas Elliot had promised him for the copyright to the volume and Elliot's extensive promotion of the *Philosophy of Natural History*, Smellie saw fit to assist McCoig and Taylor. He did so because McCoig had useful knowledge to impart about nature but lacked an author's background, and because Taylor was promising a work which Smellie would describe in his introduction to it as 'meant chiefly for the benefit of the lower class, who stand most in need of it'.[66] Taylor's project further appealed to Smellie because St Bernard's Well (the mineral spring at Stockbridge which was then just outside Edinburgh) was being restored at the expense of Lord Gardenstone whose patronage Smellie sought after his difficulties with the Earl of Buchan. Beyond this single personal motive, however, Smellie's commitment to Taylor's treatise, which seems to have involved a lengthy rewrite,[67] further testifies to Smellie's desire to see medical and natural science disseminated in many forms and made accessible to every class of reader. In assisting Taylor, Smellie was harking back to his editorial partnership with William Buchan. Once again, Smellie seized an opportunity to promote a medical work whose 'best recommendation [would] be plainness and perspicuity, without a parade of much theory'.[68]

Sadly, while Smellie delayed completing the *Philosophy of Natural History*, Elliot died suddenly of a stroke in 1790, and although this work was published later that year by the estate, Smellie's copyright fee was tied up by the lawyers until after his own death in 1795. But even in late 1789 and early 1790, Smellie was continuing to put energy into populist publications rather than into his own work, writing an introduction and printing the two volumes for Bell and Bradfute's reissue of the *History of Man, or the Wonders of Human Nature* (originally published in 1704), a work Smellie described as a 'Dictionary of Sentimental and Useful Knowledge', with a 'tendency to please the imagination, to inform the understanding, and to mend the heart'.[69] The work was popular enough to have a pirated Dublin edition in 1791 from the same firm which pirated Smellie's *Philosophy* and an edition at Perth in 1796. What attracted Smellie to such projects? Certainly not money. While he postponed his own writing in favour of this sort of cheap text, he was delaying the receipt of the greatest financial windfall of his career. The motivation here seems to have been what it was with Buchan and in all of Smellie's many projects, namely his desire to engage in public discourse, the putting of his words into the popular media, even at the expense of his professional and academic obligations. As John Murray had observed with exasperation, Smellie would set aside almost any business in order to 'write fully in a magazine or in a newspaper'[70] or, we might add, to assist the *vox populi* in any way.

When the *Philosophy of Natural History* finally appeared, the reviews were engaging and mostly positive,[71] but the sections on instinct and on the sexuality of plants caused some unease. Smellie describes instinct as 'an original quality of mind', asserting that 'all animals are, in some measure, rational beings', and that 'the dignity and superiority of the human intellect are necessary results . . . of the great variety of instincts which Nature has been pleased to confer on the species'.[72] Smellie had published this argument in the first volume of the *Transactions* of the Royal Society of Edinburgh, and it was reprinted with some notoriety.[73] A review in *The Bee* chastised Smellie for 'degrad[ing] mankind nearer to the same level with brutes', and the volume gained further public attention for its reputation of being especially accessible to women readers.[74] But Smellie's persistent opposition to Linnaeus on the sexuality of plants created most controversy and resulted in a rather nasty pamphlet by John Rotheram,[75] a doctor and medical writer employed by William Creech who by 1790 was at war with Smellie, his former printing partner, and probably took no small pleasure from spoiling Smellie's moment of greatest intellectual acclaim. Still, from what we have seen of Smellie's obsession with the public press, debate and controversy, these various print scandals surrounding the *Philosophy* seem almost necessary.

William Smellie, Radicalism and America, 1792–1795

With the death of Charles Elliot, Smellie lost his last ally in high places. In the final years of his life, he despaired of finding a publisher for the second volume of the *Philosophy of Natural History* and, typically, struggled to complete the text.[76] Many of Smellie's former correspondents were no longer replying and his letters in 1794 and 1795 to James Hutton and the banker William Forbes reveal a broken man.[77] Alexander Smellie recounts in an unpublished brief memoir that his father was quoted in the sedition trials of the 1790s from his pamphlet *On Juries* (1784), in which he had asserted the right of a juror to ignore judicial charges and follow his conscience in determining a verdict.[78] Smellie tried to revive his scheme for a new Whig paper to be called the *Scottish Chronicle*, recalling both his favourite London paper and Gavin Hamilton's failed venture which had first introduced the young apprentice to the business of dissemination and dissent. Smellie's political sentiments had shifted in sufficiently radical ways to make him decline when asked to write a public response to Thomas Paine. Smellie struggled along with the Antiquaries, becoming the Society's secretary in 1793 when the organisation was at its lowest point. He was often one of only three or four members at the monthly meeting and had to resort to keeping the natural history cabinet in his own apartments.[79] Still, he met Maria Riddell through his friend Robert Burns and kept up a regular correspondence with her that included discussions of William Godwin.[80] And he printed and encouraged his friend Peter Hill to publish Riddell's *Voyages to the Madeira and Leeward Caribbean Isles*, a rather Romantic study of nature that ranks with William Bartram in its lyricism.[81] As Smellie laboured over the second volume of the *Philosophy of Natural History*, he returned to his early interests in dreams and sleep, carrying these forward into original speculations on imagination and the will.[82] His correspondence with Maria Riddell shows him extending his theories on children and the philosophy of the mind. In his posthumous publications, we see radical ideas on the subject of early childhood development.[83] And we find Smellie in this spirit of apparent free-thinking writing to Burns and proposing that the poet compose a preface and advertisement for the second volume of the *Philosophy* at a time when Burns was himself seemingly at his most radical.[84] But that preface never materialised and Smellie was unable to find any sort of public alliance to promote himself or his ideas in Edinburgh.

But circumstances were different abroad where Smellie would become, like Priestley, something of an American sensation. His *Philosophy* was printed in Philadelphia in 1791 to no small acclaim, something confirmed by its selection as the primary natural history text on the new Harvard curriculum early in the nineteenth century. Abridged and edited by Dr. John Ware, the *Philosophy* would be reprinted scores of times in America through to 1872. In 1829, thirty-six members of the junior year at Harvard would petition to be 'permitted to study Good's *Book of Nature* or Smellie's *Philosophy of Natural History* as a substitute for Griesbach's *Greek Testament*'.[85] Furthermore, Thomas Nuttall's pioneering course in natural history at Harvard would be nicknamed 'Smellie' by adoring students. Ralph Waldo Emerson took the course; William James read the *Philosophy*; and William Bartram, the spiritual father of American environmentalism, cherished his edition of the *Philosophy*, passing along a presentation copy to his nephew James, then director of the Botanical Gardens at Philadelphia.[86] If his more conservative Scots contemporaries failed to appreciate Smellie as a proponent of change, his American cousins took his lessons very much to heart; their cultural romanticism was perhaps better suited to a naturalist who believed that our moral sense arises from the 'gradual unfolding of sympathy and benevolence which are themselves instincts' and thus necessarily 'expressions of nature'.[87]

Conclusion

William Smellie's contributions to the natural sciences, especially medicine and natural history, are remarkable in themselves and unfortunately neglected.[88] But as a printer, editor and journalist, Smellie shows us how print culture determined, shaped and ultimately politicised the spread of natural knowledge to a popular readership. From his crucial involvement in William Buchan's *Domestic Medicine* and the first edition of the *Encyclopaedia Britannica* through to his promotion of writers like Malcolm McCoig, John Taylor and Maria Riddell, Smellie was motivated by a commitment not only to the widest possible dissemination of learning but also to spreading the word as an explicit challenge to academic exclusiveness. By adopting a political posture that increasingly confronted the establishment and aggressively asserted democratic principles with regard to knowledge, especially in the natural sciences, Smellie was following in the important tradition of dissenter publishing. His attitudes and practices point back to those of John Dunton and Daniel Defoe, and his projects recall Protestant publications such as the *Athenian Mercury* (which appeared between March 1690 and June 1697), which were designed to cut across class barriers and make the specialist in any field of learning a servant of the needs of the general population. Such publishing ventures assumed that literacy necessarily nurtured a critical intelligence in the masses, making reform a natural process. The publisher Joseph Johnson and the journalist William Cobbett, not long after Smellie, would bring this tradition to its apex.

For Smellie the most effective means of disseminating learning was through periodicals and inexpensive publications. He especially favoured what we would now call popular journalism, holding magazines and newspapers in highest regard. His own attempts in these formats brought him into significant conflict with academic and political authority. This was particularly the case with the *Edinburgh Magazine and Review*. But Smellie persisted, and we can see in his career an example of how specialist knowledge was opened up to intense critical scrutiny through a new kind of periodical reviewing in the popular press and at the same time made available to a non-specialist readership.

There is still much to be done in examining how the popular press complemented and subverted learned publishing during the Enlightenment in Scotland. William Smellie is only one instance where learnedness and journalism came together in a politically combustible way in Edinburgh. Another Scot, James Tytler, who followed Smellie as editor of the second edition of the *Britannica*, shadows Smellie's career in many respects. Tytler wrote an interesting work of geography and, like Smellie, had wide familiarity with the natural sciences.[89] He also sought out popular venues to demonstrate and spread that knowledge. Over a period of several months in 1784, Tytler's attempts as a balloonist were reported almost daily in the Edinburgh newspapers, conveying to the general public a combination of sensationalism and hard scientific facts. Tytler's sponsor for the balloon experiments was the bookseller Charles Elliot, who published the first volume of Smellie's *Philosophy of Natural History* and had a hand simultaneously in popular and learned publications. Tytler continued to write extensively for Elliot's third edition of the *Britannica*, but more significantly he became one of Edinburgh's most radical journalists with the publication in 1791 of his *Historical Register*. Like Smellie, Tytler was a gifted man of science whose career took place outside the academy and who increasingly combined learning, populism and reform politics in his journalism. Recent studies by Thomas Broman and

Charles E. Clark have begun to look at the relationship between periodical print culture and the popularisation of learning.[90] But the specific question of how popular journalism contended with academic publication during the Enlightenment remains as yet unanswered.

What, then, should we take away from this case study of William Smellie? Perhaps the primary lesson is that in tracing the dissemination of ideas in intellectual history, the example of the trade in books is crucial. The book as a commodity in the Enlightenment moved rapidly, traversing continents and crossing oceans, leaving a paper trail and a narrative that combine to describe how radical ideas trickled down through society and commerce to gain acceptance. Smellie's is only one – James Tytler's is another – of the many stories of the political and social impact of the Scottish book trade.

Notes

1. In her then definitive bibliographic guide, *The British Book Trade From Caxton to the Present Day* (London, 1973), 350–55, Robin Myers lists no studies of Scottish publishing before those focused on nineteenth-century firms such as Blackwood's, Constable and Collins. Her references to eighteenth-century Scots publishers other than Foulis of Glasgow are only to the major London-based Scottish dealers, William Strahan and the first John Murray. See Warren McDougall, 'Gavin Hamilton, John Balfour and Patrick Neill' (unpublished Ph.D. thesis, University of Edinburgh, 1974). A catalogue of Hamilton, Balfour and Neill publications from 1750 to 1762 compiled by McDougall is printed in Robin Myers and Michael Harris (eds.), *Spreading the Word: The Distribution Networks of Print, 1550–1850* (Winchester and Detroit, 1990), 187–232. McDougall further cleared the way for current Scottish booktrade scholarship with his crucial study of copyright, 'Copyright Litigation in the Court of Session, 1738–1749, and the Rise of the Scottish Book Trade', *Edinburgh Bibliographical Society Transactions* 5.5 (1988), 2–31.
2. See Robert Darnton, *The Forbidden Best-Sellers of Pre-Revolutionary France* (New York and London, 1995); Kathleen M. Lynch, *Jacob Tonson, Kit-Cat Publisher* (Knoxville, TN, 1971); David Foxon and James McLaverty, *Pope and the Early Eighteenth-Century Book Trade* (Oxford, 1991); and Michael Treadwell, 'London trade publishers, 1675–1750,' *Library*, 6th series, 4 (1982), 99–134.
3. See Roger L. Emerson, 'The Social Composition of Enlightenment Scotland: The Select Society of Edinburgh, 1754–1767', *Studies on Voltaire and the Eighteenth Century* 114 (1973), 291–329; idem, 'The Philosophical Society of Edinburgh, 1737–1747', *The British Journal for the History of Science* 12 (1979), 154–91; idem, 'The Philosophical Society of Edinburgh, 1768–1783'; *The British Journal for the History of Science* 18 (1985), 255–303; 'The Scottish Enlightenment and the End of the Philosophical Society of Edinburgh', *The British Journal for the History of Science* 21 (1988), 33–66; idem, 'Scottish Universities in the Eighteenth Century, 1690–1800', *Studies on Voltaire and the Eighteenth Century* 167 (1977), 453–74; idem, *Professors, Patronage and Politics: The Aberdeen Universities in the Eighteenth Century* (Aberdeen, 1992); idem, 'Politics and the Glasgow Professors, 1690–1800', in Andrew Hook and Richard Sher (eds.), *The Glasgow Enlightenment* (East Linton, UK, 1995), 21–39; Steven Shapin, 'The Royal Society of Edinburgh: A Study of the Social Context of Hanoverian

Science' (unpublished Ph.D. thesis, University of Pennsylvania, 1971); *idem*, 'The Audience for Science in Eighteenth-Century Edinburgh', *History of Science* 12 (1974), 95–121; *idem*, 'Property, Patronage and the Politics of Science: The Founding of the Royal Society of Edinburgh', *The British Journal for the History of Science* 7 (1974), 1–41; Charles D. Waterston, *Collections in Context: The Museum of the Royal Society of Edinburgh and the Inception of a National Museum for Scotland* (Edinburgh, 1998). See also, D. C. Macarthur, 'The First Forty Years of the Royal Medical Society and the Part William Cullen Played in It', in A. Doig, J. P. S. Ferguson, I. A. Milne and R. Passmore (eds.), *William Cullen and the Eighteenth-Century Medical World* (Edinburgh, 1993).

4. Eighteenth-century Scottish creative literature before Burns is marginal at best and accounted for by a few texts in drama, poetry and fiction, including John Home's *Douglas* (1756), Robert Fergusson's *Poems* (1773) and Henry Mackenzie's *Man of Feeling* (1771). As Maurice Lindsay notes, James Thomson and Tobias Smollett both moved to London after completing their university studies in Scotland and are considered part of the English canon: 'Scotland experienced no fully-fledged Augustan age'. Lindsay describes James Beattie as 'neither much of a poet nor much of a philosopher': Maurice Lindsay, *History of Scottish Literature*, rev. ed. (London, 1992), 169–273. Perhaps the greatest poetic moment in eighteenth-century Scotland before Burns was the Ossian controversy. It is of interest to the present discussion that Macpherson's literary fraud excited Scots at the time because of its apparent historic – not its poetic – significance.

5. No Scottish writer would show a similar attitude toward the ambiguity of text and book until James Hogg's *The Private Memories and Confessions of a Justified Sinner* in 1824.

6. See Richard B. Sher's discussion of the text/book distinction with respect to Enlightenment Scotland in 'Science and Medicine in the Scottish Enlightenment: The Lessons of Book History', in Paul Wood (ed.), *The Scottish Enlightenment: Essays in Reinterpretation* (Rochester, 2000), 99–156. In their editorial introduction to *Books and the Sciences in History* (Cambridge, 2000), Marina Frasca-Spada and Nick Jardine make the point that 'the publications of the sciences provide an ideal field for exploring the most problematic parts of the communications circuit, those relating to the reading and appropriation of books' (6); in the same volume, see also James Secord, 'Progress in Print', 369–89.

7. Richard B. Sher, *Church and University in the Scottish Enlightenment: The Moderate Literati of Edinburgh* (Princeton, NJ, 1985), 151–212.

8. Michael Barfoot, 'Philosophy and Method in Cullen's Medical Teaching', in Doig *et al*, *William Cullen*, 110–32.

9. Warren McDougall's three essays provide the best contexts for the distinct character of the Scottish book trade: 'Scottish Books for America in the Mid-Eighteenth Century', in Myers and Harris, *Spreading the Word*, 21–46; 'Smugglers, Reprinters and Hot Pursuers: The Irish Scottish Book Trade, and Copyright Prosecutions in the Late Eighteenth Century', in Robin Myers and Michael Harris (eds.), *The Stationers' Company and the Book Trade, 1550–1990* (Winchester and Delaware, 1998), 151–83; and 'Charles Elliot and the London Booksellers in the Early Years', in Peter Isaac and Barry McKay (eds.), *The Human Face of the Book Trade: Print Culture and its Creators* (Winchester and Delaware, 1998), 81–96. See also Paul Wood (ed.), *The Culture of the Book in the Scottish Enlightenment* (Toronto, 2000).

10. William Zachs, *Without Regard to Good Manners: A Biography of Gilbert Stuart, 1743–1786* (Edinburgh, 1992).

11. William Zachs, *The First John Murray and the Late Eighteenth-Century London Book Trade* (Oxford, 1998), 203–4.
12. Zachs, *First John Murray*, 86.
13. In my own research, I have occasionally come across Edinburgh newspapers in which *all* the advertising in each issue for days at a stretch was derived from book trade notices. In the extant numbers of Auld and Smellie's *Edinburgh Weekly Journal* for the years 1766 through 1769, 95% of the advertisements are for books, half of which were printed by their own press. When book advertisements were not prolific, the promotion of drugs and medical treatments more than made up the difference: Zachs, *First John Murray*, 48.
14. M. E. Craig, *The Scottish Periodical Press, 1759–1789* (Edinburgh and London, 1931), 20–35.
15. Zachs, *First John Murray*, 87.
16. Richard B. Sher, 'William Buchan's *Domestic Medicine*: Laying Book History Open', in Isaac and McKay, *Human Face of the Book Trade*, 45–64; *idem*, 'The Book in the Scottish Enlightenment', in Wood, *Culture of the Book*, 43–47.
17. See Chapter 8 below and the Records of the Worshipful Company of Stationers, 9 January 1783 and 6 February 1784, Stationers Hall, London.
18. With the exception of Creech and Sibbald, none of these publishers was accorded an entry in the original *DNB*, although that astonishing oversight will be corrected in the *New DNB*. For a contemporary's account of the late eighteenth-century Edinburgh book trade, see Thomas Constable, *Archibald Constable and His Literary Correspondents*, 3 vols. (Edinburgh, 1873), 1:533–40.
19. The Bell and Bradfute material is deposited with the Edinburgh City Archives and the National Library of Scotland.
20. Both Smellie and Tytler have respectable entries in the original *DNB*, based on their careers as writers but not, however, for their contributions to the book trade.
21. Richard Sher has compiled a list of 111 medical and scientific titles published between 1746 and 1800: Sher, 'Science and Medicine in the Scottish Enlightenment', 114–127. Of the seventy-eight printed during Smellie's career (1764 to 1795), Smellie printed twenty-six (33%), including all of the richest properties such as Cullen's *First Lines* and Bell's *Surgery*. Smellie also printed at least one work of some sort for every author cited by Sher for the period 1764 to 1795.
22. Robert Kerr, *Memoirs of the Life, Writings and Correspondence of William Smellie*, 2 vols. (1811; reprinted Bristol, 1996), 1:15–18. Many of the letters quoted by Kerr are expurgated. Where discrepancies have been identified, reference will be made to the original manuscripts in the archives of the Society of Antiquaries of Scotland in the National Museums of Scotland Library, which I arranged, indexed and annotated in 1997.
23. On the Cameronians, see Richard J. Finlay, 'Keeping the Covenant: The Scottish National Identity', in T. M. Devine and J. R. Young (eds.), *Eighteenth Century Scotland: New Perspectives* (East Linton, UK, 1999), 124–26; David Stevenson, *The Covenanters: The National Covenant and Scotland* (Edinburgh, 1988), esp. pp. 70–84 on the legacy of the Covenanters among Scottish radicals. On the Freemasons, see *The Free Masons Pocket Companion* (Edinburgh, 1761), and *An Appendix to the Masons Pocket Companion* (Edinburgh, 1761); both were printed by Smellie's first business partner, William Auld, and were published in the year in which Smellie was 'passed and raised' to the degree of Master Mason. Relevant modern scholarship on Scottish Freemasonry includes: David Stevenson, *The First Freemasons: Scotland's Early Lodges and their Members* (Aberdeen, 1988), 47–49; David Murray Lyon, *History of the Lodge of Edinburgh (Mary's Chapel)*

No.1 (London, Glasgow and Dublin, 1900), 148, 213–14; Lisa Kahler, 'Freemasonry in Edinburgh 1721–1746: Institutions and Context', (unpublished Ph.D. thesis, University of St Andrews, 1998).
24. Smellie MSS, Natural History, item 8.
25. Charles Elliot once described Smellie in a letter to the London bookseller James Dodsley as 'perhaps the best Latin scholar of this country'; see Elliot to Dodsley, 13 February 1776, Elliot Letter Books, Murray Archives, London, and McDougall, 'Charles Elliot and the London Booksellers', 93.
26. Smellie MSS, Familiar Correspondence, Samuel Charteris, 1761.
27. Kerr, *Memoirs*, 1:231.
28. Kerr, *Memoirs*, 1:244–45.
29. John Gregory, *Lectures on the Duties and Qualifications of a Physician* (London, 1772), 230–36. The advertisement or preface to the 1769 edition of *Domestic Medicine* (which was apparently written by Smellie) owes much to Gregory and must have been derived, at least in part, from lecture notes. Indeed, Gregory is, perhaps, the chief stylistic as well as philosophical influence on Smellie as a writer on the natural sciences. So close are their prose styles at times that William Cullen was unable to distinguish between them and never believed that Smellie, not Gregory, had written the notorious article on the ether in the first edition of the *Encyclopaedia Britannica*: Kerr, *Memoirs*, 1:364–69. See also Smellie's life of Gregory in his *Literary and Characteristical Lives*, ed. Alexander Smellie (1800; reprinted Bristol, 1997), 1–118.
30. Works that Smellie printed for Hope include: *Termini Botanici: in essem juventus* (1767, 1770, 1778), excerpted from Linnaeus; *Genera Plantarum* (1771), also excerpted from Linnaeus; *A Catalogue of Trees and Shrubs Growing in the Botanic Garden at Edinburgh* (1775); *Index herbarum medicinalium in horto Edinburgensi sub dio cresentium* (1775); *Catalogus Arborum et Fruticum in Horto Edinensi* (1778).
31. See my analysis of the *Scots Magazine* during Smellie's editorship in 'William Smellie and the Culture of the Edinburgh Book Trade, 1752–1795', in Wood, *Culture of the Book*, 66–69.
32. Lucyle Werkmeester, *A Newspaper History of England, 1792–93* (Lincoln, NE, 1967), 37, 169.
33. Smellie MSS, Familiar Correspondence, William Tod, 1761.
34. Smellie MSS, Familiar Correspondence, William Buchan, 1765–68.
35. Copies of the *Edinburgh Weekly Journal* for 1765 to 1770 are rare. The most complete run is held in the Signet Library, with one issue (3 May 1769) in the National Library of Scotland. The subscribers' pages for the 'Proposal' which are extant among the Smellie MSS indicate that the 'Proposal' circulated in Edinburgh was not the one sent by Buchan from Sheffield but a version printed by Smellie himself. The only known complete copy of the Sheffield *Proposal* is in the Library of the Wellcome Institute in London, although it is printed in part in the *Edinburgh Weekly Journal* for 20 August 1766. In these early proposals and advertisements, the work was entitled *The Family Physician*.
36. Kerr, *Memoirs*, 1:97, 222–23; Sher, 'William Buchan's *Domestic Medicine*', 62, note 2. A more extended examination of the material from the *Edinburgh Weekly Journal* and what survives of the Edinburgh 'Proposal for Subscription' might help to support Christopher J. Lawrence's claim for Smellie as co-author of *Domestic Medicine*. Gilbert Stuart is cited by Kerr as a source for such a claim, and certainly Stuart was Smellie's constant companion from 1766 through 1770, helping write some of the articles for the *Encyclopaedia Britannica*; see C. J.

Lawrence, 'William Buchan and Medicine Laid Open', *Medical History* 19 (1975), 20–35, esp. pp. 20–21.
37. Sheffield *Proposal*, 3; *Edinburgh Weekly Journal* 3 June 1767.
38. Prefatory advertisement to William Buchan, *Domestic Medicine, or, the Family Physician* (Edinburgh, 1769), ix.
39. Bell and Macfarquhar were not as generous as Buchan. They offered Smellie a flat fee of £200 for compiling the entire *Encyclopaedia*, while Buchan paid Smellie £100 for the much lesser task of editing *Domestic Medicine*.
40. This conception of the format continued until quite late in the publication process; see a 'Proposal for Printing a Work entitled Encyclopaedia Britannica', in the *Edinburgh Evening Courant*, 20 February 1768 and the *Edinburgh Advertiser*, 11 March 1768.
41. See my discussion of the publication history of the first edition of the *Encyclopaedia Britannica* in 'William Smellie and the Culture of the Edinburgh Book Trade', 69–73.
42. Natural history and Scots law just happened to be special interests of Smellie and his close companion, Gilbert Stuart, who probably helped with the compilation.
43. Kerr, *Memoirs*, 1:92–93.
44. The dissolution of the partnership was announced in the *Caledonian Mercury* for 23 November 1771. Smellie would print over 750 medical theses and related scientific and academic publications for the University of Edinburgh during his career. This figure is derived from an examination of the English Short Title Catalogue (British Library, 1998), and the catalogues at Edinburgh University Library and the National Library of Scotland.
45. Smellie, *Literary and Characteristical Lives*, 20, 25, 397.
46. William Smellie, *The Philosophy of Natural History*, 2 vols. (Edinburgh, 1790–99), 2:350.
47. Because Murray was also the London agent for the *Edinburgh Magazine and Review* as well as publisher of the *Medical Commentaries*, he had a regular correspondence with Smellie from 1773 onwards, most of which survives in his letter books in the Murray Archives.
48. Roy Porter, 'The Rise of Medical Journalism in Britain to 1800', in W. F. Bynum, Stephen Lock and Roy Porter (eds.), *Medical Journals and Medical Knowledge* (London, 1992), 6–28, esp. pp. 16–17.
49. Smellie's correspondence with John Murray in particular is full of Murray's complaints about Smellie's unreliability as a businessman, which cost Murray dearly at times. But Murray never complains about the quality of the work – when it is eventually printed. See Murray Archives, Letter Book entries for 2 December 1773, 8 December 1774, 16 January 1775, 28 January 1775, 18 February 1775, 10 March 1775, 18 March 1775, 28 March 1775, 25 September 1775 and 11 December 1775, among others, which detail Murray's struggle with Smellie in getting the *Medical Commentaries* out. In addition, see Kerr, *Memoirs*, 1:326–34.
50. Smellie MSS, Book Trade documents, item 3; see also, Other Writing, item 6.
51. On the *Magazine and Review* from Gilbert Stuart's perspective, see Zachs, *Without Regard to Good Manners*, 63–95.
52. *Edinburgh Magazine and Review*, 5 vols. (1773–76; reprinted Bristol, 1998). Among his contributions, Smellie wrote reviews and notes on: James Campbell, *A Treatise of Modern Faulconry* (1:92–93); Robert Henry, *Revelation the Most Effectual Means of Civilizing and Reforming Mankind* (1:45); Lord Kames, *Sketches of the History of Man*, (2:430–37, 494–99, 555–57); William Worthington, *The Scripture-Theory of the Earth* (2:597–604); John Tait, *The Druid's*

Monument (2:609–10); Edward Harwood, *Of Temperance and Intemperance* (2:722); James Anderson, *Essays Relating to Agriculture* (3:320–26); Oliver Goldsmith, *An History of the Earth and Animated Nature* (3:87–92, 142–46); James Jenkinson, *A Generic and Specific Description of British Plants* (3:379–81); Thomas Hardy, *The Views which Revelation Exhibiteth* (3:375–77); James Clark, *Observations on the Shoeing of Horses* (4:489–95); John Cockburn, *A New Collection of English Prose and Verse* (4:610); James Harris, *Philosophical Arrangements* (4:426–33); William Boutcher, *A Treatise on Forest Trees* (4:496–99); Buffon, *Histoire Naturelle* (4:591–94, 649–53, 5:8–9); Thomas Martyn, *Elements of Natural History* (4:665–67); Hugh Rose, *The Elements of Botany* (4:717–23); John Innes, *A Short Description of the Human Muscles* (4:778–79); John Pringle, *Discourse on the Torpedo* (5:197–99); William Hamilton, *Observations on Mount Vesuvius* (5:250–57); Alexander Hamilton, *Elements of the Practice of Midwifery* (5:40–41).

53. *Edinburgh Magazine and Review*, 2:437, 603, 722, 4:431, 496, 5:9.
54. *Putnam's* was advertised as a 'magazine of literature, science, and art', publishing, for example, in 1854, Herman Melville's sketches of the Galapagos islands in the same issue as an examination of the work of Alfred Russel Wallace.
55. *Edinburgh Magazine and Review*, 4:723.
56. *Edinburgh Magazine and Review*, 2:432, 4:489, 666.
57. See McDougall's checklist in Ch. 8.
58. Kerr, *Memoirs* 2: 119–21; *Edinburgh Magazine and Review*, 3:89.
59. Most of the reviews of the Buffon are reprinted in Kerr, *Memoirs*, 2:122–29. Smellie's translation was still being printed in London as late as 1866 in an edition edited by Henry Augustus Chambers.
60. See *Edinburgh Magazine and Review*, 3:89 and *Natural History, General and Particular by the Count de Buffon*, trans. William Smellie, 8 vols. (Edinburgh, 1780), 1:x-xii. George Hoggart Toulmin, who got to know Smellie while he was a medical student in Edinburgh in the late 1770s, heavily plagiarised Smellie's Buffon in *The Antiquity and Duration of the World* (London, 1780). Smellie printed Toulmin's thesis in 1779, and it is likely that Toulmin had access to the Buffon translation in progress; see Dennis R. Dean, *James Hutton and the History of Geology* (Ithaca, NY, 1992), 272–74.
61. See Emerson, 'The End of the Philosophical Society', 47–63; Shapin, 'Property, Patronage, and Politics', 11–22; A. S. Bell (ed.), *The Scottish Antiquarian Tradition* (Edinburgh, 1981), 10–19; and Charles W. J. Withers, 'Natural Knowledge as Cultural Property: Disputes over the "Ownership" of Natural History in Late Eighteenth Century Edinburgh', *Archives of Natural History* 19 (1992), 289–303.
62. The content of the proposed lectures can largely be deduced from the unpublished materials collected at Smellie's death by his son Alexander for inclusion in *Literary and Characteristical Lives* and from the partial manuscripts which survive in the archives of the Society of Antiquaries of Scotland.
63. None of the commentators on this episode has paid close attention to this exchange, and the 'hints' from Buchan have never been noted. See Smellie MSS. Smellie further defied Buchan by printing in his *Account of the Institution and Progress of the Society of Antiquaries of Scotland* (Edinburgh, 1782–84) details of the charter episode which the Earl had explicitly asked Smellie to suppress (part 2:24–28).
64. Walpole to the Earl of Harcourt, 1 September 1787, in W. S. Lewis, *et al* (eds.), *The Yale Edition of Horace Walpole's Correspondence*, 48 vols. (New Haven, CT, 1937–83), 35:539.

65. See Elliot to W. S. Bennet of Rotterdam, 28 February 1788, and Elliot to Monsieur Procureur du Collège des Ecossais à Paris, 5 May 1788, in the Elliot Letter Books, Murray Archives.
66. John Taylor, *A Medical Treatise on the Virtues of St Bernard's Well* (Edinburgh, 1790), 2.
67. Smellie MSS, Book Trade Correspondence.
68. Taylor, *Medical Treatise*, 2.
69. Nathaniel Wanley, *The History of Man, or The Wonders of Human Nature*, 2 vols. (Edinburgh, 1790).
70. Murray to Smellie, 29 May 1777, John Murray Letter Books, Murray Archives.
71. See in particular *English Review* 15 (1790), 321–32, 444–54; *Critical Review* 30 (1790), 93–102; *Analytical Review* 8 (1790), 134–43.
72. Smellie, *Philosophy of Natural History*, 1:155, 159.
73. *Transactions of the Royal Society of Edinburgh* 1 (1788), 39–45.
74. *The Bee* 2 (1791), 72–77, at p. 74. The *Philosophy of Natural History* was described as being accessible to female readers in the newspaper advertisements and, in this regard, was later praised in *The Bee* 16 (1793), 280–86.
75. John Rotheram, *The Sexes of Plants Vindicated in a Letter to William Smellie* (Edinburgh, 1790). Creech spent a considerable sum on large and numerous advertisements for Rotheram's pamphlet, often set side by side with advertisements for Smellie's *Philosophy of Natural History*.
76. Smellie received what he thought an insulting offer for volume two from Bell and Bradfute and then threatened to self-publish while fruitlessly doing the rounds of other publishers; see the correspondence between John Bell and William Smellie, 22 December 1794, National Library of Scotland, Watson MSS 583, 857. Although Smellie claimed to have finished the writing by late 1794, his previous performances and his son Alexander's correspondence with Maria Riddell suggest otherwise; Smellie MSS, Riddell Correspondence.
77. Smellie MSS, General Correspondence.
78. Smellie MSS, Alexander Smellie's notes to Robert Kerr.
79. See the Minute Books of the Society of Antiquaries, 1793–95, Society of Antiquaries of Scotland Archive, National Museums of Scotland Library.
80. Smellie MSS, Riddell Correspondence.
81. Maria Riddell, *Voyages to the Madeira and Leeward Caribbean Isles with Sketches of the Natural History of these Islands* (Edinburgh, 1792). The book was reprinted in America at Salem, Massachusetts in 1803.
82. Smellie first wrote about dreams and sleep in the early 1760s, and eventually addressed the Newtonian Society on this topic in 1764. His influence is apparent in the work of Dugald Stewart, among others; see especially Dugald Stewart, *Elements of the Philosophy of the Human Mind* (Edinburgh, 1792), 320–39.
83. See Smellie MSS, Riddell Correspondence, and *Literary and Characteristical Lives*, 321–24, where Smellie revisits his theory in opposition to John Locke that 'there must be ideas *connate* at least with existence itself' (p. 322), postulating a capacity for some kind of thought by the child even in the womb.
84. Kerr, *Memoirs*, 2:355–56, and Smellie MSS, Riddell Correspondence.
85. Jeannette E. Graustein, *Thomas Nuttall, Naturalist* (Cambridge, MA, 1967), 224–25.
86. Bartram's copy is in the Library Company, Philadelphia.
87. Smellie MSS, Other Writing, item 6.
88. But see now Paul Wood's introduction to the Thoemmes Press reprint of *The Philosophy of Natural History*, 2 vols. (1790–99; Bristol, 2001) and Aaron

Garrett's introduction to the reprint of *Natural History, General and Particular*, trans. William Smellie, 3rd edn, 9 vols. (1791; Bristol, 2001).

89. Among Tytler's publications are: *Essays on the Most Important Articles of Natural and Revealed Religion* (Edinburgh, 1774); *A General History of All Nations, Ancient and Modern. Together with a Geographical Description of the Different Countries of the World* (Edinburgh, 1782); *A New and Concise System of Geography* (Edinburgh, 1788); and the polemical pamphlet, *Review of Guthrie's Geographical, Historical and Commercial Grammar* (Edinburgh, 1791). Tytler also collaborated with Thomas Salmon on *The New Universal Geographical Grammar* (Edinburgh, 1777).

90. Thomas Broman, 'Periodical Literature', in Frasca-Spada and Jardine, *Books and the Sciences in History*, 225–38; Charles E. Clark, 'Early American Journalism: News and Opinion in the Popular Press', in Hugh Amory and David D. Hall (eds.), *A History of the Book in America, Volume I: The Colonial Book in the Atlantic World* (Cambridge, 2000), 347–66.

8

Charles Elliot's Medical Publications and the International Book Trade[1]

WARREN MCDOUGALL

This chapter tells how Charles Elliot, a leading Edinburgh bookseller of the 1780s, encouraged the physician and surgeon authors of the city into print, and it opens up a hitherto unexplored area, namely the interaction between Edinburgh and Europe in the bookselling and publishing of the Scottish Enlightenment in the late eighteenth century. It describes the networks through which information about science and medicine moved, and how the Enlightenment was exported to America.

In considering the place of books and the role of book history in the Scottish Enlightenment more generally, Richard Sher has recently observed: 'We are learning that books were made by the book trade as well as by their authors, and that understanding the nature of the roles played by the most important members of the trade is one of the keys that unlocks the secrets of the Enlightenment intellectual culture. In this sense, the leading publishers . . . were among the primary enablers of the Scottish Enlightenment'.[2] In this chapter, I too have tried to place Charles Elliot and his books in wider context. Book history is enjoying something of a renaissance in contemporary scholarship: works on the culture of print, the nature of the book, and on the sciences and book history testify to this claim.[3] In an earlier groundbreaking work on the history of print culture, Lucien Febvre and Henri-Jean Martin described the early spread of printing as 'the geography of the book'.[4] My attention here to the geographical nature of Elliot's work in the book trade is concerned less with the distributional sense of the geography of the book, and more with what Robert Darnton in 1982 described as the 'communications circuit' through which printed books pass from the author to the publisher (if the bookseller does not assume that role), the printer, the shipper, the bookseller and the reader. With the reader, the circuit is complete since it is the reader, Darnton argues, who influences the author both before and after the act of composition. In these terms, what follows on Elliot's medical texts is an examination of only a part of any such communications circuit since I am less concerned with either the readership for Elliot's books or how questions of readership affected the authorship of the books for which Elliot was publisher and bookseller. My concern is not with what others have considered as the 'geographies of reading', as scientific texts were differently engaged with by different audiences but, rather, with what Adrian Johns has seen as the connections between printing, publishing, propriety and property: we must, after all, recognise books as items of economic value as much as we rightly see them as means to intellectual exchange.[5] Elliot's contemporaries certainly recognised his enabling role in what we might think of as the 'communications circuit' of Scottish Enlightenment publishing. Archibald Constable, referring to Elliot's payments to authors, said, 'The encouragement which Mr. Elliot gave to literary men was the means of producing a new spirit among the printers of Scotland'.[6] Elliot himself knew what he was about and used a colourful metaphor to describe his importance to medical authors. 'I say it without vanity', he told a medical correspondent, 'that I am the principal Man Midwife (in the literary way) here, to Man Midwives, Physicians, Apothecaries &c &c &c'.[7]

The sources for the chapter are Charles Elliot's letter books, account ledgers and other papers.[8] Their significance is that they contain the stories of hundreds of books as well as trade relationships and readership. A short-title list of medical books Elliot published and which he sold at home and abroad is added (Table 8.1): it describes the edition, shows the booksellers Elliot associated with, and gives the publication date, price, and copyright entry or payment.

Charles Elliot and medical publishing

Charles Elliot was a remarkable figure in the publishing world of the Scottish Enlightenment. He opened his bookshop in Parliament Square, Edinburgh, in 1771, when he was twenty-three, and over the next eighteen years built up a large bookselling business, in his own publications – about 600 in his career – and in stocks of other books. He was famous in his time for buying literary property – he entered more than forty copyrights at Stationers' Hall – and for not selling shares in his copyrights to other booksellers. Yet London booksellers, such as Thomas Cadell, George Robinson and John Murray sold large quantities of his books, because for particular editions he would associate with one London agent only, sharing the imprint and giving a special rate. It was a unique strategy in Edinburgh book relations with London in the eighteenth century. Elliot was bold and had flair. One of his *coups* was the selling of the *Encyclopaedia Britannica*: he bought up all the remaining copies of the second edition in 1785, and when the third edition began appearing in parts in 1788, he agreed to buy around 2,000 sets, a financial undertaking, had he lived, of around £20,000. By the time of his death in 1790, at the age of forty-one, Elliot had a booktrade network encompassing London, Scotland, the English provinces, Ireland and Europe, and he had shipped large numbers of books to America.

Elliot specialised in medical books, importing them from Europe and London, printing translations of foreign editions, and publishing the work of Edinburgh authors. There is a coherence in these titles: Elliot was deliberately forming a list of what was new, needed and would sell. This was done partly through organising an edition, and partly by encouraging authors with the payment of copy money. Elliot began publishing medical titles in 1772, with Winslow's *An Anatomical Exposition of the Human Body*, translated from the French, sharing the imprint with Alexander Donaldson, and at the beginning of the University's medical classes in November that year Elliot put out the first catalogue of his stock of medical books. His first original medical publication was a sixpenny pamphlet in 1775 by a young army surgeon, George Borthwick, on removing cataracts from the eye. By then, Elliot was attracting the interest of William Cullen as a promising medical bookseller. In December 1775, after he had obtained Cullen's advice and approval, he published the first six volumes of an 18-volume large duodecimo edition of van Sweiten's *Commentaries upon Boerhaave's Aphorisms*, finishing the edition in April 1776. It was an expensive undertaking: the printing by Colin Macfarquhar cost him £471. The book was marketed as the first comprehensive reader a medical student required, as more accurate than the London edition, and as a bargain: 3s. per volume sewed, 3s.6d. bound in calf (the subscription price was 6d a volume cheaper), or three guineas for the set bound.[9]

At the time of the Van Sweiten, Elliot was printing a pamphlet by James Lind that contains hitherto unsuspected editorial work by David Hume. Hume was Lind's friend and had recently been attempting to advance him.[10] Lind's Edinburgh medical thesis of 1768 had been published in English at London by the Dillys, as recently as 1772, as *A Treatise on the Putrid and Remitting Fen Fever, which Raged at Bengal in the Year 1762*. Lind persuaded Elliot to publish a new edition, on the grounds that the London translation had been absurd and that Hume had improved the text. These points were used in the marketing of the revised work, *A Treatise on the Putrid and Remitting Marsh Fever, which Raged at Bengal in the Year 1762*. Elliot linked it with other medical publications when he wrote to John Murray in 22 March 1776:

> In about a fortnight I shall ship for you V. Sweiten. You say 100 Copies will be sufficient. I am printing just now a New Edit of Young Dr. Gregory's Thesis, it being upon the Diseases occasioned by changes of Climate, being a Popular subject in that way and very ably hand[le]d. Make no doubt but it will sell. Also a Translation of Dr. James Lind's Thesis on the Fen fever of the East Indies in 1762. A most absurd translation of this was published by the Mr Dillys some time ago, which is the Reason the Dr wishes another for his own honour. This is done by himself and Revised, (the Language) by David Hume Esqr. Those I should think you will sell a few of. I want if possible to get them Ready to Advertise at the end of V. Sweiten. May I not send you a 100 of each.[11]

It was not true that Lind's was a new translation – the copy text was the 1772 London edition, with some factual corrections and two new footnotes inserted. The text, however, had been improved stylistically. Hume evidently took his pen to nearly every paragraph, sometimes substituting a word, altering punctuation or deleting a redundant phrase, at other times rewriting passages to make them clearer or more concise.[12] Andrew Duncan, learning of the Lind and Gregory publications while the work was going through the press, reviewed them in his amiable manner in the *Medical and Philosophical Commentaries*. For Duncan, Lind's 1768 thesis

> contains so many curious and important observations, that it was, some years ago translated into English, and published by a bookseller in London. But as the translator seems to have been unacquainted with the subject, the meaning of the author is in many places mistaken, and the translation, upon the whole, is by no means such as could be wished. It is with pleasure we can inform our readers, that the author himself has been prevailed upon to undertake the correction of it.

There would be 'several important observations by way of notes, which were neither in the original dissertation, nor in the former translation'.[13] This puffing caused Elliot to hope for double the London sales. He told John Murray in March: 'Dr Gregory & Dr Lind's small Publications are made honourable mention in the last No. of the Med.Comment., which I did not know till finished. I dare say may send you 150 or 200 of each'.[14]

John Innes's *A Short Description of the Human Muscles, chiefly as they appear on Dissection*, was published in late 1775 with a 1776 imprint. Innes, dissector to Monro II and a popular anatomy lecturer, intended that this small octavo be laid on a body as the student reader worked through the descriptions. The second edition was improved by Monro, Elliot bringing out a third in 1784, and the book was used in Edinburgh dissecting rooms for fifty years.[15] As a companion to this book, Innes produced *Eight Anatomical Tables of the Human Body*. The plates were engraved by Thomas Donaldson, after Albinus, reduced to a small quarto designed for the pocket. Elliot paid Innes £100 for the copyright of the plates and entered the property of the book at Stationers' Hall: it was a popular text which Elliot reprinted twice.

The medical books Elliot published over the next dozen years often contained plates by Edinburgh engravers, individually and collectively. For example, Donaldson (with Innes's plates), Andrew Bell and Daniel Lizars are represented in the compendium text for medical students, *A System of Anatomy* (1784 and 1787), edited by Monro's later dissector, Andrew Fyfe. Donaldson, George Cameron and John Buego engraved, while Fyfe drew for, the forty-eight plates in Monro's *The Structure and Physiology of Fishes Explained, and Compared with Those of Man and Other Animals*, as well as producing the fold-out life-size plates in Monro's *A Description of all the Bursæ Mucosæ of the Human Body*. Bell engraved the plates of Monro's own dissertation for the collection of Edinburgh medical theses, *Thesaurus Medicus*, and for William Smellie's *A Set of Anatomical Tables*. Elliot considered his medical books to be special, and at times put another kind of engraving on the title pages, a monogram of his initials, 'CE'.

Elliot entered many medical copyrights at Stationers' Hall. Authors were also given money for the property or for the rights to the edition. Elliot paid

£10 to John Brown for *Elementa Medicinæ*, and £50 to Alexander Hamilton for *A Treatise of Midwifery, Comprehending the Whole Management of Female Complaints* (Hamilton taking the payment in book purchases). Alexander Monro II was paid £42 for the plate of his father's head, which was printed in *The Works of Alexander Monro. Published by his Son*, and also sold separately. Monro received £300 for the use of the plates in *The Structure and Physiology of Fishes Explained*, while keeping their copyright. Elliot paid the prolific Andrew Duncan £50.8s for *Medical Cases, Selected from the Records of the Public Dispensary at Edinburgh*, and £20 for *Heads of Lectures on the Theory and Practice of Medicine*. When Duncan switched the publishing of his *Medical Commentaries* from London to Edinburgh, Elliot paid him copyright of £60 for each annual volume.

The contract survives in the bargain between Elliot and James Gregory. They agreed in August 1787 that Elliot would pay £250 for the right to print the third edition of Gregory's course textbook, *Conspectus Medicinæ Theoreticæ*: there would be a further payment of £50 more should there be another. Payment would be in November 1788, and, as was usual with Elliot, he would pay interest up to that time. The printer was stipulated as Murray & Cochran. Clearly with an eye to his students who would be buying the book, Gregory obtained agreement that the price of the two-volume octavo should not exceed twelve shillings in boards. Failure by either side would lead to a penalty of £200.[16]

The popularity of books on venereal disease – he also published Samuel Foart Simmons and William Nisbet – led Elliot to pay Franz Swediaur £63 for the third edition of *Practical Observations on Venereal Complaints* (1788). Elliot also paid authors for editorial work. Ralph Irving, given £25 copyright for writing *Experiments on the Red and Quill Peruvian Bark*, received £50 'for his trouble' on *The Edinburgh New Dispensatory*. The theology student and hack writer Robert Heron was paid £42 for translating from Fourcroy's *Elements of Natural History and Chemistry* (1789).

Elliot's most profitable literary property appears to have been Benjamin Bell's *A System of Surgery*. He paid Bell £600 for the copyright, and it went through several editions even as the six volumes were appearing between 1783 and 1788. The London bookseller George Robinson, who took 500 copies of the first edition, had the rights for that city, and Elliot kept him informed of the printing. Thus, on 16 February 1784, he wrote to Robinson: 'We are detained much by the author's extensive practice. We were within a sheet [of finishing volume two] 10 days ago – he was called 70 miles out of town – there kept 6 days – loss of arm – dangerous operations. This detains me for the moment, but it adds to his reputation'.[17]

In December 1786, Elliot agreed with the printer William Smellie (see also Chapter 7) on a copyright fee of 1,000 guineas for *The Philosophy of Natural History*.[18] Smellie had still to write it, but Elliot clearly viewed him as second only to William Cullen in his rank of authors. The printing was sporadic over the next three years since Smellie had trouble providing copy. Elliot was sympathetic. He told a medical correspondent in Gothenburg in October 1788: 'Smellie's Book is very much retarded & I do not even think he will be ready by the Spring – it comes hard upon a man in Trade who has a great deal to mind and [it] cannot be expected he can write fast'.[19] Elliot died before the book was finished, which has led to the supposition that Smellie, pathetically, did not get paid. His biographer, Robert Kerr, said 'Mr Smellie failed to reap the advantages of his bargain, and the fruits of his talents and literary labours: For, though the bargain was most honourably fulfilled by Mr Elliots trustees, they were legally prevented from so doing till after Mr Smellies death in 1795'.[20]

Examination of the Elliot papers show this is not the whole truth. Elliot died on 12 January 1790. On 30 January, Smellie appeared at a meeting of the Elliot trustees to say that his book was now printed but that he would not give them it unless they paid him his copyright. Elliot had already put out £688 to Smellie in bills that would be paid six, twelve and eighteen months after printing, and the trustees guaranteed he would get this money. They gave him a bond for the remainder and Smellie turned over copies of *The Philosophy of Natural History*. Smellie was in debt, however, and his creditors blocked payment of the bond. They had previously given Elliot 'arrestments' to thwart payment of the remainder of the 1,000 guineas (legally they could not arrest the bills due to Smellie), and now they served arrestments on the estate. The trustees were legally obliged to withhold the money, but they put a condition in the bond that Smellie would be paid, once he 'loosed' the obstacle, by settling with his debtors.[21] A note of money still owed by the Elliot estate in 1791 – 'Mr. Smellie or Mr. Creech, about £350' – suggests that the bookseller William Creech was the principal Smellie debtor holding up payment.[22] In 1803, eight years after Smellie's death, the arrestments were removed and the estate released £395 due to him.[23] It was a hard outcome: however, the Elliot estate did appear to pay out two-thirds of the 1,000 guineas in bills in his lifetime.

Publishing William Cullen

Cullen began buying books at Elliot's shop in 1774, where his purchases, and the copyright money he would receive for his writing, are recorded in a series of account ledger entries.[24] On 10 May 1776, Cullen dropped into the shop, bought a copy of Pomfry's *Heraldry* for six shillings, and sounded out Elliot, and, through him, John Murray of London, as to who would give the best price for selling copies of *First Lines of the Practice of Physic*, which he intended to print soon. Elliot was keen to be involved and wrote to Murray, 'It is of infinite advantage to one in my situation to be concerned in such publications'.[25] In the event, the first volume was published in 1777 by Murray and William Creech of Edinburgh: Cullen retained the copyright and was given £100 for the first edition by the publishers. Two more volumes came out in 1781 and 1783, and by 1784 Cullen had finished the fourth volume and proposed to sell the copyright of the work. He offered the whole property to Murray for the huge sum of £1,500 and was turned down.[26] Cullen justified the price to his friend Dr. Gilbert Stuart: 'Like most other Authors I may likely be disposed to put a high value upon my own work. But considering the reputation which I flatter myself it will have, and from the great number of students which I have had, and still living in the British dominions and America, I think it must have immediately an extensive sale; and as it is now to be a very complete system of the Practice of Physic, in most parts of it new and considerably improved, I hope it is to continue to be a valuable property for a long time to come'.[27] Elliot saw Cullen as central to his publishing strategy and seized the opportunity. He offered £1,200 and the professor accepted.

Elliot pointed out to a French bookseller, Amable le Roy of Lyons, that this was the most copy money to be paid in Britain for an octavo edition: previous big payments to authors, to William Robertson in particular, had been subsidised by more expensive quartos.[28] The payment defined Elliot as the Edinburgh publisher who could best compete successfully against London consortia in the payment of Scots authors. He remarked to James Keir of Birmingham: 'the London gentlemen thought it strange that a Country Bookseller (so they are pleased to call the venders of books in the Metropolis of Scotland) should outbid them in the purchase of Dr. Cullen's Practice of Physic'.[29]

For *First Lines of the Practice of Physic*, Elliot decided on a first run of 1,500 copies. Thomas Cadell of London quickly agreed to take half, as well as 500 copies of the next edition. Elliot thought at first Cullen would produce five volumes, but there would be only four, and these would retail at a typical octavo price of 6s. in boards. The trade got them at a 25% discount. The printing had to await proof correcting by Cullen and, by the end of January 1784, the edition was only just half way finished. Elliot told Cadell: 'I am sadly retarded for copy by the Doctor, but I cannot help excusing him. If ever man was a slave, he is one to the fullest degree. He makes it a rule that nothing interferes with his class, but the rest of the day he cannot call a moment his own'.[30] On 15 May 1784, Elliot shipped 750 copies of the completed edition to Cadell. The books were not long arriving in London. Elliot always made a point of loading a ship that was getting ready to leave, and he reckoned that the sailing between Leith and London took between four and eight days.

This revised version of the *First Lines* was the fourth edition, in four volumes. The third edition was only three volumes and some of it was still in the hands of the proprietors, Murray in London and Creech in Edinburgh. Cullen felt that the third edition was imperfect, fit only for cancelling, and he had no intention of bringing it up to date with a fourth volume. Murray led the outcry to get one, and Cullen reluctantly agreed that Murray and others could buy the new edition's fourth volume separately. Elliot remarked that it was not much good for the third edition, since the index did not apply and it repeated 110 pages. Once Cullen had made up his mind, there was frantic activity in Elliot's bookshop: he brought his workers in at 4 a.m. to sort out the sheets of 250 copies of the fourth volume, in order to get the material on a ship sailing to London that morning.[31]

The provision of a fourth volume to make up the third edition did not prevent Murray publishing a savage 41-page pamphlet titled *An Author's Conduct to the Public, Stated in the Behaviour of Dr. William Cullen*, in 1784. This included their correspondence about the fourth volume, with a letter from William Creech supporting Murray on the need for one. Murray said Cullen originally sold his *First Lines of the Practice of Physic* like a hawker of gingerbread, selling pieces of it for sixpence to pupils at his lecture-room door, and that he had since made great profit from the booksellers. A few months later, Murray was writing to Elliot to get the new fourth edition at a special discount, but Elliot said he had to purchase them through Cadell at the normal rates.

Elliot would distribute *First Lines* in Britain and abroad but it was not a financial success for him. When Cullen offered him a fifth volume of *First Lines* on the diseases of women and children (which in the end was not printed), Elliot told him that the most copy money he could pay was £300 for an octavo volume. He said it would take three editions, or 3,000 copies sold to the trade, to recover the £300 and pay paper and printing. He said that although he had published five editions of the four-volume *First Lines* to 1789, he had still not recovered the publishing costs or the £1,200 copy money.[32] In the mid-1780s, Cullen was researching a new edition of his *Materia Medica* and bargaining with Elliot for copyright payment. Elliot told Barrois of Paris in the autumn of 1786 that he had nearly closed with Cullen, but the author was demanding an astonishing sum, and was counting on Elliot getting a good price from the first proof sheets sent out for French and German translations.[33] Cullen had not settled by July 1787 and Elliot was warning Barrois that other Paris booksellers were interested in paying for exclusive use of the proofs.[34] In 1788, Elliot finally agreed to pay Cullen the enormous sum of £1,500 for the copyright. Elliot's financial outlay was even higher: the paper and print would cost him another £800.

Elliot was scrupulous in paying Cullen his copyright money on time. His practice was to split large payments over several stages. On 12 May 1784, Cullen was credited with £1,200 for *First Lines*. The same day he was given 602 volumes worth £158.0s.6d, and on 24 May £241.19s.6d in cash. A further instalment of £200 in cash on 8 August made up the first payment of £600.[35] In 1785, Cullen was paid £300 on 23 May, and in August given a bill for £100 payable in November; in November he was given a bill for £180 payable in March 1786. The shortfall of £20 was accounted for by various debits against Cullen, including a copy of *First Lines* for the University of Paris, at a cost of £1.4s.[36] The £1,500 copyright payment for *Materia Medica* was undertaken in three stages. On 15 April 1789, Cullen was given a bill for £500, payable in August, and a bill for £500, payable in November. Elliot made a point with Cullen and other authors of paying interest on copy money that was due, and this was honoured by his trustees after both men died early in 1790, with a payment for the period 1790 to 1791. After Cullen's shop account of around £88 was deducted, cash of £438.5s.11½d was paid to David Thomson to settle Cullen's account in July 1791.[37]

Let me now turn away from the financial aspects of Elliot's publishing of Cullen and other medical texts in Scotland to consider the international nature of the communication circuits within which Elliot worked. In so doing, my intention in looking at Elliot's network of correspondents in, first, Europe, and, secondly, in America, is to illustrate the ways in which book publishers were a part of the communications circuit by which Enlightenment knowledge moved, in print form and, of course, in private correspondence.

International Information Networks: Elliot's European 'Communication Circuit'

From the 1770s, Elliot had a number of correspondents in Europe who were graduates of the Edinburgh medical school or who were Scots merchants. He sent a list of books to Dr. John Grieve at St. Petersburg, and while saying he had little time for news, he did report on the latest medical news:

> the Med. Classes gan on as well as ever, about 325 at Dr. Monros Class this year. All our Joint Acquaintances mostly gone. I think there is no change in the Profs. Dr. Duncan has succeeded wonderfully & very deservedly in his plan. Dr. [John] Brown has sowen wonderfull dissensions amongst the students by what he calls his new Doctrines in opposition to Dr. Cullen but has fallen of late. In the list added to this youll see D. Jones Inquiry, a 6/ Book, wrote in Vindication of its Author being rejected a degree here, and in support of the favourite Theory of Joannes Brunonis as he [styles] himself in a Latin Text Book he has published.[38]

His bookselling agent at Gothenburg was the merchant David Mitchell.[39] In Holland, Elliot had a substantial trade with the Rotterdam booksellers Lambert Bennet and Charles Richard Hake. Bennet had been in Scotland for 18 months before April 1782, and before Bennet left Edinburgh, Elliot gave him lists of books he required: an Irish vocabulary printed at Paris; pocket Bibles in French and French literature; medical works in Latin; Swedish, Russian, Danish and French dispensatories. He also requested a complete set of books on medicine published on the Continent to be sent twice a year, as well as copies of the best editions of history and miscellaneous literature.[40] Bennet and Hake provided medical and scientific titles in Latin, French authors such as the Abbé Raynal, Rousseau and Voltaire, and got Elliot books from Germany. Elliot sent them Cullen's *First Lines in the Practice of Physic* when it was published in 1784 and other medical books, and kept them abreast of work in the press, such as, in 1788, Monro's *Bursæ mucosæ*, Bell's sixth volume of *A System of Surgery*, and Cullen's *Materia Medica*.[41] Peter Gosse at the Hague and S. & J. Luchtman, printers to Leyden University, traded medical and law books in Latin (which Elliot wanted for the University classes), for Elliot's medical theses in Latin and the titles of Thomas Ruddiman, which he had acquired.[42] Mr. Gregory, merchant at Campvere, was asked for a favour when a privateer seized 100 prints of the head of Dr. Cullen, and 100 sets of Hamilton's *Midwifery* in quires from a London ship heading for Leith in 1783 and carried them into Holland: Elliot asked Gregory to buy them back at the auction.[43] The Gregory dynasty of merchants in Europe – David Gregory and Son were at Dunkirk – were kinsmen of Professor James Gregory, Professor of Medicine at Edinburgh, whom Elliot published.[44]

Elliot was in contact with three booksellers in Germany. Andrew Duncan had heard from a German medical correspondent that J. Christian Dieterich of Göttingen wanted an Edinburgh bookselling contact, and he passed the letter on to his friend Elliot. Elliot exchanged catalogues and then books with Dieterich. Elliot wanted new books of merit from Germany, sent from Hamburg directly to Leith.[45] Dieterich was sent the *Thesaurus medicus* and told in 1781 that Elliot had just published the late Alexander Monro's *Works*, in a large Royal quarto of 816 pages with plates selling for £1.5s in sheets, 'which I dare say will sell very well with you and on the continent of Europe'. Dieterich could have them for 19 shillings per book in sheets if he took 100 copies: 'I dare say any of your medical Gentlemen will be able to inform you if you are a stranger [to] Dr. Monro's merit as a physician & an author'.[46] Elliot shipped books at the request of Francois de la Garde, Berlin, by way of London to Hamburg, and sent him a catalogue; Elliot's terms were 20% discount from the prices marked, and payment by bills at London in six months. 'I may be able to take some books from you in Return, and will be happy to be informed of such books you can furnish particularly in *Medicine* & *Medical Philosophy* at Vienna, Leipsick, & other parts of Germany'.[47] Elliot wrote to Jean Frederich Gleditsch, member of a well-known bookselling family, at Leipzig in 1786 at the request of William Cullen, for 16 medical titles in Latin and German for his work on the new edition of *Materia medica*: Gleditsch was to send them immediately to a merchant in Hamburg, for shipment to Elliot at Leith. 'There is a great deal expected from the work', Elliot told him: 'If you chuse to agree with me for the first sheets to be translated into either or both German and Latin I am persuaded you will find your account in it'.[48]

By the autumn of 1783, Elliot was looking for introductions to French booksellers, and one opportunity arose when a young French surgeon and physician visited Edinburgh. Dr. Jean-Joseph Sue II, a member of the Royal Academy of Surgery at Paris, had gained a medical degree at St. Andrews on 26 August 1783, and, in September that year, was in Edinburgh successfully seeking a licence and a fellowship of the Royal College of Physicians of Edinburgh, and foreign membership of the Physical Class of the Royal Society of Edinburgh, which had just been formed and where his father, also Jean-Joseph Sue, Paris surgeon and translator of Alexander Monro I, was already an honorary member.[49] Sue II made the acquaintance of a medical circle that included the surgeon Benjamin Bell, the medical lecturers John Aitken and Charles Webster, Andrew Duncan and Alexander Monro II, and he ordered books at Elliot's shop before moving on to London in mid-September. Elliot shipped Sue II's books there, along with other texts sent by Bell, Aitken and Webster. Elliot said that Monro, who had been planning to send his splendid folio on the *Nervous System* to Paris along with other books, would pass them on to Sue II at London.[50] Elliot promised to give news of medical publications at Edinburgh, and hoped Sue II would do likewise for Paris, adding: 'Pray don't forget to introduce me to the correspondence of a regular Paris bookseller'.[51] In 1784, Elliot sent Sue II a catalogue sheet listing his books, asking him to show them to any bookseller in Paris willing to exchange French titles for them.[52] By that time, though, Elliot had a publication in mind that would bring him to the notice of the French trade.

In October 1783, he wrote to A. F. Fourcroy at Paris for 25 copies in sheets of *Leçons Elémentaire Histoire Naturelle et de Chimie*, which Joseph Black wanted for the students in his chemistry lectures. Elliot explored the fastest route, and decided that Fourcroy should send them to David Gregory in Dunkirk, who would then route them through London. Elliot told Fourcroy: 'I have some thoughts of getting a translation of it made into English by a Gentleman of abilities, I dare say you would like to see the work appear in a neat English Dress, therefore if any additional hints occur, what you chuse to communicate will be very acceptable'.[53] The book was translated by Elliot's medical student nephew, Thomas Elliot, with notes by the surgeon James Russell. The work contained a complete course of chemistry lectures, and Elliot was impressed when Black insisted on having copies of the first volume for his students as it was being run off the press. Elliot offered George Robinson of London 1500 copies at subscription prices and a 10% discount, saying he was convinced there would be a general demand, 'not only from students but from private gentlemen'.[54] *Elementary Lectures on Chemistry and Natural History* was published in two volumes octavo in 1785.

The Edinburgh bookseller John Balfour, who had travelled in France, provided Elliot with another contact.[55] Elliot told Amable Le Roy of Lyons in 1783 that 'I happened to get the perusal of your Catalogue of Books from Mr. Balfour of this place. I have taken the liberty of Copying a number of Books from the Same which I am willing to take from you on Condition you take Books from my list'.[56] They agreed to pay for their transactions half in cash and half in book exchanges. A few weeks later, Elliot was writing to Le Roy and hinting at personal judgements about the worth of contemporary texts. He urged Le Roy:

> not to send the Geographie de Bushey nor any of the Sermons unless of great Character & Estimation, being an entire Stranger to their Character & Merit. if Siecle de Lous 14 & 15 is by Monsr. Voltaire [you] may send 12 [sets] instead of 6. I am amazed you have not made Choice of many more Books from my Catalogue, Such as the great Dr. Swifts works, Dean of St. Patricks, Dublin. I should think them as much esteemed as Popes, which I observe you have. There is many others equally valuable

& I would presume much esteemed in France. I have noted a few again to Remind you [and here Elliot writes down the 18-volume Swift, Sterne's *Works* in seven volumes containing *Tristram Shandy, Sentimental Journey* and *Sermons*, as well as Bailey's *English Dictionary*, new edition]. [I] could wish much to have your Liberty of adding Single Copies of a few Books by way of Specimens. Does not Books on Medicine Answer you. The Edinr Medical School is, I believe, justly esteemed the first in the world. I send many thro all Germany.[57]

Le Roy did take Edinburgh medical books together with works of English literature in return for his stock of French literature, which Elliot found cheaper than the French reprints being produced at Dublin. Elliot told him in 1784 he thought Cullen's *First Lines* 'should find its way to Lyons as it will to every corner of Germany where Dr.Cullen is better known than in France'.[58]

Elliot sold six sets of the second edition of the *Encyclopaedia Britannica* to the Versailles bookseller André de Lausin. De Lausin was in no hurry either to pay for them or to return them, and Elliot, meanwhile, said he could not sell the French books de Lausin had sent without an order, calling them waste paper and useless.[59] Elliot seems to have been aware of some of the rivalries of the French trade. When Alexander Drummond of Edinburgh toured the Paris bookshops in 1788, getting catalogues and delivering letters for Elliot, Elliot warned: 'I find Mr André of Versailles & M.Barrois are a kind of Rivals, so youll please say nothing to the one that you or I know anything of the other'.[60] At Paris, Elliot bought from Gaspard-Joseph Cuchet, sent his catalogues to N.T. Méquignon, and tried to interest Didot the younger in his books. He informed Didot that he had arranged for a copy of the *First Lines of the Practice of Physic* to be given to the Royal Medical Society of Paris, with the compliments of Dr. Cullen.[61]

European booksellers also offered Elliot first sight of proof sheets. In 1785, Pierre-Jacques Duplain of Paris proposed the work of Baron de Wenzel, but Elliot declined: 'Every one in this Country knows and gives the Baron De Wenzel every Credit as an Operator on the Eye but at the same time he is not known here as an Author, a Writer, nor as a literary Man. I must on that acct. decline your offer of the first of the work at 50 guineas; indeed I am not sure till perusal I would translate upon any terms'.[62] Elliot said he could offer Duplain or another Paris bookseller a similar offer to translate *Materia Medica* once he had finished bargaining with the author.

The Paris bookseller with whom Elliot most associated was Pierre-Théophile Barrois the younger. Elliot wrote to him in 1784: 'My friend Mr. [Dugald] Stewart, Professor of Mathematics at this University, informed me that when he was in your city last Autumn, you promised to transmit me when ready a copy of your catalogue, or list of books. I dare say you will remember the Gentleman . . . I am very anxious to settle a correspondent in Paris, that would furnish me with every book that might write for . . . if to be got by any means in Paris . . . I will at all times be much obliged to you to give me any Litterary intelligence, of whatever nature going on in your place, and if acceptable to you I will make the same return'.[63] Elliot wrote 18 more letters to Barrois before 1789. Barrois' first reply concerned Cullen's *First Lines* – he was reprinting it in Paris and wanted to ensure no further volumes were planned – but in November 1784, he put on a Paris coach a package apparently containing additional material for the translation of Fourcroy that Elliot was putting through the press. The coach arrived at the White Bear Inn, Piccadilly, where Elliot arranged for the parcel to be taken to the George and Blue Boar, Holborn, for the Edinburgh coach.[64] Thus began the use of the Paris coach for matters that could not await the normal shipping, such as the researching and production of Cullen's *Materia Medica*. Elliot wanted sent by coach '1 copy of any new book of merit, not exceeding one volume 8vo, on medicine, surgery, anatomy, chemistry, materia medica, history or entertainment'.[65] Titles coming on the coach included, in 1788, two copies of the third edition of Fourcroy's *Chemistry*, and two numbers of the *Voyage Pittoresque*.[66]

Elliot kept up a commentary on the progress of books through the Edinburgh press. In October 1784, for example, he wrote: 'I publish also this winter a large work by Dr. Monro, Professor of Anatomy, on the Natural History and Pathology of Fishes, with 46 very large Copperplates, in large folio. It will retail at about £2.2s, two guineas British, & for the accommodation of you foreigners, I mean to print a Latin Translation together with the English. I think you may be able to sell many of these'.[67] Later, in August 1785, he wrote: '[The] Edinburgh New Dispensatory will be finishd in about a month, before which I shall be glad of your order. It Contains the whole of the 1st London Pharmacopia with remarks &c and in the same order by Dr. Duncan, and will be far Superior to the former. It is, in fact, an entire new work'.[68] In 1785, Barrois took enough of Elliot's *Thesaurus Medicus Edinburgensis Novus* to have his name in the imprint.

Elliot told Barrois which of his books were worth translating. His fourth edition of *First Lines* was safe because little would be added. In 1784, Elliot also told Barrois, 'I will publish this season Vol. 3 of Bell's surgery. This is one of the first rate books on the subject. I think it would be worth your while to translate it. I have sold about 1500 copies of Vol. 1 and 2. It is already translated into German'.[69] Barrois published editions of both, translated by Edouard-Françoise-Marie Bosquillon. In 1785, however, Elliot advised Barrois against translating Cullen's *A Treatise of the Materia Medica* from the London quarto, which he said was bad, or the Dublin octavo, which was much worse. Elliot told how Cullen had promised to give him a correct edition, and this would be worth Barrois' while to translate into French.[70] While Gleditsch of Leipzig supplied the Latin and German titles during Cullen's revisions of the book, Barrois supplied the French. Sometimes Cullen wanted the books immediately. 'I am obliged to write to you this night', Elliot told Barrois in 1786, 'on Account of Dr. Cullen, who wants a Copy of each of the books above the red line sent directly by the first Dilligence from Paris to London'.[71] Cullen wanted 16 titles, including several by Antoine-Augustine Parmentier on vegetables, potatoes and bread, P. R. Vicat's *Histoire des Plantes Vénéneuses de la Suisse*, Balthasar-Georges Sage's *Analyse des Blés*, J-F. Carrère's *Bibliothèque Littéraire Historique et Critique de la Médecine Ancienne et Moderne*, R-G. Gastellier's *Des Spécifiques en Médecine*, and Chopart and Desault's *Traité des Maladies Chirurgical*, as well as a copy of Bosquillon's translation of *First Lines*. The next year Elliot again asked Barrois for a small packet on the next coach or post diligence out of Paris, including *La Pharmacie Moderne*, Bosquillon's translation of the second volume of *First Lines* if published, and the *Pharmacopoeia Helveticae* with Haller's preface: 'It is for my friend Dr. Cullen & one must submit to trouble our Correspondents on such accounts'.[72]

By August 1788, a third of the *Materia Medica* had gone through the press in Edinburgh. 'I have not yet given any person the use of the Sheets for [a French] translation and will certainly prefer you to any Stranger', Elliot told Barrois, 'but I think it is in your power to do me some service in Return. I make no doubt there will be a Considerable demand at Paris for the original, and if I prefer you, you ought to engage to take a considerable number, say from 100 to 250, at the lowest Booksellers price'.[73] Barrois actually took only 25 of the two-volume sets, but he was given the rights to the sheets as they came off the press. The batch sent in February 1789 was the third consignment and consisted of the last parts of volume one and the first two alphabets of volume two: they went by sea from Edinburgh to London, and then by coach to Paris. Barrois and Méquignon published the translation at Paris in 1789-90, making great play on the title page and the advertisement in the book that it was from the authoritative Edinburgh edition of 1789. Cullen's *Materia Medica* was used throughout Europe and North America and was a standard work for the next 40 years.[74]

Unequal International Exchange: Elliot's Book Trade with the Americas

Elsewhere I have described the Scots book trade to America and given an account of Scots book exports there in the middle of the century. The main source of letters with Scots then – to Adrian Watkins, John Balfour and Alexander Kincaid – is the correspondence of David Hall, Benjamin Franklin's printing partner. The Kincaid and Bell letter books, 1764-71, contain details of their American trade.[75] The first volume of *A History of the Book in America* has made some mention of the Scots book trade, but none of Elliot's correspondence.[76] Accordingly, this section offers new insight into the nature of the Scottish-American book trade in the later eighteenth century.

Elliot's book trade with the Americas was always one-way. It began in 1774 when he sent the *Edinburgh New Pharmacopoeia*, along with the *Edinburgh Magazine and Review*, to John Stewart, a surgeon in Grenada. The following year, a London book, Richard Brookes' six-volume *Natural History*, was sent, together with news of the Edinburgh medical school: 'The classes are always throng – no less than 22 graduates sett loose upon mankind this year from this university'.[77] Elliot had four other correspondents in Grenada and a further 15 in Jamaica.[78]

Elliot's first substantial book shipments, however, took place during the American War to British-occupied New York, to the Tory merchant Colburn Barrell in the years 1777 and 1778, and, from 1780 to 1781, to the Loyalist printers and stationers, Mills and Hicks. A later correspondent at New York, the merchant Francis Blaikie, seemed to surprise Elliot by writing in January 1784 that a cargo of Edinburgh books had not shifted, not even the medical ones, which Elliot always thought of as certain sellers. 'I am sorry,' Elliot said,

> that the books are not likely to sell as few of them can be rivalled by Irish editions, and Irish editions are so infamously incorrectly printed, that where others can be had they seldom get a preference . . . In general I must request you to sell at the best prices they will fetch after having done everything in your power to dispose of them with a profit. The expence & damage in return I cannot think of . . . Will you show the inclosed Catalogue to any of your booksellers, if there are any that can be trusted, and . . . if money or good bills are given, I will make considerable discount from the prices printed, from 15 to 20 percent.[79]

When James Rivington of New York wrote in 1784 for books and Bibles, enclosing an advance draught for £100, Elliot took the opportunity to say that he, Elliot, was the main publisher for medical books in Edinburgh and could supply him on better terms than could Thomas Cadell in London. Elliot put together six trunks of assorted books for Rivington, worth £172. They included all the sets of Rousseau that were available in Edinburgh: Elliot could get only one copy of Andrew Duncan's *Medical Commentaries* to send. Rivington was informed that a new edition of Cullen's *Nosology* was coming out.[80] There were already plenty of sets of Cullen's *First Lines* in Philadelphia and New York. The Loyalist merchant Neil Jamieson, who was staying on in New York after the war in the hope of recovering his property in Virginia, received three boxes of books, including Cullen's *First Lines*, and Jamieson also shipped Elliot's books through New York to Philadelphia before leaving in 1786.[81]

In Charleston, South Carolina, large numbers of Elliot books were imported by the merchants James and Edward Penman. Edward Penman had been Depute Collector of the Customs at Greenock, where he had organised book shipments to America for Elliot a number of times. When Penman was preparing to go to Charleston in 1780, Elliot sent him eight cases of books, four to be shipped to Mills and Hicks in New York, and four, each packed with the same titles, to be sold anywhere Penman liked in America. Elliot said they were 'the very best books', such as David Hume and William Robertson.[82] Detailed lists of books were sent separately, and are not in the Elliot papers, but Elliot makes some mention of the kinds of books they were. In July 1784 he gave the Penmans a box of new medical books and novels: in 1785 there were four large boxes, worth £178, that included amongst their contents the *Encyclopaedia Britannica*, school books, and Fanny Burney's *Cecilia* and *Evelina*. Any merchant who bought a box from the Penmans could get 15% discount on short-time credit.[83]

Another outlet for the distribution of Elliot's books were medical graduates of Edinburgh University, including Americans returning home. When Dr. John Carson returned to Philadelphia from Montrose in 1783, he took two trunks of books worth £84: one containing the latest medical books, the other a variety of titles including Elliot's *A Collection of the Most Esteemed Farces and Entertainments Performed on the British Stage*. Elliot told him the books were charged at the lowest retail prices in Edinburgh, but 'if any Gentleman take the whole or one Trunk off your hands . . . you may abate 15 to 20 per cent, always including the charges outwards'.[84] Elliot was surprised a year later to hear that Carson had not sold any books at all.[85] Elliot shared some medical school news: 'We had some amazing classes this last winter. Munro had 440 [students], he was obliged to erect two galleries, & Dr. Cullen to leave his old room and teach in the High School. We have a good deal of Americans'.[86]

One newly graduating American, Dr. John Leigh of Norfolk, Virginia, had Elliot publish his prize dissertation in Edinburgh, *An Experimental Enquiry into the Properties of Opium*, in 1786. Elliot required him and his friend, Dr. James Ramsay, also of Norfolk, to give a guarantee for the £36 cost. When they returned home, Elliot wrote to say the opium pamphlet had not sold, and he mainly blamed academic attitudes:

> I find all these Prize Dissertations are looked upon by old practitioners as the production of young men, and think more of their own consequence than to look into them, far less to buy them, & as to other students unless some particularly want to look into the subject, it is not bought . . . Besides Bark & Opium have of late been so much wrote about that the subject seems entirely laid aside.[87]

Elliot was still trying to get them to pay their debt in 1788.

Dr. Benjamin Rush of Philadelphia had warm memories of Edinburgh where he had studied under Cullen, and Elliot took advantage of this connection when he published the new fourth edition of Cullen's *First Lines* in 1784. Elliot sent Rush two boxes containing 68 four-volume sets and got him to act as a book agent, in a letter that mixed news of Cullen with hard business. He asked Rush to take a set and sign it from the author, and pass the rest on, either to Thomas Bradford or to another bookseller:

> How can I apologise for this liberty? Only Dr. Cullen's desire to have the correct copys soon with you, & before any Irish or spurious edition could appear was the causes of my troubling you . . . Dr. Cullen desires in particular manner to be remembered to you, & I can assure you his paternal regard to you is as strong as your filial duty is expressed to him.

Might I further beg of you to show the enclosed catalogue to Mr Bradford or others, & assure him I will be happy of his correspondence & without vanity may say, if he deals much in medical books, I can serve him the best terms of anyone in Britain.[88]

One of Elliot's Virginian correspondents, a Dr. Shore of Petersburg, had remarked in a letter of 1783 on a lack of books in that state. This gave Elliot the idea of setting up his nephew, William Elliot, and another young man, George Millar, in a bookshop there. Millar arrived first with an introductory letter to Shore. Elliot wrote:

> Mr Millar and . . . Mr Elliot were forming a plan of visiting your part of the world. At that time no fixed scheme was adopted by them. Your insinuation respecting the want of books in Virginia suggested to me the idea of giving them a considerable stock of books for an adventure & as a trial. Accordingly I have fitted Mr Millar with, in my humble opinion, a very useful Collection of Books, of History, Miscellanies, also of Politicks, a select Collection of Divinity, a good & large assortment of school books of every kind, & last of all in your own way, a Collection of Medical Books.[89]

Elliot advanced this new firm of Elliot and Millar £1,200 worth of books and a consignment of medicine. The cargo was insured in Scotland for 3% of the value and went from the Clyde to Hampton on the James River, and was then carried on lighters to Petersburg. In September 1784, Elliot sent off a cheerful letter to Elliot and Millar, hoping sales were going briskly in Petersburg, and despatching more goods for the store — some umbrellas, £153 worth of medicine from an Edinburgh druggist, and more medical books: Cullen's *First Lines of the Practice of Physic*, and van Swieten's *Commentaries upon Boerhaave's Aphorisms,* which, Elliot told them, 'treat fully and regularly on all . . . Diseases incidental to the Human Body, and may be more trusted than Dr. Buchan'. Also sent was Benjamin Bell's *A System of Surgery* and Andrew Fyfe's *A System of Anatomy from Monro, Winslow and Innes*: 'you may assure the medical gentlemen [they] sell amazingly'.[90] The young partners did not get along (Millar was drinking, William Elliot had a bad temper, and they quarrelled over William's debts), and the firm broke up on 15 December 1784. William took a share of the books and set off for Fredericksburg, where, in 1786, his uncle sent him another £240 worth of books in a vain attempt to rescue the business and give him a fresh start. Charles Elliot had previously sent the second edition of the *Encyclopaedia Britannica*, which was now sold out in Edinburgh.[91]

Elliot's biggest agent in America was his former clerk Thomas Dobson, whom he set up in Philadelphia. Dobson left Scotland in late October 1784 with 42 trunks containing £2,000 worth of Elliot books and some other goods. During 1785 another £1,400 worth of books arrived at Dobson's store. With all those Elliot titles, the interior of the shop must have been quite like the premises in Edinburgh's Parliament Square. Certainly, readers in Philadelphia were getting the books not that long after customers in Edinburgh. On 21 May 1785, Dobson advertised Elliot's medical publications, including Caleb Dickinson's *An Inquiry into the Causes and Nature of Fever*, Fourcroy's *Chemistry*, Irving's *Experiments on the Red and Quill Peruvian Bark* and Bell's *A System of Surgery*.[92] Dickinson was published in Edinburgh on 15 December 1784, Fourcroy and Irving on 20 December, and Bell's third volume on 8 January 1785.[93] Through Dobson, a great number of books from Edinburgh were distributed in America. But for Elliot, who did not live to see a return on his investment, the adventure was calamitous personally and financially.[94]

Conclusion

The Elliot papers offer a vivid picture of an Edinburgh bookseller working with authors, co-operating and competing with London, and trading and corresponding internationally during the Scottish Enlightenment. Yet this chapter has only touched lightly upon the resources available and upon the questions raised for Enlightenment historians concerning the authorship, production, distribution and reception of books. Further study is needed, and sources other than Elliot's papers must be incorporated in any such work. Consideration of the work and role of a major bookseller like John Balfour, for example, would extend our knowledge of European and Edinburgh connections into earlier years. Following his partnership between 1739 and 1762 with Bailie Gavin Hamilton, Balfour continued as Edinburgh University printer during a lengthy career, associating with William Smellie as well as being part of the London-Edinburgh consortium of Strahan, Cadell and Creech. Balfour was also the main supplier of French books to Edinburgh from the mid-1760s, and his knowledge of France suggests he travelled there. As a further example, the extensive collection of Bell and Bradfute ledgers covering the period 1788 to 1822 held in the Edinburgh City Archives (which could be complemented by the Bell and Bradfute collection of letters and manuscripts at the National Library of Scotland) might bear researching for questions to do with European, American and British book distribution, as well as for what they might say about other aspects of the Edinburgh book trade.[95]

It may well be, of course, that such further research has to focus on the *details* of books and the book trade. In reviewing the connections between the history of science and recent work in book history, for example, Frasca-Spada and Jardine discern a move away from 'grand pronouncements about the impact of print on the sciences' toward a more particular emphasis.[96] Not all the 'sciences' were the same. Books were not the only means by which scientific discourse was disseminated: instruments could transmit understanding, as Morrison-Low has shown in Chapter 2, and men gathered together in clubs and institutions often heard rather than read the new concerns of their peers. This is not to marginalise the book. Nor is it to marginalise the economic and social contexts in which intellectual ideas were disseminated in textual form. Quite the contrary. In a more general discussion of the place of the sciences in Enlightened Europe, it has been noted that 'Until quite recently . . . historians of the sciences have shown relatively little interest in the specific experiences of the eighteenth century, and they have encountered major problems in making precise the interactions between the Enlightenment and the natural and human philosophies of the period'.[97] Further research into books and book history in the Enlightenment may face the question, then, of how best to ally the concerns of book historians with those of historians of science in order to integrate the specific experiences of men such as Charles Elliot into those natural and human philosophies that were the substance of the books he saw into print.

Notes

1. This chapter is from work in progress on Charles Elliot, which will include an index to entries in the papers. The Elliot archive is held at John Murray Publishing, London. I thank Virginia Murray, the archivist, for her support of this project. All the Elliot letters cited here are from this archive, unless otherwise stated. I have regularised the punctuation and, sometimes, the spelling when quoting Elliot's letters. The copies were made by various hands, carelessly at times, and often without full stops: they do not reflect the standard of correctness of the original Elliot letters.
2. Richard B. Sher, 'Science and Medicine in the Scottish Enlightenment: The Lessons of Book History', in Paul Wood (ed.), *The Scottish Enlightenment: Essays in Reinterpretation* (Rochester, NY, 2000), 99–156, at p. 148.
3. See, for example, Marina Frasca-Spada and Nick Jardine (eds.), *Books and the Sciences in History* (Cambridge, 2000); Adrian Johns, 'Science and the Book in Modern Cultural Historiography', *Studies in History and Philosophy of Science* 29 (1998), 167–94; idem, *The Nature of the Book: Print and Knowledge in the Making* (Chicago and London, 1998); H. T. Mason (ed.), *The Darnton Debate: Books and Revolution in the Eighteenth Century* (Oxford, 1998); Jonathan R. Topham, 'Scientific Publishing and the Reading of Science in Nineteenth-Century Britain: A Historiographical Survey and Guide to Sources', *Studies in History and Philosophy of Science* 31 (2000), 559–612.
4. Lucien Febvre and Henri-Jean Martin, *L'apparition du Livre* (Paris, 1958). This was published in English and translated by David Gerard, as *The Coming of the Book* (New York, 1990).
5. On Darnton and the communications circuit, see Robert Darnton, 'What is the History of Books?', *Daedulus* 111 (1982), 65–83; Johns, *Nature of the Book*, 187–265. The idea of different 'geographies of reading' is discussed by James Secord with reference to the varied reception of Robert Chambers' *Vestiges* in the mid-1840s; see James A. Secord, *Victorian Sensation: The Extraordinary Publication, Reception and Secret Authorship of* Vestiges of the Natural History of Creation (Chicago and London, 2000), esp. Part Two.
6. Thomas Constable, *Archibald Constable and his Literary Correspondents*, 3 vols. (Edinburgh, 1873), 1:434.
7. Charles Elliot to Dr. H. A. Bryan, 30 April 1788, Elliot letter 4208.
8. In the Murray Archives, there are eight Elliot letter books covering the period

1774 to 1790, four account ledgers (for the period 1771 to 1790), a Trustees' book (begun after Elliot's death in 1790) and financial accounts of his estate. The letters are copies of outgoing correspondence, mostly to persons beyond Edinburgh. I have assigned each of the 4,857 entries a number. The ledgers contain the accounts and purchases of several thousand individuals, often with locations and occupations, many of them in Edinburgh, but also elsewhere in Britain, in Ireland and overseas: I cite these by ledger number and date.

9. *Edinburgh Evening Courant*, 11 December 1775; *Caledonian Mercury*, 17 April 1776.
10. Raymond Klibansky and Ernest C. Mossner (eds.), *New Letters of David Hume* (Oxford, 1954), xix, 193–95.
11. Charles Elliot to John Murray, 21 March 1776, Elliot letter 348.
12. Compare, for example, the description in the 1772 London work (pp. 9–10): 'The nature of the burying ground here did not contribute a little to spread the contagion, as, on account of that swampiness of the soil, the putrescent matter generated in the old graves kept flowing into the new ones, as fast as they were opened; so that not only the grave diggers, but even those who only attended the funerals, caught the infection; and many were seized with the disorder whilst paying the last duties to their dearest friends, who had already fallen victim to it' – with Hume's revision in the Edinburgh work of 1776 (p. 17): 'Here the burying-ground contributed not a little to spread the infection. The ground being marshy, the putrid water flowed from the old graves into the new ones, which infected the grave-diggers, and those that attended the funerals; and, from this, many were suddenly seized when performing this last duty to their deceased companions'.
13. *Medical and Philosophical Commentaries. By A Society in Edinburgh*, vol. 3 pt. 4, new edn (London 1786), 448. The Gregory review is on p. 447.
14. Charles Elliot to John Murray, 11 April 1776, Elliot letter 357.
15. *Dictionary of National Biography*, s.v. 'Innes, John'.
16. National Library of Scotland (hereafter NLS), MS. Acc. 7810, Contract between Gregory and Elliot, 1787.
17. Charles Elliot to George Robinson, 16 February 1784, Elliot letter 2928.
18. Robert Kerr, *Memoirs of the Life, Writings, and Correspondence of William Smellie*, 2 vols. (1811; Bristol, 1996), 2: 263–64.
19. Charles Elliot to Dr. Englehart, 10 October 1788, Elliot letter 4321.
20. Kerr, *Life of Smellie*, 2:295.
21. 'Sederunt Book of the Trustees appointed by Charles Elliot Bookseller in Edinburgh', 30 January and 1 March, 1790, Elliot papers. For the arrestment process, see D. M. Walker, *A Legal History of Scotland, Volume V: The Eighteenth Century* (Edinburgh, 1998), 618–19.
22. '1791 June 9th, General View of the Funds and Debts of the Late Mr Chas. Elliot', Elliot papers, p. 3.
23. 'General View of the Funds of C.E., 8th March 1803', Elliot papers.
24. Elliot ledgers, entries for Cullen: L1/255 (1774–75); L1/386 (1776–77); L2/163 (1777–82); L3/150 (1782–83); L3/387 (1784–85); L3/436 (1785–86); L4/255 (1786–88); L4/266 (1788–91).
25. Charles Elliot to John Murray, 11 May 1776, Elliot letter 387.
26. For Cullen's dealings with Murray and their subsequent dispute over *First Lines*, see William Zachs, *The First John Murray and the Late Eighteenth-Century Book Trade* (Oxford, 1998), 94–96, 191–95, and the checklist of publications.
27. William Cullen to Dr. Gilbert Stuart, 16 August 1783, Cullen Letters 1779–88, Royal College of Physicians, Edinburgh (hereafter RCPE).
28. Charles Elliot to Amable le Roy, 14 June 1784, Elliot letter 3026.

29. Elliot to James Keir, 22 June 1784, Elliot letter 3041.
30. Elliot to Thomas Cadell, 23 January 1784, Elliot letter 2913. The publishing of *First Lines* is described in 14 letters to Cadell between November 1783 and June 1784.
31. Elliot to Cadell, 21 June 1784, Elliot letter 3038.
32. Elliot to William Cullen, 3 April 1789, MS Letters to Dr. Cullen 1774–90, RCPE.
33. Elliot to P-T. Barrois, 30 October 1786, Elliot letter 3886.
34. Elliot to Barrois, 31 July 1787, Elliot letter 4027.
35. Elliot ledger L3/357, Cullen 1784–85.
36. Elliot ledger L3/436, Cullen 1785–86.
37. Elliot ledger L4/266, Cullen 1789–91.
38. Elliot to Dr. John Grieve, 27 January 1783, Elliot letter 2620. This letter was sent via the St. Petersburg merchant, William Porter, to Grieve, who was physician to the Tzarina.
39. Mitchell bought sets of John Bell's *British Poets* and *British Theatre*, and other books printed in London and sent on by Elliot, as well as the coloured mezzotints that were a rage in Edinburgh in the early 1780s. The engravings were promoted by Elliot's friend and bookselling neighbour, James Sibbald: Elliot to David Mitchell, 11 August 1780, Elliot letter 1746.
40. Elliot to Messrs. Bennet and Hake, Rotterdam, 2 April 1782 and 30 April 1782, Elliot letters 2351 and 2388 respectively.
41. Elliot to Lambert Bennet, 25 February 1788, Elliot letter 4154.
42. There are seven Elliot letters to Gosse and Pinnet and to Gosse, 1775–77, and four to the Luchtmans, 1777–82.
43. Elliot to Mr. Gregory, 14 April 1783 and 5 January 1784, Elliot letters 2676 and 2904.
44. James Gregory's ancestor, David Gregory of Kinnairdie (1627–1720), had been a merchant in Holland.
45. Elliot to J. C. Dieterich, 18 June 1779 and 1 November 1779, Elliot letters, 1366 and 1479A-B (with French translation).
46. Elliot to J. C. Dieterich, 22 June 1781, Elliot letter 2074.
47. Elliot to François de La Garde, 9 October 1785, Elliot letter 3429.
48. Elliot to J. F. Gleditsch, 12 June 1786, Elliot letter 3712.
49. I thank the Royal College of Physicians of Edinburgh for information on Sue junior: Dr. J. J. Sue's licentiate petition, 30 August 1783; RCPE Minutes, 1 Sept. and 4 Nov. 1783; History and Laws of the RCPE, List of Fellows, 5. Sue's election to the RSE was confirmed in January 1784, and his father was elected in Sept. 1783: *Transactions of the Royal Society of Edinburgh*, 1 (1788), 93–94; Clare Gaffney *et al*, typescript 'Index of Fellows of the Royal Society of Edinburgh, elected from 1783 to 1882'.
50. *Observations on the Structure and Functions of the Nervous System. Illustrated with Tables*, published at Edinburgh in 1783 by William Creech, with Joseph Johnson of London on one imprint, Cadell, Elmsley, Murray and Longman on another.
51. Elliot to Dr. Sue, 15 September 1783, Elliot letter 2805.
52. Elliot to Dr. Sue, 5 April 1784, Elliot letter 2973.
53. Elliot to M. de Fourcroy, 13 and 24 October 1783, Elliot letters 2828 and 2830.
54. Elliot to George Robinson, 29 November 1784, Elliot letter 3175.
55. Warren McDougall, 'Copyright Litigation in the Court of Session, 1738–49, and the Rise of the Scottish Book Trade', *Transactions of the Edinburgh Bibliographical Society* 5 (1988), 2–31; see esp. p. 22 for some of Balfour's French interests.
56. Elliot to Amable le Roy, 16 June 1783, Elliot letter 2743.
57. Elliot to Le Roy, 1 August 1783, Elliot letter 2774.

58. Elliot to Le Roy, 14 June 1784, Elliot letter 3026.
59. Elliot to M. André, 3 June 1788, Elliot letter 4246.
60. Elliot to Alexander Drummond, 6 August 1786, Elliot letter 3793.
61. Elliot to Didot the younger, 14 June 1784, Elliot letter 3025.
62. Elliot to Pierre J. Duplain, 17 October 1785, Elliot letter 3433.
63. Elliot to P-T. Barrois, 14 June 1784, Elliot letter 3024.
64. Elliot to George Robinson, 16 November 1784, Elliot letter 3153.1.
65. Elliot to Robinson, 12 June 1786, Elliot letter 3713.
66. Elliot to P-T. Barrois, 27 May 1788?, Elliot letter 4238.
67. Elliot to Barrois, 15 October 1785, Elliot letter 3105. The Latin translation was not printed.
68. Elliot to Barrois, 30 August 1788, Elliot letter 4306.
69. Elliot to Barrois, 15 October 1784, Elliot letter 3105.
70. Elliot to Barrois, 31 January 1785, Elliot letter 3213.
71. Elliot to Barrois, 12 June 1786, Elliot letter 3713.
72. Elliot to Barrois, 31 July 1787, Elliot letter 4027.
73. Elliot to Barrois, 30 August 1788, Elliot letter 4306.
74. A. Doig, J. P. S. Ferguson, I. A. Milne and R. Passmore (eds.), *William Cullen and the Eighteenth-Century Medical World* (Edinburgh, 1993), 38–39. For the *Materia Medica*, see also Christopher Clayton, 'William Cullen in Eighteenth Century Medicine' and R. Passmore, 'William Cullen and Dietetics', in *William Cullen*, 92–93 and 167–85 respectively.
75. Warren McDougall, 'Scottish Books for America in the Mid-Eighteenth Century', in Robin Myers and Michael Harris (eds.), *Spreading the Word: The Distribution Networks of Print, 1550–1850* (Winchester and Detroit, 1990), 21–46, and McDougall, 'Copyright Litigation in the Court of Session, 1738–49, and the Rise of the Scottish Book Trade'.
76. Hugh Amory and David Hall (eds.), *A History of the Book in America, Volume 1: The Colonial Book in the Atlantic World* (Cambridge, 2000); in this volume, see James Raven, 'The Importation of Books in the Eighteenth Century', 183–98; Calhoun Winton, 'The Southern Book Trade in the Eighteenth Century', 224–46; and James N. Green, 'English Books and Printing in the Age of Franklin', 248–96.
77. Elliot to John Stewart, 7 December 1775, Elliot letter 276.
78. Grenada: Laurance MacDowell, 1785–88; Thomas Hawkins, 1788; an Elliot kinsman, George Shortreed, 1789. Jamaica: John and James Allan, merchants, 1781–83; Dr. John Anderson, physician, 1778–83; Dr. Richard Brodie, physician, 1781–89; Thomas Brunton, 1787; James Cleland, 1781–86; William Codrington, 1786; John Farquharson, surgeon, 1775; David Gibson, land surveyor, 1776–1783; James Leslie, attorney-at-law, 1778–80; George Little, 1784–86; Daniel Macfarlane, 1775–81; Dr. George Macfarquhar, physician, 1782–83; Stevenson & Co., booksellers, 1787; Dr. William Taylor; while a Miss Spence, 1787, is indexed in the Elliot letter books but the letter to her is not copied. For the region in this period, see Roderick Cave, *Printing and the Book Trade in the West Indies* (London, 1987).
79. Elliot to Francis Blaikie, 22 March 1784, Elliot letter 2961.
80. Elliot to James Rivington, 3 July 1784 and 10 September 1784, Elliot letters 3048 and 3089.
81. Elliot to Neil Jamieson, 20 January 1785, 28 January 1785? and 13 June 1785?, Elliot letters 3199, 3209 and 3359.
82. Elliot to Edward Penman, 26 January 1780 and 2 February 1780?, Elliot letters 1569 and 1772.
83. Elliot to James and Edward Penman, 15 July 1784, 17 January 1785 and 22

August 1785, Elliot letters 3059, 3196 and *3363.
84. Elliot to Dr. John Carson, 28 June 1783, Elliot letter 2752.
85. Elliot to Carson, 22 March 1784, Elliot letter 2960.
86. Elliot to Carson, 30 July 1784, Elliot letter 3065.
87. Elliot to Dr. John Leigh, 5 November 1787, Elliot letter 4079.
88. Elliot to Dr. Benjamin Rush, 3 July 1784, Elliot letter 3049.
89. Elliot to Dr. Shore, 22 October 1783, Elliot letter 2829.
90. Elliot to Elliot & Millar, 10 September 1784, Elliot letter 3088.
91. Elliot to William Elliot, 24 July 1786, Elliot letter 3774.
92. *Pennsylvania Packet and Daily Advertiser*.
93. *Caledonian Mercury*.
94. I explore Elliot's book adventure in Philadelphia further in a forthcoming paper.
95. For the book trade in Scotland in earlier periods, see Alastair J. Mann, *The Scottish Book Trade, 1500–1720: Print Commerce and Print Control in Early Modern Scotland* (East Linton, UK, 2000).
96. Maria Frasca-Spada and Nick Jardine, 'Introduction: Books and the Sciences', in Frasca-Spada and Jardine, *Books and the Sciences*, 1–10, at p. 3.
97. William Clark, Jan Golinski and Simon Schaffer (eds.), *The Sciences in Enlightened Europe* (Chicago and London, 1999), ix.

Table 8.1: Short Title-List of Charles Elliot's Medical Publications[1]

EA = *Edinburgh Advertiser*. CM = *Caledonian Mercury*. EEC = *Edinburgh Evening Courant*. P = publication and price. Zachs = 'A checklist of Murray publications 1768–95', in *The First John Murray*.

	Date	Author	Short title	Imprint
1	1772	Charles Elliot	[A catalogue of books in physic and surgery, with the lowest prices affixed.] Not seen. P: 17 November 1772 (EA).	
2	1772	J.B. Winslow (1669–1760) trans. G. Douglas, M.D.	An anatomical exposition of the structure of the human body. 6th ed. 2 vols 8°. plates. P: noted 4 April 1772,10s. [bound] (CM).	Edinburgh: for A. Donaldson and C.E.
3	1773	Hermann Boerhaave (1668–1738)	Institutiones medicae. 8°. 498p. P: 3s. bound.	Edinburgi: C.E.
4	1775	George Borthwick, surgeon, 14th Regiment of Dragoons	A treatise upon the extraction of the crystalline lens. 8°. 38p. P: 6 Feb. 1775, 40p., 6d. [sewed] (CM 4 Feb.).	Edinburgh: for C.E
5	1775	Richard Mead (1673–1754)	The medical works. A new edition. 8°. 592p. plates. P: noted 17 May 1775, 6s. bound (CM).	Edinburgh: for A. Donaldson and C.E.
6	1776	William Cullen (1710–90)	A letter to Lord Cathcart, concerning the recovery of persons drowned, and seemingly dead. 8°. 44p. P: 1s. sewed.	Edinburgh: for C.E
7		Charles Elliot	[A catalogue of a valuable collection of books on the several branches of medicine, being the library of the late Dr George Martin physician in St Andrews . . . To which is added, A full set of all the best modern authors in physic, &c . . . which will be sold by auction . . .] Not seen. P: 20 Dec. 1775 (EA).	

	Date	Author	Short title	Imprint
8	1776	James Gregory (1753–1821)	Dissertatio medica inauguralis de morbis coeli mutatione medendis. 8°. 182p. P: by 10 April 1776, 2s. sewed (CM).	Edinburgi: C.E.
9	1776	John Innes (1739–77), dissector to A. Monro II	Eight anatomical tables of the human body; containing . . . skeletons and muscles represented in the large tables of Albinus. 4°. 52p. 8 plates [T. Donaldson]. P: 2 Nov. 1776, 6s. sewed, 6s.6d. half-bound (CM). Copyright: entered 20 July 1776 for Elliot.[2] CE paid Innes £100 for property of the plates. Zachs 127.	Edinburgh: for C.E. Edinburgh: by Balfour & Smellie, for J. Murray, London; Balfour, Drummond, Elliot and Schaw.
10	1776	John Innes	A short description of the human muscles, chiefly as they appear on dissection. 8°. 234p. P: by 20 Dec. 1775, 2s.6d. sewed, 3s. bound (EEC). Zachs 128.	Edinburgh: by Balfour & Smellie, for J. Murray, London; Balfour, Drummond, Donaldson, Elliot and Schaw.
11	1776	James Lind (1736–1812), ed. David Hume (1711–1776)	A treatise on the putrid and remitting marsh fever, which raged at Bengal in the year 1762. 8°. 70p. P: by 10 April 1776, 1s. sewed (CM). Copy at Royal College of Physicians, Edinburgh.	Edinburgh: for C.E.
12	1776	G.F. van Swieten (1700–72)	Commentaries upon Boerhaave's aphorisms concerning the knowledge and cure of diseases. 18 vols. 12°. P: v.1–6, 11 Dec. 1775, 3s. sewed, 3s.6d. bound in calf (EEC); completed 10 April 1775, £2.14s. sewed, £3.3s. bound. Zachs 138.	Edinburgh: for C.E.; sold by J. Murray, London.
13	1777 –78	Andrew Duncan (1744–1828)	Medical and philosophical commentaries. By a society in Edinburgh. Volume fifth. 8°. Four parts with separate titles. 521p. plate. P: part 20, 2 Dec. 1778, 2s; 5 vols. £1.15.6 bound (CM). Zachs 152: 1s.6d a number, 6s. per vol; variant.	London: for J. Murray; [pts 3 & 4 add J. Bell] W. Creech, C.E., and M. Drummond, Edin.
14	1777	P.J. Macquer (1718–84)	Elements of the theory and practice of chymistry. 5th ed. 8°. 646p. 6 plates. P: by 2 Nov. 1776, 6s. bound (CM).	Edinburgh: for A. Donaldson and C.E.
15	1778	Benjamin Bell (1749–1806)	A treatise on the theory and management of ulcers. With a dissertation on white swelling of the joints. To which is prefixed, an essay on the chirurgical treatment of inflamation and its consequences. 8°. 408p. plate. P: for 5 March 1778, 5s. boards (CM 4 March). ESTC n049622 describes a variant.	Edinburgh: for C.E.; sold by T. Cadell, London.
16	1778	Andrew Duncan (1744–1828)	De laudibus Gulielmi Harvei oratio. 8°. 24p. P: 6d. sewed.	Edinburgi: C.E.

	Date	Author	Short title	Imprint
17	1778	Andrew Duncan	Medical cases, selected from the records of the Public Dispensary at Edinburgh. [with De laudibus Gulielmo Harvei oratio] 8°. 386p. port. P: 4s.3d. boards. Portrait: 'Andrew Duncan M.D. A. Bell sc. Edinburgh Published by Ch. Elliot 11 May 1778'. Copyright: entered for C.E. 22 June 1778. C.E. paid Duncan £50.8s. for the property. Zachs 180.	Edinburgh: for C.E., and J. Murray, London.
18	1778	Charles Elliot	[A catalogue of books on all the different branches of medicine, with the lowest prices affixed; and all the other new books on any subject of physic published this season.] Not seen. P: 4 March 1778, gratis [to medical gentlemen] (CM).	
19	1778 –83 [-84]	Encyclopaedia Britannica	Encyclopaedia Britannica; or, a dictionary of arts, sciences, &c. 2nd ed. 10 vols. 4°. plates: A. Bell. P: vol. 1, 11 Feb 1778, 18s boards – vol. 10 [dated 1783], 6 Nov. 1784, 25s. boards; 10 vol., 360 plates, £10.10s. boards (CM). CE raises to £12 boards from April 1785.	Edinburgh: for A. Bell, C. Macfarquhar, J. Hutton; various sellers on imprint; after 22 Feb 1785 sold by C.E., and J. Robinson, London.
20	1778	H.D. Gaubius (1705?-80) trans. Charles Erskine, surgeon.	The institutions of medicinal pathology. 8°. 345p. P: 18 March 1778, 4s.3d. boards, 5s. bound (CM).	Edinburgh: for translator, sold by C.E., and T. Cadell, London.
21	1778	John Innes, ed. Alexander Monro II	A short description of the human muscles, chiefly as they appear on dissection. 2nd ed. greatly improved by Alex. Monro. 8°. 252p. P: noted 15 Nov. 1779, 2.6d. boards, [3s. bound] (CM). Copyright: entered for C.E. 18 December 1787. Zachs 190.	Edinburgh, for C.E., and J. Murray, London.
22	1778	Patrick Dugud Leslie (1705–83)	A philosophical inquiry into the cause of animal heat. 8°. 370p. P: 20 June 1779 (CM).	London: for S. Crowder, J. Robson. A. Gordon and C.E., Edin.
23	1778	Donald Monro (1727–1802), ed.	Letters and essays on the small-pox and inoculation, the measles, the dry belly-ache, the yellow, and remitting, and intermitting fevers of the West Indies. 8°. 360p. Zachs 194; 5s.	London: for J. Murray, and C.E., Edin.

	Date	Author	Short title	Imprint
24	1778–79	William Smellie (1740–1795), printer, ed.[3]	Thesaurus medicus: sive, disputationum, in academia Edinensi, ad rem medicam pertinentium, a collegio instituto ad hoc usque tempus, delectus, Gulielmo Smellio. 2 vols. 8°. 5 plates [A. Bell] P: vol. 1, 29 April 1778, 6s. boards, vol. 2, 28 Nov. 1778, 12s.6d boards for both, 14s. bound. Selected Edinburgh University medical theses 1726–58. Re-issued in 1785 as part of 4–vol Thesaurus medicus Edinburgensis novus [92]. Zachs 203.	Edinburgi: typis academicis, prostant venales apud C.E., J. Bell [v.1], G. Creech, et Londini, J. Murray.
25	1779	Benjamin Bell	A treatise on the theory and management of ulcers. 2nd ed. 8°. 440p. plate. P: 15 Nov. 1779, 5s.3d. boards, 6s. bound (CM).	Edinburgh: [by Macfarquhar and Elliot] for T. Cadell, London, and C.E.
26	1779	Matthew Dobson (d. 1784)	A medical commentary on fixed air. 8°. 204p. P: 3s. boards.	Chester: by J. Monk; sold by T. Cadell, London, C.E., Edin.
27	1779	Andrew Duncan	Medical and philosophical commentaries. By a society in Edinburgh. Volume sixth. 8°. Four parts with separate titles. 489p. P: part 2, 7 July 1779, 1s.6d. (CM). Zachs 219;1s.6d. a part, 6s. per vol; variant.	London: for J. Murray; M. Drummond, J. Bell, W. Creech, C.E., Edin
28	1779	A. von Haller (1708–77)	First lines of physiology. Translated from the correct Latin edition printed under the inspection of William Cullen. 8°. 533p. P: 15 May 1779, 6s. boards (CM).	Edinburgh: [by Macfarquhar and Elliot] for C.E.
29	1779	John Innes	Eight anatomical tables of the human body; containing . . . skeleton and muscles represented in the large tables of Albinus. 2nd. ed. 4°. 53p. 8 plates [T.Donaldson]. P: by 15 Nov. 1779, 6s. boards, 6s.6d. bound. Zachs 227.	Edinburgh: for C.E., and J. Murray, London.
30	1779	Francis Spilsbury, chemist.	Physical dissertations on the scurvy and gout. 2nd ed. 8°. 152p.	London: for J. Wilkie, C. Etherington; C.E., Edin.; J. Hoey, Dublin.
31	1780	John Brown (1735–88)	Elementa medicinae. 12°. 454p. P: for 4 May 1780, 6s. boards (CM 1 May). C.E. paid Brown £10 for the property.	Edinburgi: C.E.
32	1780	Chirurgo-Medical Society, Edinburgh.	Laws and regulations of the Chirurgo-Medical Society, Edinburgh. 2nd ed 8°. 68p.	Edinburgh: for C.E.
33	1780	Peter Degravers	A complete physico-medical and chirurgical treatise on the human eye. Translated from the French. 4°. 368p. plates.	London: for author; sold by B. Law, T. Becket, J. Robson, and C.E., Edin.

	Date	Author	Short title	Imprint
34	1780	Andrew Duncan	An account of the life and writings of the late Alexr. Monro Senr. 8°. 40p. P: 6 May 1780, 1s. sewed (CM).	Edinburgh: for C.E., and C. Dilly, London.
35	1780	William Smellie (1697–1763)	A set of anatomical tables, with explanations, and an abridgment of the practice of midwifery. 12° [+ 'few 8°'] 82p. 40 plates [A. Bell]. P: 1 March 1780, 5s. boards, 8vo size 6s. boards (CM).	Edinburgh: for C.E.
36	1780	William Smellie	A treatise on the theory and practice of midwifery. To which is now added, his set of anatomical tables. 3 vols. 12°. 40 plates. P: 1 March 1780, 9s. boards, 10s.6d. bound (CM).	Edinburgh: for C.E.
37	1781	John Aitken (d.1790)	Outlines of the theory and cure of fever, upon plain and rational principles. 12°. 100p. P: 17 February 1781,1s.6d. boards (CM).	London: T. Cadell; Edinburgh, Gordon and Murray, C.E., W. Creech, J. Bell, Balfour and sons.
38	1781	Andrew Duncan	Heads of lectures on the theory and practice of medicine. 2nd ed. 12°. 436p P: 1 Nov. 1780, 3s. boards (CM). C.E. paid Duncan £20 for the property.	Edinburgh: [by Macfarquhar and Elliot] for C.E. Edinburgh: for C.E.; sold by C. Dilly, London.
39	1781	Andrew Duncan	Medical cases, selected from the records of the Public Dispensary at Edinburgh. 2nd.ed. [With] De laudibus Harvei oratio; and An account of the life and writings of the late Alexander Monro. 8°. 428p. port. P: 4s. boards.	Edinburgh: for C.E.; sold by C. Dilly, London.
40	1781	Alexander Hamilton (1739–1802)	A treatise of midwifery, comprehending the whole management of female complaints, and the treatment of children in early infancy. 8°. 488p. Copyright: C.E. paid Hamilton £50 for the property.	Edinburgh: for J. Dickson, W. Creech, and C.E.
41	1781	Alexander Hamilton	A treatise of midwifery, comprehending the management of female complaints, and the treatment of children in early infancy. 8°. 488p. P: 18 Nov. 1780, 6s. bound [in calf] (CM). Zachs 282; copyright entry for Hamilton; 6s. boards.	London: for J. Murray; C. Dickson, W. Creech, and C.E., Edin.
42	1781	Robert Jones, M.D.	An inquiry into the state of medicine, on the principles of inductive philosophy. 8°. 394p. P: noted 29 April 1782, 5s.3d. (CM).	Edinburgh: for T. Longman, and T. Cadell, London; and C.E., Edin.

	Date	Author	Short title	Imprint
43	1781	Alexander Monro I (1697–1767), ed Alexander Monro II (1733–1817); Life by Donald Monro.	The works of Alexander Monro. Published by his son, Alexander Monro. 4°. 816p. 7 plates: Andrew Bell. Portrait engraved by Basire, after Allan Ramsay. P: 31 March 1781, £1.5s. boards (CM) Copyright: entered for C.E. 7 March 1781, assigned to him by A. Monro II; C.E. paid Monro £42 for the plate.	Edinburgh: [by Macfarquhar and Elliot] for C.E.; and G. Robinson, London. Edinburgh: [by Macfarquhar and Elliot] for C.E.
44	1781	Alexander Monro I engraving	[Portrait] Alexander Monro Senr. M.D. Professsor of Anatomy, Fellow of the Royal College of Physicians at Edinburgh, and F.R.S. Painted by Allan Ramsay Esqr. Engraved by James Basire, 1775. P: issued separately from Works. Noted 16 and 23 Nov. 1782, 5s. for proof impressions. (CM).	Published as the Act directs, for C. Elliot, Edinburgh, 1781.
45	1782	Andrew Duncan	Heads of lectures on pathology. 12°. 48p. P: 10 Dec. 1781, 6d. (CM).	Edinburgh: [by Macfarquhar and Elliot] for C.E.
46	1782	Andrew Duncan	A letter to Dr Robert Jones of Caermarthenshire. 8°. 48p. P: 8 Dec. 1781, 1s. [sewed] (CM).	Edinburgh: for C.E.; sold by T. Longman, T. Cadell, C. Dilly, London.
47	1782	Charles Elliot	[A catalogue of the new books in medicine, with the lowest prices affixed.] Not seen. P: 16 Nov. 1782.	
48	1783	William Cullen engraving	[Mezzotint engraving of Dr William Cullen, by Valentine Green, from a painting by W. Cochrane; Esq; done at the expence of the Royal Medical Society] Not seen. P. Elliot sold the T. Sommers version of this 1772 print (CM 23 Nov. 1782), bought the plate and replaced Sommers' imprint with his own. He sought to retrieve the first 100 prints from London after they were taken by privateers; 20 more were ordered 4 April 1783 [Elliot Letters 2567, 2671, 2676, 2904] P: large size, 3s. (advert in Medical Commentaries for 1787).	[C. Elliot excudebat]
49	1783–88	Benjamin Bell	A system of surgery. 6 vols 8°. 99 plates advertised [A. Bell, D. Lizars]. P: vol. 1, 23 Nov. 1782, 6s.3d. boards, 7s. bound; finished set noted 11 Oct. 1788, £1.16s.6d. boards (CM). [2nd ed. 1785–88, 3rd 1787–9, 4th ed. 1789–90, new ed. 1790.] Copyright: entered for C.E.: v. 1, 9 Jan. 1783; v.5, 17 March 1787; v.6, 6 May 1788. C.E. paid Bell £600 for the property.	Edinburgh: for C.E.; and G. Robinson, London. [v.4–5 adds G.J. and J. Robinson, v. 6 Elliot & Kay, London.]

	Date	Author	Short title	Imprint
50	1783	John Innes	Eight anatomical tables of the human body; containing . . . skeleton and muscles represented in the large tables of Albinus. 3rd ed. 4°. 53p. 8 plates. [T. Donaldson] P: 6s.6d. half-bound. Zachs 359.	Edinburgh: for C.E.; and J. Murray, London.
51	1783	Patrick Dugud Leslie	An account of the epidemical catarrhal fever, commonly called the influenza. Durham . . . June, 1782. With a letter . . . by John Clark, M.D. 8°. 106p. P: 17 May 1783, 2s. (CM).	London: for S.Crowder, J. Robson; A. Gordon, C.E., Edin.
52	1783	Alexander Monro I	The anatomy of the human bones, nerves, and lacteal sac and duct. [With] his Treatise of comparative anatomy, as published by his son. A new ed. 12°. 570p. P: about Dec. 1782 (CM 18 Nov., 7 Dec. 1782). 4s. boards.	Edinburgh: for J. and E. Balfour, and C.E.; and G. Robinson, London. [Treatise]: Edinburgh: for C.E.; and G. Robinson, London.
53	1783	Alexander Monro I	A treatise on comparative anatomy. Published by his son, Alexander Monro. A new ed. 12°. 135p. P: about Dec. 1782 (CM 18 Nov. 7 Dec. 1782). 2s. boards. Copyright: for C.E., 19 May 1783.	Edinburgh: for C.E.; and G. Robinson, London.
54	1783	Francis Spilsbury	Free observations on the scurvy, gout, diet, and remedy. 2nd ed. 8°. 176p.	Rochester: by T. Fisher; sold by G. Wilkie, Mess. Davenhill; A. Rothwell, London; C.E., Edin.; J. Magee, Dublin.
55	1784	Benjamin Bell	A treatise on the theory and management of ulcers. 3rd ed. 8°. 500p. plate. P: 23 Feb. 1784, 6s. boards , [7s. bound] (CM).	Edinburgh: for C.E.; and G. Robinson, London.
56	1784	William Cullen	First lines of the practice of physic. 4th ed. 4 vols. 8°. P: 12 May 1784, £1.4s. boards (CM). Copyright: entered for C.E., 6 Feb.19 June 1784. C.E. paid Cullen £1,200 for the property.	Edinburgh: [by Macfarquhar and Elliot] for C.E. Edinburgh: [by Macfarquhar and Elliot] for C.E.; and T. Cadell, London.
57	1784	William Cullen	First lines of the practice of physic. 2nd ed. Vol. IV. 8°. 500p. P: noted 9 June 1784, 6s. boards (CM).	Edinburgh: [by Macfarquhar and Elliot] for C.E.; and T. Cadell, London.
58	1784	William Cullen	A letter to Lord Cathcart, concerning the recovery of persons drowned, and seemingly dead. 8°. 40p. P: 1s. sewed.	Edinburgh: for C.E.; and T. Cadell, London.

	Date	Author	Short title	Imprint
59	1784	Andrew Duncan	Medical cases, selected from the records of the Public Dispensary at Edinburgh. 3rd ed. 8°. 448p. P: 5s. boards, 6s. bound.	Edinburgh: for C.E.; and G. Robinson, London.
60	1784	Andrew Duncan engraving	[Portrait.] Not seen. P: 'A very fine print of Dr Duncan, painted by Weir, and engraved by Trotter', 2s.6d (advert in Bell's Surgery, vol. 3, 1786); noted 20 Oct. 1788 (CM). Elliot ordered 250 copies from Thomas Trotter, London, 14 Aug. 1784 [Elliot Letter 3070].	[Edinburgh: C.E.]
61	1784	Thomas Addis Emmet (1764–1827)	Oratio coram societate physica. 8°. 22p. P: 6d.	Edinburgi: Macfarquhar et Elliot. C.E.; G. Robinson, Londoni; G. Gilbert, Dublini.
62	1784	John Fuller (d. 1825)	Some new hints, relative to the recovery of persons drowned, and apparently dead. 8°. 36p. P: 11 Aug. 1784, 1s. (CM).	London: for T. Cadell; and C.E., Edin.
63	1784	Alexander Hamilton	Outlines of the theory and practice of midwifery. 8°. 426p. P: 15 Nov. 1783, 5s.3d. boards, 6s. bound; bound with William Smellie's 40 Tables and explanations, 12s. (CM) Copyright: for C.E. 24 Nov. 1783.	Edinburgh: [by Macfarquhar and Elliot] for C.E.; and G. Robinson, London.
64	1784	John Innes, ed. A Monro II	A short description of the human muscles, chiefly as they appear on dissection. A new ed., greatly improved by Alex. Monro. 12°. 240p. P: 2s.6d. boards. Zachs 423.	Edinburgh: for C.E.; and G. Robinson and J. Murray, London.
65	1784	Robert Jones	An inquiry into the state of medicine, on the principles of inductive philosophy. 2nd ed. 8°. 394p.	London: for G. Robinson; and C.E., Edin.
66	1784	[A short attempt]	A short attempt to recommend the study of botanical analogy, in investigating the properties of medecines from the vegetable kingdom. 8°. 110p. P: 31 March 1784, 1s.6d. sewed (CM).	London: for G. Robinson; and C.E., Edin.
67	1784	Samuel Foart Simmons (1750–1813)	Observations on the cure of gonorrhoea, and some other effects of the venereal virus. 2nd. ed. 12°.	London: for the author; sold by J. Johnson, London; and C.E., Edin.
68	1784	William Smellie	A set of anatomical tables, with explanations, and an abridgment of the practice of midwifery. 8°. 72p., and 12°. 40 plates. P: 8° size 6s., 12° size 5s. in boards.	Edinburgh: for C.E.

	Date	Author	Short title	Imprint
69	1784	William Smellie	A treatise on the theory and practice of midwifery. To which is now added, his set of anatomical tables. 3 vols. 12°. 41 plates. P: about 16 Feb. 1784, 10s.6d. boards, 12s. Bound (CM).	Edinburgh, for C.E.
70	1784	Franz Swediaur (1748–1824)	Practical observations on the more obstinate and inveterate venereal complaints. 8°. 242p. P: 2s.6d.	London: for J. Johnson; and C.E. Edin.
71	1784	A system of anatomy, ed. Andrew Fyfe, assistant to A. Monro II	A system of anatomy from Monro, Winslow, [and] Innes. 2 vols. 8°. 16 plates [T. Donaldson, A. Bell, D. Lizars]. P: 19 Jan. 1784, 13s boards, 15s. bound (CM 24 Jan.)	Edinburgh: for C.E. Edinburgh: for C.E.; and G. Robinson, London.
72	1784	Charles White (1728–1813)	An inquiry into the nature and causes of that swelling . . . which sometimes happens to lying-in women. 8°. 92p. 3 plates.	Warrington: by W. Eyres, for C. Dilly, London; and C.E., Edin.
73	1785	Francis Balfour (fl. 1812)	A treatise of the influence of the moon in fevers. 8°. 64p. P: noted 21 Oct. 1786, 1s.6d. sewed (CM).	Calcutta: 1784; Edinburgh: reprinted, by the desire . . . of William Cullen, M.D. for C.E., Edin.; and G. Robinson, London.
74	1785–88	Benjamin Bell	A system of surgery. 2nd ed. 6 vols. 8°. plates (A.Bell, D.Lizars). P: £1.6s. boards, £2.2s. bound.	Edinburgh: [by Macfarquhar and Elliot) for C.E; and G.G.J. and J. Robinson, London. [v.5–6 adds Elliot & Elliot & Kay, London.] (CE monogram)
75	1785	Torbern Bergman (1735–84), trans. Thomas Beddoes (1760–1808)	A dissertation on elective attractions. 8°. 400p. plates. P: noted 25 March 1786, 6s. boards (CM). Zachs 456.	London: for J. Murray; and C.E., Edin.
76	1785	William Cullen	Institutions of medicine. Part I. Physiology. For the use of students in the University of Edinburgh. 3rd ed. 8°. 240p. P: 15 April 11785, 4s. boards (EA).	Edinburgh: for C.E.; and T. Cadell, London.
77	1785	T. Dewell	The philosophy of physic, founded on . . . elementary fire. 2nd ed. 8° 138p.	Marlborough: by E. Harold; for J. Murray, J. Bew, London; W. Taylor, Bath; J. Elliot, Edin.

	Date	Author	Short title	Imprint
78	1785	Caleb Dickinson	An inquiry into the nature and causes of fever. 8°. 192p. P: 15 Dec. 1784, 3s. sewed (CM).	Edinburgh: for C.E.; and G. Robinson, London.
79	1785	Andrew Duncan	Medical commentaries, for the years 1783–84. Exhibiting a concise view of the latest and most important discoveries in medicine and medical philosophy. Collected and published by Andrew Duncan. Volume ninth. 8°. 532p Zachs 468: 6s.	London: for J. Murray, and C. Dilly; W. Gordon, and C.E., Edin.
80	1785?	Andrew Duncan	Heads of lectures on the theory and practice of medicine. 3rd. ed. enlarged. 12°. P: 25 Oct. 1784 'speedily will be published', 3s.6d. boards (CM).	[Edinburgh: by Macfarquhar and Elliot]
81	1785	Charles Elliot	[A catalogue of books in the different branches of medicine; comprehending a full collection of such books as are absolutely necessary for the medical students, and also those particularly recommended by particular professors. Together with the completest assortment perhaps ever imported into this kingdom of the newest and the best German, French, and other foreign authors, on these subjects, with the lowest prices affixed.] Not seen. P: 'in a few weeks' (CM 10 November 1784).	
82	1785	Samuel Ferris (1760–1831)	A dissertation on milk. In which an attempt is made to ascertain its natural use. 8°. 224p.	[London:] by J. Abraham; sold by T. Cadell, R. Faulder, and C.E., Edin.
83	1785	A.-F. Fourcroy (1755–1809), trans. Thomas Elliot, notes by James Russell, surgeon.	Elementary lectures on chemistry and natural history. 2 vols. 8°. P: 20 Dec. 1784, 12s. boards (CM).	Edinburgh: for C.E.; and G. Robinson, London. Edinburgh: for C.E.; G. Robinson, London; W. Gilbert, Dublin.
84	1785	Alexander Hamilton	A treatise of midwifery, comprehending the management of female complaints, and the treatment of children in early infancy. 2nd ed. 8°. 336p. P: 4s. boards, 10s. in boards with William Smellie's 41 Plates and Explanations.	Edinburgh: for C.E.; G.G.J. and J. Robinson, London; W. Gilbert, Dublin.
85	1785	Ralph Irving	Experiments on the red and quill Peruvian bark. 8°. 185p. P: 20 Dec. 1784, 3s. in boards (CM). C.E. paid Irving £25 for the property.	Edinburgh: for C.E.; G. Robinson, London; W. Gilbert, Dublin.

	Date	Author	Short title	Imprint
86	1785	Samuel Foart Simmons, ed.	The London medical journal. [v. 5, 1784] 8°. 448p. P: for 5 vols 1781–84 £1.15s., single nos. 1s.6d.	London; for the editor; sold by J. Johnson; C.E., Edin.; P. Byrne, Dublin.
87	1785	Samuel Foart Simmons, ed.	The London medical journal. [v. 6, 1785] 8°. 448p.	London: for the editor; sold by J. Johnson; C.E., Edin.; P. Byrne, Dublin.
88	1785	Alexander Monro II (1733–1817)	The structure and physiology of fishes explained, and compared with those of man and other animals. royal 2°. 128p. 48 plates [G. Cameron, T. Donaldson, J. Beugo; A. Fyfe]. P: in the press 15 Dec. 1784; re-advertised 21 Oct. 1786,£2.2s. boards (CM). C.E. paid Monro £300 for the use of the plates.	Edinburgh: for C.E. Edinburgh: for C.E.; G.G.J. and Robinson, London. (CE monogram)
89	1785	William Nisbet (1759–1822)	Syllabus of a course of lectures on . . . the theory and practice of the venereal disease. 8°. 5s. 48p. P: 5s. boards.	Edinburgh: by C.E.
90	1785 ?	William Smellie	A treatise on the theory and practice of midwifery. [with] his set of anatomical tables. 41 copperplates. 3 vols. 12°. P: not seen 10s.6d. boards, 12s. bound.	[Edinburgh: for C.E.]
91	1785	William Smellie	A set of anatomical tables, with explanations, and an abridgment of the practice of midwifery. A new ed. 8°. 76p. 12°. 41 plates. 8° P: size 6s., 12° size 5s. in boards.	Edinburgh: for C.E.
92	1785	[Selected by a committee of the Royal Medical Society of Edinburgh]	Thesaurus medicus Edinburgensis novus . . . [v.1–2 title]. Thesaurus medicus: sive, disputationum, in Academia Edinensi . . . [v. 3–4 title]. 4 vols. 8°. 5 Plates [A. Bell]. P: in press 15 Dec. 1784 (CM); £1.6s. boards, £1.10s. bound. v.1–2 re-issue of William Smellie's ed. of Thesaurus medicus [24]; v.3–4 relate to medical theses 1759–85.	Edinburgi & Londoni: C.E. & G. Robinson; Dublini, G. Gilbert, Parisiis, P.T. Barrois; Viennae & Lipsiae, R. Graeffer.
93	1786	William Cullen	First lines of the practice of physic. A new edition. 4 vols. 8°. P: £1.4s. boards, £1.8s. bound.	Edinburgh: for C.E., and T. Cadell, London.
94	1786	William Curtis	Assistant plates to the materia medica: or, figures of such plants and animals as are used in medicine. 8°. plates. 16p. 20 leaves.	London: for author by Frys and Couchman; sold at his botanic garden; by B. White and son; G.G.J. and J. Robinson, London, and C.E., Edin.

	Date	Author	Short title	Imprint
95	1786	Andrew Duncan	Medical commentaries, for the year 1785. Exhibiting a concise view of the latest and most important discoveries in medicine and medical philosophy. Collected and published by Andrew Duncan. Volume tenth. 8°. 485 [2]p. P: 4 Feb. 1786, 6s. boards, £3 in boards, £3.10s. bound., for 10 vols. Zachs 518.	London: for J. Murray; C.E., Edin.
96	1786	Charles Elliot	[C. Elliot's catalogue of books in all the different branches of medicine, surgery, anatomy, chemistry, natural history, &c &c for the year 1786, with the lowest prices attached.] Not seen. P: 11 Jan. 1786, gratis to gentlemen of the faculty (CM).	
97	1786	Gentlemen of the Faculty – Charles Webster and Ralph Irving. (Revised edition of William Lewis's The new Dispensatory'.)	The new dispensatory. By gentlemen of the Faculty at Edinburgh. 8°. 752p. plates. P: due 24 Oct. 1785, 7s.6d. bound (EA 11 Oct.). 'Some copies of this work may have been sold without the word Edinburgh . . .' (EA, 1 Nov. 1775). Reissued as Edinburgh New Dispensatory to distinguish from an edition of Lewis's The new dispensatory being sold in Edinburgh Copyright: entered for C.E. 11 Nov. 1785. C.E. gave Irving £50 'for his trouble' on the book.	Edinburgh: for C.E.; G.G.J. and J. Robinson, London.
98	1786	Gentlemen of the Faculty – Charles Webster and Ralph Irving[4]	The Edinburgh new dispensatory. By gentlemen of the Faculty at Edinburgh. 8°. 783p. 6 plates on three leaves: D. Lizars. P: re-advertised 21 Oct. 1786, 734p., 7s.6d. bound (CM).	Edinburgh: for C.E.; G.G.J. and J.Robinson, London.
99	1786	A.Von Haller	First lines of physiology. Translated from the correct Latin edition printed under the inspection of William Cullen. 2 vols. 8°. P: noted 21 Oct. 1786, 7s. boards (CM).	Edinburgh: for C.E.; G.G.J. and J. Robinson, London.
100	1786	John Leigh, M.D.	An experimental inquiry into the properties of opium and its effects on living subjects. 8°. 146p. P: 28 June 1786, 2d.6d. sewed (CM).	Edinburgh: for C.E.; G.G.J. and J. Robinson, London.
101	1786	Samuel Foart Simmons, ed.	The London medical journal. [v.7 for 1786] 8°. 452p.	London: for the editor; sold by J. Johnson; C.E., Edin.; P. Byrne, Dublin.
102	1786	John Rollo (d. 1809)	Observations on the acute dysentery. 8°. 84p. P: noted 22 April 1786, 1s.6d. (CM).	London: for C. Dilly; and C.E., Edin.
103	1786	K.W. Scheele (1742–86), trans. Thomas Beddoes	The chemical essays of Charles-William Scheele. Translated from the transactions of the Academy of Sciences at Stockholm. 8°. 507p. P: noted 25 March 1786, 6s. boards. Zachs 545.	London: for J. Murray; W. Gordon and C.E., Edin.

	Date	Author	Short title	Imprint
104	1786	Thomas Skeete (1757–89)	Experiments and observations on quilled and red Peruvian bark. 8°. 400p. P: noted 25 March 1786, 6s. boards (CM). Zachs 552; 5s. boards.	London: for J. Murray; sold by W. Creech and C.E., Edin.; L. White and W. Gilbert, Dublin.
105	1786	Thomas Trotter (1760–1782)	Observations on the scurvy: with a review of the theories lately advanced on that disease; and the opinions of Dr Milman refuted from practice. 8°. 112p. P: 16 January 1786, 2s. [sewed] (CM).	Edinburgh: for C.E.; G.G.J. and J. Robinson, London.
106	1786	Franz Swediaur	Practical observations on venereal complaints. 2nd ed. 8°. 240p. P: noted 25 March 1786, 3s.6d. sewed (CM).	London: for J. Johnson; and C.E., Edin.
107	1786	Thomas Withers (1750–1809)	A treatise on the athsma. 8°. 468p.	London: G.G.J. and J. Robinson., W. Richardson; Fletchers, Oxford; Merrill, Cambridge; C.E., W. Creech, Edin.; booksellers of York.
108	1787 –89	Benjamin Bell	A system of surgery. 3rd ed. 6 vol. 8°. plates.	Edinburgh: for C.E.; C.E. & Co. [C.E. and T. Kay from v. 2],G.G.J. and J. Robinson, London.
109	1787	Benjamin Bell	A treatise on the theory and management of ulcers. 4th ed. 8°. 488p. plate.P: 18 Nov. 1786, 6s. boards (CM).	Edinburgh: for C.E.; and G.G.J. and J. Robinson, London.
110	1787	William Cullen	First lines of the practice of physic. A new ed. 4 vols. 8°. P: £1.4s. boards, £1.8s. bound.	Edinburgh: for C.E.; and T. Cadell, London.
111	1787	Andrew Duncan	Medical commentaries, for the years 1781–82: exhibiting a concise view of the latest and most important discoveries in medicine and medical philosophy. [v.8] 2nd ed. 8°. 466p. P: 6s. boards.	London: for Elliot & Kay, and C.E., Edin.
112	1787	Andrew Duncan	Medical commentaries, for the year [1786]. Decade second, vol. I. 8°. 480p. P: 'next week' , 6s. boards (CM 25 Jan 1787). Copyright: entered for C.E. 17 March 1787. C.E. paid Duncan £60 for the property of each annual volume.	Edinburgh: for C.E.; C.E. & Co., London.
113	1787	William Goldson	An extraordinary case of lacerated vagina, at the full period of gestation. 8°. 78p. Zachs 587; 1s.6d.	

	Date	Author	Short title	Imprint
114	1787	Alexander Hamilton	Outlines of the theory and practice of midwifery. A new ed. 8°. 424p. P: 20 Nov. 1786. 5s. boards; with William Smellie's Forty-one Tables and explanations, 11s. boards, 12s. bound (CM).	Edinburgh: [by Macfarquhar and Elliot] for C.E.; and G.G.J. and J. Robinson, London.
115	1787	Thomas Houlston (1745–87)	Observations on poisons; and on the use of mercury in the cure of obstinate dysenteries. A new ed. 8°. 96p. P: 1s.6d. sewed.	Edinburgh: for C.E; and Elliot and Kay, London.
116	1787	John Hunt	Observations on the circulation of the blood, and on the effects of bleeding. 8°. 92p.	London: for J. Johnson; and C.E., Edin.
117	1787	Samuel Foart Simmons, ed.	The London medical journal. [v.8 for 1787] 8°. 432p.	London: for the editor; sold by J. Johnson; C.E., Edin.; P. Byrne, Dublin.
118	1787	William Nisbet (1759–1822)	First lines of the theory and practice in venereal diseases 8°. 456p. P: 14 April 1787, 5s. boards (CM). Copyright: entered for C.E. 6 June 1787.	Edinburgh: for C.E.; C.E. and Co., G.G.J. and J. Robinson, London.
119	1787	William Smellie, ed. Alexander Hamilton	A set of anatomical tables, with explanations, and an abridgment of the practice of midwifery. A new ed . . . by Alexander Hamilton. 8°. 112p. 40 plates. P: 6s. boards.	Edinburgh: for C.E.; and Elliot & Co, London.
120	1787	A system of anatomy, ed Andrew Fyfe.	A system of anatomy and physiology; from the latest and best authors. 2nd ed. To which is added, The comparative anatomy [by Alexander Monro I]. 3 vols. 8°. 16 plates [T. Donaldson, A. Bell, D. Lizars] P: first 2 vols. announced for 21 Nov. 1786, completed 19 March 1784, 18s. boards, £1.1s. bound (CM). Copyright: entered for C.E. 6 June 1787.	Edinburgh: for C.E.; Elliot & Co. and G.G.J. and J. Robinson, London. Edinburgh: for C.E.; and G.G.J. and J. Robinson, London.
121	1787	William Withering (1741–99)	A botanical arrangement of British plants; including the uses of each species. 2nd ed. 3 vols. 8°. 19 plates.	Birmingham: by M. Swinney; for G.G.J. and J. Robinson, and J. Robson, London; J. Balfour, C.E., Edin.
122	1788	William Cullen	First lines of the practice of physic. A new ed. 4 vols 8°. P: 15 Nov. 1787, £1.4s. boards (CM).	Edinburgh: for C.E.; and C.E., T. Kay, London.
123	1788	Andrew Duncan	Medical commentaries, for the year [1787]. Decade second, vol. II. 8°. 504p. P: for 18 Jan. 1788, 6s boards (CM).	Edinburgh: for C.E., T.Kay, London; and C.E., Edin.

	Date	Author	Short title	Imprint
124	1788	Charles Elliot	[A catalogue of all the useful and modern publications on medicine and medical philosophy.] Not seen. P: 9 Feb. 1788, gratis (CM).	
125	1788	Encylopaedia Britannica Prospectus [18 vols. 1788-97]	A new edition in quarto . . . This day is published, by Elliot and Kay . . . Strand, number I. Price one shilling (to be continued weekly); also volume I. part I containing 10 numbers neatly done up in boards. 8°. P: 26 June 1788, vol. 1, part 1, 10s.6d. boards (CM).	[Edinburgh: for A. Bell and C. Macfarquhar; sold by Elliot and Kay, London; and C.E., Edin.]
126	1788	James Gregory (1753–1821)	Conspectus medicinæ theoreticæ. Ad usum academicum. Editio tertia. 2 vols. 8°. P: 16 Oct. 1788, 13s. in boards (EEC); 12s. boards 12 Nov. 1789 (CM). Copyright: C.E. paid Gregory £250 for the property; the retail price was to be 12s. in boards.	Edinburgi: C.E.; Londoni, C.E., T. Kay.
127	1788	Hippocrates, trans. John Moffat	The prognostics and prorrhetics of Hippocrates. 8°. 314p.	London: by T. Bensley; for C.E., T. Kay.
128	1788	Thomas Jameson, surgeon, Royal Navy	A treatise on diluents, and an enquiry into the diseases of the fluids of the human body. 8°. 134p. P: 14 Feb. 1789, 2s. sewed. Zachs 665; 2s.6d.	London: for the author by J. Davis; sold by J. Murray, London, and C.E., Edin.
129	1788	A. Monro II	A description of all the bursæ mucosæ of the human body. royal 2°. 60p. 10 plates [T. Donaldson, J. Beugo, G. Cameron; A. Fyfe]. P: 28 Feb. 1788, 12s. boards (CM).	Edinburgh: for C.E., T. Kay, London; and C.E., Edin. (CE monogram)
130	1788	Edward Peart (1756?-1824)	The generation of animal heat, investigated. 8° 118p.	Gainsborough: by H. Mozley; sold by J. Edwards, London, C.E., Edin.
131	1788	Samuel Stanhope Smith (1750–1819), College of New Jersey	An essay on the causes of the variety of complexion and figure in the human species. To which are added, Strictures on Lord Kames's discourse. 8°. 220p. P: 3 May 1788, 3s.6d. in boards (CM). Copyright: entered for C.E. 6 May 1788.	Philadelphia printed, Edinburgh reprinted, for C.E., and Elliot and Kay, London.
132	1788	Franz Swediaur	Practical observations on venereal complaints. 3rd ed., corrected and enlarged. 8°. 328p. P: published 'in a few days' 21 Dec. 1787, 4s. sewed. (CM). C.E. paid Swediaur £63 for the property.	Edinburgh: for C.E., T. Kay, London; C.E., Edin. (CE monogram)
133	1789	Benjamin Bell	A system of surgery. Vol. 3. 4th ed. 8°. 540p. plates.	Edinburgh: for C.E.; Elliot & Kay, and G.G. J. and J. Robinson, London.

	Date	Author	Short title	Imprint
134	1789	Benjamin Bell	A treatise on the theory and management of ulcers. A new ed. 8°. 488p. plate. P: 6s. boards, 7s. bound.	Edinburgh: for C.E.; Elliot & Kay, London.
135	1789	Joseph Black (1728–99) engraving	[Portrait] P: noted 22 Oct. 1789, 'An elegant portrait of Joseph Black, M.D., Professor of Chemistry in the University of Edinburgh. Drawn by Brown, and engraved by Beugo', 2s.6d. [for proofs](CM) 1s. to purchasers of Fourcroy (Medical Commentaries for 1789).	
136	1789	Robert Couper	Speculations on the mode and appearances of impregnation in the human female . . . By a physician. 8°. 148p. P: 11 July 1789, 2sd.6d. sewed (CM).	Edinburgh: for C.E.; Elliot & Kay, London.
137	1789	William Cullen	First lines of the practice of physic. A new ed. 4 vols. 8°. P: 13 Oct. 1788, £1.4s. boards (CM).	Edinburgh: for C.E.; C.E. & T. Kay, London.
138	1789	William Cullen	A treatise of the materia medica. 2 vols. 4°. P: 14 May 1789,£2.2s.boards (EEC). Copyright: entered for C.E. 18 April 1789. C.E. paid £1,500 for the property.	Edinburgh: for C.E.; Elliot & Kay, London.
139	1789	Andrew Duncan	An account of the life, writings, and character, of the late Dr John Hope. 8°. 32p.	Edinburgh: for C.E.; Elliot & Kay, London.
140	1789	Andrew Duncan	Heads of lectures on the theory and practice of medicine. 4th ed. 8°. 320p.	Edinburgh: for C.E.; and Elliot & Kay, London.
141	1789	Andrew Duncan	Medical commentaries, for the year [1788]. Decade second, vol. III. 8°. 507p. Copyright: entered for C.E. 31 Jan. 1789.	Edinburgh: for Elliot & Kay, London; C.E. Edinburgh.
142	1789	Andrew Duncan, ed.	The Edinburgh new dispensatory. Being an improvement upon the New dispensatory of Dr Lewis. 2nd 8°. 656p. 3 plates on six leaves [D. Lizars]. P: noted 14 May 1789, 8s. bound (EEC); 7s. boards, 8s. on a superfine paper (CM 12 Nov. 1789). Copyright: entered for C.E. 15 Nov. 1788.	Edinburgh: for C.E.; and Elliot & Kay, London.
143	1789	Charles Elliot & Thomas Kay	[A catalogue and additional list (given gratis) of their stock of books on medicine and medical philosophy.) Not seen. P: 14 May 1789 (EEC).	
144	1789	James Keir (1735–1820)	The first part of a dictionary of chemistry. 4°. 232p. P: by 12 November 1789. 10s. boards (CM).	Birmingham: by Pearson and Rollason. For Elliot & Kay, London; C.E., Edin.

	Date	Author	Short title	Imprint
145	1789	Edward Peart	On the elementary principles of nature, and the simple laws by which they are governed. 8°. 312p.	Gainsbrough: by H. Mozley; sold by J. Edwards, London, and C.E., Edin.
146	1790	Benjamin Bell	A system of surgery. Vol.2. 4th ed. 8°. 484p. 13 plates.	Edinburgh: for heirs of C.E.
147	1790	Benjamin Bell	A system of surgery. Vol. 1. A new ed. 8°. 568p. plates.	Edinburgh: for C.E.; Elliot & Kay, G.G.J. Robinson, London.
148	1790	James Clark, King's farrier for Scotland	A treatise on the prevention of diseases incidental to horses. 2nd ed. 8°. 460p. P: 2 February 1789, 7s.6d. boards (CM).	Edinburgh: for the author; sold by W. Creech, J. Dickson, P. Hill, and C.E.
149	1790	Andrew Duncan	Medical commentaries for the year [1789] Decade second. Vol. IV. 8°. 539p. P: 4 Jan. 1790, 6s. boards (EEC). Copyright: entered for C.E. 18 Dec.1789.	Edinburgh: for Elliot & Kay, London; and C.E., Edin.
150	1790	A.-F. Fourcroy, trans. mainly by Robert Heron (1764–1807)	Elements of natural history and chemistry. Translated from the last Paris edition, 1789, being the third. 3 vols. 8°. P: 24 Nov. 1789, £1.1s. boards (CM 23 Nov.). Copyright: entered for C.E. 15 Dec. 1789. C.E. paid Heron £42.	London: for Elliot & Kay; and C.E., Edin.
151	1790	Ferdinand Leber (1727–1808), ed. John Wilson, M.D.	Ferdinandi Leber . . . Prælectiones anatomicæ: editio nova. 8°. 449p. P: 4 March 1790, 5s. boards (EEC).	Edinburgi: C.E.; Londoni, Elliot & Kay.
152	1790	William Smellie, ed. Alexander Hamilton	A set of anatomical tables, with explanations, and an abridgment of the practice of midwifery. A new edition, with an entire new set of plates . . . By A. Hamilton. 8°. 112p. plates.	Edinburgh: for C.E.; and C.E and T Kay, London.
153	1790	William Smellie (1740–1795)	The philosophy of natural history. 4°. 564p. P. 4 March 1790,£1.1s. boards (EEC). Copyright: entered for heirs of C.E. 23 Feb. 1790. Smellie gave the Elliot Trustees 2,094 copies. Of his copyright of 1,000 guineas, £688 was paid out in bills 6 to 18 months after publication; the remainder was 'arrested' by his debtors until after his death.	Edinburgh: for heirs of C.E.; Elliot & Kay, T. Cadell, and G.G.J. and J. Robinson,
154	1790	Robert Walker, M.D.	An inquiry, into the small-pox, medical and political: wherein a successful method of treating that disease is proposed. 8°. 516p. Zachs 799; 6s. boards.	London: for J. Murray, and W. Creech, C.E., P. Hill, J. Elder, and J. Mudie, Edin.

1. This list is based on the *ESTC*, and draws on Elliot's advertisements in newspapers and in his medical books, and on the Elliot papers. A number of copies were inspected at Edinburgh University Library and the National Library of Scotland. For my purposes here, I have totalled the pages, rather than give pagination detail. The prices are taken from newspaper advertisements at publication, where possible; later advertisements in

Elliot's books sometimes give a different figure. Imprints and their variants are abbreviated.
2. Records of the Worshipful Company of Stationers 1554–1920. Published on Microfilm by Chadwyck-Healey Ltd., ed. Robin Myers; 'for C.E.' means that the whole share was entered for him at Stationers' Hall.
3. Attributed by Stephen Brown; i.e. the printer Smellie, not the midwifery professor Smellie as thought previously.
4. See also David L. Cowen, 'The Edinburgh Dispensatories', *Papers of the Bibliographical Society of America* 45 (1951), 85–96.

9

Reading Cleghorn the Clinician: The Clinical Case Records of Dr. Robert Cleghorn, 1785–1818

FIONA A. MACDONALD

Enlightenment thinkers set great store by empirical observation as an essential feature of physic and surgery. As Andrew Duncan I put it in the 1773 preface to his periodical, *Medical and Philosophical Commentaries*, 'medicine has long been cultivated with assiduity and attention, but is still capable of farther improvement'. For Duncan, this could best be done through attentive observation and the collection of useful facts, and 'in no age, since the revival of learning, does greater regard seem to have been paid to these particulars, than in the present'.[1] One of the main aims of Enlightenment medical publications was to bring interesting clinical cases to the attention of the medical profession as a way of improving the standards of practice.[2] The systematic recording of difficult or unusual cases imparted experience and knowledge to those who had not encountered such situations. The development of an empirical approach to therapeutic evaluation and the statistical collection of medical data for comparison were other aspects of this compilation of medical information that emerged in the later Enlightenment.[3] Several medical historians have demonstrated the importance of clinical case notes as a source of information about medical practice, clinical perception and the social dimensions of bedside consultation.[4] Nicholas Jewson has also argued that the nature of bedside practice demonstrates a great deal about the practitioner's status and professional power.[5] Critical analysis of patient case histories is now a legitimate form of medical historical investigation, and the value of preserving such documents in order to reconstruct past medical practice is also being addressed.[6]

This chapter analyses the patient case records which survive in two of the three extant private notebooks of the distinguished Edinburgh-trained physician, Dr. Robert Cleghorn, who lived from 1755 to 1821. These notebooks comprise a record of eighty-six patients seen between December 1785 and February 1818. Cleghorn's records were taken down at a period when diagnosis was still largely based upon the intensely subjective feelings of patients regarding their own illness, upon the physician's passive observation of the patient based on such things as skin colour, the appearance of the tongue or the stools, and, more rarely, upon manual examination of the patient's body.[7] According to Jewson, medicine at this time was dominated by the individual voice of the sick person and was thus symptom based.[8] Both the patient and the clinical observer have a high profile in this genus of case histories, which is dominated by the patient's narration of his symptoms and illness career, as well as by the systematic questioning, listening and analysing which was then essential to good diagnostic technique. As Steven Shapin has stressed, the reliability of patient and doctor, as sources of relevant and credible knowledge for each other, was established on the basis of trust. Reliable communication and trust in the truthfulness of others were not only regarded in the seventeenth and eighteenth centuries as integral to social order and cohesion, but for the philosophers of the Scottish common sense school, truthfulness was a prerequisite of membership of the moral community. To accept the testimony of another was morally to esteem them, and by way of knowledge, such esteem then translated into authority.[9] Not only did authority accrue to the physician in the eighteenth-century system of medical knowledge, but such was the physician's fundamental faith in the veracity of the patient's narrative that much prescribing was done by post.[10] Cleghorn's records reveal much, in addition, about the social dynamics of the professional interaction between physicians and surgeons.

Cleghorn's work is further significant in that it was part of a trend in Enlightenment medical practice towards the garnering of more objective evidence of disease at the hands, literally, of the physician. At the same time, this development began to minimise the importance of the patient's narrative of his own illness. Giovanni Battista Morgagni laid the basis for the modern understanding of disease, built around the evidence of selected case histories, in his *The Seats and Causes of Diseases investigated by Anatomy* (1761). He showed that disease left signs in the tissue of the body, through the analysis of which in postmortem examinations physicians could link the symptoms experienced by the patient when alive, and, thus, better treat the same symptoms in others.[11] The Viennese physician Leopold Auenbrugger also published his *Inventum novum* (1761), describing a form of examination called percussion, by which the body struck with fingers enabled him to establish 'by the testimony of my own senses' whether it was internally diseased by the sounds that returned to him.[12] By the 1760s, the most progressive forms of medical education, such as those at Edinburgh University, placed great emphasis on anatomical dissection and postmortem examination, all of which valued the acquisition of manual skills previously undervalued in the training of the gentleman physician.[13] Yet although these new techniques provided a general stimulus for the use of physical diagnosis, it took time for such changes to filter into general day-to-day practice.

It is generally observed, indeed, that physical diagnosis did not become routine practice until the mid-nineteenth century. Malcolm Nicolson has argued, however, that 'evidence does exist to suggest that there may have been important exceptions to the rule that the eighteenth-century physician observed and questioned at the bedside, but did not uncover or touch'.[14] Cleghorn was one such exception. Although there is no evidence in his case notes that he used percussion, he was an advocate of manual palpation and postmortem dissection and was routinely doing both in Glasgow in the 1780s.

At the heart of Cleghorn's case histories is the very personal tension he suffered as a pioneer and proselytiser of the particular form of diagnosis that involved physical manipulation of the patient's body. Himself possessed of a brusque personal manner, the evidence hints that Cleghorn did not pay exclusive attention to the subjective feelings of the patient that his peers deemed he should (many of whom did not share his modernising attitude to diagnostic technique). From the earliest days of his practice, Cleghorn relied upon his own sense of touch in forming, from his own deduction, an objective, clinical diagnosis. Like the philosophers of the Scientific Revolution and the empiricists of the Enlightenment, he affirmed the modernist belief in the authority (and, implicitly, the moral supremacy) of direct personal experience and independent verification. While not disavowing the patient's testimony, at the same time he also invested trust in his own observations as a member of the scientific and medical community.[15]

This chapter also raises questions about the local particularity of such medical issues. Adir Ophir and Steven Shapin have emphasised that science takes on its shape, meaning, reference and domain of application from the specific physical, social and cultural contexts in which it is used and practised.[16] The importance of the local nature of knowledge creation in the distinctive evolution of Enlightenment ideas and practices has been further explored by David Livingstone and Charles Withers, who make a case for locally situated knowledge-making sites which interact with each other and transfer ideas 'through practices of circulation and processes of negotiation designed to warrant the credibility of knowledge'.[17] This chapter considers the inner conflict endured by Cleghorn in making and legitimising a new form of diagnostic knowledge, locally, in his professional practice in Glasgow. Glasgow was then the only city in Britain where physicians and surgeons (and apothecaries) were admitted as members of the same corporate medical licensing body, a situation that inevitably led to greater convergence of medical and surgical ideas in the shared knowledge of the local professional medical culture.

The Physician: The Body in Question

Like many late eighteenth-century physicians, Cleghorn was wide-ranging in his professional activities, holding university appointments in materia medica and chemistry, hospital posts as physician in the poorhouse, Royal Infirmary, and Glasgow Royal Asylum, as well as concurrently maintaining a private practice. Yet, in spite of being one of the leading Glasgow medical practitioners of his day, his medical career has largely been neglected. According to the historian John M'Ure, Robert Cleghorn was sixty-six years old when he died on 18 June 1821 at Shawfield House, Rutherglen.[18] He was almost certainly christened on 4 March 1755, in Edrom in Berwick, the son of James

Cleghorn.[19] Robert went to Edinburgh to study medicine, at a time when the medical school was renowned for inculcating in its students the significance of clinical observation, and graduated MD in 1783. His thesis was entitled *De Somno* (On Sleep).[20]

Cleghorn was a member of various Enlightenment societies. In Edinburgh, he joined the Speculative Society 'for the improvement of Literary Composition and Public Speaking' in 1779. In 1783-84, he was one of the four annual Presidents of the Royal Medical Society.[21] His appointment as President is significant because of the general ethos of the Royal Medical Society in wishing to propagate research and scientific enquiry in medicine among medical students and young practitioners.[22]

In late 1785, Cleghorn took up medical practice in Glasgow, where, in 1787, he married Margaret Thomson, daughter of Andrew Thomson of Faskine. They had one child, Helen, in June 1790.[23] He was entered as a member of the Faculty of Physicians and Surgeons of Glasgow in 1786, was its President between 1789 and 1791, and its Librarian from October 1792.[24] In 1788, he was elected a Fellow of the Royal Society of Edinburgh, and, in 1792, was elected to a second stint as Secretary to the Glasgow Humane Society, presided over by David Dale, philanthropist and founder of the industrial village of New Lanark.[25] Like Dale, he also attended meetings of the Glasgow Literary Society.[26] Cleghorn lived and practised in Spreull's Land, on the north side of Glasgow's Trongate, in 1789.[27] He later lived in College Street and, in 1801, purchased part of the estate of Campbell of Shawfield, near Rutherglen, where he built Shawfield House.[28]

Cleghorn was appointed lecturer in materia medica at Glasgow University in 1788. In 1791 he switched to the lectureship in chemistry, and remained in this post until his resignation in 1817.[29] When Cleghorn took over as lecturer, the study of chemistry was in a state of flux because the research tradition founded at Glasgow by William Cullen and Joseph Black was being challenged by the revolutionary ideas of Lavoisier and the French chemists. Cleghorn's immediate predecessor in the chemistry lectureship, Thomas Charles Hope, was a convert to the French system, and it would seem that Cleghorn too taught the doctrines of Lavoisier.[30] It is unclear how much actual chemical research was carried out by Cleghorn, for he seems to have let the chemical laboratory at the University fall into disuse.[31] This is puzzling, since his clinical practice demonstrates a profound belief in the importance of practical observation in medical science. He was, however, primarily a physician, not a chemist. After retiring he indicated that he found chemistry teaching 'fretful', and had 'too long delayed the study for which I was always the best qualified. How much time have I wasted! Alas! Alas!'[32] He was nevertheless a popular lecturer favoured by his students: 'his oratory was certainly of the first order and so musical . . . that the student found himself in danger of being sung asleep'.[33] But this, according to his successor as lecturer in chemistry, Thomas Thomson, was because the lack of laboratory work allowed Cleghorn to offer 'at a very trifling expense . . . a course of lectures that charmed his hearers'.[34]

As Roger L. Emerson and Paul Wood show in Chapter 4, Glasgow spent much of the eighteenth century elaborating its institutional base, achieving this successfully because the gentry, merchants, lawyers, doctors and clergy all co-operated, with enlightened self-interest, in the improvement of the civic fabric.[35] Cleghorn was part of this civic enterprise. He was physician to the poorhouse or Town's Hospital between 1786 and 1791, and it was here that he probably developed his interest in insanity. He was also one of the original managers of the Glasgow Royal Infirmary which opened in 1794, and one of its first two physicians.[36] Later he was first physician to the Glasgow Royal Asylum for Lunatics in Dobbie's Loan in 1814. Cleghorn became a director of the Asylum, and was largely responsible 'for the enlightened attitude [shown] to its inmates'. He was an advocate of 'moral management' and, when he retired in 1819, the drop in patient numbers was seen as a direct result of the loss of his 'talents and reputation'.[37]

What of Cleghorn the man? He was, it seems, not always esteemed by many of his contemporaries. He was generally held to be the odious 'Dr. Wormwood' – the name indicative of an embittering experience – of Leonard Smith's *Northern Sketches or Characters of G*******.[38] If true, Cleghorn is caricatured as not taking his patients seriously and of indulging in breaches of clinical confidentiality:

> Wormwood presents none of that affability which every physician ought to possess in order to inspire confidence; none of that tenderness necessary to alleviate shame or to countenance delicacy. Stern and sarcastic in his manners . . . his face appears to have made a perpetual divorce from smiles. He treats his patients rather as subjects for ridicule, than relief; indulges in impertinent and superfluous interrogatories; and is not always observant of that fidelity which ought to distinguish so confidential a character as a physician.[39]

Cleghorn's most recent biographer notes only that 'whether or not Cleghorn was lacking in affability towards his patients we cannot tell'.[40] This is, strictly speaking, not true. Cleghorn is revealed by his notebooks as a somewhat patronising and conceited man. There is, however, another aspect to take into consideration in interpreting this caricature: in relying more on objective signs for diagnosis than the traditional physician, Cleghorn would have been pandering less to his clients. Further, his behaviour might have been seen as intrusive. Although Andrew Wear cites examples of seventeenth-century physicians like Christopher Wirsung and John Hall sounding the belly in dropsical cases, Stanley Reiser has argued that whereas seventeenth-century physicians might manually examine a patient, physical diagnosis was not then a major aspect of practice and 'the physician would attach far less weight to the evidence obtained by his sense of touch than to the patient's narrative and to his own visual observations'.[41] Even by the eighteenth century, many patients would have felt affronted by any physical examination, while some allowed topical, but not intimate, physical examination, like the labouring woman to whom Cleghorn was called by John Burns, in 1811, who permitted her abdomen to be touched, but 'would not be examined'.[42]

Yet 'Dr Wormwood' was also portrayed as a knowledgeable, if grasping, physician: 'fully sensible of his skill he is not less careful that it shall have its price'.[43] Cleghorn's skill as a physician is shown by the fact that medical practitioners came to him for treatment. A surgeon by the name of Holder consulted Cleghorn in 1805, when, at the age of forty, he 'began to feel an uneasy sensation in the chest particularly under the sternum, when he walked fast'.[44] Cleghorn was also respected by Dr. William Mackenzie, later a famous ophthalmologist, who, in August 1814, was Physician's Clerk at the Glasgow Royal Infirmary, where he attributed to Cleghorn all the credit '& a good deal more' for the improvement of a particular patient whom one Dr. Swan had thought past redemption.[45]

Cleghorn was also a friend of the poet Robert Burns, who addressed a letter to Cleghorn in 1796, outlining details of his recent illness during which he had been 'the victim of a rheumatic fever, which brought me to the borders of the grave. After many weeks of a sick-bed, I am just beginning to crawl about'.[46] Burns' connection with Cleghorn was probably literary and new Whiggish, if not politically radical, since Burns did much to inspire Scottish radicalism. The commercial disposition of the city gave Glasgow's Enlightenment a unique character, as is shown in Chapter 4 above. Indeed, it has been argued that the percolation of Enlightenment values into the professional and mercantile classes lay behind the city's intense political awareness. When considering that all but a handful of the professors at Glasgow University opposed the French Revolution, Cleghorn's sympathy with revolutionary politics effectively blocked his preferment to the chairs of Natural Philosophy and Medicine in 1795–96. George Jardine, the Professor of Logic and Rhetoric, who was also a Whig, wrote in January 1795 regarding Cleghorn's candidature for the chair of medicine, that:

> Dr Cleghorn, from a General conviction of his fitness, could almost have no Competitor even though it be in the Gift of the Crown – if it had not been [for] his alleged attachment to Modern Politics and his connexion with that Party here. This at present seems likely to cast the balance against him though not absolutely certain.[47]

Certainly when a former friend of Cleghorn's, Alexander Oswald, died in June 1813 of a pancreatic tumour, Cleghorn hinted at their shared political sympathies: 'Another of my early Friends gone – a man of great sense & worth, a lover of liberty'.[48]

Cleghorn not only had a sense of superiority about patients. He did not regard his fellow practitioners highly. Another of his friends, John Oswald, was treated by 'Dr M.' (probably Monteath) for ague, but was found on postmortem to have a diseased kidney and grossly enlarged prostrate. Highlighting the need to be attentive to the patient's subjective feelings, in 1800 Cleghorn pointed to the importance of structural signs of disease in the body, concluding that:

> This case is another proof of the necessity of attending to local symptoms, & to the feelings of the patient however bizarre these may appear & however contrary to nosology! Fancy or Whim or a nosological name is sufficient to check enquiry on exertion in the herd of practitioners, nay they hang a bias on the most candid & add a kind of dead weight by which the most active if not stopped are at least retarded in their course.[49]

He was one of the few denied membership of the Glasgow Medical Club, which met between 1798 and c.1814 and was known for the sociability of its meetings.[50] According to John Strang, Cleghorn was clever, but selfish, and 'had there been one man among the number who could have sat for the picture of *Dr. Wormwood*, by the sketchy limner, whose caustic pencil, during the first decade of the present century, created so much noise in Glasgow, it is certain that the Club would have been sooner entombed'.[51] It is possibly for this reason that only the merest reference is made to his death in 1821 in the Glasgow newspapers; nor does he make an appearance in Scottish biographical encyclopedias of the time.[52]

Cleghorn's Casebooks: Physical Diagnosis and Postmortem Dissection

Dr. Robert Cleghorn left three manuscript notebooks at his death. Other than some brief notes made on the early admission documents to the Glasgow Royal Asylum, and sixteen sides of case histories of lunatics that survive in a manuscript notebook of Dr. William James Fleming, these notebooks comprise the majority of his extant manuscript material.[53] There may have been other casebooks: the record of Elizabeth Halbert, on 24 March 1789, has 'vid [see] Case Book' after her name.[54] There is also a manuscript diary of an extensive Scottish tour made by Cleghorn in 1794, after the fashion of the day, to observe the manners of the people and condition of the country.[55] However, if he did make a significant contribution to the periodical press (as Alexander Duncan asserts), it was not to the Scottish medical periodical press,[56] although reference is made to 'the two cases treated at Glasgow, by Dr Cleghorn' in an account of the two-volume work on diabetes mellitus by Dr. John Rollo, Surgeon-General of the Royal Artillery.[57] The only surviving biographical notice of one of his contemporaries on which he is named as author is that of his Glasgow colleague William Hamilton; in addition, Andrew Kent claims that Cleghorn wrote the seven-page obituary of William Irvine, which is attributed to 'a worthy correspondent' in the *Medical Commentaries* of 1788.[58]

Cleghorn's notebooks are part lecture notebook, part commonplace book and part private casebook.[59] The first volume, 'begun Edinburgh January 12, 1782', is entirely a commonplace book, that is, a notebook in which miscellaneous bits of information, excerpts from books, as well as observations and ideas that appeal to the compiler are entered.[60] The second volume is a commonplace book up to page 139, at which point Cleghorn began his account of practice in Glasgow with a record of an autopsy performed, on 2 December 1785, on a ten-month-old child.[61] Thereafter, he continued to record the odd piece of information that took his eye, such as a cure for scrophula, Saunders' table of mineral waters, recipes for medicines from a variety of authors, as well as lists of the chemical constitution of different spa waters.[62] The third volume is almost entirely a casebook, apart from about fifty pages in which he noted down miscellaneous bits of history, geology and culture that interested him,[63] such as Dr. Adair's failure with 'Dr McLean's process for procuring alkalis' and a recipe for gout cordial from the family of Lord Dundas.[64]

What were Cleghorn's motives in writing these casebooks, and what epistemological issues should be taken into account in their use as historical sources? The notebooks were compiled over a period of more than thirty years, but include less than ninety case histories. This means that Cleghorn recorded an average of only three cases a year in his private notebooks. There is no way of knowing what particular merit a case had to have in order to warrant recording in his notebooks, but they must have been cases that Cleghorn perceived as contributing to the creation of an empirical foundation for knowledge. In the main they were unusual cases, as well as cases which warranted and validated the use of physical diagnostic techniques in patients' treatment or in which their disease was subsequently elucidated by post-mortem dissection. The notebooks were Cleghorn's reflexive appraisal of significant cases in his private and public practice. Internal evidence clearly shows that the cases were written up retrospectively, once the patient had been cured or had died (whereas today's case notes are written up after each consultation). There are a number of problems with the use of retrospective accounts as sources, not least that lapses of memory may have occurred and errors been introduced since some cases continued over several months, but mainly that events were inevitably subject to *post hoc* rationalisation.

A covering letter from Ebenezer Watson, dated 30 April 1882, survives in the final notebook stating that 'this book was one of several which were sent to my father from the library of Dr Cleghorn as a present for attending him on his death-bed'.[65] Watson also insisted that the book, 'entirely in Dr Cleghorn's handwriting . . . was not intended for publication'. This is probably so. Cleghorn did not publish enough to create a distinguished identity for himself in the republic of letters; he did not seek fame through writing. Rather, his reputation was established locally by himself being written about and caricatured. At least one piece of evidence indicates, however, that his notebooks had an intended readership, and that he was seeking to establish a credibility for his scientific beliefs. In an account of the illness, and subsequent post-mortem, of John Top, a 27-year-old blacksmith who had retained a piece of iron shot in his chest after being earlier wounded when a marine, Cleghorn wrote the following, although it is not clear to whom the lines were addressed:

We had formed many conjectures as to the course of the ball from the inside of the 3d to the outside of the 8th rib & they were confirmed by the appearances on dissection, which I was very anxious to have shown you but could not, for which reason I shall state them as distinctly as I can.[66]

Even if Cleghorn's notebooks had a small circulation, in writing down these cases he was helping to make the scientific knowledge of physical diagnosis and localised anatomical investigation more accessible to those who had not observed them in operation and who perhaps neither knew the patients nor necessarily agreed with his method of diagnosis. In this way, knowledge was transferred and assimilated into wider cultures of therapeutics. Trust in the casebook thus becomes part of the means by which credibility was secured for the doctor and his diagnostic ability in dealing with the patient's narrative. Casebooks from the past also serve to secure the credibility of the doctors who compiled them among readers in our own day, who must assess the reliability of casebooks as guides to illness and medical practice in history.[67]

Cleghorn's case notes begin, in a standard way, by noting background information about the patient: date of consultation, name, age, marital status, occupation and, quite often, place of residence. Religious denomination or racial origin is occasionally provided. A résumé is then given of the patients' medical history where relevant, as well as of their current illness and a description of symptoms. There then follows a description of the therapeutic treatment implemented by Cleghorn, and, in most cases, a discussion of the patients' decline and date of death. The record of postmortems subsequently performed is extremely detailed, and demonstrates considerable involvement with pathological anatomy. Only twenty-eight out of eighty-six cases do not record postmortem details, though in nine of these cases the patient died and one was murdered.

Only a handful of the case records directly refer either to infirmary or institutional patients, but it is unlikely that all of the unspecified cases are private. As such, there is not much perceptible difference between his treatment of those designated infirmary patients and private patients, although Cleghorn occasionally seems to have recorded fewer personal details such as name, age and occupation for infirmary cases.[68] It is also clear that other cases where no details are given of the patient's disease were simply opportunities for Cleghorn and his associates to dissect – the record of a prostitute, Margaret Mitchell, for example, who was anatomised in the Correction House on 23 April 1791.[69] More surprising, perhaps, is the extent of the dissection that Cleghorn was able to undertake in cases where the patient had relatives.

Cleghorn's notebooks highlight several important issues. They demonstrate that hands-on practice was beginning to infiltrate the world of the Scottish physician by the late eighteenth century. When a weaver from Calton consulted him in 1790 for pain in the gastric region, Cleghorn immediately palpated the area, finding that 'the belly was extremely retracted about the umbilicus, and in many parts was very painful to the touch. In the right hypochondrium there was a considerable hardness'. More unusually, perhaps, he felt the corpse after death (prior to postmortem), recording that the right side was remarkably livid, 'but altogether free from the hardness which I felt so remarkably before. This surprised as I expected a schirrous [cancerous] liver'.[70] Examining, in July 1801, a 52-year-old patient named Mary Millar who was much emaciated by a disease of the stomach and bowels, he commented that 'on examining the abdomen as I always do' he felt tumours around the umbilicus.[71] Some weeks after leaving the infirmary, Millar died. On opening up the body, Cleghorn noted: 'I was right as to the side of the tumour, for it lay exactly over the folds of the Ileum: but the hard tumour was form'd chiefly of omentum'.[72]

As Mary Fissell has demonstrated, through the use of postmortem dissections physicians attempted to reduce their patients' symptoms to logical patterns of physical causes and effects.[73] Yet it is difficult not to see Cleghorn as relatively unusual given the extent to which he employed physical diagnosis. He was consulted, for example, in March 1807 by Janet Craig, a 15-year-old with incessant pain in the belly, and frequent, ineffectual efforts to void urine. She had consulted three physicians before Cleghorn, two of whom had given her emetics and the third a fomentation (poultice) and diuretic – standard therapeutics for the removal of blockages. The complaint had begun eight weeks earlier: there was every sign of menstruation except the flow of blood. Cleghorn was the only doctor prepared to examine her physically. Many physicians still adhered to long-established professional tenets, not only of decency and propriety in not being prepared to touch their patients' bodies other than to feel their pulse, but also in deriding the hands-on approach that distinguished surgeons as craftsmen. But, with the mother in attendance as chaperone, Cleghorn quickly examined the vulva, where he found the hymen imperforated. He employed a specific and effective remedy: 'Next day I opend it with a Lancet when there flow'd out 1 lib. of thick high coloured blood . . . She felt immediate relief'.[74]

Cleghorn performed numerous, detailed autopsies, many, but not all, with surgeons in attendance. The Burns brothers – John and Allan – were often in attendance. This professional collaboration, primarily between Cleghorn and the elder brother, John, developed in a number of different areas. Burns was also a member of the Faculty of Physicians and Surgeons which he had joined, as a surgeon, in 1796.[75] Just as Cleghorn was keen to learn surgical skills, so John Burns was also a student in medicine, and was appointed to the office of Physician's Clerk at the Royal Infirmary on 4 May 1795, when Cleghorn was physician there.[76] Burns' esteem for Cleghorn is shown in the joint dedication of his *The Anatomy of the Gravid Uterus with Practical References Relative to Pregnancy and Labour* (1799) to him and to James Muir, surgeon.[77] The Burns brothers were the first to teach surgery, anatomy, dissection and midwifery, including obstetrics, gynaecology and paediatrics, privately in Glasgow between about 1797 and 1835.[78] It was common for a physician and a surgeon to be called together to give a prognosis on difficult cases. This afforded them a mutual protection, and many would team up together. The recently qualified Cleghorn's association with the Burns brothers gave him a tremendous grounding in general practice. His case notes even contain midwifery cases, and some notes on obstetrics.[79] He also attended cases with a number of other practitioners: James Monteath (physician after 1803); William Anderson (obstetrician); William Hamilton (Professor of Anatomy at Glasgow University); the surgeons William Couper, James? Swan, a Mr. Smart, Andrew Russell (lecturer on anatomy and surgery in the College Street School); and the physician Benjamin Watts King.[80]

The eminently practical investigations of the anatomists have been seen as one of the unqualified success stories of Enlightenment medical science. For Owsei Temkin, there was a growing convergence between the disciplines of medicine and surgery during the Enlightenment, as physicians came to realise the practical benefit of the surgeon's empirical approach to diagnosis.[81] Although Cleghorn's lecture notes frequently refer to the work of John Hunter, and that 'Johnny' as he often calls him was, by implication, something of a hero for him, there is nothing to show that the notes were taken in London.[82] It was rather his connection with the Burns brothers that facilitated his experience in performing anatomies.[83]

In November 1810 he was called to see Mr Alexander McNab, who was suffering from costive bowels and a stomach so sensitive 'that any offensive sight or smell readily excited vomiting', and who had been treated with purgatives and tonics for dyspepsia by John Burns. When the patient did not improve, his friends had a second opinion from William Anderson, who diagnosed an organic affection of the pylorus.[84] Cleghorn and Allan Burns were consulted to see if they agreed. Cleghorn found the abdomen shrunken and, on palpating it, felt it 'somewhat hard along the omental edge of the stomach, & particularly over the pylorus'. John Burns insisted there was no hardness, while his brother said 'it was the spinous process of one of the Verterbrae [sic]'.[85] Diagnosis and prognosis held no certainty in the early nineteenth century, however, even among skilled practitioners. It was usual for patients to consult several practitioners, so when the patient got no better, Dr. William Hamilton, who was clearly felt to have greater expertise, was called in from Edinburgh. Cleghorn wrote viciously in his notebook: 'He is stupid & deaf but got 40 Guineas & recommended in a distant way mercury!'[86]

In due course, the patient died, and a postmortem was performed on 7 February, when the large colon and stomach were found to be distended, and the pylorus to be 'schirrous [cancerous] for more than an inch all round'.[87] Interpreting pathological lesions, as with reading the external physical appearance, was not yet a precise science. In spite of this, John Burns continued to take the view that it was weakness of stomach increased almost to paralysis by some tartar emetic prescribed by his brother Allan, because upon examination, both he and his brother found the pylorus to be like other pylori. Cleghorn clearly found their inability to come to a common diagnosis obtuse. 'No wonder Drs should differ about internal complaints when they cannot agree even about what they see such as this on what they feel such as the pulse – I am very often 20 or 30 different from Mr Allan', he wrote, referring to the frequency of recorded pulse beats.[88]

Cleghorn seems to have taken delight in belittling, whenever he could, the 'superior' practitioners of Edinburgh, or as he once referred to them, 'the wise men of the East'.[89] When a Mr. Gartshore, a patient from Edinburgh with a heart complaint, consulted Cleghorn in November 1799, Cleghorn commented that 'his voice was very much changed for several months, a circumstance which together with the feeling about his heart, alarmed me much when others (men of deep learning & dictatorial gravity) thought the whole a whim to which indeed the patient was somewhat subject'.[90] Cleghorn is here referring to 'the Edinburgh Oracles Bell & Russell', mentioned in the account of the patient's subsequent postmortem, who had recently 'unmercifully salivated' the patient, although he had no venereal symptoms.[91] Similarly, when he performed a postmortem on a Colonel Houston on 4 October 1800, finding the liver enlarged and filled with pus, and, adhering to the liver, the stomach, which was blackened and filled with clotted blood, Cleghorn remarked sarcastically: 'This case was treated in Edinburgh as Gout! by brandy, opium & travelling!'. Although the patient had vomited blood, the vomiting was imputed to accident. Cleghorn saw him ten days before his death, and, upon examination, had felt a hardness over the heart and in other parts of the abdomen. William Dunlop and William Whyte agreed. Cleghorn commented triumphantly: 'The event shows we were right & is a new proof of the utility of examining parts'. This was a clear sign that he felt that local examination was not yet routine practice since it was something to be promoted.[92] It was only three years since the second edition of Matthew Baillie's *Morbid Anatomy of Some of the Most Important Parts of the Human Body* (1797) had emphasised the importance of postmortems for diagnostic precision and clinical practice, since this enabled the practitioner 'to detect some marked difference, by which the disease may be distinguished in the living body'.[93] A physician with an active interest in dissection and pathological anatomy, Baillie had received a classical education in Glasgow and Oxford, and then studied medicine in London under the tutelage of his uncles, William and John Hunter, in the early 1780s. Having spent seventeen years as an anatomy lecturer, Baillie is a further example of the close relationship in practice between physic and surgery that was often characteristic of elite medical practitioners in Glasgow.[94]

The use of postmortems in conjunction with physical examination began to make doctors less dependent on patients' subjective relation of their illness. In considering whether Cleghorn's attitude to his patients left much to be desired, there are a number of cases that indicate this to be an issue. Predictably, there is little in the notebooks to suggest he was guilty of breaches in confidentiality, but then he would have been unlikely to draw attention to this himself. In 1799, he treated James Angus, a patient long bedridden with consumptive complaints who finally became paralytic. The patient insisted that 'his heart seem'd to beat under the right mamma, & he was very confident in affirming that it was not always so'. Upon his death, Cleghorn performed a postmortem which showed that, indeed, the left cavity of the thorax was full of puriform fluid, and that the lungs were shrunken and hard, adhering to the mediastinum which, together with the heart, had been pushed over into the right division of the thorax by the effusion.[95] Cleghorn warned himself: 'N.B. Never slight the feelings of a Patient, even when he speaks of improbables'. Such self-censure might be taken as a sign that this was exactly what he had done.[96]

More poignant in this regard because it is one of the best, and most meticulous, examples of a case in which Cleghorn takes himself to task, is that of 'Mr Adamson a fine young Man of 17' who 'had a slight venereal complaint with which he [Adamson] tamper'd'. Cleghorn first visited Mr Adamson with John Burns on 8 February 1801. Adamson had thus far only consulted James Towers, University Lecturer in Midwifery, who probably called in his colleague, Burns, for a second opinion.[97] Cleghorn describes Adamson's condition:

> The prepuce was ulcerated with great swelling over the Glans, & urine flowd from a hole near the fraenum. The right groin was coverd with a large, ugly, deep ulcer, of most offensive smell, & penetrating below the femoral artery which was seen of a sloughly appearance, throbbing so violently that we could not look on it without horror. Both sores were coverd [sic] . . . with Pledgets of red precipitate which produced extreme irritation & pain.[98]

Constitutionally, the patient was emaciated, feverish and hectic, 'expressing the utmost anxiety & misery in every look & tone'.[99] Burns and Cleghorn began by removing the dressings, applying an antiseptic poultice of decoction of chamomile and tincture of myrrh onto the parts that were sloughing and foetid, and they then prescribed opiates, nourishment and occasional laxatives. With this treatment, the sore skinned over and seemed to heal, but was still discoloured, very tender and apt to give way. At the same time, 'an ulcer began to creep along the hairy part of the pubes very rapid in its progress & of an ugly aspect'.[100] The practitioners tried sprinkling the skin with bark of chinchona (an astringent and antiseptic) and pouring it into the ulcer and then dressing it in order to check its progress, as well as prescribing bark to the patient internally (for fever).[101]

The case is significant because it shows Cleghorn interacting with surgeons as equals, in a collegial way, administering internal and external therapies together to the patient. Nevertheless, as often happened, the practitioners' prognoses diverged radically:

> At this period we differ'd in opinion, mine being that the sore was much worse, threatening the body of the penis itself & travelling fast towards

the umbilicus, the edges being ragged, thick, with several indentations filled with green foetid pus, very irritable & painful even to the contiguous skin which was red, & plainly threaten'd in its turn. All this however appeared to me only; & it convinced me that mercury was necessary.[102]

Cleghorn was convinced both that the patient needed country air 'as he was then in a vile hole in Bell's wynd', and that calamine injured young skin when laid on its surface, and should be replaced by simple ointment (four parts olive oil to one part white beeswax). He began treating the sore with very small doses of corrosive sublimate (mercuric oxide) and decoction of sarsaparilla, which are both anti-venereals, and dressing the sores with ointment of diluted mercury nitrate rub. The progress of the sore was checked: the discharge diminished and the surrounding skin assumed a natural hue.[103] The solution was then substituted for mercury pills which the patient took for a full night and morning without the slightest salivation – the usual side effect of such treatment. On his left groin, a hardness, threatening to suppurate, stretched down to the perineum, and the patient bled occasionally on only the slightest exertion. Even so, he got short shrift from his attending practitioners: 'Of the slightest touch on those parts he complained much, but we called it hysterical, childish &c'.[104] Eventually, Cleghorn got him to the country where the treatment continued to improve the patient's condition. The sore nearly closed, and everything about it looked better except that a slight excoriation began again along the upper edge, and the redness, hardness and tenderness of the left groin increased. Nonetheless, the patient had an 'anxious look', and continued to have a parched tongue and symptoms of fever. There was also considerable haemorrhage, but since the patient did not mention its extent, Cleghorn wrote (in hindsight) that he thought it came from the surface vessels. Cleghorn indulges in the most wonderful self-ironising in this case. He anticipated Adamson's complete recovery within a week: 'En faustam prognosin [Look what a splendid prognosis!]'.[105]

On the morning of 29 June 1801, however, before 9 a.m., the woman with whom Adamson was residing came breathless to Shawfield crying that he was in the sweat of death. Attending to his own bodily needs, Cleghorn had asked: 'Was he purging? No. Bleeding? No: then he won't die & I will call after breakfasting Mr. Mylne & Craig'.[106] Unfortunately, when he attended at 10 a.m. he found the patient moribund from loss of blood that had been flowing in far greater quantity than he had been aware. The state of the patient frightened Cleghorn who dared not remove his blood-clotted dressings since he had no ligatures, and instead ran to get Mr Burns. Before they got back to the patient 'the blood had ceased to flow because it had ceased to circulate'. A few minutes later Adamson died, as Cleghorn commented, 'to the mortification of my vanity certainly, & as I think to my regret also on another score'.[107]

The following morning John Burns dissected Adamson's body. When the body was opened, it was discovered that matter from the right side had parted the tissue along the perineum to form a bag that reached across to the left side. Blood was visible from a round hole in this bag about the size of a dried pea and located near the edge of the pubic hair, but the hole had a deeper origin. The source Burns and Cleghorn discovered 'by throwing water forcibly into the aorta & a hole was found in the femoral artery between 2 & 3 inches below its passing out of the abdomen'.[108] They were essentially performing basic experiments in order to understand the practical impact of the pathological lesions on the body.

The most insightful part of Cleghorn's recording of this case is a page full of reflections at the end, in which he admits that 'the case suggests many remarks moral as well as physical'. In the Enlightenment, the quality of sympathy meant not only engaging the affection and confidence of the patient, but also being able to identify with their moral sentiments. The physician's duty to cure a patient thus became an act of moral good.[109] In this final page, Cleghorn takes himself to task for his failings in the case. In the first place, he chided himself to 'never rejoice in the misconduct of a Rival [Towers presumably] where a fellow Creature suffers by it'. It was, however, deeply unpalatable for him to accept all the blame himself. He added: 'This may apply in part to me still more to my Colleagues in the first instance with regard to our own predecessor, in the next with regard to my anticipated exultation – They differed from me but look at the event said I to myself'.[110] He rebuked himself, secondly, for having underestimated the discomfort of the patient, something that he had palpably done before: 'Never neglect or under-rate the feelings of a Patient – How often must I repeat this to myself? Poor A. could not without shrinking bear the slightest touch on the discoloured skin; we like conceited unfeeling Idiots called it Fancy!!!'.[111] Thirdly, he reminded himself that 'when a swelling rises near a great blood-vessel, with occasional haemorrhage, never neglect it again! If attention to it cannot assist the practice, it may at least direct the prognosis'.[112] He finally chided himself for not taking a sufficiently holistic view of Adamson's disease: 'count nothing on one portion of an ulcer healing while another breaks out, while the tenderness around continues unabated & the health unimproved!'[113] His lengthy account of this case and his extensive reflection on it show that he did at least try to address some of the shortcomings in bedside manner that some of his colleagues insinuated of him.

In the case of Miss M. D. who consulted him in 1817 when she began to cough after having incautiously attended her brother who died of purulent consumption, Cleghorn reproved himself for unethical behaviour. He was called to attend her with suspected whooping cough. 'Unfortunately', he comments, '& most improperly I did not insist on the Family Surgeons attending partly from resentment, chiefly that I might not be troubled with cox[comb]ry or stolidity, but I regret it & will have much reason to do so'.[114] The patient could not expectorate without vomiting, which condition was replaced by weeks of pain in her bowels and frequent stools. Her fever then increased, attended with frequent flushings. For several weeks she could not recline but sat with her head on her knees. Cleghorn, however, refused to see the obvious diagnosis, pulmonary tuberculosis, but 'still clung to hope being blinded by vanity, by vanity, the dread of interference, & the consciousness of incorrectness' that she indeed had whooping cough.[115] The patient subsequently voided a great deal of blood in a fit of coughing and died. Cleghorn sorely castigated himself:

> All the sufferings that timid sensibility, unpleasant anticipations & self reproach for repeating what I have often repented of & resolved against, are mine & will long continue tho if I may judge of the future from the past with no salutary effect on my conduct which is obviously that of a hardened Fool. Of course I opposd [sic] the dissection but being urged & threatened with a surgeon I did it myself in presence of the 2 Brothers.[116]

This example raises two important issues. Even towards the end of his career, Cleghorn still held his own opinion to be superior to that of other practitioners. His professional pride could not brook being crossed. Secondly, since Allan Burns was by this time dead, the '2 brothers' can only, in this context, refer to the patient's siblings.[117] This means that contrary to what would generally be expected in popular circles, the patient's brothers were here calling for an autopsy, though their motive was clear, namely to prove Cleghorn's diagnosis incorrect and his subsequent treatment negligent.

The postmortem showed that there were slight adhesions on the right side of the body, the left lung was plugged with mucus and the trachea filled with blood. The pericardium contained five or six ounces of fluid. The liver was large and hard and the intestines stuck to the parietes and to each other. Cleghorn queried whether this was whooping cough, but felt uncertain. He must have realised it unlikely, but he considered that his main error

> was implicitly believing it pertussis [whooping cough] & attending alone. My anxiety was little short of agony, my disappointment is such as no Fee can compensate. I feel degraded in my own estimation & I have whetted the resentment of a young active using F[orce?] against the indefatigable malice of an old B[astar]d. How little do I profit from my own experience! I have more than once since that time fallen into a similar snare, misled partly by vanity & self conceit, partly by the flattery of Patients, chiefly by my contempt of or hatred to surgeons from some of whom I receive daily proof of enmity or spite & from none any marks of confidence.[118]

Although Cleghorn here employs a Latinate diagnosis, there is little evidence in his manuscript books to support Fissell's point that doctors increasingly used it to dissociate themselves from patients' language.[119] That aside, however, the boundaries of the social and educational divide that separated the all-knowing, university-educated physicians from the craft-bound surgeons are here seen writ large.

Conclusion

This chapter has demonstrated three main things. First, it has shown the importance of case records as a source for late Enlightenment medical practice and clinical cognition in Scotland. While most patient case histories are, by their very nature, accurate records of the medical practice of their time, not all case records reveal as much about the individual's process of knowledge making and validation in a clinical context as do Cleghorn's. In these cases, he was constantly sounding the scope of his scientific beliefs.

Secondly, it has shown the significance of physical diagnosis in Cleghorn's practice in Glasgow between 1785 and 1818. While physical examination may not have been usual in Glaswegian clinical practice, it was for Cleghorn, and was so long before the routine use of physical examination in Victorian England.[120] His case notes show him to have been a physician whose practice was characterised by physical diagnosis and frequent postmortem dissection. His medical scientific identity thus legitimised and gave authority to their use in clinical practice in the West of Scotland. He knew the value of both touch and active observation. Yet, it is difficult to believe that Cleghorn was the only Scottish physician in the 1780s routinely using this degree of physical diagnosis. A systematic study of eighteenth-century Scottish physicians' extant case records, in manuscript and printed form, might establish more definitive chronological and geographical boundaries for the diffusion of physical diagnostic techniques into the country. Barbara Duden's excellent study of the casebooks of Johann Storch has shown, for example, how effectively the voice of the female patient can be found in a subtle deconstruction of the physician's case notes. Similar studies using casebooks as sources of evidence could be made in Scotland. The case notes at present available in manuscript form would be rendered more accessible by the publication of Scottish physicians' casebooks edited with contextual introductions, similar to the editions of the English physicians John Snow and William Brownrigg.[121]

Thirdly, his use of physical diagnosis, perhaps, posed something of a personal dilemma for him in his relations with his patients and with his peers, many of whom he did not regard (correctly as it turned out) to be as skilled as himself. He was an accomplished physician, as is testified by the continuity in his hospital appointments, and by the social status of those who consulted him. At the same time, his day-to-day medical practice benefited greatly from close collaboration with surgeons. Cleghorn fell between two camps. He was practising when the patient's subjective narration of his illness was still a major aspect of diagnostic practice, but he was also using a significant degree of physical diagnosis which favoured objective deduction. This must have made him an oddity. His extant case histories, while perspicacious, are not written in the pared, objective language of later nineteenth-century case records but in the extensive, rambling prose of the late eighteenth century which are the very epitome of subjective, loquacious and discursive self-indulgence. In this, he was a typical eighteenth-century physician. He even waxes lyrical in parts, recording, for example, of Thomas Adams, aged 60, a Gardener in Anderston, that he 'was leaner than the starv'd Apothecary in Hamlet'.[122]

Contemporary accounts of his alter ego, 'Dr Wormwood', hint that patients had difficulty with Cleghorn's bedside manner. He was clearly idiosyncratic and brusque, but was also capable of tenderness and empathy with his patients. Further inconvenience to Cleghorn was, perhaps, partly derived from the fact that local physicians were not yet ready to accept the significant amount of physical examination that his diagnostic technique relied upon. Thus he did not have community approval of this technique which some may have regarded as extremely crude practice for the gentleman physician. Equally, some Glaswegian surgeons (though not the Burns brothers) might have felt that he was invading their distinctive sphere of physical manipulation which may have resulted in accusations from them similar to those of the physicians. His case notes contain clear indications of difficulties in his relations with surgeons. In his pioneering *Medical Ethics* (1803), Dr. Thomas Percival certainly deplored the breakdown in traditional demarcations between practitioners in Manchester, stating that 'physicians are the only proper substitutes for physicians; surgeons for surgeons'.[123] However, he was also aware that, outside London, expediency often dictated that the smaller number of practitioners be skilled in a variety of procedures, so that traditional demarcations did not always stand. Cleghorn seems to have taken those criticisms made of him to heart and to have sought to address them, while continuing to be aware of the practical benefits of physical diagnosis. His case records are, to a great extent, a record of self-excoriation for his repeated failure to listen properly to his patients, a cardinal sin at a time when most physicians were still highly dependent on the patient's subjective feelings of illness for their diagnoses.

Is there any reason why an example of such a progressive form of diagnostic practice should be found in Glasgow as early as the 1780s, when it was not dominant elsewhere in Britain until the Victorian era? Malcolm Nicolson has argued that 'where physic and surgery were close together, socially, educationally and institutionally, one might expect to find, in the eighteenth century, forms of practice more like those which were to become dominant in the nineteenth'.[124] In this respect, Glasgow's medical licensing body was then unique in being a Faculty of Physicians and Surgeons in which physicians, surgeons and apothecaries were joined in corporate association. Surgeons in general practice in the city had been accorded an academic status in their employment as lecturers in anatomy at Glasgow University since 1714.[125] Both of these factors inevitably brought them more opportunities to fraternise than in other cities and, more importantly, to set up professional networks of co-operation and interaction between physicians and surgeons that led to a mutual exchange of skills and knowledge. In this way, Cleghorn was exceptionally fortunate in being able to team up with the Burns brothers, experts in the fields of anatomy and surgery. This fortuitous combination of circumstances in the West of Scotland resulted in a situation which encouraged, and brought on, Cleghorn's diagnostic technique.

Notes

I am most grateful for the support of the Wellcome Trust who funded a Research Fellowship at the Wellcome Trust Centre for the History of Medicine at UCL, where this chapter was written. My thanks also to: Akihito Suzuki, who discussed various aspects of patient case records with me before I began; Malcolm Nicolson, who first drew my attention to the significance of physical examination; Roy Porter, Andrew Wear and the editors, who read earlier versions of this chapter; and Vivian Nutton for providing and checking all of the Latin translations.

1. *Medical and Philosophical Commentaries* 1 (1773), v.
2. See, for example, *Medical Essays and Observations* (Edinburgh, 1733–44), and, 5 vols. *Essays and Observations, Physical and Literary*, 3 vols. (Edinburgh, 1754–71).
3. Ulrich Tröhler, *'To Improve the Evidence of Medicine': The Eighteenth Century British Origins of a Critical Approach* (Edinburgh, 2000); Lawrence I. Conrad, Michael Neve, Vivian Nutton, Roy Porter and Andrew Wear, *The Western Medical Tradition, 800 BC to 1800* (Cambridge, 1995), 465–66. For an up-to-date study of the late Enlightenment context, see Roy Porter, *Enlightenment: Britain and the Creation of the Modern World* (London, 2000), and for one of the few recent attempts at a single-volume history of medicine, idem, *The Greatest Benefit to Mankind: A Medical History of Humanity from Antiquity to the Present* (London, 1997).
4. Stanley J. Reiser, *Medicine and the Reign of Technology* (Cambridge, 1978); Edward Shorter, *Doctors and Their Patients: A Social History*, 2nd edn (Brunswick, NJ, 1991); Mary E. Fissell, 'The Disappearance of the Patient's Narrative and the Invention of Hospital Medicine', in Roger French and Andrew Wear (eds.), *British Medicine in an Age of Reform* (London, 1991), 92–109; Barbara Duden, *The Woman beneath the Skin: A Doctor's Patients in Eighteenth-Century Germany*, trans. by Thomas Dunlap (Cambridge, MA, 1991); Akihito Suzuki, 'Framing Psychiatric Subjectivity', in Joseph Melling and Bill Forsythe (eds.), *Insanity, Institutions and Society, 1800–1914: A Social History of Madness in Comparative Perspective* (London, 1999); Guenter B. Risse and John Harley Warner, 'Reconstructing Clinical Activities: Patient Records in Medical History', *Social History of Medicine* 5 (1992), 183–205. As Risse and Warner point out (p. 187), case records documenting the work of a single practitioner make it far easier to discern consistent patterns in medical practice.
5. Nicholas Jewson, 'Medical Knowledge and the Patronage System in Eighteenth-Century England', *Sociology* 8 (1974), 369–85; idem, 'The Disappearance of the Sick Man from Medical Cosmology, 1770–1870', *Sociology* 10 (1976), 225–44.
6. See, for example, Hamish Maxwell-Stewart and Alastair Tough, *Selecting Clinical Records for Long-Term Preservation: Problems and Procedures*, 2nd edn (Glasgow, 2000).
7. Reiser, *Medicine and the Reign of Technology*, 1.
8. Jewson, 'Medical Knowledge' and 'Disappearance of the Sick Man'.
9. Steven Shapin, *A Social History of Truth: Civility and Science in Seventeenth-Century England* (Chicago and London, 1994), 8–12, 38, 65.
10. For postal consultation, see Guenter B. Risse, 'Managing the Rich and Famous: William Cullen's Mail-order Physic in the Eighteenth Century', in A. Doig, J. P.

S. Ferguson, I. A. Milne, and R. Passmore (eds.), *William Cullen and the Eighteenth Century Medical World* (Edinburgh, 1993); idem, 'Doctor William Cullen, Physician, Edinburgh: A Consultation Practice in the Eighteenth Century', *Bulletin of the History of Medicine* 48 (1974), 338–51.

11. Malcolm Nicolson, 'Giovanni Battista Morgagni and Eighteenth-Century Physical Examination', in Christopher Lawrence (ed.), *Medical Theory, Surgical Practice: Studies in the History of Surgery* (London and New York, 1992), 101–34. Cleghorn's notebooks show that Morgagni was used in teaching at Edinburgh University while Cleghorn was there; see Royal College of Physicians and Surgeons of Glasgow (hereafter RCPSG), MS 20/2/1/2, 18. In 1783, Morgagni's text was being sold by the Edinburgh bookseller John Balfour; see John Murray to John Balfour, 4 December 1783, John Murray Archive (London), Letter Book 20 June 1782 to 16 January 1786, 258.
12. Reiser, *Medicine and the Reign of Technology*, 21.
13. Lisa Rosner, *Medical Education in the Age of Improvement: Edinburgh Students and Apprentices, 1760–1826* (Edinburgh, 1991), 12.
14. Malcolm Nicolson, 'Gerard van Swieten and the Innovation of Physical Diagnosis', in Ilana Löwy (ed.), *Medicine and Change: Historical and Sociological Studies of Medical Innovation* (Montrouge, France, 1992), 53.
15. Shapin, *Social History of Truth*, 16–17, 20.
16. Adir Ophir and Steven Shapin, 'The Place of Knowledge: A Methodological Survey', *Science in Context* 4 (1991), 3–21.
17. David N. Livingstone and Charles W. J. Withers (eds.), *Geography and Enlightenment* (Chicago and London, 1999), 15.
18. John M'Ure, *Glasghu Facies: A View of the City of Glasgow*, ed. J. F. S. Gordon, 2 vols. (Glasgow, 1873), 1:335; John D. Comrie, *History of Scottish Medicine*, 2nd edn, 2 vols. (London, 1932), 1:362.
19. General Register Office (hereafter GRO), Scots Origins online database, Old Parish Register (hereafter OPR) for Berwick. Cleghorn's mother's name is not given in the register. T. Gibson, *The Royal College of Physicians and Surgeons of Glasgow* (Edinburgh, 1983), 55, gives his date of birth as c.1760.
20. Cleghorn dedicated the thesis to his namesake (but not relative), George Cleghorn, the well-known Professor of Anatomy at Trinity College, Dublin; George Thomson, 'Robert Cleghorn, MD', in Andrew Kent (ed.), *An Eighteenth Century Lectureship in Chemistry* (Glasgow, 1950), 165. Cleghorn referred to his thesis some time later in November 1811: 'I have read it over and find it very well written tho' amid lassitude from fatigue, dissipation & sleeplessness. 3 other Theses have borrowd from it freely tho' generously avoiding all mention of it'. 'Adversaria Cleghorn', RCPSG MS 20/2/1/1, 201.
21. *The History of the Speculative Society, 1764–1904* (Edinburgh, 1905), 72 (Cleghorn was a member in 1779–80); James Gray, *History of the Royal Medical Society, 1737–1937* (Edinburgh, 1952), 15, 316.
22. See, for instance, the section, 'The Royal Medical Society and Medical Research', in Malcolm Nicolson and Jonathan Windram, 'Matthew Baillie Gardiner, the Royal Medical Society, and the Problem of the Second Heart Sound', *Proceedings of the Royal College of Physicians of Edinburgh*, 31 (2001), 357–67.
23. GRO Scots Origins database, OPR for Glasgow in Lanark; G. Graham Thomson, 'An Old Glasgow Family of Thomson', a paper read before the members of the Old Glasgow Club, 19 January 1903, Glasgow University Library, Mu22-e.12, no. 12, 6–7.
24. Page 139 of the second volume of his casebooks is headed 'Glasgow, December

2nd 1785', RCPSG MS 20/2/1/3; Minutes of the Faculty of Physicians and Surgeons, 1785 to 1807, RCPSG MS 1/1/1/4, fols. 44v, 70r.
25. Sheila Devlin-Thorp (ed.), *The Royal Society of Edinburgh: 100 Medical Fellows Elected 1783–1844* (Edinburgh, 1982); Hugh Frew (ed.), *Index of Fellows of the Royal Society of Edinburgh Elected November 1783–July 1883* (Edinburgh, 1984). Cleghorn was elected a Fellow on 23 June 1788.
26. W. J. Duncan (ed.), *Notices and Documents Illustrative of the Literary History of Glasgow During the Greater Part of Last Century* (Glasgow, 1831), 134. It is impossible to be specific about the date that Cleghorn joined. For the social background of the Society in the late eighteenth century, see Roger L. Emerson, 'The Enlightenment and Social Structures', in Paul Fritz and David Williams (eds.), *City and Society in the Eighteenth Century* (Toronto, 1973), 111–12.
27. *Jones's Directory; Or, Useful Pocket Companion, for the Year 1789* (Glasgow, 1789), 73.
28. Alexander Duncan, *Memorials of the Faculty of Physicians and Surgeons of Glasgow, 1599–1850* (Glasgow, 1896), 263; Comrie, *History of Scottish Medicine*, 1:362; Thomson, 'Cleghorn', 165, quoting Lanark Sasines, 4071.
29. Thomson's evidence to the Commissioners clearly states: 'in 1817, when Dr. Cleghorn resigned his situation as Lecturer in Chemistry' *Evidence, Oral and Documentary, Taken and Received by the Commissioners for Visiting the Universities of Scotland*, 4 vols. (London, 1837), 2:206.
30. James Kendall, 'Thomas Charles Hope, MD', in Kent, *Eighteenth Century Lectureship*, 158–59; Thomson, 'Cleghorn', 166–67.
31. David Murray, *Memories of the Old College of Glasgow: Some Chapters in the History of the University* (Glasgow, 1927), 190.
32. RCPSG MS 20/2/1/1, 253.
33. Gibson, *Royal College of Physicians and Surgeons of Glasgow*, 55, quoting T. Lyle, *University Reminiscences*.
34. Thomson in *Evidence, Oral and Documentary*, 2:205.
35. Emerson, 'The Enlightenment and Social Structures', 122.
36. 'Quarterly Meeting Minute Book of the Directors of the Town's Hospital, 1732–1816', Mitchell Library Rare Books and Manuscripts, MS 641983, fol. 160v. See also Fiona A. Macdonald, 'The Infirmary of the Glasgow Town's Hospital, 1733–1800: A Case for Voluntarism?', *Bulletin of the History of Medicine* 73 (1999), 64–105; Glasgow Royal Infirmary Records, 1:1787–1802, Greater Glasgow Health Board Archives, MS HB14/1/1, 13; Jacqueline Jenkinson, Michael Moss and Iain Russell, *The Royal: The History of the Glasgow Royal Infirmary, 1794–1994* (Glasgow, 1994), 37. He was also appointed physician at the Royal in 1795–96, 1798–99, 1801–2, 1805–6 and 1809–10.
37. Thomson, 'Cleghorn', 165; Iain Smith and Alan Swann, 'Medical Officers and Therapeutics, 1814–1921', in Jonathan Andrews and Iain Smith (eds.), *'Let There be Light Again': A History of Gartnavel Royal Hospital from its Beginnings to the Present Day* (Glasgow, 1993), 51, 53. His contribution to the institution was commemorated in a portrait by Sir Henry Raeburn commissioned for the asylum. G. Graham Thomson, 'Dr. Cleghorn *en voyage* (1794)', *Old Glasgow Club. Transactions* 1 (1908), 239.
38. Leonard Smith, *Northern Sketches or Characters of G******* ([London], [c.1810]). Leonard Smith is a pseudonym. The authorship of *Northern Sketches* has usually been credited to John Gibson Lockhart but, as David Murray pointed out in 1927, Lockhart would only have been 17 years old when the book was

published. Murray ascribed authorship to the accomplished poet John Finlay (1782–1810), which seems more likely: Murray, *Memories*, 207–8.
39. Quoted in Thomson, 'Cleghorn', 168. John Cowan first drew attention to this caricature in print in *Some Yesterdays with a Note upon the Development of Hospitals by Joshua Ferguson* (Glasgow, 1949), 73. Glasgow University Library currently has five copies of this book, in all of which the name Cleghorn has been pencilled in above Wormwood.
40. Thomson, 'Cleghorn', 168.
41. Andrew Wear, *Knowledge and Practice in English Medicine, 1550–1680* (Cambridge, 2000), 127–30; Reiser, *Medicine and the Reign of Technology*, 4.
42. RCPSG MS 20/2/1/1, 198.
43. Smith, *Northern Sketches*, 13.
44. RCPSG MS 20/2/1/1, 122.
45. Adam Boyd to William MacKenzie, 10 August 1814, RCPSG MS 24/2/22.
46. Burns to Cleghorn, [1796], Edinburgh University Library, MS La.II.210, fols. 2–3. Burns inscribed on the envelope: 'Burns to Cleghorn with a pretty song The letter immaterial'. This was the song 'Jane Morag'. The letter is addressed to Cleghorn at 'Saughton mills'. There was a lunatic asylum at Saughtonhall from 1800 (and possibly earlier), which, in 1816, possessed 'the peculiar advantage of being under the care of gentlemen of the medical profession': 'Third Report from the Select Committee Appointed to Consider of Provisions Being Made for the Better Regulation of Madhouses in England, 1816', *Parliamentary Papers, 1816*, Report no. 451, vol. 6, 370, 373. I am grateful to Rab Houston for the information on Saughtonhall.
47. Quoted in Roger L. Emerson, 'Politics and the Glasgow Professors, 1690–1800', in Andrew Hook and Richard B. Sher (eds.), *The Glasgow Enlightenment* (East Linton, UK, 1995), 33, citing George Jardine to Robert Hunter, 15 September 1799, Glasgow University Library MS Gen. 507, box 3.
48. RCPSG MS 20/2/1/1, 28.
49. RCPSG MS 20/2/1/1, 80.
50. See Jacqueline Jenkinson, *Scottish Medical Societies, 1731–1939: Their History and Records* (Edinburgh, 1993), 159.
51. John Strang, *Glasgow and its Clubs; or Glimpses of the Condition, Manners, Characters and Oddities of the City, During the Past and Present Century* (London and Glasgow, 1856), 301n.
52. Cleghorn is not mentioned, for instance, in Robert Chambers' *A Biographical Dictionary of Eminent Scotsmen*.
53. Andrews and Smith, 'Let There be Light Again', 60; RCPSG MS 20/2/2, which contains case notes from *c*. 1814 onwards of patients from Glasgow Asylum for Lunatics in the hand of Robert Cleghorn, with further notes by John Balmanno. This notebook contains the book plate of William James Fleming (Balmanno had an uncle by the name of William Fleming). In what follows I deal exclusively with the three manuscript notebooks left by Cleghorn.
54. RCPSG MS 20/2/1/3, 216.
55. Thomson, 'Cleghorn *en voyage*', 238. Only a few extracts of this now survive in a later secondary account. Travelling alone, Cleghorn left Glasgow on 4 August 1794, passing through Stirling, across to Dundee, and up to Peterhead, then retracing his steps south as far as Carlisle and the Lake District and from there back to Wigton, and then to Bowness which he found 'dull and ruinous', where his diary ends on 15 September, because he wished to go 'home in safety to my dear child, for whom I am now become impatient'.

56. Cleghorn made no personal contribution to the medical journals published in Scotland during the period of his practice, namely the *Medical and Philosophical Commentaries* (1773–95), the *Annals of Medicine* (1796–1804) and the *Edinburgh Medical and Surgical Journal*. His appointment as Lecturer in Materia Medica was, however, noted in the *Medical and Philosophical Commentaries*, 2nd decade, 3 (1789), 449.
57. Review of John Rollo, *An Account of Two Cases of the Diabetes Mellitus, with Remarks as they Arose During the Progress of the Cure* (London, 1796), in *Annals of Medicine* 2 (1797), 104.
58. Robert Cleghorn, 'A Biographical Account of Mr William Hamilton, Late Professor of Anatomy and Botany in the University of Glasgow', *Transactions of the Royal Society of Edinburgh* 4 (1798), 35–63; Andrew Kent, 'William Irvine, MD', in Kent, *Eighteenth Century Lectureship*, 140.
59. Unfortunately, the casebooks have not been catalogued chronologically. The first in chronological order, which extends to 254 sides of script, is RCPSG MS 20/2/1/2.
60. The Bristol physician and prolific author, Thomas Beddoes, for example, also compiled several commonplace books. Some of the medical reflections in them were extracted for an article in the main Scottish medical journal of this period; 'Extracts from Dr Beddoes's Common-Place Books', *Edinburgh Medical and Surgical Journal* 7 (1811), 184–93.
61. RCPSG MS 20/2/1/3, 139–40. There is a one-page index at the back for the commonplace entries.
62. RCPSG MS 20/2/1/3, 169, 187–88, 189–211. The cure for scrophula is from the *Gazette de Santé* for October 1786.
63. RCPSG MS 20/2/1/1 is chronologically the last in the series of casebooks, though numerically catalogued first.
64. RCPSG MS 20/2/1/1, 31–32. The final case in this notebook is unfinished and has been excluded from this study.
65. Letter inserted into the front of RCPSG MS 20/2/1/1. Ebenezer Watson graduated MD from Glasgow University in 1846. He was Professor of the Institutes of Medicine at Anderson's University (1850–76), surgeon at the Royal Infirmary (1856–60; 1866–85), and President of the Faculty of Physicians and Surgeons of Glasgow (1872–74). His father, James Watson, who was surgeon to the Royal Infirmary (1813–1814), had attended Cleghorn on his deathbed in 1821. Duncan, *Memorials*, 271, 290; Jenkinson, Moss and Russell, *The Royal*, 281.
66. RCPSG MS 20/2/1/3, 174.
67. This follows the argument in Shapin, *Social History of Truth*, 243–309.
68. See, for example, RCPSG MS 20/2/1/1, 63, for the case of a 'boy' who came to the infirmary with a bladder stone.
69. RCPSG MS 20/2/1/3, 228–29.
70. RCPSG MS 20/2/1/3, 223.
71. RCPSG MS 20/2/1/1, 88.
72. RCPSG MS 20/2/1/1, 89. The omentum is a double fold of the peritoneum that connects the stomach with the other abdominal organs.
73. Fissell, 'Patient's Narrative', 101.
74. RCPSG MS 20/2/1/1, 133.
75. RCPSG MS 20/2/1/4, fols 127r, 128r-v.
76. 'Glasgow Royal Infirmary Records, 1:1794–1812', 158–59; Duncan, *Memorials*, 266. Burns was eventually awarded a Glasgow MD in 1828.
77. The brothers were responsible for a number of important anatomical and surgical

books. A selection of John Burns' includes *Dissertations on Inflammation* (Glasgow, 1800) and *The Principles of Midwifery; Including the Diseases of Women and Children* (London, 1809), while Allan published *Observations on Some of the Most Frequent and Important Diseases of the Heart* (Edinburgh, 1809) and *Anatomy of the Head and Neck* (Edinburgh, 1811), before his death in 1813.

78. In 1809 the brothers moved their private extramural school from Virginia Street to College Street, from which time it was known as the College Street Medical School. This was the street where Cleghorn had previously lived. The new location gave the College easier access to the nearby Ramshorn Churchyard, which the students could raid for a plentiful supply of corpses: F. L. M. Pattison, *Granville Sharp Pattison: Anatomist and Antagonist, 1791–1851* (Edinburgh, 1987), 14, 241–42.

79. See, for instance, RCPSG MS 20/2/1/3, 179–80, 213; RCPSG MS 20/2/1/1, 198–202.

80. RCPSG MS 20/2/1/3, 172, 231, 245; RCPSG MS 20/2/1/1, 81, 90, 181, 189, 212, 245; Duncan, *Memorials*, 270.

81. Owsei Temkin, *The Double Face of Janus* (Baltimore and London, 1977), 489; Reiser, *Medicine and the Reign of Technology*, 19, 22.

82. See, for example, MS RCPSG 20/2/1/2, 45–47, 51, 72–74, 127, 166.

83. Significantly, the only book that he borrowed from the Hunterian Library, on 8 August 1812, was William Hunter's *Medical Commentaries* (1762): Glasgow University Library, 'Register of Books borrowed from the Hunterian Library, 1808–1852', MS MR25.

84. This is probably the surgeon William Anderson, who entered the Faculty of Physicians and Surgeons of Glasgow in 1790. John Anderson's will made him Professor Designate of Obstetrics in Anderson's College; Duncan, *Memorials*, 265.

85. RCPSG MS 20/2/1/1, 182.
86. RCPSG MS 20/2/1/1, 183.
87. RCPSG MS 20/2/1/1, 185.
88. RCPSG MS 20/2/1/1, 188.
89. RCPSG MS 20/2/1/1, 234.
90. RCPSG MS 20/2/1/1, 67.
91. RCPSG MS 20/2/1/1, 68.
92. RCPSG MS 20/2/1/1, 72.
93. Susan C. Lawrence, *Charitable Knowledge: Hospital Pupils and Practitioners in Eighteenth-Century London* (Cambridge, 1996), 308.
94. Alvin E. Rodin, *The Influence of Matthew Baillie's Morbid Anatomy: Biography, Evaluation and Reprint* (Springfield, IL, 1973), 7–9.
95. A puriform fluid resembles pus in consistency and appearance. Mediastinum is the septum formed by the juxtaposition of the two pulmonary pleurae which divides the thorax into two parts.
96. RCPSG MS 20/2/1/1, 60.
97. John Burns also gave midwifery lectures in Glasgow, and this is how they were acquainted. See, for example, Johanna Geyer-Kordesch and Fiona Macdonald, *Physicians and Surgeons in Glasgow: The History of the Royal College of Physicians and Surgeons of Glasgow, 1500–1858* (London and Rio Grande, 1999), 270.
98. RCPSG MS 20/2/1/1, 81.
99. RCPSG MS 20/2/1/1, 81.
100. RCPSG MS 20/2/1/1, 82.
101. RCPSG MS 20/2/1/1, 82; J. Worth Estes, *Dictionary of Protopharmacology: Therapeutic Practices, 1700–1850* (Canton, MA, 1990), 24, 47–49.

102. RCPSG MS 20/2/1/1, 83.
103. RCPSG MS 20/2/1/1, 82–3; Estes, *Dictionary of Protopharmacology*, 99, 117.
104. RCPSG MS 20/2/1/1, 84. Cleghorn has written 'some' immediately above the 'we', as if to lessen his own responsibility.
105. While Cleghorn undoubtedly felt that at the time, with the patient's death, he was no doubt using it ironically when he wrote his case record.
106. RCPSG MS 20/2/1/1, 84–5.
107. RCPSG MS 20/2/1/1, 85.
108. RCPSG MS 20/2/1/1, 85–6.
109. Guenter B. Risse, 'Patients and Their Healers: Historical Studies in Health Care', in Nora K. Bell (ed.), *Who Decides?: Conflicts of Rights in Health Care* (Clifton, NJ, 1982), 37, quoting John Gregory, *Observations on the Duties and Offices of the Physician* (1770), 18–19.
110. RCPSG MS 20/2/1/1, 86.
111. RCPSG MS 20/2/1/1, 87.
112. RCPSG MS 20/2/1/1, 87.
113. RCPSG MS 20/2/1/1, 87.
114. RCPSG MS 20/2/1/1, 247.
115. RCPSG MS 20/2/1/1, 248.
116. RCPSG MS 20/2/1/1, 247.
117. Allan Burns' death in 1813 was 'from a puncture got in dissection'; Geyer-Kordesch and Macdonald, *Physicians and Surgeons in Glasgow*, 407–8.
118. RCPSG MS 20/2/1/1, 250.
119. Fissell, 'Patient's Narrative', 103.
120. Roy Porter, 'The Rise of Physical Examination', in W. F. Bynum and Roy Porter (eds.), *Medicine and the Five Senses* (Cambridge, 1993), 180.
121. Jean E. Ward and Joan Yell (eds.), *The Medical Casebook of William Brownrigg, M.D., F.R.S. (1712–1800) of the Town of Whitehaven in Cumberland, Medical History* Supplement, no. 13 (London, 1993); Richard H. Ellis (ed.), *The Casebooks of Dr. John Snow, Medical History* Supplement, no. 14 (London, 1994); Duden, *Woman beneath the Skin*. Manuscript case notes of Scottish physicians are to be found in the libraries of the Royal College of Physicians and Surgeons of Glasgow, the Royal College of Physicians of Edinburgh, the National Library of Scotland, the Wellcome Library for the History and Understanding of Medicine, the Royal College of Physicians of London and the Royal Society of Medicine, London.
122. RCPSG MS 20/2/1/3, 230. Cleghorn's command of Shakespeare was, however, less good than his medicine. The 'starv'd Apothecary' makes his appearance in 'Romeo and Juliet'.
123. Thomas Percival, *Medical Ethics, or, A Code of Institutes and Precepts, Adapted to the Professional Conduct of Physicians and Surgeons* (Manchester, 1803), 59.
124. Nicolson, 'Gerard van Swieten', 57.
125. Geyer-Kordesch and Macdonald, *Physicians and Surgeons in Glasgow*, 1–35, 214–29.

10

Appealing to Nature: Geology 'in the Field' in Late Enlightenment Scotland

STUART HARTLEY

This chapter considers the role of natural knowledge gathered 'in the field' in order to understand the nature and place of geology in late Enlightenment Scotland. The question of fieldwork is also considered in relation to the presentation and discussion of field-based evidence within the social and scientific institutions of the time. In focusing on what in later decades was to become the scientific discipline or 'field' of geology, my concern is to raise questions about the hitherto rather neglected place of fieldwork in late Enlightenment science in Scotland. Several more particular themes inform what follows. I examine the nature of field enquiry in the work of Robert Jameson, Sir James Hall and Sir George Mackenzie. These men, in different ways, used the field as the testing site for debate between Wernerian neptunists and Huttonian plutonists over the primacy of either water or heat as agencies in the earth's formation. Fieldwork – being 'out there' in nature in order to secure facts with which to think theoretically – should also be seen, I suggest, as a matter of presenting material 'back here', in the social spaces and civic sites of Enlightenment institutions.

In what follows, then, I am also concerned to document something of the practices of fieldwork in the earth sciences at a key period in Scotland and in the sciences more generally. I shall also be arguing that fieldwork cannot be separated from the civic dimension of late Enlightenment science and that it was part of that historical intellectual enquiry in Scotland that was, in Christopher Berry's terms, at once 'natural' and 'civil'.[1] For other scholars, analysis of fieldwork in the earth sciences and in the physical bases to chemistry and to agricultural experimentation certainly has to be understood in relation to Enlightenment civic and cultural practices.[2]

Fieldwork also presents epistemological problems to do with the difficulties of securing reliable knowledge amidst an array of potentially overwhelming empirical data and of bringing it home, in the form of specimens, records or sketches, in ways that allow for its appreciation by others. As Lisbet Koerner has shown of Linnaeus, the field was a site for scientific endeavour and, for his students, a site for their instruction and moral guidance.[3] The field was also an arena in which the conduct of science was dependent upon the frailty of the human body, upon a reliance on the observing eye and upon the accuracy of recorded information which might make real sense only when dislocated from the field itself and brought back home to one's study or library. Knowing how science was conducted 'in the field' highlights questions to do with the knowing self, the authority of observation, the movement of knowledge and the credibility of one's claims to know. I do not pretend to do full justice to these issues in what follows. Yet signalling fieldwork as an area of scientific enquiry in the Enlightenment is to raise questions that go beyond 'the field' itself as a site for science's making and beyond the concern of any one field of knowledge such as the earth sciences.

'In the Field': Science as Fieldwork

Fieldwork has not been the subject of much attention in the history of Enlightenment science. For Henrika Kuklick and Robert Kohler, fieldwork has been marginalised in the history of science more generally because of its low status in science. Definitions of scientific rigour, they suggest, have largely been associated with the disciplinary standards of the laboratory as a 'site' for knowledge production. The field has traditionally been considered as an area of compromised work.[4] Yet as Kuklick and Kohler, their contributors and other scholars note, there are several reasons why attention should be given to the role of fieldwork in the making of natural knowledge. For Jan Golinski, for example, the field sciences ought to occupy a more important role precisely because science is shown not to be restricted to 'demarcated' locations.[5] Knowledge-producing practices are not bound to any delimited space and the field worker must be seen not as a practitioner apart from others by virtue of geographical separation but as someone whose credibility depends upon achieving what Bruno Latour has called 'translation' or 'cultural approbation'.[6] In this sense, methods applied in the field and the data collected are brought back for local use by a variety of audiences and used as 'representations' in the form of 'immutable mobiles' such as sketches, maps and specimens. Science is made by the bringing back of what lies 'beyond the laboratory walls', and the field is a crucial arena for securing natural knowledge because the uncertainty of the 'far away' and of the unknown near at hand is reduced to certainty by virtue of its classification, incorporation in current theory and by its display to others.

Dorinda Outram has shown that some prominent scientific figures, such as Georges Cuvier, were sceptical of claims to know in natural knowledge derived alone from fieldwork, and that he regarded knowledge gathered through exploration as secondary to the experiences of the sedentary naturalist. In an 1807 review of Alexander von Humboldt's *Tableaux de la Nature*, Cuvier noted how 'the field naturalist passes through at greater or lesser speed, a great number of different places, and is struck, one after the other by a great number of interesting objects and living things', but that he can only give to them 'a few instants of time and thus his observations are broken and fleeting'. For Cuvier, Outram explains, the formation of new knowledge in any continuous movement through space – the very essence of exploration – should be considered deeply suspect since 'the apparently heroic field observer lies in fact under the tyranny of the immediate'. More secure knowledge, Cuvier argued, depended upon being able to create an ordered structure out of the immediacy of the experience, upon not being tempted by nature's 'momentary vividness'.[7] Knowledge depended upon being sedentary, upon being in one's library or laboratory and upon not being mobile and bedazzled by an effulgent nature.

In addressing the place of fieldwork in Enlightenment science, we must also recognise that being in nature in order to understand it was associated with other established traditions in which nature was valued for its aesthetic and cultural significance. This is travelling in and through nature in a different sense. For many gentlemen, the Enlightenment brought with it a preoccupation with travelling, and travel had long been synonymous with notions of personal improvement and rational enlightenment.[8] It was considered highly beneficial to a young gentleman's education if his Grand Tour included observations of landscape features, the study of ruins and the art of the past. Many gentlemen undertook such pursuits, not just in Europe's great cities but also in remote, often inhospitable locations in which the interest expressed in nature was often more important than that shown in the natives or in themselves, indeed, in terms of their own self-improvement.[9]

For many late eighteenth-century figures, geology and travel were intimately connected. For Porter, such interests stemmed from an increasing conceptual and practical concern with 'strata' as a subject of investigation.[10] To make sense of strata, vast tracts of land were covered in order to connect successions of outcrops. The geological observation of landscape thus moved from three-dimensional perspectives and prospects to include conceptions of time and depth and from aesthetic sensation to scientific inference about age, process and origin. Rachel Laudan in particular has emphasised the shift from the mineralogists' concerns with single sites to the extensive examination of vast areas by earth scientists. This is not to see a straightforward distinction between uncritical description and later attempts at explanation. For Laudan, all geologists expected theory or 'system' to play a role. The issue was not whether to opt for theory or for empirical fact – it was the relationship between theory and fact and between looking and interpreting in the field that was important.[11]

It is the case, however, that by the late eighteenth century, rocks and their 'formations' identified in the field had taken over from the laboratory as sources of data for historical geology.[12] Fieldwork was understood as part of useful knowledge in terms of the practical concerns of mineralogists and, increasingly, was in use as a corroborative tool in the development of more consciously theoretical claims about the earth's origin and formative processes. For some, the field was the site of unproblematic empirical observation, the locus for a more strictly Baconian science. In the work of the Rev. Dr. John Walker, for example, fieldwork afforded the chance to record nature's bounty, not necessarily to speculate upon the causes of its diversity. As Professor of Natural History at the University of Edinburgh between 1779 and 1803, Walker was noted for lectures which did 'not deliver anything like theory, but merely a natural history of the earth'.[13] His peers were not at all of like mind. The travels and fieldwork of John Lightfoot, James Robertson, William Cullen and, in the earth sciences, of John Playfair, Sir James Hall and James Hutton to sites such as Glen Tilt, Arran, Arthur's Seat in Edinburgh and to Siccar Point all represent examples of the use of the field as sites of evidence to be used in theoretical speculation – about nature's diversity, the earth's origins or its modes of formation.

The period examined in this chapter marks something of a turning point between, on the one hand, the use of the field for empirical and utilitarian purposes and, on the other, its use in relation to geological theory. By the mid-eighteenth century in relation to what we may think of as geological fieldwork, knowledge gathering took place via largely Baconian classificatory practices and taxonomic principles. In mining and agriculture, for instance, practitioners embraced Enlightenment values of improvement based on what Walton has termed the 'Baconian project',[14] in relation, for example, to matters of land yield and to the recovery of raw materials. Yet by the mid-nineteenth century, geologists had come to attach an unprecedented importance to the methodological interrogation of 'the field' in verifying and/or in refuting 'grand theories' of the earth via, often, universally employed and standardised operational procedures based on the field-based ordering of strata and fossils. As Archibald Geikie was to put it in 1892, scientific importance attached to the use of the field as a 'site' for the production of theoretical natural knowledge was by then unchallenged:

> Let us turn from the lessons of the lecture room to the lessons of the crags and ravines, appealing constantly to nature for the explanation and verification of what is taught. And thus whatsoever may be your career in the future, you will, in the meantime cultivate habits of observation and communion with the free fresh world around you – habits which will give a zest to every journey, which will enable you to add to the sum of human knowledge, and which assuredly make you wiser and better men.[15]

This is not to suppose, however, that the changing significance of 'the field' in these terms may simply be explained as a shift from Baconian empiricism to its later place as the 'test bed' of deductive theorising. One way in which we may understand the nature of fieldwork and consider its scientific and social significance is through an empirical study of practitioners. The three men I consider briefly here each undertook excursions of a 'Grand Tour' nature and each, to varying degrees, saw and used 'the field' as a valid site for scientific rigour.

Robert Jameson, who succeeded Walker as Professor of Natural History at Edinburgh in 1804, initially favoured a philosophy of empirical observation and a rhetoric of topographic description. Latterly, however, Jameson used the field to accumulate evidence in favour of Wernerian theory: the field became his site for the corroboration of a belief held *a priori*. In contrast to him were Sir James Hall, a Huttonian theorist in whose work we can more immediately see the field being treated of as a theoretical geological 'type site' for the testing of Huttonian theory, and the more extreme Huttonian, Sir George Mackenzie, who used the field – as it happens, the igneous rocks of Iceland – almost entirely for the verificatory processes of hypothesis testing.[16]

An Enlightenment Scientist in the Field: the Case of Robert Jameson

Assessments of Jameson's work have considered his public scientific and civic roles: as museum keeper, teacher of natural history and as the 'acknowledged leader of the Scottish Wernerians' in public geological debate. Yet his activities as a field scientist, which have been relatively neglected hitherto, were also noteworthy.[17] David Allen has recently argued, for example, that 'thanks to Jameson, field teaching came to be an accepted speciality of the Edinburgh scientific curriculum'.[18] It is curious that Jameson's fieldwork has been so little studied, not least since the narratives of his excursions – which began in 1794 – show his fieldwork between that date and 1815 to incorporate many different styles and registers, namely Grand Tour rhetoric, utilitarian concerns and anecdotal accounts of travel experiences. Only after 1811 does Jameson's fieldwork incorporate in detail theoretical remarks aimed at the refutation of claims made by other men of science. In general terms, Jameson's work in the field initially demonstrates an attachment to Baconian empiricism and only latterly hints at an awareness of the role of 'the field' in providing what we may think of as 'laboratory space' for speculative earth scientists.

Jameson's commitment to practical encounter rather than to theoretical work is perhaps best exemplified in the introductory sections to his *Mineralogy of the Scottish Isles* (1800). Jameson noted:

> I fear that the theories of the formation of the earth, interesting as they are, often mislead the mind, and pervert the understanding; and those who yield to them become so involved in delusive speculations, so blind to fact and experience that, like Archimedes, they find but one thing wanting to raise worlds. Of the utility of this science there can be no question, more particularly when it is freed from the vague suppositions of the theorist.[19]

Even his most theoretically driven defence of the Wernerian doctrine of the Neptunian origin or rocks – *The Elements of Geognosy* (1808) – sought to uphold a utilitarian view of field science. Jameson referred to the work as 'a collection of the best ascertained facts respecting the aspect of the surface, and the structure of the crust of the earth', and noted that it contained 'only a few inferences, which appear to be legitimate, with regards to their mode of formation'.[20]

Unlike many of his contemporaries who travelled abroad, Jameson principally used Scotland as his 'type site', although, as I show below, the locus for his seeing the field differently in terms of theoretical speculation about the earth's formative processes was not Scotland but Saxony. Jameson's choice of localities throughout his Scottish tours owed much to his intentions to map the country fully as part of an intended mineralogy of Scotland. In 1797, and again in 1799, Jameson visited the Isle of Arran. Most of this tour is recounted as a record of factual observations, but his notebooks also contain elements written in the style of a Grand Tour, incorporating aesthetic descriptions of landscape with remarks about the scenery and the physicality of working in the field. In notes dated 14 August 1797, for example, Jameson notes that he was 'blessed with views of Jura, Isla, and Kintyre' and described the mass of islands as 'an immense colossus of stones piled up by some monstrous giant who in former times may have waged war with the gods'.[21] On Arran, Jameson was clearly aware of current theoretical issues in the earth sciences since he alluded to them, but he chose not to 'explain' the features he saw according to any theoretical criteria. For him, then, fieldwork had a descriptive purpose and a rhetorical style much like that of a traveller on the Grand Tour.

Jameson's tour of the Hebrides, undertaken between May and August 1798, was recorded in a similar style.[22] His interest in the utilitarian nature of the earth sciences was apparent in visits to a disused lead mine where calc-spar, galena and copper pyrite were present amongst the ruined workings. Unavoidably perhaps, Jameson also incorporated romantic descriptions of scenery alongside his concerns to document the mineralogy:

> Having gained the first summit, a wild scene was now before us; a rude, apparently inaccessible mountain side covered with debris of Basalt which lay scattered in wild ruin at its feet intermixed with the remains of strata of peat of here and here and there a few pools, marking the nat [sic] of the declivity of the season. The appearance of the catching lights upon the land far below, thro' the mist with difficulty we gained the summit of the hill, the rain pouring in torrents.[23]

Elsewhere he noted:

> Here a scene of the wildest grandeur is presented to the astonished eye; and near us great glens bounded by sides reddened as if [by] the powerful agency of fire and further torn by the furious torrents which collect here during storms. At a greater distance the dark, lurid Cullin Mts rising in awful majesty among black clouds convey by this tremendous covering a vast idea of their bounds and raising in our minds the most truly sublime

idea of the power and omnipotence of the framer of this wonderful globe.[24]

Jameson's description of the Western Isles was not 'Wernerian' in any strict sense of the term, nor did his observations point to any concern to gather proof in the field in favour of that theory. The lengthy descriptions and anecdotes that Bruce Hevly has described as the 'rhetoric of adventure', together with Jameson's romantic glosses, point to a natural philosopher recording the earth's features as he himself moved across them, but not explaining them. Rocks and strata were documented for their appearance rather than for what they might mean.[25]

Jameson returned to Arran in August 1799 where, in another journal, the descriptive record of local geological features was again incorporated alongside remarks upon the scenery. He also remarks upon the stratified appearance of the granite, a concept relied upon by Wernerians as evidence in support of the aqueous deposition of rock. Yet his remark is offered without recognition of any wider significance:

> Observed nothing particularly new until I came to Glen Sannicks. I however was more forcibly struck with the appearance of the granite upon several of the hills which presented a distinct stratified appearance and the strata either by being washed down by the rains seemed to surround the conical shaped summits of the mountain not unlike the leaves of an artichoke.[26]

He did, however, remark upon the use of this field evidence by others. His journals record other strata previously described by James Hutton in his work on the theory of the earth. Jameson did not comment on Hutton's interpretation, but it is clear that theoretical issues were not far from his mind. In recording what he termed 'veins of basalt' about five feet wide, Jameson mentioned Werner's theory but offered no opinions pertaining to the field evidence as 'proof':

> As to the great vein itself which I have described in the account of Arran, 1798, p. 138. I had now an opportunity of making several new observations, which will add considerable interest to the mineralogy of this curious spot. At first sight the part which is covered at high water, struck me as having more the appearance of the pitchstone being like strata inclined at from 30 and 60 but also the strata of the hard matter which bounded it on both sides having the same elevation and direction. A more complete examination demonstrated that these appearances were those of a stratified vein, as is the case with the lead vein at Argyleshire – and several similar appearances in Germany. It was upon these the celebrated Werner has founded his interesting theory of veins.[27]

In June and July of 1799, Jameson toured the Orkney Islands.[28] Aside from a single reference to Wernerian theory in a related excursion on the mainland between Cullen and Portsoy, his Orkney travels were likewise not concerned with fieldwork as the basis of theoretical explanation.

From 1802, however, it is possible to see evidence of a change in Jameson's conception of fieldwork methods and of their theoretical significance. Late in 1802, he had returned to Scotland from Freiburg where he had been a student under the tutelage of Abraham Gottlob Werner at the *Bergakademie* in Saxony. Jameson had enrolled at the academy in 1800, and he not only attended lectures there but also visited mines and travelled to Werner's chosen field sites to examine geological features.[29] Fieldwork abroad was to prove crucial to Jameson's use of the field at home.

Jameson's first Scottish excursion upon his return was to the county of Dumfries. The tone of Jameson's records and his purpose in being in the field is altogether different from his previous work. His notebooks strike the modern researcher as more professionalised, more clearly focused on the geological features and on the specifics of mining and the mineralogy of the county. Specifically, Jameson can be seen to be 'mapping' the landscape according to the system of his tutor, Werner. Lengthy descriptions in a romantic rhetoric are much less frequent. These differences are apparent in Jameson's *A Mineralogical Description of Scotland*, published in 1805, which was, essentially, an account of the geology of Dumfriesshire and which represented the first systematic geological description of any of Scotland's counties. The publication was dedicated to Werner and to his Irish mentor, the anti-Huttonian mineralogist, Richard Kirwan.[30]

It is clear, then, from Jameson's manuscript diary texts that field observations made after his time in Freiburg were concerned with explanation and employed a more rigorous terminology relating to theory-based descriptions of strata according to 'formations'. Landscape had taken on a temporal dimension with the classification of strata based on an 'age' scale. In a walk from Annan to Thornhill on 8 September 1802, for example, Jameson referred to the rocks as 'transition' (a Wernerian time period). Before 1802, he would have referred to the same feature according to a mineralogical schema as either 'slate' or 'wacke'.[31] In recording a granite vein, Jameson's language was distinctly more interpretative and was set more firmly within a nascent theoretical schema:

> The granite vein communicates with granite laying on the summit of the hill – the following sketch represents its mode of stratification. This granite I observed in great blocks on the roadside and on the side of the hills immediately after leaving Kenmore house and continues to nearly where the vein here mentioned is ... the transition strata st.7 dip N.60°.[32]

In other locations, Jameson's language explicitly articulated his explanations in Wernerian terms and did so in a more consciously precise mathematical style:

> Near the bottom of Burnswalk [there is] flinty sandstone. Does it belong to Floetz trap or the independent coal formation? The transition strata in the vicinity of Burnswalk in a rivulet that runs into a milk strata 8 dip angle. Farther down same burn str 9 Sth 65° still further down at st.8 dip sth 55°.[33]

Elsewhere, it is true, Jameson simply describes the potential apparent in local mineral deposits. He observed veins in antimony mines at Glendenning and commented upon the exploitable potential of the deposits at Leadhills and Wanlockhead. Of particular interest were the coal seams. Jameson recorded that he was 'examining the country with a view to ascertaining whether or not it contains coal'. But his utilitarian descriptions were overlaid with theoretical concerns. Jameson made observations about the coal through use of Wernerian terminology according to its age, with the 'Primitive' and 'Floetz-Trap' being Wernerian nomenclature given to different layers of rock according to their time of deposition in a universal ocean ('primitive' was the oldest):

> It frequently happens that the floetz trap formation is deposited over the independent coal formation which is still older but occurs in the same geognostical situations . . . It has to be said that wherever the old red sandstone occurs that coal is to be expected . . . the section which has frequently occurred of a tract of country on a discernible basis is composed of primitive rocks. On this it is to expected [in] the old red sandstone.[34]

This is to suggest, then, that Jameson's training at Freiburg led to a formative shift in his use of the field in geological reasoning. He still described the features of the country and was at pains to comment upon the economic utility of the mineralogy, but he also included much more than before on theoretical concerns. The field was a site of economic utility *and* a test bed for speculation upon the age and formation of the rocks themselves. Further, his Dumfries journal was less anecdotal and almost entirely devoid of topographic descriptions of landscape. Jameson's mineralogical excursions in Fife and Perthshire between 1811 and 1816 are evidence of a further marked shift toward the observation of geological features as a means to theoretical refutation.[35] His concern was by then to refute Huttonian theory, and his fieldwork records are more evidently those of a Wernerian:

> Observed a good section of what I call the fifth greenstone. The lowest part of the section is the floetz limestone of the district. It is covered with

a bed of bituminous shale which contains the beds of common clay and stone. On this rests a bed perhaps 4–8 inches thick of quartzy sandstone, this is covered by a kind of earthy greenstone – on it rests a bed of bituminous shale with a kind of hard sandstone, the upper part a bed of greenstone, the lower part brown and earthy as usual. This bed of greenstone is to be seen for a considerable distance westward in the direction of the dachy [sic] quarries.[36]

In his record of a limestone quarry associated with coal extraction at Alexander's Craig near Pettycur, Jameson made an observation that shows him opposing Huttonian theory:

The rocks on the shore from the SE end of Alexander's Craig are of trap formation. The lowest rock is basalt – in a few instances I observed below the basalt blackish coloured slate clay and mica as represented . . . the state given externally intersects with mica, bituminous shale and is sometimes in globular concretions. Below it limestone like that of Dodds head. Like bituminous shale petrifactions would externally carbonise. It is not indurated by the amygd but covers it on or in while it is in some place contained – appears rather a [?] formation with it and the idea which has some plausibility, in opposition to plutonic theory.[37]

A further indicator of Jameson's fieldwork taking a more 'theoretical' turn came in September 1813 on a visit to Glen Tilt and areas east of the River Dee. Knowing that James Hutton, Sir James Hall, John Playfair and Sir George Mackenzie had signalled the area as a 'type-site' to confirm ideas about Hutton's theory with reference to igneous intrusions and veins, Jameson went there in order to 'examine the various mineralogical aspects'. Jameson, who had hitherto refrained from recording in his field notebooks the opinions of others on mineralogical and geological matters, acknowledged Hutton's and Playfair's previous visits, and, on this occasion, offered contrary opinions to those of the Huttonians:

Remembered several of the beds of quartz in the gneiss or mica slate in our course upward. Again viewed the great bed of syenite which Dr Hutton, Prof Playfair and Sir James Hall maintain to be veins – one fact [which] is decisive again that [opinion] is that this does not rise through the faults as is the case with veins – on the contrary it has nearly the same level all along and is everywhere covered.[38]

This was an important moment to be making such claims. By 1812 the Huttonians had gained almost complete control of the Royal Society of Edinburgh, and many of its members used the society as a civic promotional vehicle for their views of the earth sciences. Jameson is here using fieldwork as a form of resistance to the civic authority of his opponents. For him, the field provided evidence that might allow one to oppose the dominant views of the plutonists. More than that, field evidence had a significance 'back home' for the social institutions in which science was the subject of civic debate. In order to offer a fuller picture of these issues, let me turn to two advocates of Huttonianism in the field.

Huttonian Science in the Field: Sir James Hall and Sir George Mackenzie

Sir James Hall is probably best known for his experiments that sought to clarify difficulties in Hutton's theory of the earth concerning the textural gradation of rocks and the action of variable cooling rates from an igneous melt. Hall was also a keen traveller in his earlier years.[39] His travels of 1781 and 1791 were, at one level, Grand Tours. But they were also more: they reveal a concern with understanding nature's works in the field to explain and to verify theories arrived at *a priori*. When in Brandenburg on 8 August 1783, he spoke of the interpretation of geological features with respect to their modes of formation:

> [There are] a quantity of large stones of various kinds of granite and roche [?] a stone with a number of round-ish parts harder than the rest that I take to be garnets. I tried to break off some but broke it all in the middle; this whole country seems to have been a vast shallow sea.[40]

Of course, like others of the time, he recorded his aesthetic appreciation of landscape, as in the following extract of his 1784 tour of the Alps:

> We saw Mont Blanc in all his majesty, I never saw his pre-eminence so strongly markt – the three peaks rose remarkably above the cloud that covered the lower parts and all the neighbouring hills . . . Sunday 11th, the place we had is a romantic spot the [mass] of stones is not above ten yards wide it is shaded by a fine oak and enclosed all round by woods and precipices on all sides.[41]

His Swiss travels were in part made sense of by local referents: in order to be understood, the unknown had to be compared to the known since, in first encounters with things foreign, memory is no secure guide to perception:

> I saw numerous stones that as far as I can judge were lavas and basalts of the purest kind, like that of Arthur's seat. I saw some several masses of quartz and roche [fenilte] but these were detached and in some places inclined in what I thought looked like volcanic matter. I came to Wapoule I found the post house built of such variety of stones that no two appeared to be alike. In general they looked like lava's and fine hard basalts of various colours and some were granit and some were pure

jasp. I broke off three pieces of the most remarkable to show to those that understand the matter or to examine myself . . . In general the country has the appearance I expect to see in a volcano. Black rocks and crossed with deep red and blue and a great quartz of clay probably from a decomposition of lavas, in some places calcareous rock.[42]

In February 1785, Hall first set eyes on Vesuvius, and made detailed accounts:

I had the opportunity of observing the formation of pumice in a gross way on this and other occasions, a quantity of matter is resolved into the state of vapour . . . at the time that the lava begins to become vitreous as one sees it sometimes in the stones that have been blown out from the volcano, those parts that have been melted into the glass if they have formed and have remained solid but near the surface the volatile matter has expanded and forced the glass out which one often finds on the stone.[43]

Later work that was to occupy Hall throughout the first part of the nineteenth century revolved around his interests in the textural gradation of rocks via the action of variable cooling rates from an igneous melt. While on his tour of France in 1791, Hall recorded having a conversation with the great French savant, Laplace, and noted:

I spoke to him of what we are doing in Scotland about the theory of the earth – he said from any astronomical observations that he was convinced the world must once have been in a fluid state by fire.[44]

For those of a Huttonian persuasion, however, fieldwork as a verificatory practice was most strongly apparent in the work of Sir George Mackenzie. For Mackenzie there was no doubt about the truth value of Hutton's theory. Mackenzie entered the field with the sole purpose of substantiating a view to which he already subscribed. Mackenzie's fieldwork in Iceland, and its publication in 1811 as *Travels in Iceland*, is perhaps the best documented and best studied aspect of what was a wide-ranging scientific life (see Chapter 11). Mackenzie turned to Iceland because its volcanic terrain offered the best opportunity to verify aspects of Hutton's theory based on the consolidation of rocks via the action of subterranean heat. His geological motives were clear from the outset: as he later remarked to the University Commissioners in 1826, 'It was with the view to ascertain the origin of the trap formation of rocks, and which was disputed between the Huttonians and Wernerians, that I went to explore Iceland'.[45]

The expedition set sail from Leith on 18 April 1810, returning in September of that year. From the outset, Mackenzie noticed that Iceland was almost entirely volcanic. He later noted: 'I saw no trace of stratification in Iceland all the rocks having been subjected to a state of perfect fusion'.[46] Central to

Mackenzie's intentions to support Huttonian thinking and to refute Wernerian theory was the discovery of obsidian and pumice under such conditions as precluded any but an igneous mode of formation:

> So intimately have Werner and his pupils interwoven the minerals in question with other rocks that if their origin shall be proved in any one instance to be volcanic, the whole of Werner's theory, as far, at least as it concerns unstratified floetz rocks, will be unhinged. The importance of a strict examination of them becomes therefore very great both to the supporters and the antagonists of the Wernerian geognosy.[47]

The search for obsidian as field evidence of a verificatory nature was of paramount importance:

> I allude to obsidian and pumice having been discovered in connection with rocks whose origin was not so apparent, they were likely to overset the great system that had been constructed by Werner who had no other resource but to deny altogether their igneous origin and to assert that they were of aqueous formation. Obsidian and pumice having been found connected with rocks, supposed by Werner to have been produced by water, is the only proof he has adduced to render of no avail the testimony of many philosophers, who have asserted from their own observations that these substances are distinctly amongst the productions of volcanoes if their origin [rocks] shall be proved in any one instance to be volcanic the whole of Werner's theory as far at least as it concerns unstratified floetz rocks will be unhinged.[48]

Mackenzie found his field evidence, and, thus, his theoretical confirmation at a site known as Torfajökull, twenty kilometres east of Mt. Hekla. Mackenzie did not hide his delight at having found the specimens:

> As if from a surface of glass, [it] delighted us with instantaneous conviction that we had attained one of the principal objects connected with the plan of our expedition to Iceland . . . Our discovery of Obsidian afforded us very great pleasure, which can only be understood by zealous geologists; and we traversed an immense and rugged mass of that curious substance with a high degree of satisfaction.[49]

Why were the rocks they discovered so significant? Mackenzie had found elements of other volcanic rocks that had embedded themselves within the obsidian. Causal processes could not be anything other than igneous:

> A remarkable and beautiful specimen the last of the series of a stream of obsidian. It is a mass of slag in a cavity in which some fusible matter has been included and reduced to the state of glass. The cavity is lined by it in

stalactitic masses and some of the matter has been drawn out to the fineness of hair. No operation of water could possibly produce these appearances.[50]

Mackenzie had found specimens that could not have been formed via the tenets of Wernerianism. Mackenzie's fieldwork had been successful in verifying *and* in refuting:

> There is no way left for Werner and his followers to evade the striking application of the facts I have ascribed to the Huttonian theory, but to deny their existence. I can hardly believe that any geologist who has paid attention to the specimens I have brought from Iceland, will shrink from the conviction of sense.[51]

In fact, the specimens did not themselves provide conclusive proof. Mackenzie attempted to use numerous interbedded tuffs and conglomerates to prove that amygdaloids were, in fact, lavas in origin. Wernerians believed amygdaloids to be aqueous in origin. Mackenzie had to concede through his own investigations that calcite could be dissolved by water charged with carbon dioxide. But he had also learned from Sir James Hall's laboratory experiments that amygdaloids could have been formed from sea-floor eruptions occurring at high pressures. The evidence of the field alone provided insufficient proof, but Mackenzie was able to show through associated evidence and, notably, from others' laboratory work, that amygdaloids were part of volcanic rather than aqueous processes. In so doing, Mackenzie both laid the foundations for current geological knowledge of Iceland and collected enough evidence to render the Wernerian theory almost untenable.[52]

What is also clear is that Mackenzie, well connected in the social and scientific institutions of late Enlightenment Scotland (see Chapter 11), used his social position to promote the values of Huttonian theory and of verificatory field science. As Steven Shapin has shown, the Royal Society of Edinburgh provided a platform for prominent Enlightenment figures to discuss the scientific issues of the day.[53] Field evidence lent weight to that body's civic concerns. Henry Holland, one of Mackenzie's assistants in Iceland, wrote a series of letters to his father both before and after the journey that cast further light upon Mackenzie's desire for wider public dissemination of theoretical geological knowledge. Holland was aware that the Icelandic specimens would be used as an anti-Wernerian 'weapon'.

Mackenzie seems to have offered his conclusive field evidence without any obvious retort from the Wernerians in the Royal Society:

> I expected some of the Wernerian party to rise here – but on this subject they all preserved a complete silence . . . When speaking of Obsidian, Sir G openly attacked the Wernerians and no reply was made.[54]

Further:

> On Monday evening Mr Allan's paper on the rocks of Salisbury Crags was read at the Royal Society to a crowded meeting – all the Wernerians were present but though several attacks were made upon their doctrines not a syllable was said by any one of the party.[55]

Mackenzie's book was no less scathing. He remarked:

> It is surely more reasonable to infer the existence of internal heat from the phenomena of volcanoes, than to believe in the Wernerian rising and falling waters of the globe without the evidence of any analogous fact whatever. It is plain, that as the Wernerians cannot prove by any analogy that the waters did actually rise and fall, or support their assumptions other than by round assertions, they cannot direct their attacks with consistency against what the Huttonians have assumed from evident and striking analogy.[56]

And, further:

> It would not be a difficult task to shew the weakness of the arguments by which the Wernerians imagine they have proved that the origin of obsidian and pumice is not igneous. They talk much of demonstration, but I must confess my utter inability to discover where anything like it is to be found either in the writings of Werner, or those of his followers. I shall not, therefore, take up the time of the reader with combating a phantom.[57]

Mackenzie's rather arrogant dismissal of Wernerian theory and, by implication, of Jameson's rather uncertain field practices, was not everywhere warmly received. The reception of Mackenzie's *Travels in Iceland* was to become, literally, a matter of public drama. For Andrew Wawn, Mackenzie's play *Gunnlaug's Saga – Properly entitled Helga or the Rival Minstrels of a Tragedy founded on an Icelandic Saga* – is more important for its contrasting claims to field evidence than it is a piece of theatre. The play was performed at Edinburgh's Theatre Royal in January 1812.[58] It was met with hostility by Wernerian geologists. Sir George wrote to Sir Walter Scott on the matter:

> The row last night I find has been complained by the Wernerians, I was left by two of my own class. I hear you would not be pleased with the manner in which the prologue was spoken – I regret that your goodness should have led you to be connected with a damned play; that is the only regret I feel. I resolved not to subject myself again to the ill nature of personal enemies and have withdrawn my play.[59]

The row over the civic performance of theory in geology derived from field evidence was mentioned by Sir Walter in a letter to the playwright Joanna Baillie:

> As for Sir Geo [sic] Mackenzie's play it was damned to everlasting redemption as elbow says and that after a tolerable fair hearing. The most mortifying part of the business was that at length even those who went as the authors friends caught the infection and laughed most heartily all the while they were applauding. The worthy Bart has however discovered that the failure was entirely owing to a set of chemists call'd Wernerians who it seems differ in their opinion concerning the cosmogony of the world from Sir George's sect of philosophers, the Huttonians. This has proved a most consolatory discovery to his wounded feelings.[60]

As Wawn has asked, why should 'a set of chemists called Wernerians' have seen fit to allow scientific controversy to enter into the theatre? The answer, I suggest, is to be found in the civic value of 'theoretically' driven field knowledge to those in Scotland who espoused Huttonian ideas. If the field site for the production of such knowledge was Iceland, the site for its consumption – indeed, for its verification *and* refutation – was the theatre. In this example, field evidence had certainly travelled, but it did not always have to be presented in its original forms – the rocks themselves – to be engaged with by the civic audiences of late Enlightenment science.

Conclusion

This chapter has argued that understanding the place of geological fieldwork as a form of scientific knowledge in late Enlightenment Scotland demands attention to the use and status of observational and descriptive traditions in the field, to questions relating to the theoretical concerns of practitioners and to differences in the audiences for science.

Hevly has certainly seen that for Victorian fieldwork 'part of the aim of the scientists as writers and explorers was for the benefit of popular audiences'.[61] This is not so readily apparent for the Scottish Enlightenment but it is clear that we should not think of the audiences for earth science as simply other natural scientists. Fieldwork in what was to become the science of geology was a matter of public utility – in mining and mineralogy, in natural history, and in the portrayal of topography. It is also clear, too, that fieldwork was central to the emergence of disciplines – scientific 'fields' themselves – and that being in the field depended upon displaying that knowledge – in texts, on the stage or by word of mouth – to others who had not had the opportunity of first-hand observation. For Jameson, the field was not, initially, a contested or problematic site. Topographical descriptions of Scotland's landscape could be accompanied by an essentially Baconian enquiry into the mineralogical utility of natural landscape or the classification of its individual features. Fieldwork was never isolated, however, from theoretical speculation as to the cause, either for the physico-theologians or for later theorists, of the earth's origin. If, for Jameson, the field was a site of reverence for the works of Providence, the field for Mackenzie was something one entered already theoretically equipped, a natural theatre for which the explanatory script was already written.

In only sketching the outlines of fieldwork in one disciplinary area, this chapter has also suggested that more needs to be known about the conduct of Enlightenment science 'in the field' in other subject domains – in botany or in natural history, for example – and more known of its treatment by audiences elsewhere. Doing science in the field was never divorced from an appreciation of the environment in which one worked: the contrast between 'scientist' as both dispassionate recorder of a compliant nature ready to yield its secrets and as a typologically different creature from the impassioned 'artist' only too aware of subjective feelings, is one drawn only in caricature.

The field is, if anything, a more difficult site for securing knowledge than the laboratory or the library. Nature yields insights from direct encounter only grudgingly. The validity and utility of knowledge gathered 'out there' may depend upon being made sense of somewhere else entirely. In effect, knowledge making depends upon bringing together the findings and practices from different sites. The field is an ambiguous place, depending as it does upon the 'translation' and impact of theory and upon the export of reliable specimens which stand, in summary form, for grander claims. As Kuklick and Kohler conclude, 'ambiguities make field sites interesting and creative work places, both for scientists and for historians and sociologists of science'.[62] Although the research task for historians of the field may be formidable because of these ambiguities, it should not deter us from attempting to understand more fully this site for knowledge making in thinking about the nature and location of science in the Enlightenment.

Notes

1. Christopher Berry, *Social Theory of the Scottish Enlightenment* (Edinburgh, 1997), 3. For more general histories of geology in this period see: Dennis R. Dean, 'James Hutton and his Public, 1785–1802', *Annals of Science* 30 (1973), 89–105; idem, *James Hutton and the History of Geology* (Ithaca, NY, and London, 1992).
2. Jan Golinski, 'Science *in* the Enlightenment', *History of Science* 24 (1986), 411–24; idem, *Science as Public Culture: Chemistry and Enlightenment in Britain, 1760–1820* (Cambridge, 1992). On lecturing in agriculture and in natural history as a civic practice of natural knowledge, see Charles W. J. Withers, 'William Cullen's Agricultural Lectures and Writings and the Development of Agricultural Science in Eighteenth-Century Scotland', *Agricultural History Review* 37 (1989), 144–56; idem, 'On Georgics and Geology: James Hutton's "Elements of Agriculture" and Agricultural Science in Eighteenth-Century Scotland', *Agricultural History Review* 42 (1994), 138–49.

3. Lisbet Koerner, *Linnaeus: Nature and Nation* (Cambridge MA, 1999); *idem*, 'Daedalus Hyperboreus: Baltic Natural History and Mineralogy in the Enlightenment', in William Clark, Jan Golinski and Simon Schaffer (eds.), *The Sciences in Enlightened Europe* (Chicago and London, 1999), 389–422. In geology, fieldwork is the subject of attention in Rachel Laudan, *From Mineralogy to Geology: The Foundations of a Science, 1650–1830* (Chicago and London, 1997). For a later period of fieldwork and geology, see Martin J. S. Rudwick, *The Great Devonian Controversy: The Shaping of Scientific Knowledge Among Gentlemanly Specialists* (Chicago and London, 1985); James A. Secord, *Controversy in Victorian Geology: The Cambrian-Silurian Dispute* (Princeton, NJ, 1986); David Oldroyd, *The Highlands Controversy: Constructing Geological Knowledge Through Fieldwork in Nineteenth-Century Britain* (Chicago and London, 1990); *idem*, *Thinking about the Earth: A History of Ideas in Geology* (London, 1996). For a partial discussion of this question for the eighteenth century, see Victor A. Eyles, 'The Extent of Geological Knowledge in the Eighteenth Century', in Cecil J. Schneer (ed.), *Toward a History of Geology* (Cambridge MA, 1969), 159–83.
4. Henrika Kuklick and Robert E. Kohler, 'Introduction', in Henrika Kuklick and Robert E. Kohler (eds.), *Science in the Field*, Osiris, 2nd series, 11 (Chicago and London, 1996), 1–16.
5. Jan Golinski, *Making Natural Knowledge: Constructivism and the History of Science* (Cambridge, 1998), 98.
6. Bruno Latour, *Science in Action: How to Follow Scientists and Engineers through Society* (Cambridge, MA, 1987).
7. Dorinda Outram, 'On Being Perseus: New Knowledge, Dislocation and Enlightenment Exploration', in David N. Livingstone and Charles W. J. Withers (eds.), *Geography and Enlightenment* (Chicago and London, 1999), 281–95.
8. This is particularly true of artistic representations of geological landscape whilst on the Grand Tour; see Martin J. S. Rudwick, *Scenes from Deep Time: Early Pictorial Representations of the Pre-Historic World* (Chicago and London, 1992); John M. Wyatt, *Wordsworth and the Geologists* (Cambridge, 1995).
9. Jeremy Black, *The British Abroad: The Grand Tour in the Eighteenth Century* (London, 1992). Observations of landscape and nature in relation to romantic sensibilities became increasingly common from the early nineteenth century, yet whilst the private diaries of scientists included writings in a romantic tone, such rhetoric was seldom reflected in their scientific publications; see Bruce Hevly, 'The Heroic Science of Glacier Motion', in Kuklick and Kohler, *Science in the Field*, 66–87.
10. Roy Porter, *The Making of Geology: Earth Science in Britain, 1660–1815* (Cambridge, 1977), 119–21.
11. Laudan, *From Mineralogy to Geology*, 101.
12. Laudan, *From Mineralogy to Geology*, 101; Porter, *Making of Geology*, 176; Secord, *Victorian Geology*, 14.
13. Harold W. Scott (ed.), *Lectures on Geology by John Walker* (Chicago and London, 1966), 181.
14. Craig Walton, 'Hume's *England* as a Natural History of Morals', in Nicholas Capaldi and Donald Livingston (eds.), *Liberty in Hume's History of England* (London, 1990), 25–52.
15. Archibald Geikie, *Geological Sketches at Home and Abroad* (London, 1882), 271.
16. On Jameson, see James A. Secord, 'The Discovery of a Vocation: Darwin's Early Geology', *British Journal for the History of Science* 24 (1991), 133–57; *idem*,

'Edinburgh Lamarckians: Robert Jameson and Robert E. Grant', *Journal of the History of Biology* 24 (1991), 1–18; Stuart Hartley, 'Robert Jameson, Geology and Polite Culture, 1796–1826: Natural Knowledge Enquiry and Civic Sensibility in Late Enlightenment Scotland' (Unpublished Ph.D. thesis, University of Edinburgh, 2001).

17. The earliest account of Jameson's fieldwork is in Jessie M. Sweet, 'Robert Jameson and Shetland: A Family History', *Scottish Genealogist* 16 (1969), 12–19.
18. David E. Allen, 'Walking the Swards: Medical Education and the Rise and Spread of the Botanical Field Class', *Archives of Natural History* 27 (2000), 335–68.
19. Robert Jameson, *Mineralogy of the Scottish Isles: With Mineralogical Observations Made in a Tour through Different Parts of the Mainland of Scotland and Dissertations Upon Peat and Kelp* (Edinburgh, 1800), 11. See also his *Outline of the Mineralogy of the Scottish Isles*, trans. W. Meuder, 2 vols. (Edinburgh, 1800).
20. Robert Jameson, *A System of Mineralogy*, 3 vols. (Edinburgh, 1804–1808), 1: Introduction (unpaginated).
21. Edinburgh University Library Special Collections, MS. Dc. 7. 126, 14 August 1797. For Jameson's published account of his Arran fieldwork, see his *Outline of the Mineralogy of the Shetland Islands and of the Island of Arran* (Edinburgh, 1798).
22. Jameson's two-volume field notebooks, 'Journal of a Tour through the Hebrides, begun 22 May 1798', are in Edinburgh University Library Special Collections, MS. Dc. 7. 127 and MS. Dc. 7. 128.
23. EUL, MS. Dc. 7. 128, 26 July 1798.
24. EUL, MS. Dc. 7. 128, 11 August 1798.
25. Hevly, 'The Heroic Science of Glacier Motion', 67. On a later example of fieldwork undertaken with no clear initial understanding as to final purpose, see Jane Camerini, 'Wallace in the Field', in Kuklick and Kohler, *Science in the Field*, 44–66.
26. EUL, MS. Dc. 7. 129, 'Notes of a Visit to Arran in August 1799'. Jameson's account of granite and veins was first published in 1802: 'On Granite', *Journal of Natural Philosophy, Chemistry and the Arts* 2 (1802), 225–33. For additional work by Jameson on this important topic, see Robert Jameson, 'On the Supposed Existence of Mechanical Deposits and Petrifactions which have been Discovered in the Newest Floetz Trap Formation', *Journal of Natural Philosophy, Chemistry and the Arts* 3 (1802), 13–20; *idem*, 'Examination of the Supposed Igneous Origin of the Rocks of the Trap Formations', *Journal of Natural Philosophy, Chemistry and the Arts* 3 (1802), 111–19.
27. EUL, MS. Dc. 7. 129, Wednesday (undated).
28. Robert Jameson, *Mineralogical Travels Through the Hebrides, Orkney and Shetland Islands and Mainland of Scotland*, 2nd edn, 2 vols. (Edinburgh, 1813). Jameson's manuscript journal of this tour is in Edinburgh University Library Special Collections, MS. Dc. 7. 130, 'Journal of My Tour Thro' the Orkney Islands begun June 8 1799'. During this tour (which lasted from 8 June until 14 July) Jameson spent time making mineralogical and geological observations on the Scottish mainland.
29. Laurence Jameson, 'Biographical Memoir of the Late Professor Jameson', *Edinburgh New Philosophical Journal* 67 (1854), 1–49.
30. Robert Jameson, *A Mineralogical Description of the County of Dumfries* (Edinburgh, 1805). For Jameson's engagement with Wernerian theory at this point, see

also Jessie M. Sweet and Charles Waterston, 'Robert Jameson's Approach to the Wernerian Theory of the Earth, 1796', *Annals of Science* 23 (1967), 81–95.
31. EUL, MS. Dc. 7. 130, 8 September 1802.
32. EUL, MS. Dc. 7. 130, 9 September 1802.
33. EUL, MS. Dc. 7. 130, 9 September 1802.
34. EUL, MS. Dc. 7. 130 (no date).
35. EUL, MS. Dc. 7. 131.
36. EUL, MS. Dc. 7. 132. As with MS. Dc. 7. 131, this notebook is unpaginated until the very end.
37. EUL, MS. Dc. 7. 132, 'Petty Cur'.
38. The mineralogical diary of the 1813 excursion is EUL, MS. Dc. 7. 133. Jameson's diary, although part of a series entitled 'Mineralogical Walks in Fife', was the only one where Fife was not the area examined. In fact Jameson explored mineralogical features in and around Perthshire, eventually returning to Edinburgh and locality including investigations near Roslin and in the Pentland Hills. The quote here is from EUL, MS. Dc. 7. 133, 23 August [1813].
39. For more on Hall's work in chemistry, see Victor A. Eyles, 'Sir James Hall, Bt. (1761–1832)', *Endeavour* 20 (1961), 210–16; *idem*, 'The Evolution of a Chemist: Sir James Hall', *Annals of Science* 19 (1963), 153–82.
40. National Library of Scotland, MS 6324, f.48.
41. NLS, MS 6325, ff.108, 120.
42. NLS, MS 6326, ff.73, 174.
43. NLS, MS 6327, f.17.
44. Sir James Hall to Laplace, 1791, NLS, MS 6332, f. 79.
45. *Evidence, Oral and Documentary, Taken and Received by the Commissioners for Visiting the Universities of Scotland*, 4 vols. (London, 1837), 4:620.
46. George Mackenzie, *Travels in the Island of Iceland during the Summer of the Year MDCCX* (Edinburgh, 1811), 361. See also John Playfair, 'Review of Sir George Steuart Mackenzie's *Travels in Iceland* (1811)', *Edinburgh Review* (1812), 416–35; Martin A. Peacock, 'The Geology of Iceland: The Pioneer Work of a Scottish Geologist', *Transactions of the Geological Society of Glasgow* 17 (1925), 185–203.
47. Mackenzie, *Travels*, 366.
48. Mackenzie, *Travels*, 366.
49. Mackenzie, *Travels*, 242–3; Peacock, 'The Geology of Iceland', 300.
50. Martin A. Peacock, 'A Contribution to the Petrography of Iceland: Being A Description of the Mackenzie Collection of 1810', *Transactions of the Geological Society of Glasgow* 17 (1925), 271–333. Peacock's analysis defines the specimens as 'Acid lavas from the post glacial formation'. The obsidian sample identified by Mackenzie is now in the Hunterian Museum, University of Glasgow, number R375.
51. Mackenzie, *Travels*, 383.
52. Peacock, 'The Geology of Iceland', 196.
53. Steven Shapin, 'Property, Patronage and the Politics of Science: The Founding of the Royal Society of Edinburgh', *British Journal for the History of Science* 7 (1974), 1–41; *idem*, 'The Audience for Science in Eighteenth-Century Edinburgh', *History of Science* 7 (1974), 95–121.
54. NLS, MS Acc. 7515, 6 February 1810 and 28 January 1811.
55. NLS, MS Acc. 7515, 10 March 1811.
56. Mackenzie, *Travels*, 364–65, see also Andrew Wawn (ed.), *The Iceland Journal of Henry Holland* (London, 1987), 149.
57. Mackenzie, *Travels*, 366.

58. Andrew Wawn, 'Gunnlaugs Saga Ormstunga and the Theatre Royal, Edinburgh, 1812: Melodrama, Mineralogy and Sir George Mackenzie', *Scandinavica* 21 (1982), 140–51. The manuscript of Mackenzie's play *Helga or the Rival Minstrels* is held in the Larpent Collection of the Huntington Library.
59. Sir George Steuart Mackenzie to Sir Walter Scott, 1812, NLS, MS 3882, f.1.
60. Wawn, 'Gunnlaugs Saga Ormstunga', 149.
61. Hevly, 'The Heroic Science of Glacier Motion', 46.
62. Kuklick and Kohler, 'Introduction', 14.

11

Late Enlightenment Science and Generalism: The Case of Sir George Steuart Mackenzie of Coul, 1780–1848

CHARLES D. WATERSTON

George Steuart Mackenzie, 7th Baronet of Coul, was born in Edinburgh on 22 June 1780, the only child of Sir Alexander Mackenzie, Major-General in the Bengal Army, and Katherine Ramsay, daughter of a Leith merchant.[1] Educated men of the late Enlightenment were expected 'to reflect on the basis of knowledge' and see science as an intellectual pursuit with direct social implications.[2] Education in Scotland therefore shunned premature specialisation so that scholars might have the breadth of learning to fulfil these expectations. For Mackenzie, one branch of learning was informed and counterbalanced by another. His writings on chemistry, agriculture, geology, archaeology and phrenology; his improvement of his estate with its social implications and consequences; his promotion of institutional scientific endeavour: all embrace features that characterised intellectual enquiry in the Scottish Enlightenment.

Yet, during Mackenzie's lifetime, the generalism of the gentleman scholar yielded to specialism, regional emphasis and professionalism. This change has been charted by many historians. In a European context, Roger Hahn introduced the concept of a second scientific revolution occurring at the beginning of the nineteenth century that involved specialisation and professionalisation.[3] Of particular relevance to Mackenzie was that specialisation in the natural sciences that Susan Faye Cannon has seen as beginning from the 1830s.[4] While J.R.R. Christie characterised the intellectual culture of the Scottish Enlightenment as generalist, he also recognised that that view was challenged by the emergence of specialisms in the early nineteenth century.[5] Roger L. Emerson, however, traced the rise of the specialist and decline of the generalist in Scottish science to well before Mackenzie's time.[6] Specialist terms may be misused, however, in describing Enlightenment scholars. To call James Hutton a geologist or Joseph Black a chemist is to ignore the agricultural and philosophical writings of the former and neglect the geological and medical interests of the latter and so diminish both. In the same way, Mackenzie's work has been described by specialists in terms of their own discipline, and they have ignored, or been unaware of, his contribution to others.

Although a full exposition of Mackenzie's work cannot be given here, this chapter signals the breadth of his accomplishments and seeks to restore something of his stature as an educated man of his period. It points to his interdisciplinary insights, and shows how his influence in one field effected change in another. The chapter is also concerned to understand how Mackenzie has been seen, both by his contemporaries and by modern scholars. Impressions of Mackenzie as a person differ. For John Prebble he was 'an ass'. Byam Shaw concluded that he was 'a man of strong opinions and cantankerous disposition'. His contemporary, Sir Charles Lyell, found him 'very gentlemanlike and intelligent and full of information'.[7] In examining the ways in which Mackenzie acted to promote the interests of science – in Edinburgh, or on his own estate, and in terms of certain subjects – the chapter also aims to shed light upon the social and intellectual significance of science in the late Enlightenment.

Mackenzie and Practical Chemistry in Late Enlightenment Scotland

At the High School of Edinburgh, until he succeeded to the baronetcy in 1796, Mackenzie was taught by Alexander Christison, classical scholar and mathematician, and the Rector, Dr Alexander Adam.[8] In December 1796, Mackenzie enrolled in the chemistry class at Edinburgh University, which was taught by Joseph Black and Thomas Charles Hope.[9] For thirty years Black, pioneer quantitative chemist, discoverer of carbon dioxide and latent heat, charismatic teacher and correspondent of Lavoisier, had occupied the chemistry chair at Edinburgh.[10] Hope, Black's former pupil, had recently returned to Edinburgh from Glasgow: like Black, he was a popular teacher. Lecture notes by one of Mackenzie's classmates for the 1796–97 session show that Black and Hope did alternating blocks of teaching.[11] Mackenzie would have heard Black deal with inflammation and with the calcination of metals in terms of Lavoisier's chemistry.[12] Hope was the sole professor when Mackenzie next attended the class in the 1798–9 session.[13]

It is unclear whether the young Mackenzie knew that these lectures were part of what was later seen as a 'revolution' in chemical theory. Hope first taught Lavoisier's ideas to his Glasgow students in 1787–88. Hope had been influenced by his friend Sir James Hall of Dunglass, who had visited Lavoisier in Paris in 1786 and delivered what is thought to have been the first account of Lavoisier's new chemistry in Britain to the Royal Society of Edinburgh in the spring of 1788.[14] Mackenzie would also have met Hall at this time. Hall had been an associate of James Hutton, whose *Theory of the Earth* transformed the earth sciences. Hall had sought proof of Hutton's contentions by experimenting with rock melts and, in 1797, following Hutton's death, he resumed these at Hope's suggestion and with his advice.[15] During 1798–99, Mackenzie found favour with Hall, for in February 1799 Sir James proposed Sir George for Fellowship of the Royal Society of Edinburgh, an advancement seconded by Hope and John Playfair, the Society's secretary. Mackenzie was elected on 24 June, two days after his nineteenth birthday.

Mackenzie's involvement with Hall in experimental science had its counterpart elsewhere. In 1799 a paper entitled 'Account of an experiment made with a view to prove whether the experiment made at the Polytechnic School, respecting the conversion of Iron into Steel by means of the diamond, is conclusive', was published by David Mushet, an employee of the Clyde Iron Works.[16] Mushet's paper criticised the experiment reported to the French National Institute by Louis Bernard Guyton de Morveau, Professor of Chemistry at the Ecole Polytechnique.[17] It has been commonly accepted that Mackenzie came to the notice of the scientific world in 1800 by obtaining 'decisive proof of the identity of diamond with carbon'. He did so, indeed, in a series of experiments on the formation of steel by the combination of diamonds with iron which he conducted in response to Mushet's 1799 paper.[18]

In order to assess Mackenzie's contribution in this context, we must note two things. First, it had been discovered in 1786 that cast iron, wrought iron and steel differed only in their carbon content.[19] Secondly, there had been a rapid change in understanding the nature of diamond. Guyton had been experimenting on diamond since 1785. In a series of experiments in 1797–98, he burned diamond in oxygen in the focus of a lens and discovered that the diamond combined with the oxygen in fixed proportions.[20] By 1799 Guyton 'regarded diamond as the only pure form of carbon'.[21] If diamond is 'carbon', French chemists thought it should be possible to convert soft iron into steel by cementation with diamond. The successful experiment was conducted at the Ecole Polytechnique, of which Guyton was then acting director, by J.-F. Clouet, J. J. Welter and J. N. P. Hachette and reported to the French National Institute by Guyton in August 1799.[22] Mushet read of the experiment in the November issue of Tilloch's *Philosophical Magazine*, which printed Mushet's criticism of it in a later issue of the same volume. Mushet contended that in the French experiment the carbon necessary for converting the iron into steel had not come from the diamond but from the solution of carbon in caloric permeating the porous materials of the earths of the vessels. He believed that his own experiments had produced steel without diamond from wrought iron heated in this way in contact with sand only.

Many of these French chemists were known personally to Edinburgh chemists and respected by them. Black, who was born in Bordeaux, was in touch with his French counterparts.[23] Hall had toured Europe between 1783 and 1786 and met many leading scientists, notably Lavoisier and his associates in Paris in 1786.[24] Hope visited France in 1788. It is likely, therefore, that they had faith in the experiment of the chemists of the Ecole Polytechnique and wished to verify and extend the experimental evidence and so refute Mushet's criticism of it. Mackenzie was at hand, newly fledged from Hope's class and of independent means, and we may speculate that they asked him to undertake the experiments as an appeal to his scientific capacity, or perhaps as an act of social patronage. In any event, Mackenzie first experimented on diamonds at the end of 1799, sometimes in the presence of Sir James Hall, Lord Webb Seymour and Dr. Robert Kennedy.[25] Kennedy was a chemist who taught chemistry classes at Edinburgh and worked closely with Hall.[26] Following

Kennedy's early death in 1803, Mackenzie wrote of the practical help he had received from 'my deceased friend Dr Kennedy, in the art of mineral analysis'.[27] Hall was by then doing his rock melts and using 'the great reverberating furnace at Mr Barker's iron foundry'.[28] In his diamond work, Mackenzie used similar experimental methods to Hall's and also employed Barker's furnace.

Mackenzie repeated the French experiment with complete success at a temperature of 150 Wedgwood (greater than 1150–1250C).[29] The same treatment of iron, without the diamond, did not afford steel, as Mushet had alleged in his 1799 paper. Experiments were then undertaken to show that carbon did not penetrate the vessels in strong fire, as Mushet had supposed. Mackenzie also reduced oxide of iron by diamond powder. He presented the results of his 'Experiments on the Combustion of the Diamond' to the Royal Society of Edinburgh on 3 February 1800 and submitted the paper to *Nicholson's Journal of Natural Philosophy, Chemistry and the Arts* on 1 May.[30] It is alleged that Mackenzie made free use of his mother's diamonds in these experiments.[31] This story may be apocryphal, but Mackenzie, concerned about the purity of the diamond used, certainly noted that 'All the diamonds used in these experiments were cut and polished'.[32]

Mushet left the Clyde Iron Works in 1800. In 1801, while founding the Calder Iron Works, he discovered the blackband ironstone that he showed was capable of being worked economically.[33] In 1802, when he returned to the diamond controversy, he accepted that diamond was a form of carbon but required more rigorous proof of its union with iron to form steel and questioned the nature of the products of Mackenzie's experiments.[34] By then, Mackenzie's conclusions had been generally accepted and Sir George did not reply to Mushet's criticisms.

Mackenzie, Agricultural Science and Improvement

Sir George reached his majority in 1801 and assumed responsibility for his estates in mid- and wester Ross. In the same year he was admitted to the Highland Society. In June 1802 he married Mary, daughter of Donald Macleod of Geanies, a well-known improver and later Sheriff of Ross-shire.[35] To fit him for his responsibilities as a landowner, Mackenzie returned to Edinburgh University in December 1802 to attend Andrew Coventry's class in Agriculture and Rural Economy, and Alexander Monro's Anatomy class.[36] Coventry's syllabus covered the practices of plant and animal husbandry and estate management and his class was popular among those with a direct involvement in agriculture – young landowners, the sons of practical farmers and lawyers responsible for the management of estates.[37] Mackenzie's knowledge of anatomy stood him in good stead when applied to animal husbandry. Coventry was developing his own thoughts on that subject when Mackenzie attended his class. Coventry deplored 'Promiscuous Breeding' and in his published work on the subject wrote: 'select the best and breed from them'.[38]

It has been claimed that in eighteenth-century agriculture the absence of science meant that knowledge was unorganised, diverse and unreliable.[39] Mackenzie sought to apply Enlightenment ideas, and science in particular, to the advance and consolidation of Scotland's husbandry. While still a student, he proposed to Archibald Constable, the Edinburgh publisher, that he should produce an anatomical atlas to remedy the ignorance of farmers in treating sick or injured animals. To be written by a skilled anatomist, it should have:

> tables of the Anatomy of all animals used in rural affairs – describe the parts most liable to disease – the morbid appearance of such – the external symptoms occasioned by them & the means most likely to remove the disease . . .
> These tables would also show the direction & situation of blood vessels, by which Farmers would be enabled in the case of external injury, to find & take up any important channel of blood which may be hurt or be ruptured.[40]

The suggestion was not taken up, but it illustrates Mackenzie's early concern that scientific knowledge – in this case anatomy – should yield practical outcomes in animal husbandry.

In the following year, he again urged the application of science – now chemistry – to a practical problem in agriculture. Soil chemistry and plant nutrition had long been a concern of agriculturalists and Mackenzie's Professor of Agriculture, Andrew Coventry, had been influenced in this field by such men as Lord Kames, Francis Home and William Cullen. Liming of fields had become widespread and farmers were 'unknowingly bestowing their money, time and labour on a disguised substance, the half of which is sand or clay'. A correspondent to *The Farmers' Magazine* had offered directions for assessing the quality of lime by dissolving equal quantities of the samples in hydrochloric acid 'which may be obtained from any laboratory' – that which leaves least sediment when dissolved being the best.[41] Sir George responded that the determination was not so easy. First the acid must be pure and free of sulphuric acid otherwise insoluble sulphates will be precipitated. Secondly, limestone may contain soluble impurities. The amount of residual sediment after solution was not itself reliable: 'Without a regular analysis by a skilful chemist, the real value of limestone cannot be ascertained'. He followed his claim with an analysis of marl 'to show such of your readers as may not have studied chemistry the intricacy even in the analysis of a mineral whose decomposition requires less dexterity than perhaps any other'.[42] Mackenzie thus urged refinement within extant practices in agricultural chemistry: practical experimentation was geared to socially useful ends.

Mackenzie's major contribution to agriculture, however, is as a leading thinker in livestock management. Following the American Revolution, fine Spanish Merino wool was being bought in preference to British wool. Sir

Joseph Banks had been convinced of the merits of Merino wool by French manufacturers contacted through William Eden, later Lord Auckland, as early as 1780. Banks wished to promote British self-sufficiency in wool and, with the aim of improving the quality of British wool, obtained royal approval to introduce Merino sheep into England with Kew Gardens as the experimental base for the Royal Flock. It was not until he was appointed ambassador to the Spanish court in 1789, however, that Auckland could explore the possibility of obtaining Merino sheep for Banks. Auckland's plans for procuring them obtained royal approval in 1790 and, in October of the following year, a flock of Merinos arrived in England. This was the nucleus of the royal breeding flock at Kew from which sheep were sent to breeders to establish strains and crosses that would flourish in Britain.[43] To promote this improvement, Banks published an article on extending Merino sheep into all parts of Great Britain which was printed in *Nicholson's Journal* in 1800 only four months after Mackenzie's diamond paper appeared there.[44] In common with many contemporaries, Sir George appreciated the importance of cross- and selective sheep breeding and the recognition of 'quality characteristics'. Carter has argued that he purchased Merino sheep from the King's flock in 1807 because of the published reports of Banks and Caleb Hillier Parry in the south and, in Scotland, by the experience of Sir James Montgomery of Stobo and General William Robertson of Lude.[45]

Mackenzie's *Treatise on the Diseases and Management of Sheep* was published in 1809, with an *Appendix, containing documents exhibiting the value of the Merino Breed of Sheep, and their progress in Scotland*.[46] For Carter, this work is important because it was the first to gather together the reports on the Royal Flock printed by Banks between 1800 and 1809.[47] Animal husbandry had more than economic significance. Its scientific importance was then in the discerning eye and understanding of the breeder rather than in the practices of contemporary scientists. Darwin later wrote that the importance of selective breeding and the skill of the breeder 'consists in the great effect produced by the accumulation in one direction, during successive generations, of differences absolutely inappreciable by an uneducated eye . . . Not one man in a thousand has accuracy of eye and judgement sufficient to become an eminent breeder'.[48]

In his last address to the Royal Society of Edinburgh, delivered on 5 April 1847, Mackenzie drew on his lifetime's experience as a breeder to deny the Lamarckian belief in the fixity of acquired characters and the orthogenetic view of an inherent tendency to progressive development. A report of the address states:

> His object was to shew there was no analogy in nature, rendering a prospective law of progressive development from lower to higher types probable . . . in all cases in which variation took place among domestic animals and vegetable productions, the varieties, though they might be regarded as improved or new, were not permanent, but required selec-

tion and care to preserve them . . . The stronger, because cultivated (not newly developed), intellect and energy of [the races of] Europe are subduing or extirpating the inferior and weaker races in other parts of the world; but the substitution of a better race in this manner is not progressive development.[49]

This was written eleven years before Darwin first proposed the mechanism of natural selection in evolution or the survival of the fittest.

The perfection of the Cheviot sheep for the Highland economy has been described as 'a technological change in no less a sense than the great mechanical changes in contemporary industry'.[50] In 1839, Patrick Sellar called his Cheviots 'machines for converting highland pastures into wool and mutton'.[51] Sir John Sinclair 'convinced the world of the superior quality and higher value of Cheviot wool' in calculating that by means of Cheviots, 'instead of £300,000 of lean cattle, the Highlands could export £600,000 worth of mutton and £3,600,000 of woollens'.[52] From the 1790s, Cheviot sheep had proved successful in other parts of Scotland but had not yet been established on the hills of Ross-shire.[53] Sir George pioneered their introduction into that county. In 1808 he bought a large stock of Cheviots for 'an extensive sheep walk which he has lately taken into his hands in Ross-shire' and hoped that his example would be followed.[54]

In his *General View of Agriculture of the Counties of Ross and Cromarty* (1810), Mackenzie emphasised the 'skill and patient attention which are absolutely necessary for the good management of a sheep stock'.[55] He wrote of the remarkable results selection could have on a single breed and considered also cross-breeding with Merino sheep. This was only beginning to be understood but already Mackenzie knew that acquired characteristics would not persist without selection. Although wool quantity was increased from the first cross, 'a careful selection of breeding yews [is] necessary, in order to bring the wool of a whole flock as near to equality as possible. If no selection is made, the whole wool will soon degenerate'.[56]

The changed social conditions following the failure of the Jacobite risings presented Highland landlords with a stark choice, either to cling to the old ways, like Stewart of Garth, and suffer financial ruin, or to survive by applying Enlightenment ideas of science and economics to improve landholdings.[57] Mackenzie's reputation as an improver, for good or ill, has been widely recognised. Richards and Clough saw him as 'one of the most robust and uncompromising of improvers' whose *General View* not only 'propagated the best thinking in agricultural practice but was a substantial description and indictment of the older forms of life in the Highlands, and a prescription for a new society and economy'.[58] His influence, particularly on sheep rearing in the Highlands, was acknowledged by Patrick Sellar, the largest sheep farmer in the country who, according to Richards, was 'a keen follower of George Culley and Sir George Mackenzie in their efforts to improve technical knowledge of sheep rearing'.[59] Prebble's less charitable view is that Mackenzie's writings on sheep were heeded more than those of Sir John Sinclair because 'he did not confuse his readers with philanthropy or concern for the Highlander'.[60]

Mackenzie's contemporaries, Patrick Sellar (1780–1851) and James Loch (1780–1855), attended Edinburgh University when he did. Although seldom classmates, they probably knew one another during these years.[61] They gained a shared vision of economic and social improvement for the Highlands based on scientific principles which required engrossing small arable cultivations into farms of 100 acres or more and the development of extensive sheep walks in upland areas with wintering pasturage on lower ground. They believed that scientific and economic legitimacy would justify social policy and accepted that a consequence would be the displacement of a large section of the indigenous population. Mackenzie and Sellar, both with Highland origins, remained uncompromising practitioners of this doctrine. In their advocacy of it, their arguments and even their language have striking similarities. The case for Highland improvement was sophisticated and drew upon many sources. The improvement of arable cultivation by manuring and crop rotation, and of sheep stocks by selective breeding, were obvious scientific benefits. The views of Adam Smith and Malthus on economics were also powerful influences. For Mackenzie, a more immediate inspiration was Thomas Douglas, Lord Selkirk, who had preceded him at Edinburgh University.[62]

Selkirk was probably known to Mackenzie socially, being a brother-in-law of Mackenzie's friend and scientific associate Sir James Hall. The first edition of Selkirk's famous *Observations on the Present State of the Highlands of Scotland with a View of the Causes and Probable Consequences of Emigration* was published in 1805. It was written to arrest the dispersal of displaced Highlanders to foreign lands where 'they were lost not only to their native country, but to themselves as a separate people'.[63] He hoped that they might be brought together in Britain's colonies where they would be useful to the nation and able to preserve their customs and language. Selkirk argued that subsistence farming in the Highlands resulted from the clan system under which the value of land was reckoned 'not by the rent it produced, but by the men whom it could send into the field' as fighting men. The subdivision of holdings meant that arable possessions yielded hardly enough for the subsistence of the occupiers who, in poor seasons, depended upon the chief's generosity. Little 'regular industry' was required to work the small portions of arable land, which encouraged indolence. Much land was fit only for pasturage.[64]

Selkirk's view of the over-populated Highlands was Malthusian. Malthus' *Essay on the Principle of Population* (1798) asserted that uncontrolled reproduction will always outstrip food supply, with war, famine and disease checking population growth. Philanthropy would not cure the misery of the poor so long as these conditions persisted. Misery followed bad years and the generosity of the chiefs only perpetuated the tenants' distress. From at least the 1720s and more so after 1745, when the Highlands were disarmed, the pecuniary interests of the proprietors became paramount. Higher rents could be paid when arable land was worked more efficiently by the merger of small farms. As Selkirk put it, enlargement 'throws [farms] into the hands of men of education and efficient capital, who, by following improved modes of cultivation, increase productiveness of the soil'. But since enlargement made many people redundant, who might then become an unacceptable burden on the proprietors, a great part of the present inhabitants of the Highlands must in one way or another seek for means of livelihood totally different from those on which they have hitherto depended'.[65] Selkirk's book was subjected to the strictures of tenant Highlanders. In the Lowlands, it was well received. Sir Walter Scott believed it traced 'the political and economical effects of the changes' in Scotland 'with great precision and accuracy'.[66] Mackenzie agreed and shared the opinion of many highland landowners that it 'would be the means of bringing a right understanding on the subject of Highland populations'.[67]

Mackenzie's *General View*, which was prepared for the Board of Agriculture and published in 1810, draws upon Selkirk's analysis and conclusions and applies them, with his own observations, to Ross and Cromarty. Echoing Malthus and Selkirk, Mackenzie noted the melancholy effect of recent crop failure in the over-populated Highlands and went on to state his conviction that 'the true value of land is to be found, not in the number of ignorant and idle people, who can contrive to live upon it, but in the number of cattle and sheep, and in the quantity of corn it can produce'.[68] He estimated that half of the Highland population would suffice to cultivate the soil properly, and thereby to support themselves and their families. 'What is to be done with the other half, is a question of far greater difficulty in the solution than people in general are aware of'.[69] The result of overpopulation and overstocking was poverty. Mackenzie described Highland living conditions in an unpopular but well-known passage:

> The Highland tenantry are universally ill accommodated. They live in the midst of smoke and filth; that is their choice. Wherever farms have been laid out on a proper scale, and are occupied by substantial and well educated men, we find the farm-houses and offices handsome and commodious . . . The present race of Highland tenants will yet find themselves much happier, and more comfortable, in the capacity of servants to substantial tenants, than in their present position.[70]

Mackenzie stressed the importance of adequate capital and skill for the economic welfare of the Highlands:

> The Highland tenants have neither capital nor skill, and therefore cannot be considered fit to improve the country. It is for the interest of proprietors, and of the public, that the soil should be made to yield all that it possibly can afford, that it may support a large and industrious population . . . it being necessary for the introduction of a better system,

to engross a number of small farms, a great many [Highlanders] are consequently thrown out of their possessions, and have only the alternative of starving, or leaving the country where no employment can be found for them . . . It cannot be denied that unless employment can be found, there is no occasion for people . . . The humane feelings, which were excited by the prospect of distressing an indolent and almost useless population cannot be blamed . . . Caution in every change is not only proper but necessary; but to keep reason aloof is absurd.[71]

Late eighteenth-century improvers looked to Smith's doctrine of the division of labour for a solution to society's ills. Initially, Mackenzie shared these views and believed they offered hope for the superfluous population of the Highlands 'Our only means of making the natives industrious, is to make them depend on each other for many of the necessaries of life . . . If a few can be prevailed upon to study different trades, and commence business, many will follow their example'.[72] He went on to advocate the establishment of manufactures in villages to provide employment, but was disillusioned by his later experience of running a woollen mill in Inverness which, although close to the source of its raw materials, had to use machinery to be competitive and so employed few people. Eventually, he thought fisheries offered the best prospect for employment.

Mackenzie's adoption of rational principles in the management of his tenants certainly incurred their wrath. Dr. John Mackenzie remembered that 'On the Coul estate Sir George had evicted every crofter, and they were a host – over thirty in Rogie alone'. Alexander Mackenzie, author of *The History of the Highland Clearances* (1883), was himself evicted as an infant from the township of Tournaid (Tournaig) on Mackenzie's lands in Wester Ross.[73] Yet in distant Edinburgh it was possible to take another view:

> Sir George Mackenzie was an active and exemplary country gentleman, a kind and liberal landlord, who abated rent in unfavourable seasons (remarkable on occasion of the potato failure), respected the elective franchise as the sacred right of his tenants, and practically acted on the principle that the estate should, at all times, find employment for its population. No part of the Highland Destitution Fund, lately subscribed by the public, was required on the estates of Sir George Mackenzie.[74]

The Coul estate probably made no claim on the Fund because its remaining 'industrious' people were employed, the 'useless population' having been removed by Sir George long since.

It is tempting to suggest that Mackenzie used his estate as a social laboratory. While this might be argued for the Sutherland estates where, because they were backed by the Leveson-Gower fortune, a choice of strategies was possible and, indeed, discussed, Mackenzie's policy was determined by financial necessity. It is no surprise to learn that 'Sir George was always short of money': he had expensive tastes.[75] He probably saw scientific improvement as the only means of making his estate viable and capable of yielding the income he desired.

Mackenzie in the Field: Iceland and Huttonian Geology

In his inaugural lecture as first Professor of Geology at the University of Edinburgh, Archibald Geikie spoke of Hutton, Playfair and Hall as leaders of the Scottish 'school' of geology. 'As worthy but less celebrated associates', continued Geikie, 'we must not omit to add the names of Mackenzie, Webb Seymour and Allan.'[76] Writing to the English geologist Henry De la Beche in 1836, Mackenzie claimed to be one of 'the only two disciples of the old Huttonian School still alive' – the other was Thomas Charles Hope, Mackenzie's former chemistry professor.[77] Mackenzie became a Huttonian while a chemistry student. His professor, Joseph Black, was one of Hutton's oldest friends and staunchest supporters. His other associates were Hope, James Hall, Webb Seymour and Robert Kennedy, all enthusiastic Huttonians. In 1802 John Playfair's *Illustrations of the Huttonian Theory of the Earth* was published and Mackenzie acknowledged that its clarity was a further influence.[78]

The history, content and influence of James Hutton's *Theory of the Earth* have been exhaustively studied.[79] Its deductive reasoning was characteristic of a major strand of Scottish Enlightenment thinking.[80] Like Guettard and Desmarest, Hutton believed that basalt crystallised from molten volcanic lava but he developed ideas about the earth's internal heat and pressure (Plutonism) that went beyond those of the volcanists. From the evidence of baking, veining and disruption of pre-existing rocks that are found in contact with granite and greenstone, he concluded that these must be unerupted molten rocks produced by subterranean heat. He deduced also that earth processes seen at work today, such as running water, if given enough time, would have enormous erosive powers. His theory envisaged cyclicity combining these creative and destructive elements. Erosion of a former landscape would lead to sedimentation and, on burial, these sediments would be subjected to the earth's internal heat and pressure that would consolidate them into rock strata and subsequently uplift them as a new landscape. Following publication of his theory by the Royal Society of Edinburgh, Hutton and his friends sought 'proofs and illustrations' in the field. Their observations were recorded in Hutton's manuscript of the third volume of his *Theory of the Earth*, which remained unpublished until 1899.[81] The original drawings, made for Hutton in the field, were not rediscovered until 1968 and were eventually published in 1978.[82] These made clear the extent of Hutton's journeys and the accuracy of his geological observations. After Hutton's death, the search for field evidence was continued by his followers to counter those opposed to the theory (see Chapter 10). Among the objectors were the disciples of A. G. Werner, such as Robert Jameson, who believed in the serial precipitation of all rocks from a primaeval sea (Neptunism). The appointment of Jameson, a pupil of Werner, to the Chair of Natural History in 1804 meant that many students within the University of Edinburgh were influenced against Huttonian geology, and pro- and anti-Huttonian factions developed within the city.[83]

Several works describing field relationships at home and abroad that supported Huttonian theory were published. Mackenzie's book, *Travels in the Island of Iceland* (1811), was one such. *Letters on Iceland* by Uno von Troil, who had accompanied Sir Joseph Banks' expedition to the island in 1772, had been published in English in 1780 and, as Sir George noted, had awakened 'the curiosity of science to that neglected, but remarkable country'.[84] A chance meeting in 1807 with an Icelandic medical student, Olafur Loptsson, revived that interest in Mackenzie.[85] As Hartley has shown in Chapter 10, Sir George saw Iceland, with its active volcanoes, hot springs and geological relationships, as offering a natural laboratory where observation would put Huttonian theory to the test. A visit was projected in 1809 and, writing to William Hooker in October of that year, Banks mentioned that he had advised 'the cargo of Scotsmen', with Mackenzie at their head, to wait another season because of the ongoing Icelandic insurrection under Jorgen Jorgensen.[86] So it was that on 18 April 1810, Mackenzie set sail from Leith with Henry Holland and Richard Bright, accompanied by Loptsson as interpreter, on what Wawn has described as 'the first significant British expedition to Iceland since that of John Thomas Stanley in 1789'.[87]

Holland and Bright had been school friends in Bristol and students at Edinburgh. They were later to become famous: Sir Henry Holland FRS as physician to the Queen; Dr. Richard Bright as the renowned physician whose name is associated with the renal disease he first described. Both were interested in geology but tensions arose in the field because their inductive approach did not accord with Mackenzie's enthusiasm for deduction and theory.[88] Holland kept a journal of the expedition that Mackenzie used in his chapter on 'Mineralogy' in *Travels in the Island of Iceland*. Wawn has shown that Mackenzie drew heavily on Holland's journal in that chapter.[89] The chapter is presented as a justification of Huttonian geology and Holland, in letters to his father, objected to the tendentious way Mackenzie had misrepresented his geological observations.[90] For the second edition (1812), Holland insisted that a clear distinction be drawn between the recorded observations, for which he was to be given credit, and what he regarded as the controversial interpretations of those observations for which Mackenzie was to bear full responsibility.[91]

During their four months in Iceland, they explored Gullbringusysla, the Snaefellsnes peninsula and Rangarvallasysla. They found much to interest them. They puzzled over the origin of the different types of lava that they saw, but in other cases the field evidence was more easily interpreted. As Hartley notes, they found clear evidence for the igneous origin of obsidian and pumice in a lava seen north of Hekla where obsidian graded upwards into pumice and slag (see Chapter 10). Bright's paper to the London Geological Society suggests that he may have been the first to realise the significance of this field evidence.[92] They saw black vitreous coatings along the walls of veins or dykes due to chilling of the intruded molten rock and noted veins of greenstone and basalt cutting lavas. They studied the relationships of the stratified beds of tuff.[93]

In one important respect, Mackenzie was led to query Hutton's belief in the adequacy of the erosive powers of earth processes seen at work today to explain the dramatic field evidence of sudden erosion which he saw in Iceland. Towards the conclusion of his chapter on mineralogy, Mackenzie wrote:

> To consider the slow operations of the atmosphere, or rivers, as sufficient for shaping out huge mountains, and forming stupendous precipices, which are known to defy the most violent external attacks the destructive agents of nature can make, is a poor resource, either for the philosophers who can raise and sink the waters of the ocean as fancy may prompt, or for those who have seen the effects of the earthquake and the volcano, and can appreciate the power of subterraneous heat.[94]

'Thus', comments Dean, 'in defending Hutton, Mackenzie jettisoned a great part of his theory.'[95] Sir George's belief that something greater than the 'slow operations of the atmosphere, or rivers", was required to account for the Icelandic evidence of rapid erosion was supported at the Royal Society of Edinburgh in 1812 by his friend Sir James Hall. Hall postulated a cataclysmic torrent flowing across Scotland from west to east to account for much erosion and the striated rock surfaces, erratic blocks, and drift deposits which cover the country and directional topographic features such as crag and tail.[96]

In the summer of 1812, Thomas Allan (also cited by Geikie as a worthy associate of the 'Scottish school of geology') accompanied Mackenzie on an expedition to the Faroes to see whether 'in a Trap Country, where no trace of external volcanoes existed, anything similar to the peculiar features of the rocks of Iceland was to be found'.[97] Mackenzie's account of the expedition again supported the theory of 'the ingenious Hutton [which] explains every phenomenon of the mineral region' and contrasted with Allan's more objective account.[98] They described from the Faroes many geological features that had been seen in Iceland. Not least of the achievements of the Iceland and Faroes expeditions were the important mineral and rock collections made. Lyell spoke of Mackenzie's 'magnificent collection of mineralogical treasures' from Iceland.[99] Sir George gave Icelandic minerals to the Hunterian Museum and divided his rocks between the Museum of the Royal Society of Edinburgh and the Natural History Museum of the University of Edinburgh.[100]

Mackenzie continued to contribute papers on various geological topics and in 1835 joined the controversy concerning the origin of the 'Parallel Roads' of the glens of Lochaber and the possible former glaciation of Scotland. In 1817, Macculloch interpreted the Parallel Roads of Glen Roy as successive shore lines of a lake which had once filled the valley to the height of the highest terrace and had subsided step by step. In the following year, Sir Thomas Dick-Lauder also interpreted the terraces of Glen Roy and other parts of Lochaber as lake shorelines.[101] Neither provided a satisfactory explanation for the existence or disappearance of the lake. Mackenzie agreed that the Parallel Roads were shorelines and saw their origin in Hall's cataclysmic flood whose waters were ponded up by detrital barriers caused by currents and eddies as they passed over the varying topography of the land.[102] He stuck to this idea in the face of various later explanations including the now generally accepted view of Agassiz that they are the product of an ice-dammed lake. Mackenzie accepted the former existence of glaciers in Scotland but invoked the cataclysm to explain it.[104] Cunningham noted that, in his review of Forbes' classic paper on the former existence of glaciers in the Skye Cuillins (1846), Mackenzie rejected the possibility of any significantly colder climate in the past, and thought that their glaciation occurred when the Coolin stood at a greater elevation, a reversion to the ideas of Hutton and Playfair in relation to the Alps.[105]

Mackenzie and the Science of the Mind

In this chapter, I have considered how the work of men like Black, Hope, Hall, Kames, Francis Home, Coventry, Hutton and Playfair was tested, carried forward and applied by their successors such as Mackenzie. In contrast, it is instructive to consider how an idea, although initially attractive, was recognised as unscientific because the technology required to test it experimentally was not available. Phrenology was such a 'science'.

When F. J. Gall traced a connection between cranial anatomy and psychology, it was of interest to many Scots since, during the eighteenth century, Enlightenment philosophers had wrestled with the implications of man's understanding of himself in body and mind.[106] Mackenzie became a phrenologist when Gall's disciple J. G. Spurzheim spent almost six months in Edinburgh in 1816–17 demonstrating dissections of the brain and lecturing.[107] No sooner had Mackenzie's interest been aroused than he made his views known. Thus, 'for amusement and exercise', he wrote *An Essay on Some Subjects connected with Taste* and read most of it to meetings of the Royal Society of Edinburgh in the 1816–17 session.[108] He believed that a study of the comparative development of parts of the brain associated with 'particular manifestations of talent or character' would subject the discussion of taste to scientific test and so remove it from the arbitrary views of moral philosophers. Such views were anathema to philosophers of the Common Sense school such as Thomas Reid and Dugald Stewart.[109] In an article published in 1816, Stewart wrote: 'Is there no Arbuthnot now to chastise the follies of our Craniologists'.[110] This was taken personally by Mackenzie who wrote two long letters in defence of phrenology to the ageing philosopher in 1821.[111] If the tone of Sir George's references to phrenology in the *Essay* was tentative, his commitment was unqualified in his *Illustrations of Phrenology*, one of the earliest works on the subject in the English language.[112]

As a Fellow of the Royal Society of London and President of the Physical Section of the Royal Society of Edinburgh, Mackenzie's support was welcomed by the phrenologists in their struggle to establish their scientific credentials. Attacks by Sir William Hamilton and Alexander Monro *tertius* on the anatomical foundations of phrenology proved the turning point in the phrenological debate.[113] Hamilton's papers were read before the Royal Society of Edinburgh between 1825 and 1829, but not until the last of these had been given did the Society allow Mackenzie the right of reply.[114] This he did in addresses in December 1829 and January 1830.[115] T. C. Hope took the chair at the second address and his summing-up defined the position of the majority of Edinburgh's scientific men. After acknowledging the existence in the brain of distinct organs that differ in appearance, substance and structure, he went on:

> it is a study truly philosophical and strictly physiological, to investigate the special use of each of these organs, and the particular mental function to which each was subservient . . . without doubt, both physical and metaphysical science would profit greatly from successful enquiries into the uses of these multifarious and finely constructed organs in the interior of the brain . . . so far as he knew, [phrenologists] had not ascertained the function performed by any one of them.[116]

Phrenologists then changed the emphasis from the scientific to the social implications of their subject and many scientists left their ranks. Mackenzie, however, continued to write and speak in its support and, in doing so, damaged his scientific standing.

Mackenzie and the Civic Promotion of Science

Convinced of natural knowledge's potential for social and cultural improvement, Mackenzie worked to promote the natural sciences in Scotland. He saw learned societies as important in this respect and regarded enlightened laws, an active museum and effective publication as factors that distinguished such bodies. He believed that societies should be outward looking, have international links and facilitate representation to political decision makers. His drive and initiative were valued in the many societies to which he belonged. Only a partial indication can be given here of ways in which his work has been of lasting significance.

Mackenzie was elected to the Society of Antiquaries of Scotland in 1798 and to the Royal Society of Edinburgh in 1799 when both were suffering decline. Sir David Brewster later cited Sir George among those who 'made arrangements' to correct the situation at the Royal Society. For Stevenson, Mackenzie was the catalyst for change at the Antiquaries.[117] With others, Mackenzie did much to re-establish the museums of both and, by promoting mutual co-operation in their management, overcame what ill feeling remained from the time of their foundation.[118] In 1811 he convened a committee whose report assisted the establishment of the Royal Society's own museum.[119] He was elected President of that Society's Physical Section in 1812, an office that he held for ten years. As a Vice-President of the Antiquaries, he was instrumental in the adoption of new statutes and bye-laws in 1815.[120] He suggested writing to the clergy of Scotland inviting notices of antiquities in their parishes, a scheme which was implemented in the north of the country.[121] Having corresponded with learned men in Denmark and Norway, he 'expressed his opinion of the importance of establishing a connexion with learned Societies abroad'. Count Bedemar was elected an Honorary Member of the Society of Antiquaries of Scotland in 1814 and, in the following year, Sir George was elected a Member of the Committee of Antiquities of the Royal Literary Society of Copenhagen. He edited unpublished papers in the Antiquaries' possession for its *Transactions*.[122] Later he contributed his own archaeological researches, especially on vitrified forts in the north of Scotland. He had published on this subject in the *Edinburgh Encyclopaedia* and in a letter to Sir Walter Scott on the fort of Knockfarril.[123] His communications to the Antiquaries on the topic formed part of a report prepared for the Society by Samuel Hibbert for whom Mackenzie was the most able exponent of the theory that vitrification was due to beacon fires, not intentional activity.[124] In his later years, Mackenzie used his position as an office-bearer of the Royal Society of Edinburgh and the Scottish Society of Arts as a power base in the long-running dispute over the slow progress of the Ordnance Survey in Scotland. He wrote to his friend, Sir George Clerk of Penicuik MP, secretary to the Treasury, to ask that Clerk use his good offices with the Chancellor of the Exchequer, Sir Robert Peel, to propose a proper grant for the survey.[125]

Mackenzie was a prime mover in establishing the Edinburgh Astronomical Institution. At a public meeting in 1811, he proposed that an address by John Playfair on the merits of such an institution be inserted in the newspapers and subscribers be invited.[126] As chairman of an interim committee, he wrote personal letters inviting subscriptions stressing the usefulness of an Institution 'the want of which has long been a reproach to the Capital of Scotland'.[127] His own name appears on the list of First Class Subscribers, and in 1812 he was elected Vice-President. He convened the sub-committee that prepared the Laws of the Institution adopted in 1814.[128] He worked for the establishment of its observatory by continued fundraising and sought financial help from the government through Lord Melville.[129] He negotiated with the Town Council about the bounds of the observatory on Calton Hill and laid its foundation stone.[130] Wanting the observatory to be well provided, he donated equipment to that end.[131] Following Playfair's death, Mackenzie was elected President of the Institution in 1819 and held office for four years. When George IV visited Edinburgh in 1822, it was Mackenzie who drafted the Institution's loyal address and, while attending the king as Vice-President of the Royal Company of Archers, passed it to the Secretary of State.[132] Peel replied that the king was 'graciously pleased to permit the Observatory on the Calton Hill to be styled The Royal Observatory of King George the Fourth', later abbreviated to 'the Royal Observatory of Edinburgh'. In 1834, the Institution made over its observatory to the University of Edinburgh on condition that it be made a public establishment. The Government created the public office of Principal Observer, which was combined with that of the Regius Professor of Astronomy, the holder of the joint posts becoming 'The Astronomer Royal for Scotland'.[133]

The cross-fertilisation of Mackenzie's astronomical and horticultural interests is apparent in his writing to Sir Joseph Banks in 1815 on the shape of a glass forcing house. In 1809, Banks had declared that 'The public have still much to learn about Hot-houses'.[134] Early volumes of the *Transactions of the Horticultural Society of London* were full of papers on greenhouse heating and the proper pitch of glass roofs and, in 1814, Mackenzie had given a paper on an economical hot-house to the Caledonian Horticultural Society.[135] Banks published Mackenzie's communication. In Elliott's view, 'the real break through came in 1815, with Sir George Mackenzie's paper recommending a spherical roof, parallel to the dome of the heavens, as the means of obtaining optimum light for plants at every time of day'. In a postscript, Mackenzie suggested that the glass semi-dome be constructed in two parts on rollers, 'in the manner of an observatory' to allow it to be opened and so expose the plants to the full influence of the sun.[136]

A new society that Mackenzie supported, but which lasted for only ten years, was the Northern Institution for the Promotion of Science and Literature, based in Inverness. As a Vice-President, Sir George presided at the Institution's first meeting (in March 1825) and offered a gold medal for the best essay on the State of Society in the Highlands in 1745, and on the progress which had since been made.[137] He presented it to John Anderson WS, author of the history of the Frasers and one of the secretaries of the Society of Antiquaries, in October 1826.[138] The winning essay, with Mackenzie's address on that occasion, was published in 1827.[139]

Mackenzie was admitted to the Phrenological Society of Edinburgh in 1820 and drafted the laws which, after amendment, were adopted in 1821.[140] He was the main force in establishing the Society's museum which is today one of the largest surviving phrenological collections in the world.[141]

Conclusion

This chapter has suggested that the Enlightenment generalist was no mere dilettante. Study of Mackenzie demonstrates that the state of contemporary science allowed an educated man to contribute in many fields. Mackenzie's achievements have been indicated, and literature cited, showing that his contributions have been considered significant by historians of science in many disciplines. As knowledge and specialisation increased, interdisciplinary perceptions became less the province of the generalist and more that of learned bodies. Mackenzie stood between these extremes. As a generalist, he was drawn to interdisciplinary societies, such as the Royal Society of Edinburgh and the Northern Institution. Yet he demonstrated his support also for specialist bodies such as the Edinburgh Astronomical Institution. He recognised science as an international undertaking and worked for the free exchange of ideas through effective publication and personal interchange.

Mackenzie also believed that science should be applied for the benefit of society and he used the improvement of his estate as a practical example to others. He was not alone in thinking that the short-term social consequences could not be allowed to stand in the way of progress towards long-term social benefit. His support for the Astronomical Institution was not only to encourage astronomy but for its potential to be useful. He believed also that, through the application of mental science, taste and ethics could be removed from arbitrary discussion and subjected to scientific test and thus prove useful in matters such as criminality. We should not now criticise Sir George Steuart Mackenzie for his partial understanding of the many issues with which he was involved. For example, scanning techniques which now allow observation of the reaction of different parts of the living brain to differing stimuli have reactivated the social and ethical questions posed by the phrenologists in the nineteenth century. Understood in the context of his time, Mackenzie was a committed advocate of useful science. He was a practical experimenter, in chemistry and in sheep breeding as well as with the 'industrious qualities' of the Highland peasantry. He was sufficiently committed to theoretical debate in the earth sciences to undertake arduous fieldwork in order to confirm others' accounts. Generalist, then, he may have been, but along with Mackenzie's desire to promote scientific understanding as a whole was his engagement to promote the application of science in particular practices which indicates the more specialist ways in which scientific activity was a part of the intellectual life of late Enlightenment Scotland.

Notes

1. Baptismal Register, June 1780, Lady Yester Church, Edinburgh.
2. Steven Shapin, 'Science', in David Daiches (ed.), *The New Companion to Scottish Culture* (Edinburgh, 1993), 275–76.
3. Roger Hahn, *The Anatomy of a Scientific Institution: The Paris Academy of Sciences, 1666–1803* (Berkeley and London, 1971).
4. Susan F. Cannon, *Science in Culture: The Early Victorian Period* (New York, 1978), esp. Ch. 4.
5. John R. R. Christie, 'The Rise and Fall of Scottish Science', in M. Crosland (ed.), *The Emergence of Science in Western Europe* (London and Basingstoke, 1975), 111–26.
6. Roger L. Emerson, 'The Scottish Enlightenment and the End of the Philosophical Society of Edinburgh', *The British Journal for the History of Science* 21 (1988), 33–36.

7. John Prebble, *The Highland Clearances* (London, 1963), 35; C. Byam Shaw, *Pigeon Holes of Memory, The Life and Times of Dr. John Mackenzie, 1803–1886* (London, 1988), 201; Katherine M. Lyell (ed.), *Life, Letters, and Journals of Sir Charles Lyell, Bart.*, 2 vols. (London, 1881), 1:156.
8. National Library of Scotland (hereafter NLS), MS 14298, ff.1–65. Alexander Christison (1753–1829) became Professor of Humanity at Edinburgh University in 1806. Pupils from his High School class of 1787 to 1791 formed a club which met annually for fifty years from 1798 and which Sir George Mackenzie attended on ten occasions between 1799 and 1830.
9. Edinburgh University Medical Matriculation Register, 1796–97.
10. R. G. W. Anderson, *The Playfair Collection and the Teaching of Chemistry at the University of Edinburgh, 1713–1858* (Edinburgh, 1978), 19–33.
11. Edinburgh University Library Special Collections (hereafter EUL), MS Gen.48D.
12. W. P. Doyle, 'Black, Hope and Lavoisier', in A. D. C. Simpson (ed.), *Joseph Black, 1728–1799: A Commemorative Symposium* (Edinburgh, 1982), 43–46. Doyle traces Black's gradual acceptance of Lavoisier's theory of combustion from about 1783–84. Black corresponded with Lavoisier in 1789–90 and was fully converted by 1790. His letter of affirmation to Lavoisier was printed in *Annales de Chimie* in March 1791.
13. Edinburgh University Medical Matriculation Register, 1798–99.
14. V. A. Eyles, 'The Evolution of a Chemist: Sir James Hall', *Annals of Science* 19 (1963), 153–82; Neil Campbell and R. Martin S. Smellie, *The Royal Society of Edinburgh, 1783–1983: The First Two Hundred Years* (Edinburgh, 1983), 22.
15. Dennis R. Dean, *James Hutton and the History of Geology* (Ithaca, NY, and London, 1992), 89–92.
16. David Mushet, 'Account of an Experiment made with a View to prove whether the Experiment made at the Polytechnic School, respecting the conversion of Iron into Steel by means of Diamond, is Conclusive', *Tilloch's Philosophical Magazine* 5 (1799), 201–5.
17. J. R. Partington, *A History of Chemistry*, 4 vols. (London, 1961–1970), 3:517; W. A. Smeaton, 'L. B. Guyton de Morveau, A Bibliographical Study', *Ambix* 6 (1957), 18–34; idem, 'Guyton de Morveau', in Charles C. Gillespie (ed.), *Dictionary of Scientific Biography*, 18 vols. (New York, 1970–90), 5:600–4.
18. *Dictionary of National Biography*, s.v. 'Mackenzie, Sir George Steuart'.
19. See 'Lettre a M. De La Metherie par M. Hassenfratz' and 'Lettre de M. De Morveau, sur la théorie de la conversion du Fer en Acier et fur la Plombagine', *Observations sur la Physique* 29 (1786), 210–21 and 308–12 respectively, and L. B. Guyton de Morveau, 'Acier', in L. B. Guyton de Morveau, et al, *Encyclopédie Méthodique; Chymie, Pharmacie et Métallurgie*, 6 vols. (Paris, 1786–1815), 1:420–51.
20. L. B. Guyton de Morveau, 'Sur la combustion du diamant', *Annales de Chimie* 31 (1799), 72–112 and in *Tilloch's Philosophical Magazine* 5 (1799), 55–61, 174–88.
21. Partington, *History of Chemistry*, 3:531.
22. L. B. Guyton de Morveau, 'De la conversion du fer doux en acier fondu par le diamant', *Annales de Chimie* 31 (1799), 328–36 and in *Tilloch's Philosophical Magazine*, 5 (1799), 89–93.
23. Guyton had written to Black in 1787 about his new chemical nomenclature as mentioned, but the name is misquoted as 'Mr Guyot', in a letter from John Robison to James Watt, 25 February 1800, in Eric Robinson and Douglas McKie (eds.), *Partners in Science: Letters of James Watt and Joseph Black* (London, 1970), 339.

24. Among Lavoisier's associates were Claude Louis Berthollet, Gaspard Monge, Antoine François de Fourcroy, Pierre Simon de La Place (see Chapter 10) and Guyton de Morveau, who joined them in adding notes on Richard Kirwan's *Essai sur le Phlogistique* (Paris, 1788). The book was given to Thomas Charles Hope in Paris to take to Hall in Edinburgh by Madame Lavoisier in the year of publication; see Eyles, 'Evolution of a Chemist', 167-70, 176.
25. George S. Mackenzie, 'Experiments on the Combustion of the Diamond', *Nicholson's Journal of Natural Philosophy, Chemistry and the Arts* 4 (1800), 103-10; an abstract of this paper appears in *Transactions of the Royal Society of Edinburgh* 5 (1805), 11-14.
26. Dean, *James Hutton*, 91-2. In December 1798, Kennedy read his paper 'Chemical Analysis of Three Species of Whinstone and Two of Lava' before the Royal Society of Edinburgh.
27. George S. Mackenzie, 'Experiments on the Analyzation of Marl', *The Farmer's Magazine* 5 (1804), 271.
28. James Hall, 'Experiments on Whinstone lava', *Nicholson's Journal of Natural Philosophy, Chemistry and the Arts* 4 (1800), 8-18, 56-65. Alex Barker's Edinburgh foundry in middle Leith Walk appears in the *Edinburgh Directory* from 1797 to 1803.
29. I am grateful to Alison Morrison-Low for drawing my attention to Josiah Wedgwood's papers on his pyrometer; see Josiah Wedgwood, 'An Attempt to make a Thermometer for measuring the higher Degrees of Heat, from Red Heat up to the strongest that Vessels made of Clay can support', *Philosophical Transactions* 72 (1782), 305-36; idem, 'An Attempt to compare and connect the Thermometer for strong Fire, described in Vol. LXXII of the Philosophical Transactions with the common Mercurial Ones', *Philosophical Transactions* 74 (1784), 358-84; idem, 'Additional Observations on making a Thermometer for measuring the higher Degrees of Heat', *Philosophical Transactions* 76 (1785), 390-408. The 1784 paper (p. 370) gives a table of supposed equivalent temperatures on the Wedgwood and Fahrenheit scales which shows that 130'W is the temperature at which cast iron melts. Because the scale is nonlinear against any thermodynamic scale, it is not possible to extrapolate to the Celsius equivalent of 150'W except that it exceeded the fusion temperature of cast iron. In the present context, it is of interest that Hall was already using a Wedgwood pyrometer in his rock fusion experiments.
30. Mackenzie, 'Experiments on the Combustion of the Diamond', 103-10.
31. Mrs. Margaret Maria Gordon, *The Home Life of Sir David Brewster* (Edinburgh, 1869), 213.
32. Mackenzie, 'Experiments on the Combustion of the Diamond', 103.
33. *Dictionary of National Biography*, s.v. 'Mushet, David'. Blackband was not exploited until the coming of the hot blast smelting process.
34. David Mushet, 'Remarks upon Several Experiments made to prove the conversion of Iron into Steel by means of the Diamond', *Tilloch's Philosophical Magazine* 11 (1802), 289-94.
35. Ian R. M. Mowat, *Easter Ross, 1750-1850: The Double Frontier* (Edinburgh, 1981), 35, 76.
36. Edinburgh University Literary and Philosophical Matriculations Register; Medical Matriculations Register.
37. Ian J. Fleming and Noel F. Robertson, *Britain's First Chair of Agriculture at the University of Edinburgh, 1790-1990* (Edinburgh, 1990), 26-29.
38. Andrew Coventry, *Remarks on Livestock and Relative Subjects* (London, 1806).

39. Rosalind Mitchison, *Agricultural Sir John: The Life of Sir John Sinclair of Ulbster, 1754–1835* (London, 1962), 103.
40. Sir George Mackenzie to Archibald Constable, Edinburgh 12 January 1803, NLS, MS 673 ff.255–56.
41. A. S., 'On ascertaining the Quality of Lime', *The Farmers' Magazine* 5 (1804), 27–29.
42. George S. Mackenzie, 'Thoughts on the analyzation of Lime', *The Farmers' Magazine* 5 (1804), 265–68, and, in the same volume, 'Thoughts on the analyzation of Marl', 269–72.
43. John Gascoigne, *Science in the Service of Empire: Joseph Banks, the British State and the Uses of Science in the Age of Revolution* (Cambridge, 1998), 44, 105–6, 116.
44. Joseph Banks, 'A Project for extending the Breed of Fine-Woolen Spanish Sheep, now in the Possession of his Majesty, into all Parts of Great Britain, where the growth of finer Clothing Wools is found to be profitable', *Nicholson's Journal of Natural Philosophy, Chemistry and the Arts* 4 (1800), 289–93.
45. H. B. Carter, *Sir Joseph Banks* (London, 1988), 467.
46. George S. Mackenzie, *Treatise on the Diseases and Management of Sheep with Appendix, containing Documents exhibiting the Value of the Merino Breed of Sheep, and their Progress in Scotland* (Inverness, 1809).
47. Carter, *Banks*, 467.
48. Charles Darwin, *On the Origin of Species by means of Natural Selection*, 2nd edn. (London, 1860), 29.
49. George S. Mackenzie, 'Remarks on the Hypothesis of Progressive Development in the Organic Creation', *Proceedings of the Royal Society of Edinburgh* 2 (1847), 130–31.
50. Eric Richards, *A History of the Highland Clearances. Volume 1: Agrarian Transformation and the Evictions, 1746–1886* (London and Canberra, 1982), 190.
51. Eric Richards, *Patrick Sellar and the Highland Clearances* (Edinburgh, 1999), 312.
52. Richards, *Highland Clearances*, 1:189, and Mitchison, *Agricultural Sir John*, 110.
53. Malcolm Gray, *The Highland Economy, 1750–1850* (Edinburgh, 1956), 86–88.
54. *Inverness Journal*, 22 April 1809, quoted in Richards, *Highland Clearances*, 1:202; George S. Mackenzie, *General View of the Agriculture of the Counties of Ross and Cromarty* (London, 1810), 222.
55. The date of publication is often given as 1813. A limited number of copies appear to have been published in 1810 and are so dated on the title page (see, for example, the two copies in Edinburgh University Library Special Collections). It is probable that much of the print remained unbound and was issued with a new title page dated 1813 when funds became available. The chairman of the Board of Agriculture, Sir John Sinclair, had obtained additional funding for his General Report on Scotland and for the County reports from the Prime Minister in 1812.
56. Mackenzie, *General View*, 218–19, 222.
57. Major General David Stewart of Garth, author of *Sketches of the Character, Manners, and present State of the Highlanders of Scotland* (Edinburgh, 1822).
58. Eric Richards and Monica Clough, *Cromartie: Highland Life, 1650–1914* (Aberdeen, 1989), 184.
59. Richards, *Sellar and the Highland Clearances*, 297.
60. Prebble, *Highland Clearances*, 36.
61. Edinburgh University Matriculation Registers. Mackenzie and Loch attended James Finlayson's Logic class in 1796–97 and were together in Hope's Chemistry

class in 1798–99. Mackenzie and Sellar attended David Hume's Scots Law class of 1798–99. Otherwise their classes did not coincide.
62. Mackenzie, *General View*, 295.
63. Lord Selkirk [Thomas Douglas], *Observations on the Present State of the Highlands of Scotland with a View of the Causes and Probable Consequences of Emigration* (London, 1805), 3.
64. Selkirk, *Observations*, 11, 13, 14, 16–19.
65. Selkirk, *Observations*, 36.
66. For example, see Robert Brown, *Strictures and Remarks on the Earl of Selkirk's Observations on the Present State of the Highlands of Scotland, with a View of the Causes and Probable Consequences of Emigration* (Edinburgh, 1806); Walter Scott, *Waverley or 'Tis Sixty Years Since* (Edinburgh, 1829), Ch. lxxii.
67. Mackenzie, *General View*, 295.
68. Mackenzie, *General View*, 85, 295.
69. Mackenzie, *General View*, 85.
70. Mackenzie, *General View*, 73–74.
71. Mackenzie, *General View*, 294–95.
72. George S. Mackenzie, *Letter to the Proprietors of Land in Ross-shire* (Edinburgh, 1803), 17.
73. Richards and Clough, *Cromartie*, 185; Byam Shaw, *Pigeon Holes*, 196; Eric Richards, *A History of the Highland Clearances, Volume 2: Emigration, Protest, Reasons* (London and Canberra, 1985), 78–79.
74. Address by James Simpson to the Phrenological Society, 11 December 1848, EUL, MS Gen. 608/1.
75. John Mackenzie, in Byam Shaw, *Pigeon Holes*, 198.
76. Archibald Geikie, 'The Scottish School of Geology', in his *Geological Sketches at Home and Abroad* (London, 1882), 291.
77. Dennis R. Dean and T. Sharpe, 'Letter from an Old Huttonian, Sir George Mackenzie to Henry De la Beche, 11 April 1836', *The Edinburgh Geologist, Magazine of the Edinburgh Geological Society* 33 (1999), 32.
78. George S. Mackenzie, *Travels in the Island of Iceland, during the Summer of the Year MDCCCX* (Edinburgh, 1811), 360.
79. Dean, *James Hutton*; James Hutton, 'Theory of the Earth; or an Investigation of the Laws observable in the Composition, Dissolution, and Restoration of the Land upon the Globe', *Transactions of the Royal Society of Edinburgh* 1 (1788), 209–304; idem, *Theory of the Earth, with Proofs and Illustrations*, 2 vols. (Edinburgh, 1795).
80. George E. Davie, *The Democratic Intellect* (Edinburgh, 1961), 189ff.
81. James Hutton, *Theory of the Earth, with Proofs and Illustrations, Volume 3*, ed. Archibald Geikie (London, 1899).
82. Gordon Y. Craig, Donald B. McIntyre and Charles D. Waterston, *James Hutton's Theory of the Earth: The Lost Drawings* (Edinburgh, 1978).
83. Robert Jameson, *Elements of Geognosy* (Edinburgh, 1808); see also the facsimile reprint with an introduction by Jessie M. Sweet and a foreword by G. W. White, *The Wernerian Theory of the Neptunian Origin of Rocks* (New York, 1976).
84. Mackenzie, *Travels*, vii. Von Troil, later Archbishop of Uppsala, accompanied Banks to Iceland in 1772. For Banks' Icelandic involvements, see A. Agnarsdottir, 'Sir Joseph Banks and the Exploration of Iceland', in R. E. R. Banks et al, *Sir Joseph Banks: A Global Perspective* (Kew, 1994), 31–48.
85. Mackenzie, *Travels*, viii–ix; Andrew Wawn (ed.), *The Iceland Journal of Henry Holland, 1810* (London, 1987), 21, 86.

86. Carter, *Banks*, 464. Mackenzie had hoped that his friend Thomas Allan would join his expedition in 1809: Mackenzie, *Travels*, x. Mackenzie asked Holland to join the expedition at a meeting of the Royal Society of Edinburgh in February 1810: Henry Holland to his father, 6 February 1810, NLS, MS 7515.
87. Wawn, *Iceland Journal*, xi.
88. Holland and Bright both gave accounts of the geology of Iceland to the London Geological Society before Mackenzie's *Travels in Iceland* was published. Holland's 'Sketch of the Mineralogy of Iceland' was read in his absence by Bright on 17 May 1811, and Bright read his own paper 'On the Obsidian & other Mineral Products of Iceland' in July 1811. Both papers exist in manuscript in the Mineral Department of the Natural History Museum, London.
89. Holland's journal, together with an account of the background to the expedition, was first published in full in Wawn, *Iceland Journal*; Mackenzie, *Travels*, xiii.
90. In January 1812, a few months after the publication of his book on Iceland, Mackenzie's play *Helga* was produced at the Theatre Royal (see Chapter 10). Young Wernerians are said to have packed the house and, despite a prologue by Sir Walter Scott and epilogue by Henry Mackenzie, they booed it off the boards: Dennis R. Dean, 'Scott and Mackenzie: New Poems', *Philological Quarterly* 52 (1973), 265–73; Andrew Wawn, 'Gunnlaugs saga ormstungu and the Theatre Royal Edinburgh 1812: Melodrama, Mineralogy and Sir George Mackenzie', *Scandinavica* 21 (1982), 139–51.
91. Andrew Wawn, personal communication.
92. Mackenzie, *Travels*, 368–70; R. M. Kark and D. T. Moore, 'The Life, Work, and Geological Collections of Richard Bright, MD (1789–1858)', *Archives of Natural History* 10 (1981), 119–51.
93. Mackenzie, *Travels*, 372, 382–83, 385–88.
94. Mackenzie, *Travels*, 393.
95. Dean, *James Hutton*, 156.
96. James Hall, 'On the Revolutions of the Earth's Surface', *Transactions of the Royal Society of Edinburgh* 7 (1815), 139–211.
97. George S. Mackenzie, 'An Account of Some Geological Facts Observed in the Faroe Islands', *Transactions of the Royal Society of Edinburgh* 7 (1815), 213.
98. Thomas Allan, 'An Account of the Mineralogy of the Faroe Islands', *Transactions of the Royal Society of Edinburgh* 7 (1815), 229–67.
99. Lyell, *Life of Lyell*, 1:156.
100. J. Laskey, *A General Account of the Hunterian Museum* (Glasgow, 1813), 47; Charles D. Waterston, *Collections in Context: The Museum of the Royal Society of Edinburgh and the Inception of a National Museum for Scotland* (Edinburgh, 1997), 157–58; Martin A. Peacock, 'The Geology of Iceland: Pioneer Work of a Scottish Geologist', and his 'A Contribution to the Petrography of Iceland: Being a Description of the Mackenzie Collection of 1810', *Transactions of the Geological Society of Glasgow* 17 (1925), 185–203 and 271–333 respectively.
101. John Macculloch, 'On the "Parallel Roads" of Glen Roy', *Geological Society Transactions* 4 (1817), 314–92; Thomas Dick-Lauder, 'On the Parallel Roads of Lochaber', *Transactions of the Royal Society of Edinburgh* 9 (1823), 1–64.
102. George S. Mackenzie, 'On the Theory of the Parallel Roads of Glen Roy', *London and Edinburgh Philosophical Magazine and Journal of Science* 7 (1835), 433–36; idem, 'An Attempt to Classify the Phenomena in the Glens of Lochaber with those of the Diluvium, or Drift', *Edinburgh New Philosophical Journal* 44 (1847), 1–12.
103. This was first published in *The Scotsman*, 7 October 1840.
104. George S. Mackenzie, 'On the Most Recent Disturbance of the Crust of the

Earth, in Respect of its Suggesting an Hypothesis to Account for the Origin of Glaciers', *Proceedings of the Royal Society of Edinburgh* 1 (1841), 364–65, and also in the *Edinburgh New Philosophical Journal* 33 (1842), 1–9.
105. F. Cunningham, *James David Forbes, Pioneer Scottish Glaciologist* (Edinburgh, 1990), 193; George S. Mackenzie, 'A Few Remarks suggested by Professor Forbes's Description of the Effects of Glacial Action among the Cuchullin Hills, and Mr Maclaren's Views of the Facts observed by him at the Gareloch', *Proceedings of the Royal Society of Edinburgh* 2 (1846) 65–67; James D. Forbes, 'Notes on the Topography and Geology of the Cuchullin Hills in Skye, and on the Traces of Ancient Glaciers which they present', *Edinburgh New Philosophical Journal* 40 (1846), 76–99.
106. Nicholas T. Phillipson, 'Scottish Enlightenment', and Steven Shapin, 'Science', in Daiches, *New Companion to Scottish Culture*, 297 and 276–77 respectively.
107. Roger Cooter, *Phrenology in the British Isles: An Annotated Bibliography* (Metuchen, NJ, and London, 1989), 217, which has a bibliography of Mackenzie's phrenological works; idem, *The Cultural Meaning of Popular Science: Phrenology and the Organisation of Consent in Nineteenth-Century Britain* (Cambridge,1984); M. H. Kaufman, 'Phrenology – Confrontation between Spurzheim and Gordon – 1816', *Proceedings of the Royal College of Physicians Edinburgh* 29 (1999), 159–70; Geoffrey N. Cantor, 'The Edinburgh Phrenological Debate: 1803–1828', *Annals of Science* 32 (1975), 195–218; idem, 'A Critique of Shapin's Social Interpretation of the Edinburgh Phrenological Debate', *Annals of Science* 32 (1975), 245–56; Steven Shapin, 'Phrenological Knowledge and the Social Structure of Early Nineteenth Century Edinburgh', *Annals of Science* 32 (1975), 219–43.
108. George S. Mackenzie, *An Essay on Some Subjects connected with Taste* (Edinburgh, 1817).
109. Cantor, 'Edinburgh Phrenological Debate', 218.
110. Dugald Stewart, 'Note to Preliminary Dissertation exhibiting a General View of the Progress of Metaphysical, Ethical, and Political Philosophy, since the revival of Letters in Europe', in the Supplement to the 4th and 5th editions of the *Encyclopaedia Britannica* (Edinburgh, 1816).
111. Report Book of the Phrenological Society, Session 1821–22, EUL, MS Gen. 608, 47–57. This was published after Dugald Stewart's death in the *Phrenological Journal* 7 (1831–2), 303–9.
112. George S. Mackenzie, *Illustrations of Phrenology, with Engravings* (Edinburgh, 1820).
113. Kaufman, 'Phrenology', 163.
114. Shapin, 'Phrenological Knowledge', 231 n.34.
115. George S. Mackenzie, 'Essay on the Principles of Phrenology . . . read to the Royal Society of Edinburgh, January 1830', *Phrenological Journal* 6 (1829–30), 332–43, 355–62.
116. T. C. Hope, *The Scotsman*, 6 February 1830, and the *Phrenological Journal* 6 (1829–30), 362.
117. David Brewster, 'Presidential Address', *Proceedings of the Royal Society of Edinburgh* 5 (1865), 321–22; R. B. K. Stevenson, 'The Museum, its Beginnings and its Development', in A. S. Bell (ed.), *The Scottish Antiquarian Tradition* (Edinburgh, 1981), 55, 58.
118. Steven Shapin, 'Property, Patronage and the Politics of Science: The Founding of the Royal Society of Edinburgh', *British Journal for the History of Science* 7 (1974), 15–35.
119. Waterston, *Collections in Context*, 45–46.

120. Society of Antiquaries of Scotland (hereafter SAS), Council Minutes of the Society of Antiquaries, 14 March 1814.
121. SAS, Council Minutes, 10 and 24 January, 1 and 14 March 1814.
122. SAS, Council Minutes, 11 December 1813; Sir George Mackenzie to Archibald Constable, 22 February 1815, undated but from 1815 and undated, NLS, MS 673, ff.267–68, 271–72, 275. These papers are contained in volume 2, part 1 of *Archaeologia Scotica – Transactions*, which was published in 1818. According to a manuscript note in the copy in the library of the Royal Museum of Scotland, the date of the completed volume is 1822. At the time, Mackenzie was negotiating with Constable on behalf of the Caledonian Horticultural Society, offering back stock for sale and stating the intention of that Society to publish a new part each six months; see Mackenzie to Constable, 2 May 1815, NLS, MS 673, ff.269–70.
123. George S. Mackenzie, 'Forts, Vitrified', in David Brewster (ed.), *Edinburgh Encyclopaedia*, 9 (Edinburgh, 1815), 523–28; George S. Mackenzie, *A Letter to Sir Walter Scott, containing Observations on Vitrified Forts* (Edinburgh, 1824).
124. Samuel Hibbert, 'Observations on the Theories which have been proposed to explain the Vitrified Forts of Scotland', *Transactions of the Antiquarian Society of Scotland* 4 (1857), 169.
125. Sir George S. Mackenzie to Sir George Clerk of Penicuik, undated, National Archives of Scotland, GD 18/3406; cited by kind permission of Sir John Clerk of Penicuik.
126. Royal Observatory of Edinburgh Library (hereafter ROE), MS A.2.3 (Playfair's Address) and MS A.1.3, Minute Book 1 (1811–31) of the Edinburgh Astronomical Institution, citation from p. 1.
127. George S. Mackenzie to the Hon. Gilbert Elliot, 24 February 1812, NLS MS 12226, f.1.
128. ROE, MS A.1.3, Minutes of 26 November and 3 December 1814, Minute Book 1 (1811–31), 35, 37.
129. ROE, MS A.1.3, Minute Book 1 (1811–31), 301–8.
130. ROE, MS A.1.3, Minute Book 1 (1811–31), 41; *The Scotsman*, 2 May 1818.
131. ROE, MS A.1.3, Minute Book 1 (1811–31), 30, 31, 33. Between 1812 and 1814 he presented a Gregorian telescope, a lucernal microscope, an asimuth compass and a solar microscope.
132. ROE, MS A.1.3, Minute Book 1 (1811–31), 99; H. A. Brück, *The Story of Astronomy in Edinburgh* (Edinburgh, 1983), 10–11; J. Balfour Paul, *The History of the Royal Company of Archers* (Edinburgh, 1875), 182–83. I am grateful to Captain J. D. B. Younger for his letter of 8 December 1997 confirming that Sir George Mackenzie was admitted to the King's Body Guard for Scotland, Royal Company of Archers on 3rd October 1801.
133. Brück, *Story of Astronomy*, 17–18.
134. Joseph Banks, 'On the Forcing-Houses of the Romans, with a List of Fruits cultivated by them, now in our Gardens', *Transactions of the Horticultural Society of London* 1 (1812), 147–56.
135. George S. Mackenzie, 'Description of an Economical Hot-house', *Memoirs of the Caledonian Horticultural Society* 2 (1819), 55–59. For this, he was awarded a gold medal. Royal Botanical Gardens of Edinburgh (hereafter RBG), Royal Caledonian Horticultural Society Minute Book 1 (1809–1820), Meeting of 5 December 1815.
136. George S. Mackenzie, 'On the Form which the Glass of a Forcing-House ought to have, in order to receive the greatest possible Quantity of Rays from the Sun',

Transactions of the Horticultural Society of London 2 (1818), 171–77; B. Elliott, 'The Promotion of Horticulture', in Banks, *Sir Joseph Banks*, 125.
137. *Inverness Courier*, 24 March 1825.
138. *Inverness Courier*, 1 November 1826.
139. The Institution assembled a valuable museum but financial support was not enough and the collections passed to the directors of the Inverness Academy who settled its debts; see J. Barron, *The Northern Highlands in the Nineteenth Century* 2 vols. (Inverness, 1907), 2: xli-xliii.
140. EUL, Minute Book of the Phrenological Society 1, 26 February and 4 May 1821.
141. Cooter, *Phrenology in the British Isles*, 217; M. H. Kaufman and N. Basden, 'Items relating to Dr. Johann Spurzheim (1776–1832) in the Henderson Trust Collection, formerly the Museum Collection of the Phrenological Society of Edinburgh: With an abbreviated Iconography', *Journal of Neurolinguistics* 9 (1995–96), 304.

12
Afterword: New Directions?

CHARLES W. J. WITHERS AND PAUL WOOD

Understanding the importance of science and medicine in the Enlightenment will depend upon *what* and *who* is studied: a particular subject or person; national and international exchanges of books, instruments and ideas; or local practices and institutional differences. It will depend, too, upon *why* and *how* we look at Enlightenment science, for Scotland and elsewhere. For contemporaries, the unity of the sciences was a guiding belief of Enlightenment thinkers in Scotland. In one sense, they saw this unity in terms of the exchange of ideas between individual sciences. Dugald Stewart observed that 'the modern discoveries in astronomy and in pure mathematics, have contributed to bring the art of navigation to a degree of perfection formerly unknown. The rapid progress which has been lately made in astronomy, anatomy and botany, has been chiefly owing to the aid which these sciences have received from the art of the optician'.[1] In another, and perhaps deeper, sense they followed David Hume in seeing this unity as rooted in the study of human nature: 'Even *Mathematics, Natural Philosophy, and Natural Religion*, are in some measure dependent on the science of MAN; since they lie under the cognizance of men, and are judg'd of by their powers and faculties'. As Hume further noted, 'There is no question of importance, whose decision is not compriz'd in the science of man; and there is none, which can be decided with any certainty, before we become acquainted with that science'.[2]

This book has documented something of what Hume and Stewart would have recognised as science, including medicine, in the Scottish Enlightenment. The essays collected here have paid attention to what particular sciences were understood to do, to what individual men of science did and to the specifics of scientific conduct and exchange. It is our hope, moreover, that the essays have affirmed rather than diminished the claims of contemporaries as to the importance of natural knowledge in the Enlightenment, and that they have done so by enriching our understanding of what science and medicine were and what they meant for their various audiences. Our general intention has been neither to privilege science and medicine above and beyond the social and intellectual worlds of which they were part, nor to assume that the essays offered here provide an exhaustive treatment of our topic. We have, however, sought to build upon and extend established work in the conviction that the study of science and medicine ought to form a part of the analysis of the Enlightenment, in Scotland and elsewhere.

What emerges from these essays, then, is not a single definitive answer to the question 'What was the nature of science and medicine in Enlightenment Scotland?', so much as more specific questions and answers to do with given discourses, particular people and certain locations and scientific practices. Morrison-Low suggests that, until very late in the eighteenth century, Scottish science largely depended upon instruments made outside Scotland and that we must be careful to recognise differences between the cultural, economic and practical value of instruments as well as between the different markets and audiences for them. Withers argues that social factors rather than instrumental capabilities contributed to the public credibility of those map makers who drew upon mathematical knowledge in drawing their maps of the enlightened nation. Emerson and Wood discern geographical, social and other differences within Glasgow's enlightened 'scientific community', as well as emphasising the role of the university and the particular character of that mercantile city as explanatory features of the nature of Glasgow's enlightened science. For Grabiner, the utility of the mathematical sciences was, in Maclaurin's world, differently apparent in maps, molasses and mortality tables. For Guerrini and for Macdonald, the probing instruments of medical enquiry were not expensive technical artefacts. They were one's eyes and ears and, importantly, one's hands as physicians like George Cheyne and Robert Cleghorn observed, conversed and, in ways not before done, hesitatingly touched their way to diagnoses and, in time, to new medical practices. For Brown and for McDougall, ideas in Enlightenment science and medicine were disseminated through the material worlds of booksellers in which personal politics, print runs, postage rates and the speed of coaching services must be taken into consideration in determining how ideas travelled. Hartley shows how fieldwork mattered, both as the means to a practical understanding of Scotland's mineralogical resources and as a site of theoretical debate between adherents of opposing explanatory views in the earth sciences. Waterston's account of George Steuart Mackenzie shows how one late Enlightenment figure sought to make sense of wider scientific concerns and the practical demands of estate management; for Mackenzie, Enlightenment science was grounded, literally, in the soils of Ross-shire and in the igneous rocks of Iceland.

Collectively, these essays have raised questions to do both with the level or scale on which we may study science and medicine in the Enlightenment, and with the discursive practices that demarcate and connect them. It may be helpful to distinguish between different levels of historical analysis, dealing with individuals, particular institutions, given cities, national engagements with science and medicine and with questions of international collaboration. These essays have shown such levels to be connected by a range of intellectual, social and material practices: by the exchange of instruments, books and ideas; by the movement of people as well as of ideas; by the emergence of print cultures for science and for medicine; by the testing of ideas in different ways – through experimental method in the laboratory, by direct encounter in the field, in private conversation and even through theatrical performance.

What brought these levels and practices together? How may we best understand the Enlightenment worlds in which science and medicine were undertaken – by individuals who were also members of institutions, readers of books, purchasers of instruments? One answer rests in that shared belief in the unity of the sciences that we have noted. A second answer has to do with a further shared belief, namely in the utility of knowledge for Scotland's different publics. A third has to do with the intrinsic sociability of the Enlightenment and in the ways science was part of public culture.[3] In these senses, what characterised the production, dissemination and disputation of science and medicine in Enlightenment Scotland was, in essence, no different from what was occurring elsewhere.

In her afterword to a collection of essays on the sciences in Enlightenment Europe, Lorraine Daston suggests, for example, that the enlightened commitment to public utility throughout Europe should be thought of as subdivided. Rather than focus alone on the question of 'material improvements', we should recognise also the sense in which 'public utility might refer to edification in the form of physico-theology, anthropomorphized natural history, public demonstrations on the wonders of anatomy and electricity, or even lectures on the intricacies of the new analysis'.[4] In the Scottish context, the idea of utility underlying Maclaurin's interest in mathematical precision and the measurement of the state's capacity in the barrel or on the map should not straightforwardly be equated with that idea of utility for best medical practice which demanded that men like Cleghorn develop a more tolerant bedside manner by listening more and talking less. For others, utility rested not in the content of books but in hearing public lectures, or in purchasing better instruments. Utility as differently understood may demand, then, that we pay more attention to the different audiences and markets for enlightened science and medicine.

If it is the case, as Daston further noted, that 'the enlightened sciences were central to the project of the Enlightenment not because of their achievements in theory and application, but because they claimed to reveal the true identity of nature, heroine of all narratives',[5] then we would hope that attention continues to be paid in the future to the nature, place and conduct of science and medicine in Enlightenment Scotland. Individual authors here offer suggestions as to the possibilities for further work. While it is not our intention either to be prescriptive or proscriptive about the future historiographical development of the field, several related 'new directions' suggest themselves as worthy of attention. The first has to do with the different ways of scientific knowing. The second concerns the geographies of scientific and medical knowledge. The third has to do with books, the book trade and print culture. Finally, in proposing these topics as possible new research agenda for the case of enlightened science in Scotland and for its different audiences, we would also urge that moves to further understanding should not lose sight of the world beyond Scotland.

Different Ways of Knowing

In his *Ways of Knowing* (2000), John Pickstone offers a novel approach to the history of science, technology and medicine. His approach has four key features: a long timescale (the history of the last 300 years); a broad range in which the histories of science, technology and medicine are linked to other human histories; understanding science-technology-medicine as having constituent elements – '*ways of knowing*' as he terms it; and, finally, seeing these ways of knowing 'as forms of *work* related to various ways of making and mending'.[6] Pickstone distinguishes four ways of knowing: natural history, analysis, experimentalism and 'technoscience' (which largely falls outwith the period of Enlightenment science and is thus irrelevant to our concerns here). Natural history is, essentially, about classification and taxonomy. Analysis seeks knowledge by dissection. Experimentalism, which Pickstone sees as 'a set of practices chiefly institutionalised and theorised from the mid-nineteenth century', builds upon analysis in order to promote synthesis and to extend understanding of the systematic production of novelty.[7]

There is not the space here to discuss Pickstone's approach in detail, but it is nonetheless suggestive in thinking about the ways in which what passed for credible scientific and medical knowledge in the Scottish Enlightenment was created. How far should we understand Enlightenment science to be classificatory and taxonomic and how far was it analytic or experimental? Medicine was, in one context, literally concerned with dissection. But analytic techniques also depended upon reading and feeling and representing surface features – of the bodies of Scots as well as of the mapped landscape – and in working inferentially from what could be seen, touched and measured to what lay beneath and was unobservable.

It is not strictly correct, of course, to regard the culture of experimentalism as alone a nineteenth-century set of practices.[8] Sir James Hall's laboratory experiments in proof of Huttonian theory demonstrate a private attempt to replicate large-scale natural processes under controlled conditions. Many other Enlightenment scientists experimented. Virtually all of Francis Home's *Principles of Agriculture and Vegetation* (1757), for example, was taken up with reports upon seeds, different soil conditions and in testing different plant-soil-growing conditions as 'analogues' of Scottish agriculture. It was because 'the art of husbandry' had not been subject to what Home termed the 'princyples of science' that Scotland's agriculture was in the poor state it was: 'This art is, in general, carried on by those whose minds have never been improved by science, taught to make observations or draw conclusions, in order to attain the truth'.[9] The practical concerns that men such as William Cullen, Joseph Black and James Hutton had with experimentation in philosophical chemistry might contradict Pickstone's point about experimentalism. Yet it is not transparently clear in all of the work understood as Enlightenment science that descriptive and classificatory practices took second place to analysis and the analytic method.

New Directions?

Attention to 'ways of knowing' might also prompt us to ask *how* what was known *was* known? Fieldwork, for example, was a post-Enlightenment means of student instruction.[10] For the relatively few who were students in universities, Enlightenment science came through direct instruction, conversation with professors and fellow students, and observation. But not everybody could see for themselves. Taking seriously what one heard in a lecture class, read in a book or had relayed by a third party through conversation demanded that criteria of credibility and trust had to be established, even if only implicitly. As Steven Shapin has shown, establishing trust in those who undertook science was a crucial part of the move to empirical rigour and of 'modern' scientific methodology from the late seventeenth and early eighteenth centuries. Trust in science and in the scientist is no less important today.[11] How was trust in the claims of enlightened men of science and physicians engendered in eighteenth-century Scotland? How did people come to believe what they were told? Answering such questions arguably requires attending to the seemingly prosaic nature of ways of knowing: hearing, seeing, reading a book or being told by others.

As Peter Burke shows, understanding the nature of knowledge demands that we know how knowledge was professed, established, classified, controlled and sold as well as trusted.[12] Burke's thesis can usefully be applied to the case of science and medicine in Enlightenment Scotland. The enlightened in Scotland, for instance, wrote to one another extensively. Much of our knowledge of science and medicine in the Enlightenment is gleaned from surviving personal letters. Yet as Andrea Rusnock has argued regarding the Royal Society between 1700 and 1750,

> As a scientific method, correspondence raises a number of issues concerning the social relations of collaborators. While it is appealing to regard correspondence networks as channels of communication among more-or-less equals – a vision of a democratic science or Baconian fact-gathering – such was not the case in the eighteenth century. Hierarchy, centralization and social status shaped the form and content of science by correspondence.[13]

If this was true of London, was it so of enlightened men of science in Edinburgh, Glasgow and Aberdeen?

Geographies of Knowledge

Several of the contributors raise questions to do with the geographical dimension of knowledge in the Enlightenment. Morrison-Low, Withers and Grabiner variously identify differences between and within the urban audiences and markets of the Scottish Enlightenment as topics that would repay further attention. Emerson and Wood extend our understanding of the geographical origins of Glasgow's scientific community and offer the view that it is not appropriate to think of *the* Glasgow enlightenment, but of different enlightenment*s* within that city. Such claims have obvious comparative potential. It is also clear that different sites were differently employed in the making and witnessing of Enlightenment science and medicine: the university with its teaching rooms and regulatory practices, the analytic space of the anatomy theatre, and the classificatory display spaces of the botanical garden and the museum.

These claims accord with what others have seen as a 'geographical turn' in the history of science.[14] This work has been marked by a move away from a belief in science's universalist and transcendental nature towards an acceptance of its local and relativist nature. This, in turn, has demanded that knowledge claims be understood to secure their credibility 'not through absolute standards but through the workings of *local* causes operating in contexts of judgement'.[15] Attention has therefore been paid to particular sites of knowledge making: coffee houses, lecture rooms, the library, the public house and especially the laboratory.

The laboratory has since the seventeenth century been seen as 'the pre-eminent site for making knowledge in the experimental sciences'. It has been regarded as such

> because it straddles the realms of private seclusion and public display . . . On the one hand, the laboratory is a place where valuable instruments and materials are sequestered, where skilled personnel seek to work undisturbed, and where intrusion by outsiders is unwelcome . . . On the other hand, what is produced there is declaredly "public knowledge"; it is supposed to be valid universally and available to all.[16]

Students of eighteenth-century Scotland might ask whether this summary description fits what is known of laboratories and their use in the Scottish context, not least since empirical studies of the laboratory as a scientific site have focused more on the seventeenth and the nineteenth centuries and have relatively neglected the Enlightenment period.

There are also connections to be made between sites and with ways of knowing. Managing the tensions in the laboratory between the private and the public realm also involves making social distinctions concerning who had access to natural knowledge and who was to be trusted in the making of it. In turn, issues of trust arise regarding the discursive forms in which knowledge moved in and out of sites of knowledge production, such as written reports, mathematical and chemical formulae and verbal accounts. What, in short, were the languages of Enlightenment science? How did they travel between different sites? It is just such questions that have prompted interest in how knowledge travels. As Shapin has put it, 'we need to understand not only how knowledge is made in specific places but also how transactions occur between places'.[17] In several ways, then, we need to know more about the place of Enlightenment science and medicine as a geographical matter: of how ideas moved between sites, of the movement of practitioners and audiences between different spaces of knowledge making and of the constitutive role of those different spaces.

Books, the Book Trade and Print Culture

As Brown and Macdougall here suggest, books were essential instruments for both the already and the would-be enlightened. Macdonald also sees a key role for the book in medical practice, albeit of a rather different sort from printed and published works, for Robert Cleghorn used his casebooks to consolidate his doubts, directives and diagnoses. Their chapters thus extend recent claims concerning the place of the book in the histories of medicine and of science.[18] For Enlightenment Scotland, Richard Sher has stressed the international nature of the country's book culture, the importance of studying book culture in relation to modes of oral expression and, not least, of treating book culture 'as a broad-based phenomenon, involving the complex interplay of authors, members of the book trade, books themselves (considered as texts, artefacts, and commodities), and their readers, and as one of the central components of the culture of the literati that lay at the heart of the movement known as the Scottish Enlightenment'.[19]

These are important points to note for future research on the place of science and medicine in the culture of the book in Enlightenment Scotland. Recognising with Sher the connections between them, in future it may also be worth making clear in further study the distinctions within and between the categories 'book', 'book trade' and 'print culture'. These themes come together in relation to such issues as production, ownership and readership. They thus connect with different ways of knowing and with the geographies of knowledge hinted at above, but they are also capable of more precise separation. Was there, for instance, a market for children's books in science? Were there books of science and medicine aimed at women? How did the periodical literature as opposed to the printed book promote and debate public interest in science? Since print culture is not just textual but visual, can we discern developing or different traditions of scientific draughtsmanship in the representation of science and of medicine? Ludmilla Jordanova has made important advances, for example, in understanding the visual image as a form of Enlightenment print culture, particularly in the medical world.[20] How did Scotland's enlightened men of science and physicians portray themselves?

In advancing these and related questions about books, the book trade and print culture, it may be helpful also to distinguish between the social history and historical geography of book *ownership*, and changes in the nature of science and medicine evident from what we might term the *genealogy* of books. As Richard Yeo has shown, the scientific dictionary and the encyclopaedia were distinct features of Enlightenment book culture. But by the end of the eighteenth century, major encyclopaedias were no longer the classificatory texts they had been in the Enlightenment. They 'were now the province of experts contributing long disciplinary treatises, some of which served as opportunities for specialist controversy rather than as careful public statements of the present consensus in various fields'.[21]

To such questions of ownership and of genealogy, we might add those to do with translation and the practices of reading. Recent work by Nicolaas Rupke and James Secord, albeit dealing with a later period, is suggestive here. Rupke has shown how the translation into Dutch and German of Robert Chambers' *Vestiges of the Natural History of Creation*, first (and anonymously) published in English in 1844, was used to invest the text with new meanings which subtly reflected the political beliefs of different scientific communities within the countries concerned. In a longer study of Chambers' *Vestiges* and its British reception, Secord has identified different 'geographies of reading' of the text between the polite and aristocratic circles of genteel society in London, and the merchants and clerics of Liverpool.[22] In terms of eighteenth-century Scotland, we need to ask questions about how the works of foreign authors translated – in the geographical and epistemological senses of that term – into enlightened circles and how Scottish science and medicine were understood elsewhere.

Thomas Reid and the Potato: Interpreting Science, Medicine and Enlightenment

Among the surviving manuscripts of Thomas Reid, there is a single sheet with, on one side, an algebraic problem, and, on the other, a geometrical problem and a short note in Reid's hand. The note reads 'April 24 1758 Planted the Northermost Row of my Pittatoes of Small Seed uncut the two next, of the largest seed Cut the fourth of Midling Seed Cut. And Same day planted 1700 Cabbage at 7 Score to the hundred'.[23]

Knowing that Reid cultivated potatoes and that, earlier that year, he had carried out experiments on their chemical properties, is of interest, for it was only in the mid-eighteenth century in Scotland that the potato began to be a part of the staple diet of ordinary Scots rather than a luxury food item. To whom Reid's working with potatoes in April 1758 is of interest depends, of course, upon wider considerations. Most historians of the Enlightenment know Reid as a leading Scottish and European philosopher, not as a 'tattie howker'. Reid, no doubt rightly, has no place in the history of eighteenth-century Scottish agriculture. Particular interests and disciplinary preoccupations will determine at the outset whether one is more interested in the potato (the historian of Scottish life and diet), the man and his ideas (the historian of the Enlightenment), the soil, or, even, in knowing what sort of instrument was used to work the ground (the historian of technology). But to see Reid as *only* one thing and not another and hence to miss the connections between Enlightenment ideas and the material conditions in which they were rooted, is, we would argue, to undertake a limited sort of historical enquiry.

Understanding the place and nature of science and medicine in the Scottish

Enlightenment will not be advanced as it might be, we contend, if all that is studied is simply the conjunction of science and medicine and Scotland in the Enlightenment. There are limits, of course, to all forms of scholarly enquiry. Yet historians and others with an interest in these topics would do well to consider more exactly the interrelations between the Enlightenment and other sorts of historical evidence, namely the social, political and constitutional histories of eighteenth-century Scotland. This is not to say that it is not happening at present, only to urge that it happen more systematically in the future.[24] We hope, too, that researchers might in future look less at Scotland in isolation. In developing and extending the arguments discussed here, they might also examine Scottish science and medicine in the wider context of Enlightenment in its British and international dimensions.[25]

Notes

1. Dugald Stewart, *Elements of the Philosophy of the Human Mind* (Edinburgh, 1867), 9.
2. David Hume, *A Treatise of Human Nature*, eds. David Fate Norton and Mary J. Norton (Oxford, 2000), 4. This point about the unity of the sciences and about the science of man as the study of human reason is emphasised by Alexander Broadie in discussing 'Science and the Enlightenment' in his *The Scottish Enlightenment* (Edinburgh, 2001), 186–218.
3. On science and public culture in the Enlightenment, see Jan Golinski, *Science and Public Culture: Chemistry and Enlightenment in Britain, 1760–1820* (Cambridge, 1992); Paul Wood, 'Science, the Universities and the Public Sphere in Eighteenth-Century Scotland', *History of Universities* 13 (1994), 99–135; Charles W. J. Withers, 'Towards a History of Geography in the Public Sphere', *History of Science* 37 (1999), 45–78.
4. Lorraine Daston, 'Afterword: The Ethos of Enlightenment', in William Clark, Jan Golinski and Simon Schaffer (eds.), *The Sciences in Enlightened Europe* (Chicago and London, 1999), 495–504, at p. 496.
5. Daston, 'Ethos of Enlightenment', 503.
6. John Pickstone, *Ways of Knowing: A New History of Science, Technology and Medicine* (Manchester, 2000), 5.
7. Pickstone, *Ways of Knowing*, 13.
8. For a seminal discussion of this issue, see Thomas S. Kuhn, 'Mathematical versus Experimental Traditions in the Development of Physical Science', in his *The Essential Tension: Selected Studies in Scientific Tradition and Change* (Chicago and London, 1977), 31–65.
9. Francis Home, *The Principles of Agriculture and Vegetation* (Edinburgh, 1757), 5.
10. David E. Allen, 'Walking the Swards: Medical Education and the Rise and Spread of the Botanical Field Class', *Archives of Natural History* 27 (2000), 335–67.
11. Steven Shapin, *A Social History of Truth: Civility and Science in Seventeenth-Century England* (Chicago and London, 1994); on trust and credibility in modern science, see Jane Gregory and Steve Miller, *Science in Public* (Cambridge, MA, 1998).
12. Peter Burke, *A Social History of Knowledge: From Gutenberg to Diderot* (Cambridge, 2000).

13. Andrea Rusnock, 'Correspondence Networks and the Royal Society, 1700–1750', *British Journal for the History of Science* 32 (1999), 155–69, at p.169.
14. For a review of literature on this topic, see Crosbie Smith and Jon Agar (eds.), *Making Space for Science: Territorial Themes in the Shaping of Knowledge* (London, 1998). For a particular discussion of the topic in relation to Scotland, see Charles W. J. Withers, *Geography, Science and National Identity: Scotland since 1520* (Cambridge, 2001).
15. Adir Ophir and Steven Shapin, 'The Place of Knowledge: A Methodological Survey', *Science in Context* 4 (1991), 3–17, at p.6.
16. Jan Golinski, *Making Natural Knowledge: Constructivism and the History of Science* (Cambridge, 1998), 84.
17. Steven Shapin, 'Placing the View from Nowhere: Historical and Sociological Problems in the Location of Science', *Transactions of the Institute of British Geographers* 23 (1998), 3–12, at p.7.
18. Marina Frasca-Spada and Nick Jardine (eds.), *Books and the Sciences in History* (Cambridge, 2000).
19. Richard B. Sher, 'Science and Medicine in the Scottish Enlightenment: The Lessons of Book History', in Paul Wood (ed.), *The Scottish Enlightenment: Essays in Reinterpretation* (Rochester, NY, 2000), 99–156, at p.148.
20. Deanna Petherbridge and Ludmilla Jordanova, *The Quick and the Dead: Artists and Anatomy* (London, 1997); Ludmilla Jordanova, *Defining Features: Scientific and Medical Portraits, 1660–2000* (London, 2001).
21. Richard Yeo, *Encyclopaedic Visions: Scientific Dictionaries and Enlightenment Culture* (Cambridge, 2001), 278.
22. Nicolaas Rupke, 'Translation Studies in the History of Science: The Example of Vestiges', *British Journal for the History of Science* 33 (2000), 209–22; James A. Secord, *Victorian Sensation: The Extraordinary Publication, Reception, and Secret Authorship of Vestiges of the Natural History of Creation* (Chicago and London, 2000).
23. Aberdeen University Library, MS 2131/5/II/2.
24. It is noteworthy how, for example, in T.M. Devine and J. R. Young (eds.), *Eighteenth Century Scotland: New Perspectives* (East Linton, UK, 1998), only two essays (in a total of seventeen) explicitly address the Scottish Enlightenment. In contrast, David Allan gives much more attention to 'Belief' and to 'Ideas' in his *Scotland in the Eighteenth Century* (London, 2002).
25. It is interesting that, in his latest work on the subject, Roy Porter sees the Enlightenment both as a *British* phenomenon and as the basis of the creation of the modern world: Roy Porter, *Enlightenment: Britain and the Creation of the Modern World* (London, 2000).

Index

Aberdeen, 6, 18, 24, 34, 36, 38, 40–41, 62, 63, 71, 75n, 79, 80, 82, 95, 104, 11, 113n, 125n, 147, 172, 173, 174, 176, 178, 179, 186, 331
   Castlehill Observatory, 40
   enlightenment, 6, 8, 110
   Episcopalian enlightenment, 186
   Philosophical Society, 6, 101, 107, 126n
   politics, 173
   religion, 173, 187n
   St. Nicholas Church, 174
   scientific community, 81, 82
   See also King's College, Aberdeen, and Marischal College, Aberdeen
Aberdeenshire, 24, 70
achromatic lenses, 30
Ackermann, Silke, 45n
Ackworth, 195
Acts of Parliament of Scotland, 27
Adair, James, 114n
Adair, James Makittrick, 115n, 262
Adair, John, 27
   A True and Exact Hydrographical Description of the Sea Coast and Isles of Scotland, 27
Adam, Dr. Alexander, 302
Adam, Robert and James, 37
Adam, William, 118n
Adams, Ian, 50n 77n
Adams, Thomas, 272
Adamson, Mr., 267–69
Addison, W. Innes, 76n, 121n
Adie, Alexander, 22, 27, 28, 36, 41, 45n
Adie family, 23, 37
Adie, John, 28
Adie, Patrick, 28
Agar, Jon, 73n, 336n
Agassiz, Jean Louis, 314
Agnarsdottir, A., 322n
Agnew, Sir Andrew, 27
agriculture, 58, 197, 296n, 301, 304–8, 330, 334
Ainslie, John, 59, 65–68, 69, 70, 72
   Comprehensive Treatise on Land Surveying, 66

The Gentleman and Farmer's Pocket Book, 65–66
'Scotland Drawn and Engraved', 70
air pumps, 21, 29, 35, 39, 44n, 53n
Aitken, John, 224, 225
Albinus, Bernard, 218
Albury, W. R., 162n
d'Alembert, Jean le Rond, 55, 73n
   Encyclopédie, 55
Alexander's Craig, 289
algebra, 150, 166n, 197
Alison, Rev. Francis, 156
Allan, Alexander, 40
Allan, David, 113n, 116n, 336n
Allan, Thomas, 294, 311, 313, 323n
Allen, David E., 284, 298n, 335n
Allen, Elias, 24, 26, 46n
Alps, 290, 314
Alston, Charles, 195
America, 8, 34, 71, 84, 216, 221, 223, 232
   Central, 91
   instrument trade in, 27
   North, 91, 228
American Revolution, 305
Amory, Hugh, 214n, 236n
Amsterdam, 24
analysis and synthesis, 158–59, 161
anatomy, 81, 86, 92, 96, 97, 122n, 197, 264, 265, 272, 305, 314, 327, 329
Anderson, John, 7, 38, 39, 52n, 75n, 81, 85, 86, 87, 93, 95, 97–100, 101, 102, 103–4, 105, 106, 111, 113n, 114n, 119n, 120n, 121n, 122n, 123n, 127n, 128n, 278n
   Anderson's University, 98–100
   his rain gauge, 105
Anderson, John, WS, 317
Anderson, P. J., 117n
Anderson, R. G. W., 10n, 38, 43n, 45n, 46n, 51n, 52n, 319n
Anderson, William, 99, 265, 278n
Anderson's University (1828), 277n
Anderston, 101
Andrewes, William J. H., 44n

Andrews, Jonathan, 129n, 275n, 276n
Anglicanism, 147
Angus, 82
Angus, James, 266
Annan, 287
Anne (Queen of Britain), 180, 184
annuities, 143, 155–58
Aphir, Adir, 336n
apothecaries, 257, 272
Arbroath, 46n
archaeology, 301, 316
Archer, Mr., 128n
Argathelians, 93, 148, 149
Argyll, 37, 66–67, 70, 82, 286
Aristotelianism, 60
Aristotle, 183
arithmetic, 60, 148
Armstrong, Andrew and Mostyn, 67
  A Scotch Atlas, 67
Arnot, Hugo, 202
Arran, 283, 285, 286
Arrowsmith, Aaron, 59, 68–69, 70, 72, 77n, 78n
  Memoir Relative to the Construction of the Map of Scotland, 68–69, 70
Arthur, Archibald, 85, 105, 123n, 128n
artisans, 79, 81, 82, 83, 84, 94
Ash, Marinell, 48n
astrolabes, 24, 28, 46n
astronomy, 6, 17, 27, 56, 60, 61, 62, 63, 81, 86, 92, 93, 95, 105–6, 126n, 151, 161, 291, 316–317, 327
  aberration of light, 127n
  astronomical clocks, 40
  shape of the earth, 143, 151–52
Atkinson, Thomas, 129n
Atlantic
  Ocean, 8
  world, 8, 107
Auchtermuchty, 45n
Auenbrugger, Leopold, 256
  Inventum novum, 256
Auld, 196, 198, 209n
Ayrshire, 27, 70, 82

Baconianism, 79, 80, 152, 160, 282–83, 284, 295
Baillie, Joanna, 295
Baillie, Matthew, 86, 266
  Morbid Anatomy, 266
Bain, John, 65
balances, 39, 45n, 53n
Balfour, John, 193, 194, 196, 198, 225, 228, 232, 274n
Balmanno, John, 99, 115n, 121n, 276n
Baltimore, 34

Banffshire, 34, 70, 82
Banks, Sir Joseph, 67, 306, 312, 317, 321n, 322n
Banks, R. E. R., 322n
Barclay, William, 33
Barfoot, Michael, 14n 15n, 107, 128n, 208n
Barker, Mr., 304
Barnes, Barry, 10n
barometers, 21, 22, 37, 38, 44n, 53n, 93, 126n
Barrell, Colburn, 229
Barrois, Pierre-Théophile, the younger, 222, 226, 227, 228, 235n, 236n
Baron, J., 326n
Barrow, G. W. S., 113n
Bartram, James, 205
Bartram, William, 205
Bath, 180, 182
Bath waters, 181
Beattie, James (naturalist), 60
Beattie, James (moralist), 208n
Beddoes, Thomas, 102, 103, 124n, 277n
Bedemar, Count, 316
Bedford, Hilkiah, 46n
Bedini, Silvio, 27–28, 30, 47n
Beighton, Henry, 149
Bell, A. S., 76n, 212n, 324n
Bell and Bradfute, 193, 194, 203, 209n, 213n, 232
Bell, Andrew, 196, 197, 211n, 218
Bell, Benjamin, 193, 209n, 219, 224, 225, 266
  A System of Surgery, 193, 209n, 219, 223, 227, 231, 232
Bell, John, 193, 213n, 228
Bell, Nora K, 279n
belles lettres, 89, 101
Bendall, Sarah, 50n
Bengal, 56
Bennet, Lambert, 223, 235n
Bennet, W. S., 213n
Bennett, J. A., 43n, 50n
Bennion, Elisabeth, 45n
Benson, K. R., 188n
Benthamism, 111
Berg, Maxine, 52n
Berkeley, George (Bishop of Cloyne), 150–51, 154, 158, 160, 166n
  The Analyst, 150–51
Berlin, 224
Bermingham, Ann, 48n
Bernoulli, Daniel, 158
Bernoulli, Jakob, 157
Berry, Christopher J., 74n, 280, 296n
Berthollet, Claude Louis, 320n

Index

Berwick, 257, 274n
Bewley, Christine, 122n
Bird, John, 38, 51n
Birkbeck, George, 41, 82, 100
Birmingham, 23, 85, 86, 221
  Reference Library, 43n
  Soho Works, 39
Black, Jeremy, 297n
Black, Joseph, 1, 3, 17, 28, 38, 41, 81, 82, 85, 86, 94, 97, 102, 103, 107, 118n, 120n, 122n, 125n, 225, 258, 301, 302, 303, 311, 314, 319n, 330
Black, William, 60
Blackie, William, 37
Blackwood's Magazine, 200
Blaeu, Johannes, 24, 45n, 54
  Atlas novus, 24
Blaeu, Willem, 45n
Blaikie, Francis, 229, 239n
Blondel, C., 44n
Bloor, David, 10n
Board of Agriculture, 309, 321n
Board of Longitude, 29, 30, 61
Board of Ordnance, 29
Board of Trustees for Fisheries and Manufactures in Scotland, 19, 40, 148, 154
Bolle, Bert, 51n
Bologna, 90
Boney, A. D., 13n, 116n, 120n, 129n
Bordeaux, 82, 303
Borthwick, George, 217
Boscovich, Roger Joseph, 167n
Bosquillon, Edouard-Françoise-Marie, 227, 228
Boston, 34
Boswell, James, 202
botany, 67, 81, 86, 96, 98, 198, 296, 327
Boulton, Matthew, 39
Bourignon, Antoinette, 177, 179, 180, 185, 188n
Bowler, Peter, 188n
Bowness, 276n
Boyd, Adam, 276n
Boyer, Carl B., 166n, 170n
Boyle, Robert, 92, 173, 174
  Boyle Lectures, 91, 178
Bradford, Thomas, 230, 231
Bradley, Rev. James, 123n
Brandenburg, 290
Bravo, Michael, 73n
Breslau, 156
Brewer, John, 15n, 29, 44n, 48n, 74n, 168n

Brewster, Sir David, 28, 37, 315, 324n, 325n
  A New Treatise on Philosophical Instruments, 37
Bright, Dr. Richard, 312, 323n
Brisbane, Matthew, 89–90, 116n
Brisbane, Thomas, 96, 115n, 120n
Brisbane, Sir Thomas Makdougall, 28
Bristol, 45n, 182, 277n, 312
Britain, 143, 221, 222, 232, 257, 272, 302, 306
  as a consumer society, 29
  audience for science in, 32
  enlightenment in, 335
  growth of industry in, 86
  Hanoverian, 56
  North, 154
  north coast of, 152
  provincial science in, 111
  religion in, 173
British
  Admiralty, 59
  army, 30
  colonies, 85, 221, 308
  Isles, 30
  navy, 30
  state, 29–30
British Linen Company, 154
Broadie, Alexander S., 74n, 335n
Broman, Thomas, 206, 214n
Brookes, Richard, 228
  Natural History, 228
Brown, James, 102
Brown, John, 198, 223
  Elementa Medicinae, 219
Brown, Joyce, 46n
Brown, Olivia, 50n
Brown, Robert, 322n
Browning, Andrew, 116n
Brownrigg, William, 271
Brück, Hermann A., 48n, 325n
Bruland, Kristine, 52n
Brumberg, Joan Jacobs, 189n
Bryan, Dr. H. A., 233n
Bryce, Alexander, 58, 74–75n
Bryden, D. J., 2, 10n, 14n, 22, 25, 27, 28, 29, 38, 42, 42n, 44n, 45n, 46n, 47n, 48n, 49n, 50n, 51n, 52n, 53n, 121n, 127n
Bryson, Robert, 30
Buchan, William, 193, 195, 196–97, 198, 200, 202, 203, 205, 211n
  Domestic Medicine, 193, 195, 196–97, 198, 205, 210n
Buchanan, James, 92, 93
Buchanan, M. S., 129n

Buchwald, Jed Z., 162n, 163n, 168n
Budenz, Julia, 162n
Buego, John, 218
Burke, Edmund, 55, 73n
Burke, John G., 51n
Burke, Peter, 331, 335n
Burnett, James, Lord Monboddo, 104
Burnett, J. E., 48n
Burney, Fanny, 230
  Cecilia, 230
  Evelina, 230
Burns, Allan, 120n, 264, 265, 266, 270, 272, 278n, 279n
Burns, John, 99, 120n, 122n, 260, 264, 265, 267, 268, 269, 272, 277n, 278n
  The Anatomy of the Gravid Uterus, 264
Burns, Robert, 191, 204–5, 208, 260, 276n
Burnswalk, 288
Bute, 82
Butler, J. A. V., 128n
Butt, John, 52n, 121n, 122n
Butts, Robert E., 162n
Buxton, 182
Bynum, W. F., 13n, 14n, 279n

Cable, John A., 48n, 52n 118n
Cadell, Thomas, 194, 216, 221, 229, 232, 235
Caithness, 82
calculus, 145–46, 150–51, 159
Calder, Angus, 48n
Calder Iron Works, 304
Calder, Jenni, 43n
Cameralism, 59, 75n
Camerini, Jane, 298n
Cameron, George, 218
Cameronians, 194, 209n
Campbell, A. M., 52n
Campbell, Archibald, of Knockbuy, 165n, 166m
Campbell, Archibald, 1st Earl of Ilay and 3rd Duke of Argyll, 5, 33, 37–38, 63, 86–87, 92–93, 95, 105, 109, 114n, 117n, 143, 148, 154, 165n
Campbell, C. A., 117n
Campbell, Rev. Colin, 145, 147, 163n, 164n, 165n
Campbell, John, 2nd Duke of Argyll, 86, 93, 148
Campbell, Neil, 319n
Campbell, Rev. Neil, 119n
Campbell, R. H., 12n, 43n, 76n, 117n, 125n, 130n

Campbeltown, 66
Campvere, 224
Candlish, James, 128n
Cannon, Susan Faye, 301, 318n
Cant, R. G., 19, 39, 43n, 45n, 52n, 76n
Cantor, G. N., 2, 10, 11n, 14n, 118n, 166n, 324n
Capaldi, Nicholas, 297n
Caribbean, 84
Carlisle, 66, 276n
Carlyle, Rev. Alexander, 122n, 123n, 149, 166n
Carmichael, Gershom, 91–92, 117n, 147
Carmichael, Patrick, 87
Carolina, 90
Caroline (Queen of Britain), 35
Carrère, J.-F., 228
  Bibliothèque Littéraire Historique et Critique, 228
Carson, Dr. John, 230, 237n
Carstares, Rev. William, 90
Carter, H. B., 306, 321n, 323n
Carter, J. J., 12n, 165n, 187n, 188n
Cartesian dualism, 183
Cartesianism, 60, 92, 147, 150, 151, 158, 159, 176, 178
cartography, 7, 14
Cassini, Jacques, 167n
Catherine (Tzarina of Russia), 87, 88
Catholicism, 8, 173
Cave, Roderick, 236n
Chalmers, Rev. Thomas, 108
Chambers, Henry Augustus, 212n
Chambers, Robert, 276n, 334
  Vestiges of the Natural History of Creation, 233n, 334
Chandrasekhar, Subrahmanyan, 167n
Charles II (King of England and Scotland), 27
Charleston (South Carolina), 34, 229
Charteris, Rev. Samuel, 194, 195
Checkland, S. G., 114n
chemistry, 1, 3, 4, 6, 17, 19, 34, 40, 57, 74n, 81, 86, 93, 94, 96, 97, 98, 99, 102, 103, 109–10, 111, 199, 257, 258, 301, 302–4, 305, 318, 321, 330, 334
Cheyne, George, 8, 172, 174, 178–86, 328
  The English Malady, 180
  An Essay of Health and Long Life, 180
  An Essay on the Gout, 180
  Essay on Regimen, 182–84
  Natural Method of Cureing the Diseases of the Body, 172, 183

Index

Philosophical Principles of Natural Religion, 178–80, 183
   on diet, 180, 184, 185
Chitnis, Anand, 4–5, 11–12n
   The Scottish Enlightenment, 4
Chopart, François, and Desault, Pierre-Joseph, 228
   Traité des Maladies Chirurgical, 228
chorography, 24
Christianity
   Biblical exegesis and natural philosophy, 177
   Biblical prophecy, 183
   doctrines of, 145, 160
   Scripture, authority of, 150
   varieties of, 173
Christie, J. R. R., 2–3, 4, 5, 9, 10n, 11n, 12n, 18, 42n, 51n, 94, 113n, 118n, 166n, 301, 318n
Christison, Alexander, 302, 319n
chronometers, 21, 29, 61
Church of Scotland, 4, 58, 155–58, 174, 178, 194
   General Assembly, 155, 178
   Moderate Party, 98, 110, 147
   Popular Party, 95, 98
   Scottish Ministers' Widows Fund, 57, 155–58, 169n, 170n
   Synods, 155
Clackmannan, 70, 82
Clairaut, Alexis-Claude, 167n
Clark, Charles E., 207, 214n
Clark, James, 26–27, 47n
Clark, John (instrument maker), 32
Clark, John (land surveyor), 65
Clark, Peter, 123n
Clark, William, 44n, 126n, 169n, 187n, 237n, 297n, 335n
Clarke, Dr. Samuel, 91
Clarke, T. N., 44n, 45n, 47n, 48n, 50n, 52n
Clason, Patrick, 119n
Clayton, Christopher, 236n
Clayton, Daniel, 56, 74n
Cleghorn, George, 274n
Cleghorn, Helen, 258
Cleghorn, James, 257–58
Cleghorn, Robert, 9, 102, 103, 105, 124n, 255–79, 328, 329, 333
   and physical examination, 9, 256–57, 262, 263–64, 266, 271–72
   and postmortem examinations, 257, 262–70, 271
   career, 257–61
   De Somno, 258
   his character, 257, 259–61, 272

   his politics, 260
   notebooks of, 9, 255, 261–72, 333
Cleland, James, 121n, 129n
Clerk, Sir George, of Penicuik, 316, 325n
Clerk, Sir John, 118n, 171n
Clerk-Maxwell, Sir George, 111
Clercq, P. R. de, 43n
clergymen, 83, 84
Clifton, Gloria, 45n
clocks, 21, 29, 34, 40, 46n
Close, C., 48n
Clouet, J.-F., 303
Clough, Monica, 307, 321n, 322n
Clow, Archibald, 130n
Clow, Nan L., 130n
Clyde (river), 231
Clyde Iron Works, 303, 304
Cobbett, William, 206
Cochrane, Archibald, 9th Earl of Dundonald, 104
Cockburn, Henry, 122n
Cohen, I. Bernard, 144, 162n, 163n, 168n, 170n
Cole, Benjamin, 40
Cole, Humphrey, 23, 24
Colley, Linda, 48n, 187n
Collins, John, 25
Colquhoun, Patrick, 111
Columbia University, 88
Commission for the Annexed Estates, 58, 65, 125n, 154
Commissioners for making Roads and building Bridges in the Highlands of Scotland, 69
compasses (navigational), 21, 29, 39, 325n
Comrie, John D., 274n, 275n
Conduitt, John, 158, 165n
Connor, Robin, 23, 45n, 50n, 52n
Conrad, Lawrence I., 273n
consensus, ideal of, 153–55, 168n
Constable, Archibald, 305, 321n, 325n, 216
Constable, Thomas, 209n, 233n
constructivism, 7, 143
consumerism, 29
Cook, Harold J., 76n
Cooper, Richard, 74n
Cooper, Thomas, 107, 128n
Cooter, Roger, 168n, 324n, 326n
Copenhagen, 87, 316
   Royal Danish Academy of Sciences and Letters, 87
   Royal Literary Society, 316
Copernicanism, 60, 106

Copland, Patrick, 36, 40–41, 75n, 110
Corss, James, 25, 46n, 81
Cosens, Nicholas, 45n
Cosh, Mary, 51n
cosmography, 60, 71
cosmology, 60
Couper, Rev. James, 87, 107, 128n
Couper, William, 99, 128n, 265
Coutts, James, 116n, 118n, 119n, 120n, 127n, 129n
Coventry, 21
Coventry, Andrew, 304, 305, 314, 320n
Cowan, John, 276n
Cowan, Robert, 99, 100n
Craig, ?, 268
Craig, Gordon Y., 322n
Craig, Janet, 264
Craig, M. E., 209n
Craig, Robert, 27
Craig, W. S., 117n
Craige, John, 163–64n
  Theologiae Christianae Principia Mathematica, 163–64n
Craik, E. M., 115n
Crawford, Adair, 82, 114n, 124n, 127n, 128n
Crawford, John, 96
Crawford, Patricia, 190n
Crawforth, M. A., 43n
credibility, 7, 328, 330, 331, 332
Creech, William, 193, 202, 204, 209n, 213n, 220, 221, 222, 232
Crichton, 28, 39, 41, 45n, 52n, 99
Critical Review, 65
Crosland, M., 11n, 318n
Cross, John, 62, 99, 121n
Cruickshank, William, 86
Cruikshanks, Stewart, 129n
Cuchet, Gaspard-Joseph, 226
Cuillin Mountains, 285
Cullen, 287
Cullen, William, 283
Cullen, Dr. William, 1, 3, 8, 81, 84, 85, 86, 94, 96, 97, 102, 103, 104, 107, 109, 114n, 118n, 120n, 122n, 123n, 193, 195, 199, 210n, 217, 219, 220–23, 224, 226, 227, 228, 230, 234n, 235n, 258, 305, 330
  First Lines of the Practice of Physic, 193, 203, 209n, 220–22, 223, 226, 227, 228, 229, 230, 231, 234n, 235n
  Materia Medica, 222, 223, 224, 226, 227, 228
  Nosology, 229

Culley, George, 307
Cumberland, Richard, 92
Cumming, Alexander, 30, 37, 39
Cunningham, Andrew, 14n
Cunningham, F., 324n
Cunninghame, John, 123n
Cupar, 23
Cuvier, Georges, 281–82

Daiches, David, 51n, 318n, 324n
Dale, David, 258,
Dalgleish, George, 45n
Dalrymple, David, Lord Hailes, 111
Darien Company, 90, 91
Darling, Robert, 64, 66
Darnton, Robert, 191, 207n, 215, 233n
Darwin, Charles, 306, 307, 321n
  theory of evolution by natural selection, 307
Daston, Lorraine, 143, 157, 170n, 329, 335n
Davenport, Robert, 25–26
Davie, George E., 2, 11n, 162n, 322n
Davies, Roger, 45n
Davis, Dorothy, 43n
Davis, Herbert, 161n
Davis, John W., 162n
Dean, Dennis R., 14n, 212n, 296n, 313, 319n, 320n, 322n, 323n
Dee (river), 289
Dee, P. R., 119n
Defoe, Daniel, 206
Deism, 91, 150, 160, 173
De la Beche, Henry Thomas, 311
Delano-Smith, Catherine, 56, 74n
De Luc, Jean André, 104, 126n
Demainbray, Stephen, 33
De Moivre, Abraham, 157, 168n
Denholm, James, 62
  An Historical and Topographical Description of the City of Glasgow, 62
Denina, Carlo, 114n
Denmark, 316
Derby, 35
Desaguliers, J. T., 33, 149, 168n
DesBrisay, Gordon, 187n
Descartes, René, 147, 151, 173, 174, 175, 178
  theory of epigenesis, 175
  theory of vortices, 150
Desmarest, Nicolas, 311
Devine, T. M., 112n, 130n, 209n, 336n
Devlin-Thorp, Sheila, 13n, 275n
diagnosis, techniques of, 255–57, 259–60, 262,70, 328

Index

Dick I, Robert, 19, 38, 86, 94–95, 118n, 119n, 122n, 127n
Dick II, Robert, 86, 87, 95, 103, 105, 119n
Dickinson, Caleb, 232
  An Inquiry into the Causes and Nature of Fever, 232
Dick-Lauder, Sir Thomas, 314, 323n
Diderot, Denis, 55, 191
  Encyclopédie, 55
  Le Neveu de Rameau, 191
Didot, Pierre-François, the younger, 226, 236n
Dietrich, J. Christian, 224, 235n
Dilly, Charles and Edward, 217
Ditton, Humphrey, 119n, 149
dividing engines, 30
Dobson, Thomas, 231–32
Dodsley, James, 210n
Doig, A., 11n, 12n, 14n, 15n, 112n, 118n, 208n, 236n, 273n
Donald, Thomas F., 114n
Donaldson, Alexander, 193, 216
Donaldson, Thomas, 218
Doncaster, 34
Donnachie, Ian, 48n
Donovan, Arthur L., 1–2, 10n, 14n, 118n, 120n, 125n
Donovan, Robert Kent, 79, 111, 112n
Dorret, James, 65, 66, 69
Douglas, James, 86
Douglas, James, 14th Earl of Morton, 152, 153, 167n
Douglas, John, 86
Douglas, Thomas, Lord Selkirk, 308, 309, 322n
  Observations on the Present State of the Highlands, 308
Douglass, George, 63
  An Appeal to the Republic of Letters, 63
  The Art of Drawing in Perspective, 63
  The Elements of Euclid, 63
Dow, Derek A., 13n
Dow, J. B., 169n
Doyle, W. P., 319n
Dragoni, Giorgio, 48n
Drayton, Richard, 75n, 77n
Drew, Henry, 38, 94, 118n
Drumlanrig Castle, 46n
Drummond, Alexander, 226, 236n
Drummond Castle, 46n
Drummond, George, 18, 35
Dryden, John, 191
Dublin, 34, 86, 203, 226, 227
  See also Trinity College, Dublin

Duden, Barbara, 271, 273n, 279n
Dudley, Hugh A. F., 13n
Dumbarton, 81
Dumbartonshire, 82, 99
Dumfriesshire, 82, 99, 287
Duncan, Alexander, 120n, 121n, 261, 275n, 277n, 278n
Duncan, Andrew, 192, 198, 201, 217, 218, 219, 224, 227, 229, 255
  Medical Cases, 219
  Medical and Philosophical Commentaries, 192, 198, 201, 211n, 218, 229, 255
Duncan, W. J., 117n, 124n, 128n, 129n, 275n
Dundas, Henry, 1st Viscount Melville, 70, 86, 100, 262
Dundas, Robert Saunders, 2nd Viscount Melville, 316
Dundee, 24, 46n, 276n
Dunkirk, 224, 225
Dunlop, A. Ian, 169n
Dunlop, William, 266
Dunlop, Rev. William, 90, 104, 116n
Dunn, John, 28
Dunton, John, 206
  Athenian Mercury, 206
Duplain, Pierre-Jacques, 226, 236n
Durham, 26
Dwyer, John, 112n

East Indies, 152
East Lothian, 35
Eastwood, 91
Eddy, M. D., 14n
Eden Treaty (1786), 109
Eden, William, Lord Auckland, 306
Edinburgh, 2, 4, 5–6, 9, 18, 19, 22, 23, 24, 26, 27, 28, 30, 32, 33, 34, 35–37, 38, 40, 41, 47n, 62, 63, 64, 71, 75n, 79, 80, 81, 84, 85, 86, 95, 102, 104, 108, 110–11, 111n, 113n, 179, 186, 192, 199, 200, 201, 205, 206, 212n, 216, 218, 219, 220, 221, 223, 224, 225, 226, 227, 228, 229, 230, 231, 232, 235n, 255, 265, 266, 299n, 302, 303, 310, 314, 320n, 331
  Arthur's Seat, 283, 290
  audience for science in, 6
  The Bee, 204
  book trade, 193, 215–54
  Botanical Gardens, 203
  Caledonian Horticultural Society, 317, 325n

Caledonian Mercury, 198, 211n
Calton Hill, 316
connections with Europe, 215, 223–28, 232
Edinburgh Advertiser, 198
Edinburgh Astronomical Institution, 316–17, 318
Edinburgh Chronicle, 196
Edinburgh City Archives, 232
Edinburgh Encyclopaedia, 316
Edinburgh Evening Courant, 198, 203
Edinburgh Magazine and Review, 198–201, 206, 211n, 228
Edinburgh New Pharmacopoeia, 228
Edinburgh Review (1802), 111, 200
Edinburgh Weekly Journal, 196–97, 198, 209n, 210n
enlightenment, 8, 110, 113n
Faculty of Advocates, 22
Greyfriars Churchyard, 162n, 164n, 194
High School, 194, 302, 319n
Honourable the Society for the Improvement in the Knowledge of Agriculture, 4
Mary's Chapel Masonic Lodge, 194
National Library of Scotland, 232
New North Church, 155
New Town, 19
Old Town, 194
Parliament Square, 216, 232
periodicals in, 192
Philosophical Society, 5, 35–36, 58, 75n, 81, 85, 87, 92, 114, 123n, 143, 153–54, 155, 168n, 201, 202
Phrenological Society, 317
Pleasance, 194
Ramsay's Land, 64
Rankenian Club, 143, 155, 168n
Royal College of Physicians, 89, 92, 192, 224
Royal Company of Archers, 316
Royal Medical Society, 258
Royal Observatory, 30, 317
Royal Society of Edinburgh, 2, 6, 13n, 35, 63, 65, 81, 87, 110, 114n, 127n, 130n, 192, 201–2, 224, 258, 289, 293, 294, 302, 304, 306, 311, 313, 314, 315–16, 318, 323n
  museum, 315–16
  Transactions, 204
St. Bernard's Well, Stockbridge, 203
Salisbury Crags, 294
scientific community, 81, 82, 110
Scottish Mint, 27

Scottish Society of Arts, 316
Society for the Importation of Foreign Seeds and Plants, 5
Society of Antiquaries, 59, 64–68, 110, 197, 199, 200, 201–2, 204, 315–16
  and mapping, 65–68
  as a centre of geographical calculation, 64–68
  museum, 64–65, 315
  Transactions, 65, 316
Stair's Close, 63
Surgeons' Company, 88
Surgeons' Hall, 97
Theatre Royal, 294, 323n
Tolbooth Kirk, 155
Town Council, 25, 33, 66, 148, 165n, 316
See also University of Edinburgh
Edney, Matthew, 56, 73n
Edrom, 257
Egerton, Judy, 44n
electrical machines, 35
electricity, 95, 105, 329
Elliot, Charles, 8, 9, 193, 194, 199, 201, 203, 204, 206, 210n, 215–54
  American trade, 228–32
  and medical publishing, 216–20
  and William Cullen, 220–23
  career, 216
  A Collection of the Most Esteemed Farces, 230
  correspondents in Grenada and Jamaica, 236n
  The Edinburgh New Dispensatory, 227
  European contacts, 223–28
Elliot, Hon. Gilbert, 325n
Elliot, Sir Gilbert, Lord Minto, 152
Elliot, James, 66
Elliot, Thomas, 225
Elliot, William, 231, 237n
Elliott, B., 326n
Ellis, Richard H., 279n
Emerson, Ralph Waldo, 205
Emerson, Roger L., 5, 6, 7–8, 9, 12n, 33, 37, 43n, 49n, 50n, 51n, 75n, 77n, 112n, 113n, 114n, 116n, 117n, 130n, 164n, 165n, 168n, 173, 187n, 191, 202, 207n, 212n, 259, 275n, 276n, 318n, 328, 331
Encyclopaedia Britannica
  first edition, 193, 196, 197–98, 205, 210n, 211n
  second edition, 193, 206, 216, 226, 230, 231
  third edition, 193, 203, 206, 216

Index

engineering, 100, 168n
England, 20, 23, 33, 34, 42, 56, 82, 84, 89, 91, 100, 108, 191, 306
  aristocracy, 180–86
    and childbirth, 181
    diet, 182, 184
  enlightenment, 8, 59
  north of, 33, 130n
  provincial, 8, 30, 32, 130n
Englehart, Dr., 234n
English Civil War, 174
English Review, 192
Enlightenment, 269, 329
  and books, 233
  and medicine, 3–6, 255, 327–35
  and quantification, 36
  and religion, 8, 172–73, 186
  and science, 3–6, 7, 296, 327–35
  communications circuit in, 216, 223
  empiricism in, 257
  experimentalism in, 330
  geometric spirit of, 58
  historical geography of, 72
  levels of historical analysis of, 328–29
  ways of knowing in, 329–31, 332
  See also Scottish Enlightenment
The Enlightenment in National Context, 4
Epicureanism, 178
Episcopalianism, 8, 173, 174, 177
Equitable Insurance Company (London), 157, 170n
Erskine, David, 11th Earl of Buchan, 64, 67, 99, 114n, 123n, 197, 200, 201–2, 203, 212n
Erskine, John, 11th Earl of Mar, 86
Estes, J. Worth, 278n, 279n
ethics, 89
Euclid, 107
  Elements, 107
Europe, 8, 20, 23, 33, 42, 84, 89, 143, 147, 154, 172, 216, 223, 224, 228, 232, 233, 282, 301, 303, 329
Exchequer, 149
Excise, 19, 56–57, 148–49, 153, 168n
Eyles, V. A., 166n, 297n, 299n, 319n

Falkirk, 47n
Fara, Patricia, 44n
Farber, Paul L., 188n
Far East, 91
farmers, 82
The Farmers' Magazine, 305
Faroe Islands, 313
Fauvel, John, 166n
Faversham Arsenal, 121
Febvre, Lucien, 215, 233n

Feldman, Theodore S., 126n
Ferguson, Adam, 3
Ferguson, James, 32, 34, 40, 49n
Ferguson, J. P. S., 11n, 12n, 14n, 15n, 112n, 118n, 208n, 236n, 274n
Fergusson, Robert, 208n
Field, J. V., 43n
fieldwork, 9, 56, 63, 68, 75n, 280–300, 311–14, 318, 328, 331
  and geological theory, 283
  and self-improvement, 282
  and travel, 282, 283
  and useful knowledge, 282, 284, 285, 288, 295
  as Baconian science, 282–83
Fife, 82, 288, 299n
Figlio, Karl, 189n
Finlay, John, 276n
Finlay, Major John, 87, 109, 115n, 121n, 129n
Finlay, Richard J., 209n
Finlayson, James, 321n
Fishlock, D., 44
Fissell, Mary, 264, 270, 273n, 277n, 279n
Flanders, 23
Fleming, Ian J., 320n
Fleming, William James, 261, 276n
Flinn, Michael W., 169n
Flood, Raymond, 166n
Florence, 90
Florida, the, 90
fluxions, 148, 150–51, 153
Folkes, Martin, 147, 149, 164n, 165n, 166n, 167n
Forbes, Eric G., 13n, 56, 73n, 114n
Forbes, J. D., 314, 324n
Forbes, John, of Corse, 173
Forbes, William, 204
Forfar, 23
Forsythe, Bill, 273n
fortification, 95, 168n
Foulis, Andrew, 92, 123n
Foulis, Robert, 92, 104, 123n
Fourcroy, Antoine François Comte de, 227, 235n, 320n
  Elementary Lectures on Chemistry and Natural History, 225, 227, 232
  Elements of Natural History and Chemistry, 219
  Leçons Elémentaire Histoire Naturelle et de Chimie, 225
Foxon, David, 191, 207n
Frame, James, 123n
France, 56, 109, 125n, 191, 226, 232, 291, 303

provincial academies, 107
wool manufacture, 306
Frängsmyr, Tore, 50n, 73n, 126n
Franklin, Benjamin, 228
Frasca-Spada, Marina, 208n, 214n, 232, 233n, 237n, 336n
Free Church of Scotland, 155
Freemasonry, 33, 194, 209n
Freer, Robert, 84, 102, 103
Freiburg, 287, 288
  Bergakademie, 287
French, Roger, 273n
French Revolution (1789), 98, 260
Frew, Hugh, 275n
Fritz, Paul, 11n, 130n, 275n
Fullarton, William, 27
Funk, G., 165n
Fusoris, Jean, 24
Fyfe, Andrew, 218
  A System of Anatomy, 218, 231
Fyffe, J. G., 51n

Gall, F. J., 314
Garde, François de la, 224, 235n
Garden, Francis, Lord Gardenstone, 203
Garden, George, 8, 172, 173–78, 179, 180, 185, 186, 187n, 188n
Garden, James, 174
Gardner I, John, 39, 45n
Garnett, Thomas, 82, 100, 122n, 124n
Garrett, Aaron, 213–14n
Gartshore, Mr., 266
Gascoigne, John, 77n, 321n
Gasking, Elizabeth, 173, 177, 187n, 188n
  Investigations into Generation, 173
Gastellier, R.-G., 228
  Des Spécifiques en Médecine, 228
gauging, 153, 154, 161, 168n
Gavine, David, 14n, 48n, 49n, 76n, 127n, 129n
Geikie, Archibald, 283, 297n, 311, 313, 322n
Gemini, Thomas, 23
gender, 6, 8, 62, 182, 183
generation, theory of, 8, 172–90
  animalculism, 172, 175, 176–77
  epigenesis, 174–75, 176
  metamorphosis, 175, 176
  pre-existence, 175, 176, 178–79
  preformationism, 172, 174, 175, 176–77, 179, 180, 182, 183, 186
  spermatozooism, 175, 176–77, 183
  spontaneous generation, 176
  ovism, 175, 177
geography, 7, 54–78, 119n, 126n, 206
  and civic discourse, 57
  and polite sociability, 57
  as national self-knowledge, 54
  audiences for, 54, 71
  collaboration in, 68
  development of, 57
  in France, 68
  teaching of, 54, 57, 59–64, 71
  See also maps
geology, 6, 9, 104, 126n, 262, 280–300, 311–14
  and religion, 296
  fieldwork and, 9, 280–300, 311–14
  neptunism, 280, 284, 312
  plutonism, 280, 290, 311
  Scottish 'school' of, 311, 313
  uniformitarian, 149
geometry, 54–78, 144
  ancient, 150, 151, 158–59, 168
  authority of, 150
  teaching of, 59–64
George III (King of Britain), 33, 37
George IV (King of Britain), 316
Germany, 223, 224, 226, 286
Geyer-Kordesch, Johanna, 13n, 88, 115n, 117n, 120n, 121n, 122n, 129n, 278n, 279n
Gibb, Andrew, 112n
Gibson, Alexander, 75n
Gibson, John, 97, 120n
Gibson, T., 274n, 275n
Gillespie, Charles C., 162n, 319n
Glanvill, Joseph, 90
  Sadducismus Triumphatus, 90
Glasgow, 6, 7–8, 9, 18, 26, 27, 28, 34, 36, 39, 41, 43n, 45n, 62, 71, 75n, 79–142, 148, 153, 186, 257–72, 302, 331
  Academy, 62
  agricultural society, 107
  Anderson's Institution, 39, 97–100, 101, 109, 110, 121n, 122n
  Anderston, 272
  Bell's Wynd, 268
  Calton, 263
  Chamber of Commerce, 86, 109
  chemical society, 108
  City land surveyor, 39
  College Street, 258, 265, 278n
  Considerable Club, 87, 101
  Correction House, 263
  Dunlop Street, 62
  enlightenment, 6, 8, 79–142, 260, 332
  evangelicalism, 79, 110
  Faculty of Physicians and Surgeons, 88, 96, 97, 115n, 120n, 129n, 258, 264, 272, 277n, 278n

Index 347

Foulis Press, 38, 207n
Gaelic community, 98, 113n
Garnethill Observatory, 62, 108
Graeme Street, 108
Hodge Podge Club, 87
Humane Society, 258
Ingram Street, 62
Ingram Street Gaelic Chapel, 121
John Street, 62
King Street, 121n
linen trade, 109
Literary and Commercial Society, 80, 108
Medical Club, 87, 261
medical culture, 9, 257, 272
molasses trade, 153
Philosophical Society, 62, 63, 80, 87, 107
poorhouse, 257
Ramshorn Churchyard, 278n
Royal Botanic Institute, 108
Royal Infirmary, 108, 124n, 257, 259, 260, 264, 277
Royal Lunatic Asylum, 108, 257, 259, 261
scientific community, 80–88, 110, 130–42, 328, 331
Scruton's laboratory, 97, 99
Society for Promoting Astronomical Science, 108
Spreull's Land, 258
Stirling Library, 108
Technical College, 39
tobacco trade, 109
Town Council, 108, 114n
Town's Hospital, 6, 88, 96, 101, 108, 120n, 259
Trongate, 258
Virginia Street, 278n
See also University of Glasgow, University of Strathclyde
Glassford, John, 104, 109
Gleditsch, Jean Fredrich, 224, 227, 235n
Glen Roy, 'parallel roads' of, 313–14
Glen Sannicks, 286
Glen Tilt, 283, 289
globes, 28, 37
  celestial, 60
  makers of, 63
  terrestrial, 60
  use of, 61, 62, 71
Glorious Revolution (1688), 27, 88, 174
Glynne, Richard, 37, 51n
Godlewska, Anne, 56, 68, 73n, 74n, 77n
Godwin, William, 205

Goldsmith, Oliver, 201
  History of the Earth, 201
Golinski, Jan, 6, 13n, 15n, 44n, 73n, 74n, 118n, 126n, 169n, 187n, 237n, 281, 296, 297n, 335n, 336n
Good, Gregory A., 73n
Goodison, Nicholas, 22, 44n
Gordon, Daniel, 19
Gordon, J. F. S., 274n
Gordon, John, 96
Gordon, Margaret Maria, 320n
Gordon, Robert, of Straloch, 24
Gosse, Peter, 223, 235n
Gothenburg, 220, 223
Göttingen, 224
Gough, Richard, 54, 55, 67, 77n
  British Topography, 67
  Essay on the Rise and Progress of Geography, 54
Grabiner, Judith V., 7, 8, 14n, 15n, 57, 162n, 166n, 167n, 168n, 328, 331
Graaf, Regnier de, 177
Graham, George, 30, 152, 167n
Graham, James, 1st Duke of Montrose, 86
Graham, James, 2nd Duke of Montrose, 86
Graham, James, 3rd Duke of Montrose, 86, 100
Graham, John, 120n
Graham, Patrick, 104, 105
Graham, Thomas, 28
Grahame, Alexander, 123n
Grand Tour, 32, 282, 285, 290
Grattan-Guiness, I., 167n
Graustein, Jeanette E., 213n
Gray, James, 274n
Gray, John M., 115n
Gray, Malcolm, 321n
Gray, Thomas, 191
Green, James N., 236n
green sickness, 180, 185
Greenberg, John L., 162n, 167n
Greenock, 39, 45n, 81, 229
  McLean Museum, 45n
Greer, Frederick, 123n
Gregory, Mr., of Campvere, 224, 235n
Gregory, Charles, 19, 39
Gregory, David (Edinburgh and Oxford professor), 153, 164n, 168n
Gregory, David, and Son (merchants), 224, 225
Gregory, David, of Kinnairdie, 235n
Gregory I, James (St. Andrews and Edinburgh professor), 24–25, 27–28, 60

Gregory II, James (Edinburgh professor), 155
Gregory, Dr. James, 202, 217, 218, 219, 224, 235n
   Conspectus medicinae theoreticae, 202, 219
Gregory, Jane, 335n
Gregory, John, 104, 195, 198, 200, 210n, 279n
   A Comparative View, 104
Grenada, 228–29
Grieve, Dr. John, 223, 235n
Guerrini, Anita, 8, 14n, 16n, 187n, 188n, 189n, 190n, 328
Guettard, Jean-Étienne, 311
Guicciardini, Niccolò, 14n, 162n, 167n
gunnery, 93, 95, 119n
Gutteridge, G. H., 73n

Haakonssen, Knud, 112n
Habsburg Empire, 56
Hachette, J. N. P., 303
Hackmann, W. D., 43n, 44n, 48n
Haddington, 82
Hague, the, 223
Hahn, Roger, 301, 318n
Hake, Charles Richard, 223, 235n
Halbert, Elizabeth, 261
Hales, Stephen, 149
Hall, A. Rupert, 10n, 165n
Hall, David, 228
Hall, David D., 214n, 236n
Hall, Sir James, 9, 280, 283, 284, 289, 290–91, 293, 299n, 302, 304, 308, 311, 313, 314, 320n, 323n, 330
Hall, John, 259
Haller, Albrecht von, 228
Halley, Edmond, 147, 149, 156, 161–62n, 164n
Hamburg, 224
Hamilton, 47n, 121n
Hamilton, Alexander, 219, 224
   A Treatise of Midwifery, 219, 224
Hamilton, Balfour and Neill (publishers), 191, 194, 232
Hamilton, Sir David, 180, 184
Hamilton, Frederick, 123n
Hamilton, Gavin, 196, 204
Hamilton, Henry, 130n
Hamilton, James, 5th Duke of Hamilton, 86
Hamilton, Robert (Aberdeen professor), 60–61
Hamilton, Robert (Glasgow physician), 96, 115n
Hamilton, Thomas, 85, 87, 102, 123n
Hamilton, William (Edinburgh physician), 265
Hamilton, William (Glasgow professor), 261, 265
Hamilton, Sir William, 315
Harcourt, George Simon, 2nd Earl of Harcourt, 212n
Hardie, D. W. F., 113n, 121n
Harding, Dr., 189n
Hardy, Thomas, 200
Hardy, William, 30
Hare, D. J. P., 169n
Harris, John, 149
Harris, Michael, 207n, 208n, 236n
Harrison, J. A., 49n
Harrison, John, 30, 61
Hartley, Stuart, 9, 298n, 312, 328
Harvard University, 36, 205
Harvey, William, 174–75, 176, 177
   De generatione, 174
Harwood, Edward, 199
Hastings, Lady Ann, 182
Hastings, Lady Elizabeth, 181, 182, 189n
Hastings, Lady Frances, 182
Hastings, Lady Margaret, 182
Hastings, Selina, 190n
Hastings, Selina, Countess of Huntingdon, 181–82, 184, 185, 189n
Hastings, Theophilus, Earl of Huntingdon, 181, 183, 184, 189n
Hauksbee, Francis, 32, 40, 53n
Havart, William, 26
Hawick, 66
Hay, Denys, 116n
Hearne, George, 40, 53n
Heath, Thomas, 36
Hebrides, 125n, 285
Heidelberger, Michael, 170n
Heilbron, J. L., 50n, 56, 73n, 126n, 167n
Helden, A. van, 44n
Henderson, Ebenezer, 49n
Henderson, G. D., 186, 187n
Henry, Joseph, 109
Henry, Robert, 200
Heron, Robert, 219
Hevley, Bruce, 286, 295, 297n, 298n, 300n
Hibbert, Samuel, 316, 325n
Highland Destitution Fund, 310
Highland Society, 304
Higton, Hester, 46n
Hill, Peter (Edinburgh bookseller), 193, 205

Index 349

Hill, Peter (instrument maker), 37
Hills, R. L., 43n 52n
history, 5, 89, 90, 116n, 191, 223, 262
  civil, 57, 61, 71, 80, 89, 98, 280
  four stages theory of, 4, 57
  natural, 4, 6, 8, 58, 61, 67, 71, 74n, 80, 89, 91, 93, 97, 98, 103–4, 126n, 193, 198, 199–200, 205, 211n, 280, 295, 296, 296n, 329, 330
  of the book, 7, 8–9, 215
  Scottish narrative, 192
A History of the Book in America, 228
History of Science, 1
Hobbes, Thomas, 160
Hodge, M. J. S., 11n, 118n
Hodgson, James, 32
Hogg, James, 208n
Holbrook, Mary, 45n, 46n, 51n, 52n
Holland, 60, 84, 223, 224, 235n
Holland, Sir Henry, 293, 312, 323n
Home, Francis, 305, 314, 330, 335n
  Principles of Agriculture and Vegetation, 330
Home, Henry, Lord Kames, 104, 126n, 200, 201, 305, 314
  Sketches of the History of Man, 200
Home, John, 208n
Hont, Istvan, 117n
Hook, Andrew, 11n, 12n, 15n, 112n, 113n, 207n, 276n
Hooker, William, 312
Hope, John, 195, 198, 210n
Hope, T. C., 102, 103, 104, 107, 115n, 125n, 258, 302, 303, 311, 314, 315, 320n, 321n, 324n
Horseley, John, 149
Hounslow Heath, 33
Houston, Colonel, 266
Houston, Robert, 82, 114n
Howarth, Richard J., 44n
Howse, Derek, 75n, 128n
Hull, 34
  Society for the Purpose of Literary Information, 34
humanism, 80
Humbie, 35
Humboldt, Alexander von, 281
  Tableaux de la Nature, 281
Hume, Baron David, 322n
Hume, David, 3, 94, 158, 160, 193, 217, 229, 327, 335n
Hume, Patrick, 147
Hunter, ?, 123n
Hunter, John, 84, 86, 265, 266
Hunter, Michael, 116n, 188n

Hunter, Robert, 276n
Hunter, William, 6, 84, 86, 92, 97, 266, 278n
  The Anatomy of the Human Gravid Uterus, 6
Hurlet Alum Works, 121n
Hutcheson, Francis, 3, 79, 92, 101, 146, 147, 150, 161, 164n
Hutton, General Charles, 66
Hutton, James, 9, 14n, 104, 105, 111, 126n, 127n, 149, 161, 204, 283, 286, 289, 301, 302, 311, 313, 314, 322n, 330
  Theory of the Earth, 302, 311
Huttonianism, 280, 284, 288, 289, 290–95, 312–13, 330
Huygens, Christiaan, 158
hydrometer, 39
hydrostatics, 95

Iceland, 284, 291–93, 295, 312–13, 323n, 328
  Gullbringusyslsa, 312
  Mount Hekla, 292, 312
  Rangarvallasysla, 312
  Snaefellsnes peninsula, 312
  Torfajökull, 292
Ignatieff, Michael, 117n
improvement, 2, 4, 5, 6, 57, 80, 89, 143, 149, 154, 283, 304–11, 315, 318, 329
  agricultural, 6, 13n, 104–5, 304–8
  cultural, 2, 315
  economic, 2, 308–11
  personal, 282
industrialists, 81, 83, 85, 87, 109–10
Ingamells, John, 114n
Inkster, Ian, 47n, 48n, 49n, 111n
Innes, John, 218
  Eight Anatomical Tables of the Human Body, 218
  A Short Description of the Human Muscles, 218
instruments
  astronomical, 38, 105
  demonstration apparatus, 19, 28, 29, 32–35, 36, 38–41, 92
  drawing, 37
  makers of, 2, 19, 23, 35–41, 63, 81, 93
  markets for, 7, 18, 27–32, 328
  mathematical, 7, 20–21, 22, 24, 29, 94
  medical, 22–23
  musical, 17
  navigational, 21
  optical, 20, 21

precision, 20
scientific, 2, 7, 17–53, 94, 165n, 232, 328
  definition of, 17, 20–21
  surveying, 21, 37
Inverary, 37
Inverness, 310, 317
  Academy, 326n
  Northern Institution for the Promotion of Science and Literature, 317, 318, 326n
Inverness-shire, 82
Iona, 125n
Ireland, 30, 91, 184, 216
irreligion, 173
Irvine I, William, 85, 97, 102, 103, 104, 105, 120n, 124n, 125n, 126n, 127n, 128n, 261
Irvine II, William, 124n, 125n, 126n, 127n
Irving, Ralph, 219, 232
  Experiments on the Red and Quill Peruvian Bark, 219, 232
Isaac, Peter, 15n, 208n, 209n
Isla, 285
Italians, itinerant, 22
Italy, 65, 90

Jackson, Gordon, 112n, 130n
Jackson, Thomas, 87
Jacob, Margaret C., 49n, 172–73, 187n
Jacobitism
  the '15, 307
  the '45, 58, 307
Jacyna, L. S., 13n
Jamaica, 36, 38, 229
James, William, 205
Jameson, Laurence, 298n
Jameson, Robert, 9, 280, 283–90, 295, 296, 298n, 299n, 311, 312, 322n
  The Elements of Geognosy, 284
  A Mineralogical Description of Scotland, 287
  Mineralogy of the Scottish Isles, 284
Jamieson, George, 47n
Jamieson, Neil, 229, 236n
Jarden, George, 38
Jardine, George, 97, 107, 108, 260, 276n
Jardine, Nick, 208n, 214n, 232, 233n, 237n, 336n
Jedburgh, 65, 66
Jeffray, James, 102, 103, 120n, 122n
Jeffreys, Thomas, 65, 66, 68
Jenkinson, Jacqueline L. M., 13n, 129n, 275n, 276n, 277n

Jessop, T. E., 166n
Jewson, Nicholas, 255, 273n
Johns, Adrian, 215, 233n
Johnson, William, 70
Johnstone, William, of Caskieben, 24
Johnstone, W. T., 52n
Johnstoun, John, 92, 96, 97, 117n, 119n, 167n
Jones, Jean, 14n, 51n
Jones, Peter, 12n, 43n, 51n, 112n
Jordanova, Ludmilla, 6, 7, 15n, 333, 336n
Jorgensen, Jorgen, 312
journalism, 192, 195, 206, 207
Jura, 285

Kahler, Lisa, 210n
Kain, Roger, 56, 74n
Kark, R. M., 323n
Kaufman, M. H., 324n, 326n
Keir, James, 221, 235n
Keith Grammar School, 34
Keller, Eve, 177, 188n
Kelso, 66
Kendall, James, 124n, 125n, 275n
Kenmore House, 287
Kennedy, Dr. Robert, 303–4, 311, 320n
Kent, Andrew, 120n, 124n, 125n, 261, 274n, 275n, 277n
Kepler's laws, 144, 163n
Kerr, Robert, 193, 196, 209n, 210n, 211n, 212n, 220, 224n
Kew Gardens, 67, 306
Kidd, Colin, 113n
Kilmarnock, 81
Kincaid, Alexander, 193, 228
King, Benjamin Watts, 265
King, John, 40, 41
King's College, Aberdeen, 60, 113n, 115n, 125n, 117n, 173, 174
  curriculum, 174, 175, 176
Kinross-shire, 82, 99
Kinsley, James, 123n, 166n
Kintyre, 67, 285
Kirkcaldy, 34
Kirkwall, 36
Kirwan, Richard, 287, 320n
Kitchen, Thomas, 67
  Geographia Scotiae, 67
Klibansky, Raymond, 234n
Knibb, Joseph, 46n
Knight, Isabel, 74n
Knockfarril, 316
knowledge
  credibility and, 7, 55, 70, 328, 330, 331, 332

Index

different ways of knowing, 329–31
Enlightenment map of, 54, 71, 89
factors affecting growth of, 1, 2, 9–10n
medical, 1, 9, 256
natural, 1, 5, 6, 179, 186, 296n, 315
public, 154
sites for consuming, 295
sites for creating, 55, 88, 101, 257, 281, 296, 332
socially situated nature of, 1, 257
useful, 95, 100, 152, 195, 199, 318, 329
Knowles, Edward, 61
Knox, James, 109, 121n
Koerner, Lisbet, 280, 297n
Kohler, Robert, 281, 296, 297n, 298n, 300n
Konvitz, Josef W., 73n
Krüger, Lorenz, 170n
Kuhn, T. S., 81, 113n, 335n
Kuklick, Henrika, 281, 296, 297n, 298n, 300n

Lake District, 276
Lamarck, J. B., 306
 theory of fixity of acquired characteristics, 306
Lanark, 81
Lanarkshire, 82
Lancashire, 21
Landes, David S., 44n
Landsman, Ned C., 79, 111, 112n
Lang, Gilbert, 101, 123n
Langford, Paul, 187n
Langlands, John, 66–67
languages
 Danish, 223
 Dutch, 334
 English, 217, 334
 French, 222, 223, 227
 Gaelic, 82
 German, 222, 224, 227, 334
 Greek, 89
 Hebrew, 90, 93
 Irish, 223
 Latin, 89, 98, 223, 224
 modern, 89
 oriental, 90
 Russian, 223
 Swedish, 223
Laplace, Pierre-Simon de, 167n, 291, 299n, 320n
Laskey, J., 323n
Latour, Bruno, 76n, 281, 297n
Laudan, L. L., 158, 162n, 170n

Laudan, Rachel, 282, 297n
Lausin, André de, 226, 236n
Lavoisier, Antoine-Laurent, 39, 103, 258, 302, 303, 319n, 320n
Lavoisier, Mme. Marie-Anne-Pierrette, 320n
law, 98
 Scots, 4, 98, 211n, 322n
Law, John, of Ballarnock, 91, 94, 117n
 Calendarium lunae perpetuum, 91
Lawrence, C. J., 2, 10n, 13n, 210n, 274n
Lawrence, Susan C., 278n
lawyers, 82, 83
Leadhills, 288
Leclerc, Georges Louis, Comte de Buffon, 104, 197, 199, 201
 Histoire Naturelle, 201
lecturing
 extramural, 18, 28, 32, 40, 61–63, 81, 96–97, 100
 itinerant, 32, 33, 34
 public, 58, 59, 94, 149
Lee, R. A., 46n
Leechman, Rev. William, 97, 119n
Leeds, 35
Leeuwenhoeck, Anthony van, 175, 176, 177
Legendre, Adrien-Marie, 167n
Leibnitz, G. W., 91, 150, 158, 159
Leigh, Dr. John, 230, 237n
 An Experimental Enquiry into the Properties of Opium, 230
Leighton, John M., 129n
Leighton, Robert, 173
Leipzig, 224, 227
Leith, 221, 224, 291, 301, 312
Lenman, Bruce P., 116n
Lennox, Charles, 3rd Duke of Richmond and Lennox, 119n, 121n
Leslie, John (Aberdeen professor), 123n
Leslie, John (Edinburgh professor), 28
Levack, Iain D., 13n
Lewis, G. Malcolm, 73n
Lewis, Judith Schneid, 181, 189n, 190n
Lewis, W. S., 212n
Leybourne, William, 46n
Lhuyd, Edward, 89, 91
Liddel, Duncan, 23–24
Liddesdale, 66
light
 aberration of, 127n
 wave theory of, 2
Lightfoot, John, 283
Lincoln, 35

Lind, James, 217–18
  A Treatise on the Putrid and Remitting Fen Fever, 217
  A Treatise on the Putrid and Remitting Marsh Fever, 217
Lindeboom, G. A., 188n
Lindqvist, Svante, 49n
Lindsay, Hercules, 122n
Lindsay, Ian G., 51n
Lindsay, Maurice, 208n
Lindsay, William, 123n
Linnaeus, Carl, 104, 197, 198, 204, 210n, 280
Liverpool, 45n, 334
Livingston, Donald, 297n
Livingstone, David, 73n, 74n, 257, 274n, 297n
Lizzars, Daniel, 218
Lloyd, J. T., 51n
Loch, James, 308, 321n
Lochaber, 'parallel roads' of, 313–14
Lochnaw Castle, 27
Lock, Stephen, 211n
Locke, John, 91, 92, 213n
Lockhart, John Gibson, 275n
logic, 80, 89, 91, 92, 321n
London, 19, 20, 23, 25, 26, 28, 30, 32, 33, 34, 35, 36, 37, 39, 40, 41, 46n, 60, 84, 85, 86, 92, 106, 111, 147, 157, 165n, 179, 185, 192, 195, 199, 216, 217, 218, 219, 220, 221, 224, 225, 227, 228, 229, 232, 235n, 266, 272, 331, 334
  as centre of instrument making in Britain, 42
  booksellers, 216
  Christ's Hospital, 119n
  Clerkenwell, 21
  Fleet Street, 40
  Geological Society, 313, 323n
  George and Blue Boar, Holborn, 227
  hospitals, 84
  London Chronicle, 195, 204
  Mint, 23
  Royal Greenwich Observatory, 29, 30
  Royal Society, 32, 34, 35, 36, 57, 58–59, 63, 74n, 87, 147, 149, 154, 168n, 173, 174, 315, 331
    Philosophical Transactions, 63, 91, 105, 106, 173, 174, 177
  St. Paul's Cathedral, 33
  Society of Arts and Manufactures, 33
  Soho Square, 67
  Stationers' Hall, 216, 218
  Tower, 23
  Transactions of the Horticultural Society of London, 317
  White Bear Inn, Piccadilly, 227
Loomes, Brian, 45n
Lopston, Olafur, 312
Loudon, John, 115n
Love, John, 96
Low Countries, 23
Löwy, Ilana, 274n
Luce, A. A., 166n
Lucretius, 161n
Lunan, Charles, 41
Luton Hoo, 33
Lux, David S., 76n
luxury, 184, 185
Lyell, Sir Charles, 302, 313
Lyell, Katherine M., 319n, 323n
Lyle, T., 275n
Lynch, Kathleen, 191, 207n
Lyon, David Murray, 209n
Lyons, 221, 225, 226

Macarthur, D. C., 208n
Macaulay, James M., 52n 113n
McCoig, Malcolm, 203, 205
  Flora Edinburgensis, 203
McConnell, Anita, 44n, 48n
Macculloch, John, 313–14, 323n
McCullogh, Kenneth, 40
Macdonald, Angus, 45n, 50n, 52n
Macdonald, Fiona A., 9, 13n, 88, 115n, 117n, 120n, 121n, 122n, 129n, 275n, 278n, 279n, 328, 333
McDougall, Warren, 8, 191, 193, 207n, 208n, 212n, 235n, 236n, 328, 333
McElroy, D. D., 124n
Macfarlane, Alexander, 17, 38, 105, 127n
Macfarquhar, Colin, 196, 197, 211n, 217
MacGibbon, David, 47n
MacGregor, James, 109, 110
McIntosh, Rev. Angus, 121n
Macintosh, Charles, 85, 99, 109, 110, 111, 113n, 115n, 121n, 129n
Macintosh, George, 85, 109, 110, 113n, 121n, 128n, 129n, 130n
MacIntyre, Donald, 166n, 171n, 322n
McKay, Barry, 15n, 208n, 209n
McKay, Margaret M., 125n
McKendrick, Neil, 29, 48n
Mackenzie, Alexander, 310
  The History of the Highland Clearances, 310
Mackenzie, Sir Alexander, 301

Index

Mackenzie, Sir George Steuart, 9, 280, 284, 289, 291–5, 296, 299n, 300n, 301–26, 328
  An Essay on Some Subjects connected with Taste, 314
  as a Huttonian, 311–13
  as an improver, 304–11
  as a multi-competent intellect, 9, 301, 317–18
  as a promoter of the sciences, 315–17
  General View of Agriculture of the Counties of Ross and Cromarty, 307, 309
  Gunnlaug's Saga, 294–95, 323n
  his estates, 310
  Illustrations of Phrenology, 315
  on the nature of diamond, 303–4
  on phrenology, 314–15
  on sheep breeding, 305–8, 318
  Travels in Iceland, 291, 294, 312
  Treatise on the Diseases and Management of Sheep, 306
Mackenzie, Henry, 111, 208n, 323n
Mackenzie, James Stuart, 66, 75n
Mackenzie, Dr. John, 310, 322n
Mackenzie, Murdoch, 36, 59
Mackenzie, William, 260, 276n
McKie, Douglas, 42n, 53n, 125n, 127n, 130n, 319n
Mackie, J. D., 51n, 52n, 117n, 118n
Maclaurin, Colin, 4, 7, 8, 19, 28, 34, 35, 36, 57, 74–75n, 91, 117n, 119n, 143–71, 328, 329
  An Account of Sir Isaac Newton's Philosophy, 146, 148, 149–52, 153, 158
  'De viribus mentium Bonipetus', 145–46
  Geometrica organica, 147, 165n
  graduation thesis, 146–47
  Memorial to the Honourable Commissioners of Excise, 148, 149
  on tides, 152
  Treatise of Algebra, 148
  Treatise of Fluxions, 146, 148, 149–52, 153, 158
Maclaurin, Rev. John, 119n
McLaverty, James, 191, 207n
Maclean, Dr. John, 128n, 129n, 262
MacLeod, Christine, 52n
MacLeod, Donald, 304
MacLeod, Finlay, 50n
MacLeod, Mary, 304
McNab, Alexander, 265
Macpherson, James, 208n
M'Ure, John, 257, 274n

McVey, Andrew, 123n
magnetism, 95
Maitland, Richard, Earl of Lauderdale, 46n
Malebranche, Nicholas, 176
  Recherche de la vérité, 176
Malpighi, Marcello, 176, 177
Malthus, Thomas R., 156, 157, 308, 309
  Essay on the Principle of Population, 308
Manchester, 86, 272
  Literary and Philosophical Society, 107
Mann, Alastair J., 237n
maps, 7, 24
  and Enlightenment, 55–57
  and mapping networks, 59
  and the 'quantifying spirit', 56
  and the state, 56
  as survey, 56
  conjoined with texts, 68–69
  in the Scottish Enlightenment, 57–59
  making of, 7, 24, 54–78, 328
  representational practices and, 54–55
  See also geography
Marischal College, Aberdeen, 19, 23–24, 36, 40–41, 60, 66, 110, 115n, 125n, 143, 147, 165n, 174
  curriculum, 174, 175, 176
Marke, John, 46n
Marshall, John, 96
Marshall, Peter J., 73n
Marshall, Robert, 87
Martin, Benjamin, 32, 34, 49n
Martin, Henri-Jean, 215, 233n
Martine, George, 38, 128n
Mary II (Queen of England and Scotland), 174
Maskelyne, Nevil, 58–59, 75n, 106
Mason, H. T., 233n
materia medica, 81, 86, 97, 102, 257, 258
mathematical practitioners, 18, 28
mathematics, 6, 7, 8, 57, 74n, 81, 86, 90, 92, 95, 106–7, 115n, 199, 327, 328
  applied, 168
  applied to ethics, 145–46
  authority of, 149
  in France, 68
  teaching of, 60, 61, 62, 71, 89, 165n
Matthews, W. M., 113n, 114n
Maupertuis, Pierre-Louis Moreau de, 107, 151, 158
Maxwell, Henry, 123n

Maxwell, Sir John, 91
Maxwell-Stewart, Hamish, 273n
mechanics, 86, 92, 95
  laws of, 146
  principle of least action, 146
Mechanics' Institutes, 34
Medical Commentaries, 261
medicine, 1, 2, 3, 4, 5, 6, 7, 8, 19, 80, 81, 86, 96–97, 98, 99–100, 102–3, 191, 193, 195, 199, 200, 205, 215–54, 265, 327, 328, 329
  authority in, 256
  clinical, 9, 192, 255, 271
  clinical case notes, 255, 263, 271
  empirical method in, 255, 329
  medical theory, 6, 14n, 96–97
  physician-patient relationship, 9, 255–56
  professionalisation of, 5
  relation to philosophy, 6, 128n
Meek, Ronald L., 73n
Meikle, Henry W., 122n
Meikleham, William, 99, 100, 105
Melling, Joseph, 273n
Melvill, John, 123n
Melvill, Thomas, 85, 87, 101, 105, 106, 122n, 123n
Melville, Herman, 212n
Menzies, William, 58
Méquignon, N. T., 226, 228
merchants, 79, 81, 82, 83, 84
metaphysics, 89, 107
meteorology, 105, 126n, 127n
method
  empirical, 5, 255
  of analysis and synthesis, 158–59
  scientific, 2, 331
micrometers, 36
microscopes, 21, 28, 29, 32, 325n
Middle Ages, 146
midwifery, 81, 86, 97, 100, 103, 122n, 184, 264, 265, 278n
military service, 82, 83
Millar, Andrew, 193
Millar, George, 231, 237n
Millar, James, 61, 106, 107, 119n, 122n, 123n, 126n, 128n
Millar, John, 3, 97, 122n
Millar, Mary, 264
Millar, Richard, 102–3
Millburn, J. R., 48n, 49n, 50n
Miller and Adie (instrument makers), 28
Miller, David Philip, 13n, 77n
Miller I, John, 36, 38

Miller II, John, 36–37, 41
Miller, Karl, 122n
Miller, Steve, 335n
Mills and Hicks (printers and stationers), 229
Mills, Stella, 117n, 163n, 164n, 165n, 166n, 167n, 168n, 169n, 171n
Milne, I. A., 11n, 12n, 14n, 15n, 112n, 118n, 208n, 236n, 274n
Milne, James, 23
mineralogy, 37, 282, 285, 288, 295, 328
mining, 295
Miss M. D., 269–70
Mitchel, John, 93
Mitchell, Sir Andrew, 167n, 169n
Mitchell, David, 223, 235n
Mitchell, Ian, 75n
Mitchell, Margaret, 263
Mitchison, Rosalind, 112n, 321n
Moir, D. G., 45n, 74n, 75n, 77n
Molland, A. George, 45n
Money, John, 74n, 130n
Monge, Gaspard, 320n
Monro I, Alexander, 224
Monro II, Alexander, 218, 219, 223, 224, 227, 230
  A Description of all the Bursæ Mucosæ of the Human Body, 218, 223
  Observations on the Structure and Functions of the Nervous System, 225, 235n
  The Structure and Physiology of Fishes Explained, 218, 219, 227
  The Works of Alexander Monro, 219, 224
Monro III, Alexander, 304, 315
Monroe, Joseph, 129n
Mont Blanc, 290
Monteath, James, 99, 120n, 260, 265
Montgomery, Sir James, 306
Montrose, 27, 30
Moor, James, 85, 92, 122n
Moore, D. T., 323n
Moore, James, 79, 112n, 117n
Moore, John, 84, 87
Moore, John N., 47n
Moray, 82
More, Henry, 173
Morgagni, G. B., 256, 274n
  The Seats and Causes of Diseases investigated by Anatomy, 256
Morgan, John, 39
Morgan, W., 170n
Morrell, J. B., 2, 10n, 18, 42n, 43n, 47n, 52n, 111n
Morris, Andrew, 97, 120n

Index

Morrison-Low, A. D., 7, 15n, 43n, 44n, 45n, 46n, 47n, 48n, 49n, 50n, 51n, 52n, 232, 328, 331
mortality tables, 156, 168n, 328
Morton, A. G., 14n
Morton, A. Q., 33, 49n
Morveau, Louis Bernard Guyton de, 303, 319n, 320n
Moss, Michael S., 13n, 129n, 275n, 277n
Mossner, Ernest C., 234n
Mountaine, William, 74n
Mowat, Ian R. M., 320n
Moyes, Henry, 34–35
Muir, James, 52n, 113n, 119, 121n, 127n
Muir, James (Glasgow surgeon), 264
Muirhead, Lockhart, 104, 125n
Mullett, C. F., 189n, 190n
Murdoch, Alexander, 165n
Murdoch, John, 114n
Murdoch, Patrick, 74n, 152, 155, 167n, 168n, 169n, 170n
Mure, Baron William, of Caldwell, 86, 120n
Murray and Cochran (printers), 219
Murray, David, 117n, 118n, 121n, 125n, 126n, 127n, 128n, 129n, 275n
Murray, John, 192, 193, 194, 198, 199, 204, 207n, 211n, 213n, 216, 217, 218, 220, 221, 222, 234n, 274n
  An Author's Conduct to the Public, 222
Musgrave, William, 174
Mushet, David, 303, 304, 319, 320n
Musson, A. E., 113n, 130n
Myers, Robin, 207n, 208n, 236n
Mylne, ?, 268
mysticism, 172, 177, 179, 186

Nairn, 82
Napier, John, 25
Nash, Richard, 164n
National Museums and Galleries, Merseyside, 47n
natural jurisprudence, 92
natural theology, 92, 143, 160, 178, 179, 201, 296
  argument from design, 146–47
  final causes, 201
navigation, 21, 60, 61, 62, 71, 74n, 95, 119n, 154, 168n
Neilson, J. B., 116n, 119n
Netherlands, 21

networks, 63, 68–70, 71–72
  correspondence, 223
  print, 215
Neve, Michael, 273n
Newcastle-upon-Tyne, 34, 86
Literary and Philosophical Society, 34
New Lanark, 258
newspapers, 192
  Glasgow, 261
  Scottish, 61
Newton, Sir Isaac, 8, 91, 92, 106, 143, 144, 145, 147, 149, 150, 151, 154, 158, 159, 161, 161n, 162n, 163n, 164n, 167n, 168n, 170n, 173, 174, 178
  absolute space and time, 150, 158
  analysis of planetary motion, 144–45
  authority of, 150
  Enumeratio linearum tertii ordinis, 165n
  hypotheses non fingo, 144
  Newtonian style, 143–71
    definition of, 144
  Opticks, 144, 158, 159
  Principia, 144–45, 148, 150, 151, 158, 159, 161n, 163n, 164n, 170n
  theory of gravitation, 144, 146–47
Newtonianism, 6, 24, 32, 33, 34, 40, 60, 106, 143–71, 178
  popularisation of, 154, 161
Newtonians, 2, 149, 179
New York, 229
Nichols, John, 76n, 77n
Nicholson's Journal of Natural Philosophy, Chemistry and the Arts, 304, 306
Nicol, William, 34–35
Nicolson, Malcolm, 256, 272, 274n, 279n
Nicolson, William (Bishop of Carlyle), 91
Niddrie Marischal, 47n
Nisbet, William, 219
Noble, Daniel, 123n
North Ronaldsay, 36
Northern Lighthouse Board, 19
northern passage, search for, 152
Norton, David Fate, 335n
Norton, Mary J., 335n
Norway, 316
Nottingham, 35
Nutton, Vivian, 273n

obstetrics, 98, 264, 265
O'Donoghue, Yolande, 48n
Ogborn, Miles, 74n, 168n

Ogilvie, William, 103, 125n
O'Gorman, Frank, 187n
ÓGráda, Cormac, 190n
Olby, Robert C., 162n
Oldroyd, David, 297n
Olson, Richard, 2, 10n, 171n
  Scottish Philosophy and British Physics, 2
Ophir, Adir, 257, 274n
optics, 6, 35, 95, 106, 327
Ordnance Survey, 30, 316
Orkney and Shetland, 82
Orkneys, 36, 287
orreries, 28, 40
Osler, Margaret J., 188n
Ossianic controversy, 208n
Oswald, Alexander, 260
Oswald, John, 260
Oughton, General Sir James, 111
Oughtred, William, 26, 46n
Outram, Dorinda, 281, 297n

Paine, Thomas, 204
Paisley, John, 96
Paris, 24, 36, 98, 222, 223, 224, 225, 226, 227, 228, 302, 303, 320n
  Académie des Sciences, 57, 143
  École Polytechnique, 303
  French National Institute, 303
  hospitals, 84
  National Convention, 98
  Royal Academy of Surgery, 224
  Royal Medical Society, 226
  See also University of Paris
Parmentier, Antoine-Augustine, 228
Parot, F., 44n
Parry, Caleb Hillier, 306
Parry, Martin L., 77n
Partington, J. R., 319n
Passmore, R., 11n, 12n, 14n, 47n, 112n, 118n, 208n, 236n, 274n
Paterson, James, 22
Paton, George, 66–68
Patoun, William, 123n
patronage, 5, 7, 34, 37, 58, 59, 66, 68, 72, 86–87, 92–93, 143, 147–49, 154
Pattison, F. L. M., 278n
Paul, J. Balfour, 325n
Peacock, Martin A., 299n, 323n
Pederson, Kurt Møller, 14n
Peebles, 23
Peel, Sir Robert, 316
Pemberton, Henry, 168n
Penman, Edward, 229–30, 236n
Penman, James, 229–30, 236n
Pennant, Thomas, 201

Pentland Hills, 194, 299n
Percival, Thomas, 272, 279n
  Medical Ethics, 272
periodicals, 192, 195, 206, 261, 333
Perrin, Carleton E., 14n, 125n
Perry, Ruth, 189n
Perth, 203
Perthshire, 46n, 58, 70, 288, 299n
Peterhead, 276
Petherbridge, Deanna, 336n
Petty, Sir William, 155
Pettycur, 289
La Pharmacie Moderne, 228
Pharmacopoeia Helveticae, 228
Philadelphia, 34, 156, 192, 205, 229, 230, 231–32
  Botanical Gardens, 205
  First Presbyterian Church, 156
Phillipson, N. T., 2, 4, 5, 11n, 12n, 51n, 324n
philosophy, 60
  common sense, 111, 256, 314
  moral, 3, 5, 80, 176, 191, 200
  natural, 5, 17, 19, 80, 81, 89, 91, 92, 93, 94, 95, 103, 106, 159, 160, 172, 173, 174, 177, 179, 191, 199
  audience for, 95
  relation to science and medicine, 6
phrenology, 301, 314–15, 318
physicians, 82, 84, 86, 89, 96, 110, 215, 256, 257, 264, 271
physico-theology, 296, 329
physics, 151
  general, 89
  special, 89
Physiocrats, 59, 75n
physiology, 96, 102
Pickstone, John, 330, 335n
  Ways of Knowing, 330
Pinkerton, John M., 44n
Pitcairne, Dr. Archibald, 92, 164n, 179
Pittock, J. H., 12n, 165n, 187n, 188n
Platonism, 183
  Cambridge, 173, 175
Playfair, John, 75n, 166n, 283, 289, 299n, 302, 311, 314, 316
  Illustrations of the Huttonian Theory of the Earth, 311
Plumb, J. H., 29, 48n
Plummer, Andrew, 3
pneumatics, 95
pneumatology, 89
polite
  conversation, 63
  letters, 5, 80, 191
  sociability, 57, 62, 329

Index

politeness, 89, 95, 97, 110, 111
political arithmetic, 57, 155, 168n
political economy, 80, 101
Pont, Timothy, 24, 54, 70
Ponting, Betty, 45n
Poovey, Mary, 146, 157, 164n, 169n
Pope, Alexander, 143, 161n, 191, 225
Porter, Roy, 8, 11n, 13n, 14n, 15n, 43n, 44n, 48n, 50n, 73n, 111–112n, 187n, 198, 211n, 273n, 279n, 282, 297n, 336n
Porter, Theodore M., 143, 157, 169n
Porter, William, 235n
Portsoy, 287
potatoes, 334
Prebble, John, 302, 307, 319n, 321n
Prentiss, Joseph, 14n
Presbyterianism, 173, 174, 194
Prescot, 21
Price, Richard, 157, 169n, 170n
 Observations on Reversionary Payments, 157
Priestley, Joseph, 102, 107, 205
Princeton University, 34, 88
prisms, 35
probability theory, 156, 157
professionalisation, 9, 97, 301
 of medicine, 5
professors, 82, 89
progress, 57
Prussia, 56
psychology, 314
Ptolemy, 57
 Ptolemaic system, 60
public sphere, 57, 59, 88
Pumfrey, Stephen, 44n, 49n, 168n
Purdie, James, 122n
Putnam's Monthly Magazine, 200, 212n
Pyle, Andrew, 116n, 188n

quadrants, 29, 30, 36, 38, 40, 46n, 47n
quadrivium, 60
Quakerism, 8, 173

Raeburn, Sir Henry, 275n
Ramsay, Allan, 114n
Ramsay, Dr. James, 230
Ramsay, John, 75n, 195
Ramsay, Katherine, 301
Ramsden, Jesse, 30, 40
Rankine, Alexander, 105, 127n
Raven, James, 236n
Raynal, Abbé Guillame-Thomas-Francois, 223
reading, 63, 334
Reid, John S., 12n, 40–1, 50n, 52n, 53n

Reid, Thomas, 7, 60, 63–64, 75n, 76n, 85, 97, 102, 103, 106–7, 111, 125n, 126n, 127n, 128n, 130n, 166n, 314, 334
Reid, Thomas (clockmaker), 30
Reill, Peter Hanns, 77n
Reiser, Stanley, J., 259, 273n, 274n, 276n, 278n
religion, 101, 173, 176, 186
 in eighteenth-century Britain, 173
 personal, 173, 177, 179
Renfrewshire, 82, 90
Renwick, John, 15n, 77n
Renwick, Robert, 115n
representation, 6, 7, 333
Restoration, the, 174
Revel, Jacques, 74n
rhetoric, 89
Richards, Eric, 307, 321n, 322n
Richardson, Samuel, 183, 185, 190n
Richardson, William, 123n, 124n
Riddell, Maria, 204–5, 206, 213n
 Voyages to the Madeira and Leeward Caribbean Isles, 205
Rider, Robin, 50n, 73n, 126n
Rigaud, S. P., 123n
Risse, Guenter B., 13n, 14n, 273n, 274n, 279n
Ritchie, James, 126n
Rivington, James, 229, 236n
Roberts, Philip, 189n
Robertson, James, 283
Robertson, John, 11n, 12n, 113n
Robertson, Noel F., 320n
Robertson, General William, 306
Robertson, Rev. William, 3, 73n, 111, 221, 229
Robinson, Eric, 42n, 53n, 125n, 127n, 130n, 319n
Robinson, George, 216, 219, 225, 234n, 235n, 236n
Robison, John, 17, 23, 36, 61, 86, 87, 92, 97, 103, 106, 117n, 118n, 119n, 120n, 125n, 127n, 128n, 319n
Rodin, Alvin E., 278n
Roe, Shirley, 175, 188n
Roger, Jacques, 188n
Rollo, John, 261, 277n
Ronan, Colin, 44n
Roscoe, John, 45n
Roscoe, Robert, 45n
Rosenberg, Nathan, 20, 43n
Roslin, 299n
Rosner, Lisa, 13n, 274n
Ross, A. J., 70

Ross and Cromarty, 82, 309
Ross, Andrew, 116n
Ross, Rev. Andrew, 123n
Ross, George, 119n, 123n
Ross-shire, 304, 307, 328
Ross, Thomas, 26, 47n
Rotherham, John, 204, 213n
Rotterdam, 223
Rousseau, George S., 73n
Rousseau, Jean-Jacques, 191, 223, 229
Rowley, 36
Roxburghshire, 99
Roy, Amable le, 221, 225–26, 234n, 235n, 236n
Roy, General William, 58–59, 65, 69, 78n
  See also Scotland, Military Survey of
Royal Bank of Scotland, 148, 154
Ruat, William, 123n, 127n
Ruddiman, Thomas, 223
Rudwick, Martin J. S., 297n
Ruestow, Edward G., 188n
Rupke, Nicolaas, 334, 336n
Rush, Dr. Benjamin, 230, 237n
Rusnock, Andrea A., 169n, 331, 336n
Russell, Andrew, 265
Russell, C. A., 51n
Russell, Iain F., 13n, 129n, 275n, 277n
Russell, James, 225, 266
Russell, John, 47n
Russell, John L., 75n
Russia, 65
Rutherglen, 257, 258
  Shawfield House, 257, 258, 268
Ryan, W. F., 43n

St. Andrews, 3, 82, 128n
St. Lawrence River, 61
St. Petersburg, 36, 87, 88, 223, 235n
  Economic Society, 87
  Imperial Academy of Sciences, 88
Sage, Balthasar-Georges, 228
  Analyse des Blés, 228
Sageng, Erik L., 162n, 163n, 164n, 165n, 166n, 167n
Salmon, Thomas, 214n
Santucci, Antonio, 15n
Saunderson, Nicholas, 168n
Saxony, 56, 285, 287
Scarborough, 182
Schaffer, Simon, 44n, 50n, 126n, 143, 162n, 169n, 187n, 237n, 297n, 335n
Schiehallion, 58, 106, 128n
Schneer, Cecil J., 297n
Schweizer, Karl W., 12n, 49n

science, 1, 2, 3, 4, 5, 6, 7, 327
  and the public sphere, 6
  as public culture, 6, 154
  authority in, 68
  authority of, 149
  civic dimension of, 280, 295
  fieldwork in, 9, 280–300
  in relation to philosophy and medicine, 6
  of man, 4, 6
  of the mind, 314–15
Scientific Revolution, 5, 29, 60, 257
  in Scotland, 20
  second, 301
Scotland, 20, 30, 82, 91, 108, 148, 173, 179, 216, 221, 223, 231, 285, 287, 291
  Astronomer Royal for, 30, 317
  audience for popular science in, 161
  book trade in, 191–93
  central, 26
  education in, 301
  famine in, 184
  glaciers in, 314
  Highlands, 65, 307, 308–11, 317
  importation of instruments to, 23, 328
  landed elites in, 2
  Lowlands, 309
  mapping of, 54–78
  Military Survey of, 58, 65, 68, 69, 75n
  National Archives of, 22, 193
  National Museums of, 29, 38
    'Brass & Glass', 29
  Newtonianism in, 6, 144, 148
  north of, 58, 95, 316
  north east of, 8, 186
  northern and western isles of, 59, 153
  religious factionalism in, 173
  Royal Society of, 5
  sites of knowledge production in, 191
  taxation in, 153
  teaching of geography and geometry in, 59–64
  urban elites in, 2
  vitrified forts in, 316
  west of, 71, 86, 95, 271, 272
  western isles of, 286
Scots Magazine, 165n, 195–96, 198, 210n
Scott, Harold W., 297n
Scott, J. F., 162n
Scott, R., 166n
Scott, Sir Walter, 294, 295, 300n, 309, 316, 322n, 323n

Index

Scott, W. F., 169n
Scott, W. R., 121n
Scottish Enlightenment, 36, 70, 232, 301, 327
  and historical truth, 36
  and the culture of the book, 8–9, 191–214, 215–54, 329, 333–34
  and Enlightenment internationally, 335
  and the popular press, 206
  and religion, 173
  as an oral culture, 191
  audiences for science in, 295, 327, 329, 331
  communications circuit in, 216, 223
  cultural role of mathematics in, 143–71
  end of, 9
  experimentalism in, 330
  export of, 8, 215, 223–32
  definition of, 72, 79–80, 111, 172, 173, 186, 318
  geographies of scientific production in, 59, 329, 331–33
  historiography of, 3–6, 11n, 18
  levels of historical analysis of, 328–29
  markets for instruments in, 32–35, 41, 328
  patronage and, 33–34
  place of geography in, 54
  public sphere of, 6, 55, 61, 70, 329
  role of the generalist in, 301, 317–18
  role of natural knowledge in, 318, 327–35
  unity of the sciences in, 327, 329
  usefulness of knowledge in, 329
  ways of knowing in, 329–31, 332
Scottish universities, 2, 4, 5, 6, 18, 19, 32, 33, 36, 37, 41, 80, 98, 155, 158, 191, 192
  curricula of, 3, 19, 59, 60, 80, 88–89
  cursus philosophicus, 89
  Parliamentary Commission of Visitation (1690), 80
  Parliamentary Commission of Visitation (1826), 291
  professorial system, 3
  science and religion in, 147
Scougal, Henry, 173–74, 177, 187n
  Life of God in the Soul of Man, 173
Scougal, Patrick, 173
Scruton, John, 99, 121n
Scruton, William, 99, 121n
Secord, J. A., 49n, 208n, 297n, 334, 336n
Sellar, Patrick, 307, 308, 322n

sensibility, 186
Seymour, Lord Webb, 303, 311
Shakespeare, William, 279n
Shapin, Steven, 2, 6–7, 10n, 15n, 18, 28, 42n, 44n, 47n, 75n, 77n, 113n, 130n, 143, 168n, 169n, 191, 207–8n, 212n, 256, 257, 273n, 274n, 277n, 293, 299n, 318n, 324n, 331, 335n, 336n
Sharp, L. W., 116n
Sharpe, T., 322n
Shaw, C. Byam, 302, 319n, 322n
Shaw, John Stuart, 165n
sheep, 305–8
  Cheviot, 307
  Merino, 306, 307
  Royal Flock, 306
Sheffield, 23, 35, 195, 196
Shepherd, Christine M., 12n, 75n, 117n, 188n
Sher, Richard B., 7, 12n, 15n, 43n, 79, 101–2, 111, 112n, 113n, 124n, 193, 196, 207n, 208n, 209n, 210n, 215, 233n, 276n, 333, 336n
Shetlands, 152
Shore, Dr., 231, 237n
Short, James, 28, 30, 33, 35–36, 37, 39, 40, 41
Short, Thomas, 40
Shorter, Edward, 273n
Shortland, Michael, 166n
Sibbald, James, 193, 209n, 235n
Sibbald, Sir Robert, 5, 27, 54, 55, 68, 70, 72, 72n, 77n, 89, 90, 116n
  Scotia Illustrata, 27
Siccar Point, 283
Silverthorne, Michael, 117n
Simmons, Samuel Foart, 219
Simpson, A. D. C., 11n, 23, 43n, 44n, 45n, 46n, 47n, 48n, 50n, 51n, 52n, 118n, 319n
Simpson, James, 322n
Simson, 85, 86, 87, 92, 95, 101, 105, 106, 107, 117n, 119n, 122–23n, 127n, 128n, 148, 166n
Sinclair, George, 22, 90, 91, 95, 116n, 119n
  Hydrostaticks, 22
  Satan's Invisible World Discovered, 90
Sinclair, Sir John, 57, 65, 68, 99, 157, 307, 308, 321n
  Statistical Account of Scotland, 57, 65, 68, 157
Sinclair, Robert, 90, 91, 116n
Sisson, Jonathan, 40
Skene, David, 63–64

Skene, George, 60
Skinner, Andrew S., 12n, 43n, 76n, 117n, 125n
Skye, 314
Slater, Terence R., 77n
Sloane, Sir Hans, 89, 91
Smart, Alistair, 114n
Smart, Mr., 265
Smart, R. N., 113n, 114n
Smart, Rev. William, 62
Smeaton, W. A., 319n
Smellie, Alexander, 194, 204, 212n, 213n
Smellie, R. Martin S., 319n
Smellie, William (printer and natural historian), 8, 76n, 191–214, 219–20, 232
  and the dissemination of learning, 193, 195, 197, 198, 200, 201, 203–4, 206
  and the *Edinburgh Magazine and Review*, 198–201, 206
  and the *Encyclopaedia Britannica*, 197–98
  and the Society of Antiquaries, 201–2, 204
  as editor of the *Scots Magazine*, 195–96, 210n
  equates liberty with suspicion, 194, 200
  his politics, 193, 201, 204–5, 206
  life and career of, 191–214
  on dreams, 202, 205, 213n
  on instinct, 202, 204
  On Juries, 204
  on sexuality of plants, 198, 204
  on usefulness of knowledge, 195, 199
  Philosophy of Natural History, 201, 202–4, 205, 206, 219–20
  reviews of scientific works, 199–200, 211–12n
  Scottish Chronicle, 204
  Thesaurus Medicus, 201, 218, 224, 227
  translation of Buffon's *Histoire Naturelle*, 201, 212n
  versus systems, 200
Smellie, William (surgeon), 86
  A Set of Anatomical Tables, 218
Smith, Adam, 3, 122n, 161, 308
  division of labour, 310
Smith, Annette, M., 75n
Smith, Crosbie, 73n, 336n
Smith, Denis, 52n, 127n
Smith, Diana C. F., 50n, 75n
Smith, G. E., 144, 163n
Smith, Iain, 129n, 275n, 276n

Smith, John, 44n
Smith, John (Cambridge Platonist), 173
Smith, Leonard (pseudonym), 259, 275n, 276n
  Northern Sketches or Characters of G******, 259–60
Smith, Robert, 35, 36, 168n
  A Compleat System of Opticks, 35, 36
Smitten, Jeffrey R., 112n
Smollett, Tobias, 208n
Smyton, Andrew, 24
Snedden, Ian, 116n, 117n
Snow, John, 271
Somerville, Andrew, 26, 46n, 47n
Somerville, Thomas, 50n
Sorrenson, Richard, 29–30, 32, 48n, 50n, 74n, 167n
specialisation, 3, 5, 9, 80, 97, 108, 113n, 301, 317–18
Spinoza, Benedict de, 158
Spurzheim, J. G., 314
Squadrone, 93, 147, 149
Stanley, John Thomas, 312
Stansfield, Dorothy A., 124n
Staynred, Philip, 45n
Sterne, Laurence, 191
  Tristram Shandy, 191
Steuart, David, 66
Steuart, Robert, 91, 118n
Stevenson, Alexander, 87, 102, 124n
Stevenson, David, 209
Stevenson, R. B. K., 315, 324n
Stevenson, Sarah, 51n
Stewart, Major General David, 307, 321n
Stewart, Dugald, 149, 213n, 226, 314, 324n, 327, 335n
Stewart, John (Aberdeen professor), 60, 66
Stewart, John (Edinburgh professor), 36
Stewart, John (surgeon), 228, 236n
Stewart, Larry, 32, 49n, 74n, 118n, 149, 154, 160, 166n, 168n, 169n, 170n
Stewart, M. A., 15n, 79, 112n, 115n, 164n, 170n, 187n
Stewart, Matthew, 35, 61, 92, 114n, 149
Stimson, Alan, 46n
Stirling, 81, 276n
Stirling, James, 167n, 168n
Stirlingshire, 36, 82
Stobhall, 46n
Stobie, James, 70
  map of Perth and Clackmannan, 70
Stone, Edmund, 63
  Some Reflections ..., 63
Stone, Lawrence, 11n, 184, 190n

Stock, J. T., 45n
Storch, Johann, 271
Stott, Rosalie M., 13n
Strahan, Andrew, 194
Strahan, William, 193, 194, 199, 207n, 232
Strang, John, 114n, 123n, 129n, 261, 276n
Stuart, Gilbert, 192, 197, 198, 199, 210n, 211n, 221, 234n
Stuart, John, 3rd Earl of Bute, 33, 40, 86
Sue I, Jean-Joseph, 224
Sue II, Dr. Jean-Joseph, 224–25, 235n
Suggett, Martin, 47n
sundials, 20, 26–27, 28, 91
surgeons, 9, 82, 84, 86, 96, 99, 215, 256, 257, 264, 271
surgery, 97, 98, 100, 122n, 264, 265, 272
surveying, 21, 61, 62, 95, 119n
surveys, 68, 104, 125n
 parochial, 64–65, 155, 157
Sutherland, James, 116n
Sutton, Henry, 46n, 47n
Suzuki, Akihito, 14n, 273n
Swan, William, 28, 40
Sweden, 33
Swammerdam, Jan, 175, 176, 177, 188n
Swan, Dr., 260, 265
Swann, 275n
Swediaur, Franz, 219
 Practical Observations on Venereal Complaints, 219
Sweet, Jessie M., 298n, 299n, 322n
Sweiten, G. F. van, 217
 Commentaries upon Boerhaave's Aphorisms, 217, 231
Swift, Jonathan, 191, 225
Swinbank, Peter, 42n, 52n, 118n
sympathy, 269

Tandberg, J. G., 49n
Taylor, E. G. R., 45n, 46n
Taylor, John, 203, 205, 213n
 A Medical Treatise on the Virtues of St. Bernard's Well, 203
teaching, sites of, 62, 63
Teich, Mikulá, 11n, 43n
telescopes, 21, 29, 33, 36, 40, 53
 Gregorian, 24, 28, 33, 35, 91, 325
Temkin, Owsei, 265, 278n
Tennant, Charles, 85, 109
Tennant, E. W. D., 121n
Teviotdale, 66
Thackray, Arnold, 8, 15n, 111, 112n, 113n

theodolites, 21, 30
theology, 161, 172, 174, 176
thermometers, 22, 38, 39, 41, 93, 128n
Thoday, A. G., 22, 44n
Thom, Rev. William, 71, 78n, 95, 119n
Thomson, Andrew, of Faskine, 258
Thomson, David, 222
Thomson, George, 125n, 274n, 275n, 276n
Thomson, G. Graham, 274n, 275n, 276n
Thomson, James, 208n
Thomson, John, 54, 55, 59, 70, 72, 73n, 78n
 Atlas of Scotland, 54, 70
Thomson, John (Edinburgh professor), 118n, 120n, 123n, 125n
Thomson, Margaret, 258
Thomson, Thomas, 40, 52n, 115n, 258, 275n
Thorkelin, G. J., 52n
Thornhill, 287
Thorold, John, 118n
Thrower, Norman, 165n
Tilloch, Alexander, 129n, 303
 Philosophical Magazine, 303
Tod, William, 195
Top, John, 262
Topham, Jonathan R., 233n
Torrens, Hugh, 77n
Tough, Alastair, 273n
Toulmin, George Hoggart, 212n
Towers, James, 82, 108, 267
Towers, John, 120n
Trail, Robert, 104, 126n
Trail, William, 92, 103, 117n, 125n, 128n
Treadwell, Michael, 191, 207n
Trevor-Roper, Hugh, Lord Dacre, 3–4, 5, 11n
Triewald, Mårten, 33, 118n, 149
trigonometry, 61, 71
Trinity College, Dublin, 274n
Tröhler, Ulrich, 273n
Troil, Uno von, 312, 322n
 Letters on Iceland, 312
Troughton, Edward, 40
trust, 9, 169n, 331
 and historical records, 9, 263
 and the use of instruments, 7
 in geography, 7, 68
 in medicine, 256, 257, 263
 in natural history, 7
Turnbull, George, 161
Turnbull, H. W., 46n, 162n, 167n
Turner, A. J., 43n, 44n, 45n, 46n, 48n

Turner, G. L'E., 43n, 45n, 48n, 49n, 50n, 52n
Tweddle, Ian, 14n
Tweedie, Charles, 162n, 167n
Tytler, James, 193, 206, 207, 209n, 214n
Historical Register, 206

Ulman, H. Lewis, 13n, 126n
Union, Act of (1707), 2, 4, 5
United States of America, 156
University of Cambridge, 35
  Whipple Museum of the History of Science, 47n
University of Edinburgh, 19, 27, 36, 38, 60, 61, 75n, 84, 87, 90, 110, 115n, 143, 147, 148, 149, 192, 194, 195, 201, 211n, 217, 223, 226, 230, 232, 274n, 283, 284, 302, 304, 308, 312, 317, 319n
  Academy of Physics, 2
  curriculum, 60, 148, 165n, 284
  Leslie Affair (1805), 113n
  medical school, 2, 6, 192, 195, 201, 223, 228, 230, 256, 258
  medical theses, 196, 201, 217, 218, 223
  Natural History Museum, 313
  Newtonian Society, 195, 198, 201, 213n
  Science Studies Unit, 1
  Speculative Society, 258
  student societies, 2
University of Glasgow, 17, 19, 37–39, 61, 79, 81, 82, 84, 85, 88–97, 98, 99, 100, 102, 107, 108, 109, 110, 111, 115n, 120n, 122n, 125n, 127n, 129n, 144, 146, 257, 258, 260, 266, 272, 277n
  chemical society, 107, 121n, 128–29n
  chemical laboratory, 258
  essay prizes, 61, 98
  experimental philosophy course, 95
  Hunterian Museum, 116n, 299n, 313
  Literary Society, 80, 81, 87, 98, 101–7, 108, 123n, 124n, 125n, 258
  Macfarlane Observatory, 127n
  medical school, 96–97
  Medico-Chirurgical Society, 108
  observatory, 105
  physick garden, 96
  professorial system, 93–94
  proposed Academy, 98
  student theological societies, 101
  Visitation Commission (1726–27), 93, 94, 120n

University of Leyden, 27, 84, 115n, 120n, 223
  S. and J. Luchtman (University printers), 223, 235n
University of Lund, 33
University of Oxford, 266
University of Paris, 222
University of Rheims, 84, 120n
University of St. Andrews, 19, 24–25, 39–40, 60, 61, 115n, 224
  observatory, 24
University of Strathclyde, 39, 51n
University of Utrecht, 84
Uppsala, 36
Ure, Andrew, 28, 39, 62, 115n
Ure, David, 99, 121n

Vancouver, George, 56
'vapours', the, 180
Veitch, James, 37
venereal disease, 219
Verona, 65
Versailles, 226
Vesuvius, Mount, 291
Vickery, Amanda, 181, 189n
Vienna, 224
Vilant, Nicholas, 61
The Elements of Mathematical Analysis, 61
Virginia, 229
  Fredericksburg, 231
  Hampton, 231
  James River, 231
  Norfolk, 230
  Petersburg, 231
virtue, 57
virtuosi, 5, 79–80, 101, 108, 110, 172, 174
  ideal of the virtuoso, 88–97, 98, 110
Voltaire, François Marie Arouet de, 223, 225
Voyage Pittoresque, 227
Vream, 32

Walker, Rev. John, 14n, 87, 98, 104, 106, 121n, 125n, 128n, 202, 282–3
Wallace, Alfred Russel, 212n
Wallace, Rev. Robert, 57, 155–57, 169n
  A Dissertation on the Numbers of Mankind, 156
  Various Prospects of Mankind, Nature and Providence, 156
Wallis, P. J., 168n, 170n
Walpole, Catherine, 180–81, 185, 189n
Walpole, Horace, 202, 212n

Index

Walpole, Robert, 148, 149, 170n
Walters, Alice N., 76n
Walton, Craig, 283, 297n
Wanley, Nathaniel, 213n
   The History of Man, or the Wonders of Human Nature, 203, 213n
Wanlockhead, 288
Wapoule, 290
Ward, Jean E., 279n
Warden, David, 61
Ware, John, 205
Warner, D. J., 43n
Warner, John Harley, 273n
watches, 21–22
Waters, D. W., 46n
Waterston, Charles D., 9, 13n, 191, 208n, 299n, 322n, 323n, 324n, 328
Watkins, Adrian, 228
Watson, Ebenezer, 262, 277n
Watson, James, 277n
Watt, James, 6, 17, 28, 38–39, 41, 43n, 52n, 63, 84, 103, 105, 109, 110, 114n, 122n, 122n, 124n, 125n, 127n, 319n
Watt, James (surgeon), 82
Wauchopes of Niddrie, 47n
Wawn, Andrew, 294, 295, 299n, 300n, 312, 322n, 323n
Wear, Andrew, 259, 273n, 276n
Webster, Rev. Alexander, 155–57
   Account of the Number of People in Scotland, 155–56
Webster, Charles, 224, 225
Webster, Diana C. F., 50n
Wedgwood, Josiah, 109, 320n
weights and measures, 23, 36, 39, 106
   Imperial standard, 39
   Scots pint, 23, 26
Weiss, Leonard, 44n
Weiss-Amer, Melitta, 189n
Welch, Edwin, 189n
Welter, J. J., 303
Wenzel, Baron de, 226
Werkmeester, Lucyle, 210n
Werner, Abraham Gottlob, 286, 287, 292, 293, 311, 312
   Wernerianism, 280, 284, 286, 288, 291, 292, 293, 294, 295
Wess, J. A., 33, 49n
Westfall, Richard S., 162n, 165n, 168n, 173, 187n
West Indies, 82
West Lothian, 82
Wester Ross, 310
Westminster Confession, 174

Whatley, Christopher, 48n
White, Edwin J., 45n
White, G. W., 322n
White, John, 45n
Whiteside, D. T., 170n
Whitman, Anne, 162n
Whitney's cotton gin, 109
Whittington, Graeme, 75n
Whitton, 33
Whyte, William, 266
Widmalm, Sven, 56, 58, 73n, 75n
Wigton, 276n
Wigtownshire, 27
William III (King of England and Scotland), 174
William Augustus, Duke of Cumberland, 35
Williams, David, 11n, 130n, 275n
Williams, Glyndwyr, 73n
Williams, John, 65, 68, 77n
Williams, M., 44n
Williamson, James, 92, 106, 107, 127n
Wills, Virginia, 75n
Wilson, Adrian, 190n
Wilson, Alexander, 38, 83, 85, 86, 87, 93, 102, 105–6, 107, 123n, 126–27n, 128n
Wilson, John, 64, 76n
   Trigonometry ..., 64
Wilson, John, of Hurlet, 99, 121n, 129n
Wilson, Patrick, 82, 105–6, 122n, 126–27n, 128n
Wilson, Robin, 166n
Windram, Jonathan, 274n
Winslow, J. B., 216
   An Anatomical Exposition of the Human Body, 216
Winton, Calhoun, 236n
Wirsung, Christopher, 259
Wise, M. Norton, 169n
Wishart, John, 60
Withers, Charles W. J., 6, 7, 12n, 13n, 14n, 15n, 28, 47n, 48n, 72n, 73n, 74n, 75n, 76n, 77n, 113n, 115n, 116n, 119n, 121n, 125n, 126n, 169n, 212n, 257, 274n, 296n, 297n, 328, 331, 335n, 336n
Wodrow, James, 123n
Wodrow, John, 96, 115n, 120n
Wodrow, Robert, 90–91, 116–17n
women, 32, 62, 95, 100, 119n, 149, 161, 172–90, 204, 333
Wood, C. G., 127n
Wood, M., 46n
Wood, Paul, 6, 7–8, 9, 11n, 12n, 13n, 15n, 16n, 18, 20, 43n, 45n, 49n,

51n, 52n, 53n, 72n, 74n, 76n, 112n, 113n, 115n, 116n, 117n, 119n, 120n, 124n, 125n, 127n, 128n, 130n, 165n, 166–67n, 170n, 187n, 208n, 209n, 210n, 213n, 233n, 259, 328, 331, 335n, 336n
Woodward, John, 89, 91
Worster, Benjamin, 32
Worthington, William, 200
Wray, E. M., 52n
Wright, John P., 15n, 164n
Wright, Joseph, of Derby, 21, 44n
Wright, Peter, 99, 121n

Wright, Robert, 99
Wyatt, John M., 297n
Wynne, Henry, 46n

Yell, Joan, 279n
Yeo, Richard, 333, 336n
Yewsdale, 66
York, 35, 45n
Yorkshire, 195
Young, David, 40
Young, J. R., 209n, 336n

Zachs, William, 192, 193, 208n, 209n, 211n, 234n